Introduction to Spacecraft Thermal Design

Develop a fundamental understanding of heat transfer analysis techniques as applied to Earth-based spacecraft with this practical guide. This essential text is written in a tutorial style and provides a how-to manual tailored for those who wish to understand and develop spacecraft thermal analyses. Providing an overview of basic heat transfer analysis fundamentals such as thermal circuits, limiting resistance, MLI, environmental thermal sources and sinks, as well as contemporary space-based thermal technologies, and the distinctions between design considerations inherent to room temperature and cryogenic temperature applications, this is the perfect tool for graduate students, professionals, and academic researchers.

Dr. Eric A. Silk is a lecturer in the Aerospace Engineering Department at the University of Maryland, College Park with 22 years of engineering experience in thermal system design and analysis for space-based systems.

Cambridge Aerospace Series

Editors: Wei Shyy and Vigor Yang

1. J. M. Rolfe and K. J. Staples (eds.): *Flight Simulation*
2. P. Berlin: *The Geostationary Applications Satellite*
3. M. J. T. Smith: *Aircraft Noise*
4. N. X. Vinh: *Flight Mechanics of High-Performance Aircraft*
5. W. A. Mair and D. L. Birdsall: *Aircraft Performance*
6. M. J. Abzug and E. E. Larrabee: *Airplane Stability and Control*
7. M. J. Sidi: *Spacecraft Dynamics and Control*
8. J. D. Anderson: *A History of Aerodynamics*
9. A. M. Cruise, J. A. Bowles, C. V. Goodall, and T. J. Patrick: *Principles of Space Instrument Design*
10. G. A. Khoury (ed.): *Airship Technology, Second Edition*
11. J. P. Fielding: *Introduction to Aircraft Design*
12. J. G. Leishman: *Principles of Helicopter Aerodynamics, Second Edition*
13. J. Katz and A. Plotkin: *Low-Speed Aerodynamics, Second Edition*
14. M. J. Abzug and E. E. Larrabee: *Airplane Stability and Control: A History of the Technologies that Made Aviation Possible, Second Edition*
15. D. H. Hodges and G. A. Pierce: *Introduction to Structural Dynamics and Aeroelasticity, Second Edition*
16. W. Fehse: *Automatic Rendezvous and Docking of Spacecraft*
17. R. D. Flack: *Fundamentals of Jet Propulsion with Applications*
18. E. A. Baskharone: *Principles of Turbomachinery in Air-Breathing Engines*
19. D. D. Knight: *Numerical Methods for High-Speed Flows*
20. C. A. Wagner, T. Hüttl, and P. Sagaut (eds.): *Large-Eddy Simulation for Acoustics*
21. D. D. Joseph, T. Funada, and J. Wang: *Potential Flows of Viscous and Viscoelastic Fluids*
22. W. Shyy, Y. Lian, H. Liu, J. Tang, and D. Viieru: *Aerodynamics of Low Reynolds Number Flyers*
23. J. H. Saleh: *Analyses for Durability and System Design Lifetime*
24. B. K. Donaldson: *Analysis of Aircraft Structures, Second Edition*
25. C. Segal: *The Scramjet Engine: Processes and Characteristics*
26. J. F. Doyle: *Guided Explorations of the Mechanics of Solids and Structures*
27. A. K. Kundu: *Aircraft Design*
28. M. I. Friswell, J. E. T. Penny, S. D. Garvey, and A. W. Lees: *Dynamics of Rotating Machines*
29. B. A. Conway (ed): *Spacecraft Trajectory Optimization*
30. R. J. Adrian and J. Westerweel: *Particle Image Velocimetry*
31. G. A. Flandro, H. M. McMahon, and R. L. Roach: *Basic Aerodynamics*
32. H. Babinsky and J. K. Harvey: *Shock Wave–Boundary-Layer Interactions*
33. C. K. W. Tam: *Computational Aeroacoustics: A Wave Number Approach*
34. A. Filippone: *Advanced Aircraft Flight Performance*
35. I. Chopra and J. Sirohi: *Smart Structures Theory*
36. W. Johnson: *Rotorcraft Aeromechanics vol. 3*
37. W. Shyy, H. Aono, C. K. Kang, and H. Liu: *An Introduction to Flapping Wing Aerodynamics*
38. T. C. Lieuwen and V. Yang: *Gas Turbine Emissions*
39. P. Kabamba and A. Girard: *Fundamentals of Aerospace Navigation and Guidance*
40. R. M. Cummings, W. H. Mason, S. A. Morton, and D. R. McDaniel: *Applied Computational Aerodynamics*

41. P. G. Tucker: *Advanced Computational Fluid and Aerodynamics*
42. Iain D. Boyd and Thomas E. Schwartzentruber: *Nonequilibrium Gas Dynamics and Molecular Simulation*
43. Joseph J. S. Shang and Sergey T. Surzhikov: *Plasma Dynamics for Aerospace Engineering*
44. Bijay K. Sultanian: *Gas Turbines: Internal Flow Systems Modeling*
45. J. A. Kéchichian: *Applied Nonsingular Astrodynamics: Optimal Low-Thrust Orbit Transfer*
46. J. Wang and L. Feng: *Flow Control Techniques and Applications*
47. D. D. Knight: *Energy Disposition for High-Speed Flow Control*
48. Eric A. Silk: *Introduction to Spacecraft Thermal Design*

Introduction to Spacecraft Thermal Design

ERIC A. SILK
University of Maryland

CAMBRIDGE
UNIVERSITY PRESS

University Printing House, Cambridge CB2 8BS, United Kingdom

One Liberty Plaza, 20th Floor, New York, NY 10006, USA

477 Williamstown Road, Port Melbourne, VIC 3207, Australia

314–321, 3rd Floor, Plot 3, Splendor Forum, Jasola District Centre, New Delhi – 110025, India

79 Anson Road, #06–04/06, Singapore 079906

Cambridge University Press is part of the University of Cambridge.

It furthers the University's mission by disseminating knowledge in the pursuit of
education, learning, and research at the highest international levels of excellence.

www.cambridge.org
Information on this title: www.cambridge.org/9781107193796
DOI: 10.1017/9781108149914

First published 2020

A catalogue record for this publication is available from the British Library.

Library of Congress Cataloging-in-Publication Data
Names: Silk, Eric, author.
Title: Introduction to spacecraft thermal design / Eric Silk.
Description: New York : Cambridge University Press, 2020. | Series: Cambridge aerospace series |
Includes bibliographical references and index.
Identifiers: LCCN 2019049044 (print) | LCCN 2019049045 (ebook) | ISBN 9781107193796 (hardback) |
ISBN 9781108149914 (epub)
Subjects: LCSH: Space vehicles–Thermodynamics. | Heat–Transmission. | Space vehicles–Cooling. |
Low temperature engineering.
Classification: LCC TL900 .S54 2020 (print) | LCC TL900 (ebook) | DDC 629.47–dc23
LC record available at https://lccn.loc.gov/2019049044
LC ebook record available at https://lccn.loc.gov/2019049045

ISBN 978-1-107-19379-6 Hardback

This work is dedicated to the two women that delivered me to my undergraduate educational institution (freshman year) where this journey began: my mother Rachel Silk and my aunt Maple Skinner.

Contents

Figures

Tables

Nomenclature

A	Area [m^2]
A	Amps
Alb	Albedo, fraction of Earth-reflected solar energy
A_p	Projected surface area [m^2]
$B_{i,j}$	Fraction of gray body radiation transfer and absorption between surface i and j
Bo	Bond number $\left[g(\rho_l - \rho_v) \cdot l_c^2/\sigma_l\right]$
C	Thermal conductance [W/K], Vacuum line conductance [Pa · m^3/s]
CC	Compensation chamber
°C	Degrees Celsius
CHF	Critical heat flux [W/m^2]
COP	Coefficient of performance
CTE	Coefficient of thermal expansion [K^{-1}]
D	Diameter [m]
$\dot{E}$	Rate of energy transfer [W]
E	Energy [J]
E_b	Blackbody emissive power [W/m^2]
E_{mod}	Modulus of elasticity [MPa]
$E_{\lambda b}$	Blackbody spectral emissive power [W/m^2 · μm]
Eff	Effectiveness [0 ⇔ 1.0]
EIR	Earth IR energy [W/m^2]
$F_{i,j}$	View factor
$\hat{F}_{i,j}$	Fraction of gray body radiation transfer between surface i and j
$\mathfrak{F}$	Radiative transfer factor
FOM	Figure of merit
G	Irradiation [W/m^2]
G_λ	Spectral irradiation [W/m^2 · μm]
H	Magnetic field intensity [A · N_{turns}/m]
HP	Heat pipe
Hz	Hertz
I	Intensity [W/sr · m^2], Current [A]
I_b	Blackbody intensity [W/m^2]
I_λ	Spectral intensity [W/sr · m^2 · μm]
IR	Infrared
J	Radiosity [W/m^2]

J	Joules
J_e	Current density [A/m^2]
J_λ	Spectral radiosity [W/m^2 · μm]
$\boldsymbol{K}$	Permeability [m^2]
K	Kelvin
KE	Kinetic energy [J]
km	Kilometers
L	Length vector [m]
L,*l*	Length [m], Honeycomb panel length [m]
L_o	Lorenz number 2.45×10^{-8} [W · Ohms/K^2]
LHP	Loop heat pipe
M	Million
N	Newtons, Number
NCG	Non-condensable gas
O	On the order of
OP	Orbital period [minutes]
Q	Energy [J]
$\dot{Q}$	Heat transfer [W]
$\dot{Q}''$	Heat flux [W/m^2]
P	Pressure [kPa], Power [W]
PRT	Platinum resistance thermometer
R	Thermal resistance [K/W], Radius [m]
R	Radius on virtual hemisphere
Re	Reynolds number [$\rho \cdot d_h \cdot Vel/\mu$]
R_e	Electrical resistance [Ohms, Ω]
R_{Flow}	Fluid flow resistance [m^{-1} · s^{-1}]
R_m	Radius of molecule [m]
RRR	Residual resistivity ratio
S	Shape factor [m]
S_y	Yield strength [MPa]
S.P.	Specific power
STC	Strength-to-thermal-conductivity ratio [MPa/W/m · K]
T	Temperature °C or K
T	Honeycomb panel thickness [m], Tesla [N/A · m]
T_c	Transition temperature for super-conducting state [K]
TC	Thermocouple
Torr	0.133322 kPa
TM	Thermistor
TP	Transmission probability
U	Overall heat transfer coefficient [W/m^2 · K]
USD	United States dollars
UV	Ultraviolet
V	Voltage
V	Volts

Vel	Velocity [m/s]
Vol	Volume [m^3]
$\dot{Vol}$	Volume flow rate [m^3/s]
w, W	Width [m], Honeycomb panel width [m]
W	Watts [J/s]
$\dot{W}$	Work input [W]
Wb	Weber [T · m^2]
We	Weber number $\left[\rho_v \cdot Vel_{avg}^2 \cdot d_h/\sigma_l\right]$
X_{leak}	Leak rate [Pa · m^3/s]
$\dot{Z}$	Throughput [Pa · m^3/s]
a	Semi-major axis
c_o	Speed of light [2.9979×10^8 m/s]
c_p	Specific heat [J/kg · K]
cm	0.01 meters
dA	Differential surface area [m^2]
d_h	Hydraulic diameter [m]
d_s	Sphere diameter [m]
dB	Decibels
$d\Omega$	Differential solid angle [sr]
e^-	Charge of an electron [-1.602×10^{-19} Coulombs]
f	Friction factor
g	Earth gravitational acceleration [9.8 m/s]
$\dot{g}$	Energy generation [W]
$\dot{g}'''$	Energy generation per unit volume [W/m^3]
h	Height [m]
	Fluid enthalpy [kJ/kg], Planck's Constant [6.6256×10^{-34} J · s]
h_{conv}	Convection coefficient [W/m^2 · K]
h_f	Head loss [m], Liquid phase enthalpy [kJ/kg]
h_{fg}	Enthalpy of vaporization [kJ/kg]
h_s	Solid phase enthalpy [kJ/kg]
h_{sf}	Enthalpy of fusion [kJ/kg]
k	Thermal conductivity [W/m · K]
k_B	Boltzmann's Constant [1.381×10^{-23} J/K]
k_{eff}	Effective thermal conductivity [W/m · K]
kg	Kilograms
kPa	1,000 N/m^2
l_c	Characteristic length [m]
lbs	Pounds
$\dot{m}$	Rate of mass change [kg/s], Mass flow rate [kg/s]
m	Mass [kg]
m	Meters
mil	1/100th of an inch
mV	millivolts

mK	milli-Kelvin
$\hat{\mathbf{n}}$, n	Surface normal direction
$\dot{q}$	Heat transfer [W]
$\dot{q}''$	Heat flux [W/m^2]
r	Radial length [m]
r_b	Average bubble radius [m]
r_p	Pore radius [m]
s	Seconds
s	Entropy [J/kg · K]
s_{arc}	Arc length
sr	Steradians
t	Time [seconds]
th	Thickness [m]
x	Length [m], Fluid quality [0 ⇔ 1.0]
x_{MSR}	Melt-to-solidification ratio [0 ⇔ 1.0]
y	Length [m]
z	Longitudinal length [m]
Ξ	Balance equations coefficient matrix
Φ	Balance equations solution vector
Φ_v	Viscous dissipation
Δ	Difference in value
Λ	Thomson coefficient [V/K]
Ω	Solid angle [sr], Ohms
Π	Peltier coefficient [V]
Γ	Seebeck coefficient [V/K]
α	Absorptivity
β	Beta angle
δ	Ribbon thickness [m], Error
δ_F	Foil thickness [mils]
η	Efficiency
θ	Azimuthal angle [Degs, °], Inclination angle [Degs, °]
ε	Emissivity
ε^*	Effective emissivity
γ	Solar constant to surface normal angle [Degs, °]
λ	Wavelength [μm], Lambda point for transition of He into the superfluid state
λ_{MFP}	Mean free path [m]
ξ_{load}	Duty cycle for environmental thermal loading
ρ	Density [kg/m^3], Reflectivity
ρ_e	Electrical resistivity [Ω · m]
ρ_l	Liquid density [kg/m^3]
ρ_v	Vapor density [kg/m^3]
ω	Honeycomb cell size [m]

τ	Transmissivity
ϕ	Cylindrical circumferential direction, Circumferential angle [Degs, °]
ϕ	Porosity [0 $\Leftrightarrow$ 1.0], Electrostatic voltage [V]
μ_l	Liquid viscosity [kg/m · s]
μ	Product of universal gravitational constant and mass of earth [3.98603×10^{14} m^3/s^2]
μ-g	Microgravity
μ_{J-T}	Joule–Thomson coefficient [K/kPa]
μ_o	Magnetic permeability [Wb/A · m]
μm	Microns [10^{-6} m]
π	3.14159265359
σ	Stefan Boltzmann constant [5.67×10^{-8} $W/m^2 \cdot K^4$], Honeycomb ribbon extension factor
σ_{dc}	DC electrical conductivity [$\Omega^{-1} \cdot m^{-1}$]
σ_e	Electrical conductivity [$\Omega^{-1} \cdot m^{-1}$]
σ_l	Liquid surface tension [N · m]

Subscripts

Adia	Adiabatic
C,c	Cold
CC	Compensation chamber
CHF	Critical heat flux
E	Endpoint
EIR	Earth-IR
F	Foils
FB	Film boiling
FE	Finite element
Fermi	Fermi-based value
H,h	Hot, High
HL	Heat leak
HX	Heat exchanger
Iso	Isothermal
L	Honeycomb panel height,Low
LHP	Loop heat pipe
LRL	Liquid return line
MHF	Minimum heat flux
MLI	Multi-layer insulation
NC	Natural convection
OFF	Off operating state
ON	On operating state
ONB	Onset of nucleate boiling
P	Packing
PCM	Phase-change material
PEL	Peltier
PV	Photovoltaic
RNB	Rapid nucleate boiling
S	Shape, Solar
T	Temperature, Honeycomb panel thickness
Thom	Thomson effect
TS	Thermal strap
VL	Vapor line
W	Honeycomb panel width
a	Apogee

a,b,c	Nodal position in planar grid
abs	Absorption, absorber
actual	Actual value
avg	Average
b	Blackbody, Perigee
boil	Boiling
cap	Capillary
comp	Compressor
con	Condenser
cond	Conduction
couple	Couple level
eff	Effective
elec	Electrical
emb	Embedded
ent	Entrainment
end	Endpoint
env	Environmental
evap	Evaporator
exit	At outlet location
final	Final or end state
g	Generated, Gravitational
h	Convection, Isenthalpic
high	High value, high end
i	Inner
in	Inflow/Ingoing
init	Initial
inlet	Entrance
int	Interface
isen	Isentropic
joule	Joule heating
k	Conductivity
l,liq	Liquid
lat	Lattice
link	Thermal coupling
load	Heat load value
low	Lower value
lower	Lower end
m	Mean
max	Maximum
melt	Value during melting
min	Minimum
mod	Module level
net	Net difference
o	Outer
out	Outwards, rejected, outflow/outgoing

p	Projected
par	Particle
$para$	Parallel
pri	Primary
$pump$	At mechanical pump
r	Real, Radial direction
rad	Radiator, radiation
$recup$	Recuperator
ref	Reflected
rej	Rejected
s	Isentropic value
sat	Saturation
sec	Secondary
ser	Serial
$sink$	Circuit cold temperature location
sol	Solid
st	Stored
sup	Superheat
$surf$	Surface
tot	Total
tp	Triple point
$tran$	Transition point
$trans$	Transmitted
$turb$	Turbine
$upper$	Higher end
v, vap	Vapor
$wall$	At wall location
$waste$	Excess, non-usable
$wick$	Wick structure
$wire$	Wire/s
x	Cartesian x direction
y	Cartesian y direction
z	Cartesian z direction, Cylindrical longitudinal direction
λ	Spectral
$2\text{-}\Phi$	Multiphase
θ	Cylindrical circumferential direction
∞	Infinite distance
$+$	Positive value
$-$	Negative value

1 Introduction

1.1 Background

Over the course of the past several years, I have had the opportunity to teach spacecraft thermal design to both undergraduate and graduate students. In researching texts for my first year of teaching the subject, I quickly found that there was no textbook available that methodically walked the student through the process of thermal design and analysis of space-based systems. Standard heat transfer texts simply did not provide detailed content addressing thermal design considerations and constructs for analysis of space-based systems: contemporary engineering texts only provided a cursory overview of space-based conduction and radiation phenomena. Texts that explicitly addressed the topic of spacecraft thermal design did so in the context of an overall systems perspective, with thermal as one of several chapters dedicated to different subsystems. Other chapters covered topics such as mechanical, guidance navigation and control, electrical, etc. Last but not least there was a lack of detailed treatment of techniques and methodologies developed from fundamental heat transfer, thermodynamics and fluid dynamics principles. In creating the lecture content for my course, I found this limited the conveyance of important subject matter to students. Ultimately, the solution for providing course participants with a coherent stream of lecture content was to combine information in previously published texts on the subject with information I deemed important. Information was collected in written form as class notes and provided to students as the primary text for the course. Ironically, while it is some years since I became a lecturer in aerospace, the absence of a single introductory text for students on spacecraft thermal design persists. Arguably, this is a key void in the literature associated with space systems. This volume aims to fill that void with teaching materials covering the topic of spacecraft thermal design. I hope that the concepts and principles conveyed in this book will inspire early-career practitioners and students enrolled in space-based courses, and enable them to gain a fundamental understanding of the thermal environment in space, as well as the basic skills required to perform spacecraft thermal design and analysis.

1.2 Why Is Space Important?

While this text focuses on the thermal aspects of the space environment and the design of systems that mitigate potentially negative effects associated with the extreme temperatures experienced in space, a greater realization often sparks our interest in this topic: that space is important and highly relevant to our lives as citizens of planet Earth. However, in the face of contemporary societal challenges (e.g., the race towards a cure for cancer, solving world hunger and preserving the environment) one might ask: "Why is space important?" In discussions of the need for increased national and/or international investment to accelerate the goals of the space program, this question is often posed with cynical overtones, implying that other vital and immediate concerns are more worthy of investment than space-based pursuits. However, with a little contemplation, the relevance of space in all our lives is painstakingly apparent. To begin with, we reside on planet Earth, which resides in space as a living planet and does not display the attributes of what many would classify a "dead" planet, such as Mars. Many space researchers believe that Mars was once hospitable to carbon-based life forms similar to those found on Earth today. Much of the contemporary exploration of the Martian environment has therefore been dedicated to validating the presence of remnant forms of life on the surface and in the soil. Whichever side of the debate regarding life on Mars individuals may find themselves on, it is abundantly clear that the present environmental conditions on the Martian landscape are not conducive to human life (nor carbon-based life forms in general). Another way to interpret this is to state that Mars has poor planetary health. How is this relevant to Earth and its inhabitants? While our planet is today a living planet, "today" is an operative term. If Mars once contained life and somehow lost it, identifying the events that led to Mars' present state may help educate the human family about what we need to do on Earth today to avoid similar pitfalls, and the near-term and long-term challenges of maintaining our planet in a living state.

The human family faces many important challenges. Three that have plagued societies since the foundation of civilization (although the list is not exhaustive) are war, disease and pestilence and famine. Although famine is not usually acknowledged as a threat to contemporary first-world countries, the reduction in their quantity of farmland, combined with increasing global demands for food production, could easily tax the existing food supply chain in years to come. A fourth challenge that has arisen in recent years is that of changing weather patterns and natural disasters, confirmed via phenomena such as extreme drought in the western United States, the melting of polar ice shelves, increased frequency of flooding events along America's east coast (resulting from the La Nina/El Nino weather cycle off the west coast of Africa) and continual record-setting summer-time temperatures worldwide. While these changing weather patterns have been closely coupled to the concept of global warming, scientists have yet to unanimously agree upon the causes. The debate regarding whether or not global warming is manmade continues to rage, although the evidence that it is occurring is irrefutable.

Human beings living in environments that have been subject to chemical pollutants have been shown to routinely develop adverse health conditions. Environmental pollutants have shown similar effects in animal, insect and plant life. By the early 2000s there was a noticeable reduction in bee populations throughout the United States. Less bees means less pollination capability for crops and ultimately lower food yields. Left unaddressed, a chain of events like this could lead to destabilization of the food table, which could reduce the ability to sustain society. (For a detailed assessment of the bee crisis, see the documentary *More than Honey*.) While it is undoubtedly true that the planet will remain regardless of the outcome of these challenges, the key question is whether or not human life can be sustained after these events.

What about our neighborhood in space?

If we examine our planet in the context of the larger universe, it can easily be seen that we live in a dangerous neighborhood. By January 2020, NASA's Center for Near Earth Object Studies had identified 21,699 NEAs (near-earth asteroids). When an object like the Chelyabinsk meteor of 2013 (approximate size 20 m) enters the Earth's atmosphere, the resultant land strike can devastate human populations in the area surrounding the impact zone. In addition, the Sun, which sustains life on our planet, is a decaying star; approximately 1.5 billion years remain before it burns out. While we do have the option of improving local conditions in order to sustain ourselves in the Earth's environment, even in the best-case scenario of not being adversely affected by challenges internal to the planet, human existence on Earth is term limited. One day we will be forced to leave. If we don't invest the time and effort into developing enabling technologies and a higher understanding of the universe that will allow us to efficiently transition the human race from our planet to another when the time comes for departure, we may not be ready. Extinction would be a sad conclusion to the human family. Thus, space is highly relevant and vital to our future as a species. If we succeed in identifying a new home conducive to human life, technical challenges that will need to be addressed in preparation for our transport to "new Earth" include:

- Heavy lift capability/reusable space vehicles
- Survivability in space (must be addressed in order to get to our intended celestial destination)
  - μ-g effects (long term)
  - Radiation dosage outside the Earth's magnetic field
- Logistics of transport to "new Earth"
  - Earth's present population is approximately 7.5 billion
- Reduction in launch costs
  - Average cost is US$10,000/lb. of launch mass [1]
  - One trip to the International Space Station costs approximately US$58 million.

Members of the space community are charged with working towards answers for today's challenging spacecraft thermal problems that will benefit society in the future. This text is designed to help refine the reader's craft in the technical area of spacecraft thermal design. Welcome aboard and enjoy the reading.

1.3 Space-Based Thermal Energy Analysis Constructs

Spacecraft thermal design presents many problems associated with thermal and/or temperature control of space platform components. Most problems will require temperature control of an embedded component dissipating thermal energy remote from the platform's thermal interface with the space environment. The overall challenge then becomes maintaining temperature and getting the heat out. In Figure 1.1, which illustrates this problem, the source term for the energy to be dissipated by the remote component is $\dot{g}$. Using this construct, there is a thermal conductance coupling (C_{tot}) between the temperature-controlled item at temperature T_i and the radiator surface (where heat rejection from the space platform takes place), as well as a radiative coupling between the temperature-controlled item and the interior surfaces of the structure within its field of view (FOV). The heat-flow path between the temperature-controlled item and the heat rejection location can consist of multiple components, assembly of which will have an effective total thermal conductance. The total thermal conductance term in Figure 1.1 has units of [W/K]. The variable A is the area. Successful operation of the temperature-controlled item will typically require it to be maintained within high and low temperature values. Thermal analysis and design of the whole system, in other words, heat transfer analysis, ensures that the temperature requirements are satisfied while operational in the space environment.

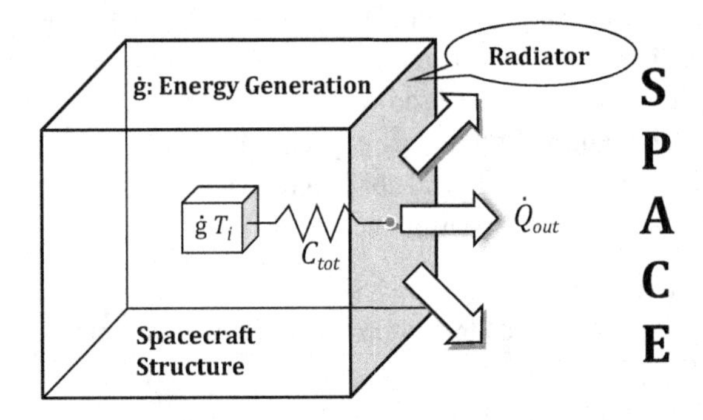

Figure 1.1 Heat flows onboard a simplified space structure

1.4 Units

Heat transfer is defined as the transfer of thermal energy that takes place as a result of temperature differences. In our study of heat transfer and related basic thermodynamics topics, it is important to establish conventions for common variables that allow us to seamlessly interleave the heat transfer concepts and constructs with those commonly used in thermodynamics. In general, thermodynamics deals with energy transfer. Examples of ways that energy is transferred include work, heat and kinetic energy. Regardless of the method of energy transfer, the base unit for energy often used in thermodynamics is the joule [J]. Both work and heat use joules. Heat transfer deals with the rate of thermal energy transfer (i.e., dQ/dt or $\dot{Q}$), which has units of [J/s] or Watts [W]. For equations defining heat transfer mechanisms, heat transfer texts typically use the variable Q and attribute units of [W] to it, as opposed to $\dot{Q}$. For consistency, $\dot{Q}$ (which has units of [W]) will be used throughout this text to define the heat transfer rate. Table 1.1 provides a summary of the base SI units used in the text. Kelvin will be used as the primary unit for temperature. However, Celsius will be used interchangeably for some calculations.

Table 1.1 Base units of measure

Quantity	Unit	Symbol
Length	Meter	m
Mass	Kilogram	kg
Time	Second	s
Temperature	Kelvin	K
	Celsius	°C
Heat	Joules	J
Heat transfer rate	Joules per second	W

1.5 Fundamental Heat Transfer Mechanisms

The majority of this text will focus on conduction and radiation, both of which are highly applicable to the space environment. Note, however, that radiative heat transfer has also been known to dominate convection for surfaces in thermal communication with ambient at suborbital altitudes. While convective heat transfer outside a platform in space is non-existent, boiling and condensation processes, which have been widely investigated, are the basis for heat transfer in several technologies frequently used in thermal control system architectures on space-based platforms. Since boiling and condensation are typically treated as convection based, the constructs used in this text will rely on all three basic mechanisms in heat transfer theory: conduction, convection and radiation.

1.5.1 Conduction

Fourier's Law for conduction (equation 1.1) is the relationship used to model thermal energy exchange between two locations in solid media.

$$\dot{Q} = \frac{kA(T_1 - T_2)}{\Delta x} \tag{1.1}$$

where

k: Thermal conductivity [W/m · K]
A: Heat exchange surface area [m^2]
Δx: Length of conducting path [m]
T_1, T_2: Temperature at each location [K]

The rate of thermal energy transfer $(\dot{Q})$ is equal to the product of the thermal conductivity (k), the cross-sectional area for heat flow ($A = \Delta y \cdot \Delta z$) and the temperature difference at the two discrete measurement locations ($T_1 - T_2$), all of which is divided by the length (Δx) over which heat flow occurs. Based on the direction of the heat flow, in the example of the solid experiencing heat transfer shown in Figure 1.2, $T_1 > T_2$. It is important to note that while the value for the thermal conductivity has temperature dependence (as is the case for all thermophysical properties) it is shown in equation 1.1 as a constant. The acceptance of the thermal conductivity as a constant is generally applicable for steady-state room temperature scenarios (i.e., 270K $\leftrightarrow$ 370K). In cryogenic applications, the temperature dependence of this property is important for determining accurate predictions of heat flows at temperature. This will be revisited and expanded upon in Chapter 7.

In the Fourier's Law conduction equation, the thermal conductivity is the constant of proportionality for the product of the A/L ratio and the temperature gradient. From a physical perspective at standard-size scale, it is the rate at which energy is transferred via diffusion through the media per unit length, per unit temperature. From a solid-state physics perspective, materials are atoms configured into a lattice structure (example in Figure 1.3). The lattice structure contains both immobile electrons (bound at the atoms) and mobile electrons (also known as conduction electrons) that have transport capability through the structure [2]. Conduction-based thermal energy transfer through a material is a function of conduction electron mobility (through the material), as well as vibrations of the material's lattice itself (otherwise known as phonons) [2–4]. The total thermal conductivity is the sum of these two individual contributions (as shown in equation 1.2).

$$k_{tot} = k_e + k_{lat} \tag{1.2}$$

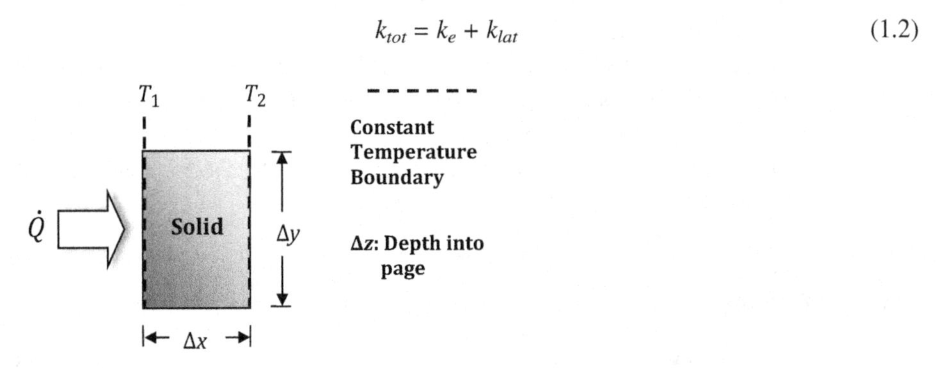

Figure 1.2 1-D heat flow through a solid

For pure metals, the contribution to k_{tot} from the conduction electrons has a much greater magnitude than that from phonons. However, as the amount of impurities in a metal increases, the phonon contribution will also increase. For non-metals, k_{tot} is primarily determined by the phonon level of the lattice structure. Both k_e and k_{lat} can be modeled using kinetic gas theory [2–6]. For pure metals, the electron component of the thermal conductivity (k_e) is inversely proportional to the electrical resistivity (ρ_e) of the material. The equation which captures the relationship between these two properties is the Wiedemann–Franz Law [7], shown in equation 1.3.

$$\rho_e = \frac{L_o}{k} T \tag{1.3}$$

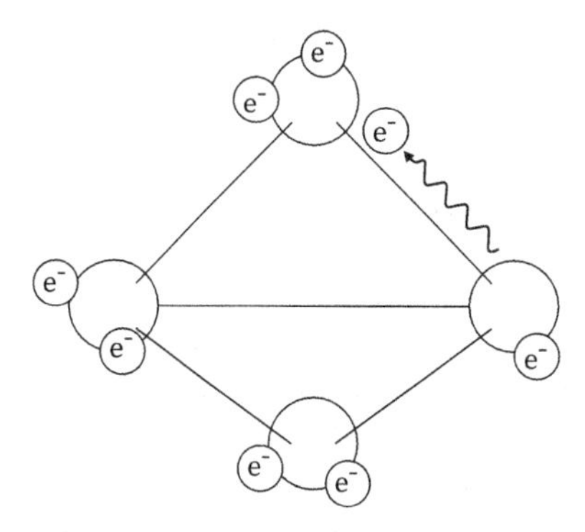

Figure 1.3 Molecular lattice

where

ρ_e: Electrical resistivity [Ohms · m]
L_o: Constant of 2.45×10^{-8} [W · Ohms/K^2]
k: Thermal conductivity [W/m · K]
T: Temperature [K]

In the Wiedemann–Franz relationship, the electrical resistivity is equal to the temperature divided by the thermal conductivity times a constant of proportionality (L_o). The Wiedemann–Franz Law is typically applicable to pure metals at temperatures from 4 K to 300 K and is used in cryogenics to determine thermal conductivity at low temperatures when the electrical resistivity is known and the thermal conductivity is not [7]. However, experimental studies have shown that the Wiedemann–Franz Law can have reduced accuracy in predicting the thermal conductivity of particular metals (e.g., phosphor bronze) in the cryogenic temperature regime [8]. When applying this law attention should be paid to the material in question and the segment of the temperature regime being designed for.

1.5.2 Convection

The fundamental relationship used to model thermal energy exchange between a fluid (either liquid or vapor phase) and a surface is Newton's law of cooling. This relationship is shown in equation 1.4 and illustrated in Figure 1.4.

$$\dot{Q} = h_{conv} A \cdot \left(T_{surf} - T_l\right) \tag{1.4}$$

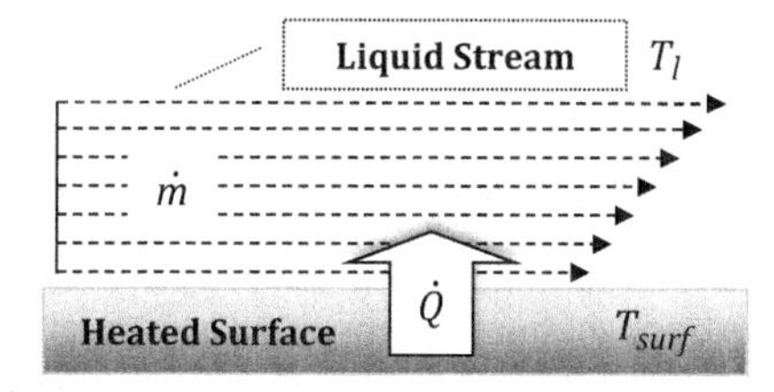

Figure 1.4 Heat transfer resultant from fluid flow over a heated surface

where

h_{conv}: Convection coefficient [W/m^2 · K]
A: Heat exchange surface area [m^2]
T_{surf}: Temperature of heat exchange surface [K]
T_l: Temperature of bulk fluid stream [K]

Depending on the direction of the temperature gradient between the fluid and the surface, the fluid is either heating or cooling the surface. In the example in Figure 1.4, $T_{surf} > T_l$. Thus, the fluid is providing cooling to the surface. In equation 1.4, the rate of heat transfer $(\dot{Q})$ is equal to the product of the heat transfer coefficient (h_{conv}), the temperature difference $T_{surf} - T_l$ and the surface area of the fluid contact (A). The heat transfer coefficient is the constant of proportionality for the process. Furthermore, h_{conv} is a function of the thermophysical properties of the fluid and the flow. In order for Newton's law of cooling to be applicable, the fluid present must satisfy the continuum hypothesis. The continuum hypothesis is valid if the mean free path (i.e., the average distance for molecule-to-molecule collisions) is significantly less than the characteristic length of travel to the bounding surface [9]. The equation used to determine the mean free path is

$$\lambda_{MFP} = \frac{k_B T}{2^{\frac{5}{2}} \pi R_m^2 P} \tag{1.5}$$

where

k_B: Boltzmann's Constant, 1.3805 × 10^{-23} [J/K]
T: Temperature of bounding surface [K]
R_m: Radius of molecule [m]
P: Pressure [N/m^2]

Otherwise, a molecular dynamics-based approach must be used to determine the rate of heating and/or cooling on the surface.

1.5.3 Radiation

The Stefan–Boltzmann Law (equation 1.6) is the fundamental relationship used to model radiative thermal energy rejection from any surface that is at a temperature above the zero point (i.e., 0 K).

$$\dot{Q} = \varepsilon \sigma A T_{surf}^4 \tag{1.6}$$

where

ε: Emissivity $0 < \varepsilon < 1.0$
σ: Stefan–Boltzmann constant, 5.67 × 10^{-8} [W/m^2 · K^4]
A: Heat exchange surface area [m^2]
T_{surf}: Temperature of heat exchange surface [K]

Figure 1.5 shows a surface radiating thermal energy away from it. In the Stefan–Boltzmann Law the heat transfer rate is the product of the emissivity (ε), the Stefan–Boltzmann constant (σ), the surface area emitting energy away (A) and the fourth power of the temperature (T_{surf}). The emissivity is an indicator of how efficiently the surface in question emits thermal energy relative to an ideal surface (i.e., $\varepsilon = 1.0$).

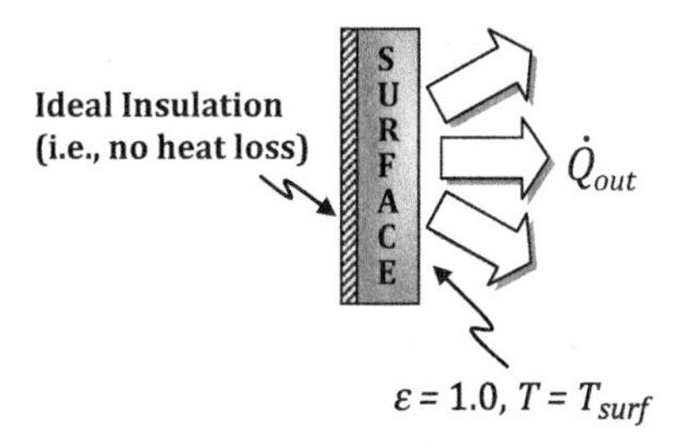

Figure 1.5 A surface radiating thermal energy away

The heat transfer rate's dependence upon the fourth-order temperature suggests that the phenomenon inherent to the process is fundamentally different than those observed in the other two mechanisms (i.e., conduction and convection). The Stefan–Boltzmann Law establishes the analytical relations for radiative thermal energy exchange between multiple surfaces, and will be examined in detail in Chapter 3.

1.6 The Energy Balance

One of the key concepts in mechanical engineering is the law of conservation, which can be applied to mass, momentum and energy. In the heat transfer context, it is also called the energy balance. The principle of conservation of energy is often used to bookkeep heat flows, and the transfer or storage of thermal energy. Throughout the text we will be performing steady-state and transient analysis of thermal circuits containing multiple individual components, using CV (control volume) theory. Figure 1.6 illustrates the energy balance where thermal energy exchange is occurring across non-adiabatic surfaces.

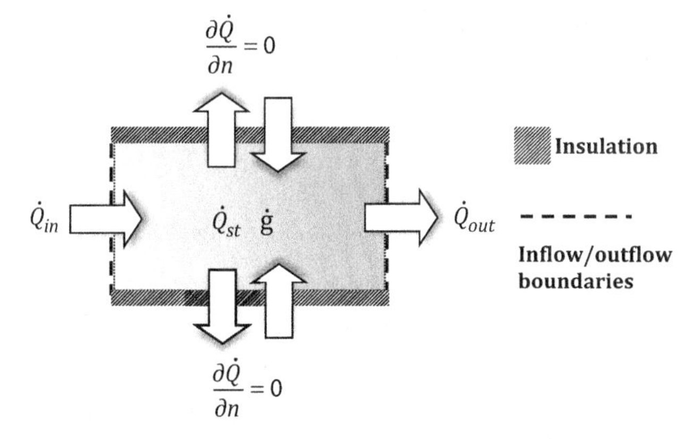

Figure 1.6 Control volume undergoing thermal energy exchange, storage and generation

The corresponding energy balance relationship is shown in equation 1.7.

$$\dot{Q}_{in} + \dot{g} - \dot{Q}_{out} = \dot{Q}_{st} \tag{1.7}$$

This equation states that the rate of thermal energy transfer into the CV plus the rate at which energy is generated ($\dot{g}$) internal to the CV minus that which is exiting the CV is equal to the rate at which thermal energy is stored internally. When no energy is generated, this equation simply reduces to

$$\dot{Q}_{in} - \dot{Q}_{out} = \dot{Q}_{st} \tag{1.8}$$

Once a system has reached equilibrium the energy storage term goes to zero, leaving

$$\dot{Q}_{in} = \dot{Q}_{out} \tag{1.9}$$

The energy balance can be relied upon as a sanity check to confirm that all energy in a system is accounted for in the temperature solution determined.

1.7 Supplemental Resources

A seasoned spacecraft thermal engineer may often perform thermal analysis of an instrument or system with a high node count (30,000–50,000). Thermal models with such a high node count are referred to as a DTM (detailed thermal model). DTM models require computationally intensive software tailored to the space community, preferably with preprocessor and post-processor capability built in. Thermal Desktop® is an example of such a software package. However, the examples and problems in this text are designed for either calculation by hand or solution of an RTM (reduced thermal model) using a standard equation solver. The combined mode heat transfer problems found in the latter part of the text, for example, require limited nodalization schemes (15 nodes or less). Both Matlab and EES (Engineering Equation Solver) are viable solvers for these problems. Internet resources such as the Thermophysical Properties for Fluid Systems section of NIST's Webbook, as well as their online Cryogenic Materials Property Database, are reliable reference sources for thermophysical properties data. Finally, online video content displaying boiling phenomena and the behavior of fluids in the microgravity environment will be referenced in the Chapter 5 discussion on multiphase heat transfer conductance technologies.

References

[1] Larson, W.J., and Wertz, J.R., 1999, *Space Mission Analysis and Design*, Microcosm Press, Portland

[2] Kittel, C., 2005, *Introduction to Solid State Physics*, 8th edn, John Wiley & Sons, Inc., New York, NY

[3] Rosenberg, H.M., 1988, *The Solid State*, Oxford University Press

[4] Ziman, J.M., 1960, *Electrons and Phonons*, Oxford University Press

[5] Ekin, J.W., 2006, *Experimental Techniques for Low Temperature Measurements*, Oxford University Press

[6] Incropera, F.P., and Dewitt, D.P., 1990, *Fundamentals of Heat and Mass Transfer*, John Wiley & Sons

[7] Wiedemann, G., and Franz, R., 1853, *Annalen der Physik und Chemie*, Vol. 89, p. 530

[8] Tuttle, J., Canavan, E., and DiPirro, M., 2009, "Thermal and Electrical Conductivity Measurements of CDA 510 Phosphor Bronze," 2009 Space Cryogenics Workshop, Couer d'Alene, ID

[9] Kaviany, M., 1998, "Flow in porous media," *The Handbook of Fluid Dynamics*, edited by Johnson, R.W., CRC Press, Boca Raton, FL, Chap. 21, pp. 21.35–21.36

Problems

1.1 A CV has an energy inflow of 40 W and an energy outflow of 25 W. What is the amount of energy being stored in the CV?

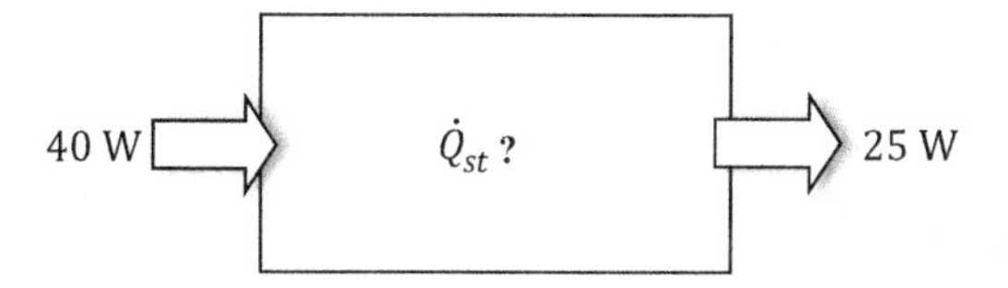

Figure P1.1 Control volume with energy storage

1.2 A CV has thermal energy transfer on four sides. On the left there is a thermal energy flux of 50 W/m^2. Heat transfer into the CV is 10 W from both the top and bottom surfaces. If the surface area undergoing flux-based loading on the left is 0.25 m^2 and $\dot{Q}_{st} = 0$, what is the heat transfer out of the CV?

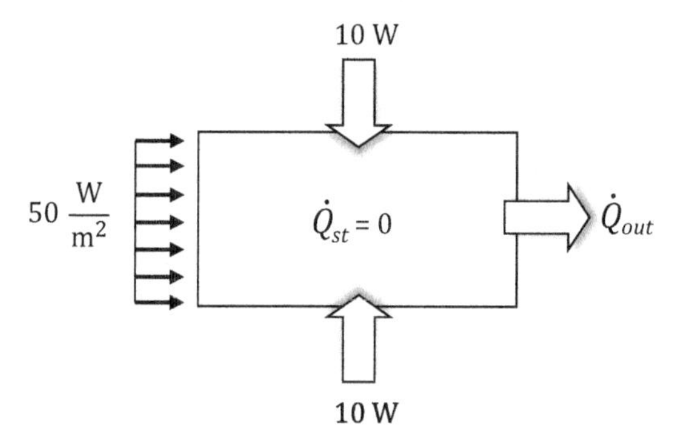

Figure P1.2 Control volume with a heat flux boundary condition

1.3 A plane wall of 2024 aluminum ($k_{Al,\,2024} = 177$ W/m · K) has a height of 0.4 m, a width of 0.2 m and a length of 0.75 m. If 100 W of thermal energy is applied to one end of the wall that is at a temperature of 285 K, assuming 1-D conduction, what is the temperature on the opposite end of the wall?

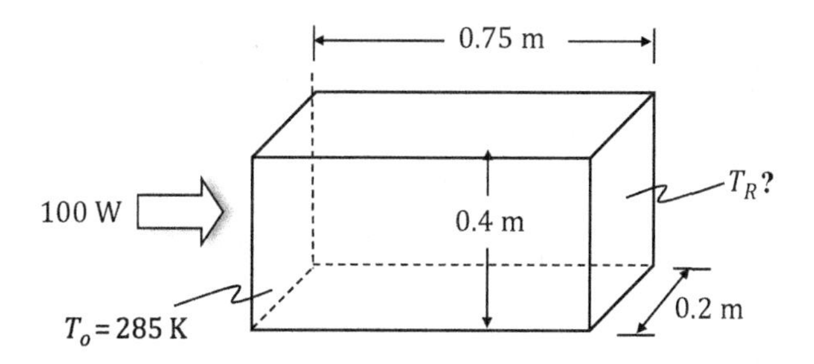

Figure P1.3 Rectangular control volume with unknown boundary temperature

1.4 Chilled water is flowing over a 1.0 m^2 surface that has a temperature of 80°C.
i) If the convection coefficient is 1.5 $W/m^2 \cdot K$ and the stream temperature is T_{surf} = 30°C, what is the thermal energy transferred from the surface to the stream?

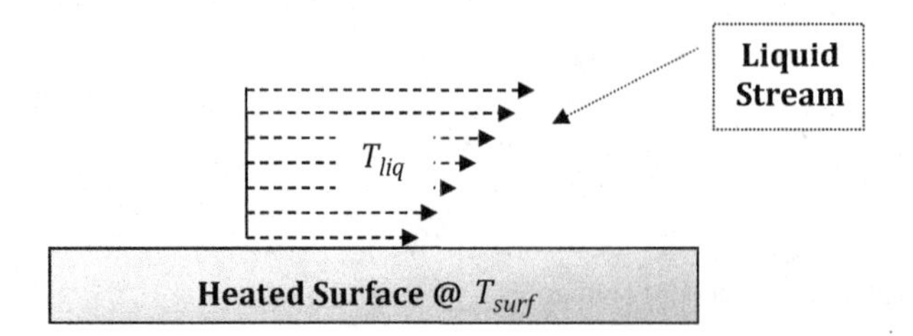

Figure P1.4 Convective heat transfer from a heated surface

ii) How many joules are transferred to the stream over a time of 120 seconds?

1.5 What is the radiative heat transfer from a 0.5 m^2 surface with an emissivity of 0.89 that is at a temperature of 180 K?

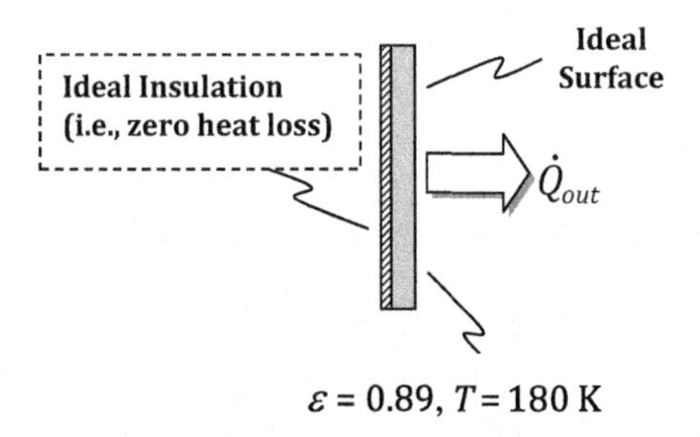

Figure P1.5 Radiating heat transfer surface

1.6 A wire made of platinum has a diameter of 0.2 mm and a resistivity of 11×10^{-8} Ohms · m.
i) What is the thermal conductivity across the wire at a temperature of 293 K?
ii) How does the calculated value in part a) compare to the tabulated value provided by www.matweb.com?

1.7 A system of air molecules are at a temperature of 263 K and a pressure of 20 kPa.
i) Assuming that the diameter of an air molecule is 4.0×10^{-10} m, what is the mean free path of the system of air molecules?
ii) Given that the molecules in part a) are contained in a cubic box measuring 50 cm × 50 cm × 50 cm, does the continuum hypothesis apply to heat transfer phenomena based on the state conditions of the air in the box?

2 Conduction Heat Transfer Analysis

2.1 Introduction

Thermal conduction is one of the two primary heat transfer mechanisms present on space-based platforms. The goal of this chapter is to establish proficiency in the design and interpretation of thermal conductance resistance networks. The geometries studied are assemblies and/or sub-assemblies with multiple components of variable shapes and sizes.

2.2 1-D Conduction

If we derive the energy equation in Cartesian coordinates for a differential element in *dx*, *dy* and *dz*, the final form applicable for a solid experiencing energy generation is shown in equation 2.1.

$$\rho c_p \frac{\partial T}{\partial t} = k \frac{\partial^2 T}{\partial x_i^2} + \dot{g}''' \text{ where } i = \hat{i}, \hat{j}, \hat{k} \tag{2.1}$$

The LHS of the equation contains the transient term which comprises the product of the thermal storage capacity per unit volume (i.e., the product of the density ρ and the specific heat c_p) and the first-order partial derivative of temperature with respect to time (t). The RHS contains the product of the thermal conductivity (k) and the second-order partial derivate of temperature with respect to a specific Cartesian direction (denoted by i), all of which is added to the energy generation per unit volume ($\dot{g}'''$), which has units of W/m^3. In this equation, the thermal conductivity is assumed as being constant and isotropic. Without energy generation the relation reduces to equation 2.2.

$$\rho c_p \frac{\partial T}{\partial t} = k \frac{\partial^2 T}{\partial x_i^2} \tag{2.2}$$

Example 1 The Plane Wall In the 1-D plane wall problem (Figure 2.1), heat is being transferred in the direction of the primary dimension (i.e., the positive x direction). This inherently fosters an isothermal assumption in the y direction for any given x location. Given that $T_1 > T_2$, what will be the temperature as a function of x spanning the length $x = 0$ to $x = L$ at steady state?

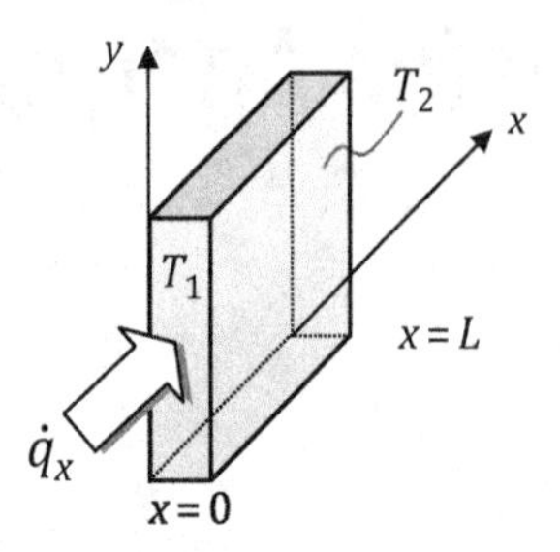

Figure 2.1 Heat flow through a 1-D plane wall
Source: Adapted with permission of John Wiley & Sons, Inc. from figure 1.3, *Fundamentals of Heat and Mass Transfer*, Bergman, T., Lavine, A., Incropera, F.P., and Dewitt, D.P., Vol. 7, 2017, reference [1]; permission conveyed through Copyright Clearance Center, Inc.

Solution

To answer this question we must derive a temperature function dependent upon the variable x. Equation 2.2 can be used as the governing equation, since the problem is one-dimensional with no energy generation. However, the problem is also at steady state which implies that the transient term goes to zero. Thus, our governing equation reduces to

$$k\frac{d^2T}{dx^2} = 0 \tag{2.3}$$

If we divide through by the thermal conductivity (k), this can be recast as

$$\frac{d^2T}{dx^2} = 0 \tag{2.4}$$

In order to get $T(x)$ we must integrate equation 2.4 twice w.r.t. x. This gives the following temperature function with unknown constants of integration c_1 and c_2.

$$T(x) = c_1 x + c_2 \tag{2.5}$$

From Figure 2.1 we can determine that the boundary conditions and the corresponding temperature functions at the boundaries in the x direction are

$$T = T_1 \ @ \ x = 0; \ T_1 = c_1 \cdot 0 + c_2 \tag{2.6}$$

$$T = T_2 \ @ \ x = L; \ T_2 = c_1 \cdot L + c_2 \tag{2.7}$$

Since c_1 and c_2 are unknown they must be determined in order to achieve a general form for the temperature function applicable throughout the x domain. Using equation 2.6 we can determine that $c_2 = T_1$. The new value for c_2 can then be placed in equation 2.7, giving

$$c_1 = \frac{T_2 - T_1}{L} \tag{2.8}$$

Having solved for both constants of integration, they can now be placed in equation 2.5 to recast the temperature function as

$$T(x) = (T_2 - T_1)\frac{x}{L} + T_1 \tag{2.9}$$

What does the thermal circuit for this plane wall problem look like? Since $T = T(x)$, the heat transfer relation for Fourier's law for conduction can be defined in differential form as

$$\dot{q}_{in} = \dot{q}_{out} = -kA\frac{dT}{dx} \tag{2.10}$$

If we apply the derivative w.r.t. x to the temperature function defined in equation 2.9, the relation for heat transfer becomes

$$\dot{q}_{in} = \left[\frac{kA}{L}\right](T_1 - T_2) \tag{2.11}$$

The thermal circuit for heat flow through the plane wall can be illustrated as Figure 2.2.

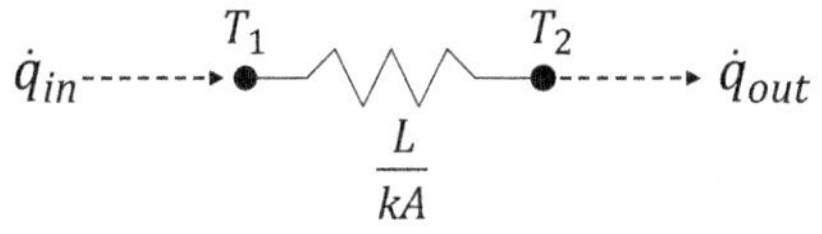

$\dot{q}_{in}$ ‑‑‑‑‑‑‑→ T_1 ●—⁄\/\/\—● T_2 ‑‑‑‑‑‑‑→ $\dot{q}_{out}$

$\frac{L}{kA}$

Figure 2.2 Thermal circuit for heat flow through a 1-D plane wall

The heat flow through the wall experiences a resistance as it propagates between the bounding surface temperatures of T_1 and T_2. The resistance is defined as

$$\frac{L}{kA} \tag{2.12}$$

and has units of [K/W]. It is inversely proportional to the conductance (i.e., resistance = 1/conductance) which is defined as kA/L. The conductance is the factor multiplied by the temperature gradient in equation 2.11. The A term in the resistance and conductance relations represents the cross-sectional area for the heat flow path. The A/L ratio, which is dependent upon the geometry experiencing heat flow, is also known as the shape factor (S).

2.2.1 Series Resistance

Example 2 The Composite Wall Assume two consecutive plane walls of different materials (i.e., different thermal conductivity values) are experiencing heat flow in the direction of the primary dimension (x). Material 2 is bounded on the side opposite material 1 by a flowing fluid (material 3) with a known convection coefficient (h_{conv}) and free stream temperature T_∞ (Figure 2.3). For a common cross-sectional area (A), what is the heat flow through each material and the thermal circuit?

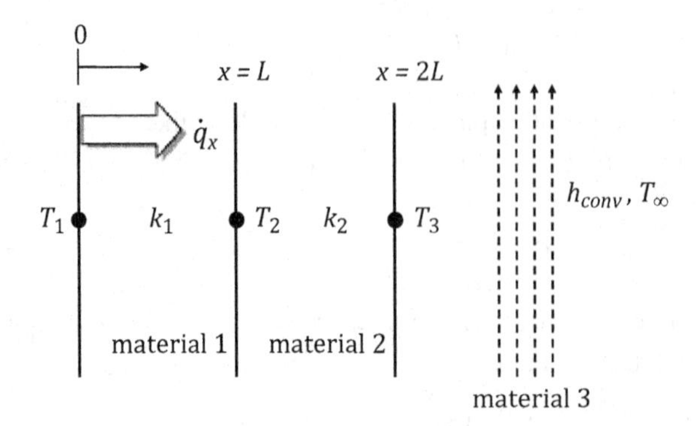

Figure 2.3 Composite wall structure experiencing convective heat transfer along one boundary

Solution

Based on the direction of heat flow imposed, it can be assumed that $T_1 > T_2 > T_3 > T_\infty$. Applying Fourier's Law to materials 1 and 2 gives the following equations

$$\dot{q}_1 = \frac{k_1 A}{L}(T_1 - T_2) \tag{2.13}$$,

$$\dot{q}_2 = \frac{k_2 A}{L}(T_2 - T_3) \tag{2.14}$$

Newton's law of cooling gives the heat transfer from material 2 to material 3 as

$$\dot{q}_3 = h_{conv}A(T_3 - T_\infty) \tag{2.15}$$

The thermal circuit is illustrated in Figure 2.4. Each of the resistances has been placed, in series, between the temperatures bounding the material in question. The direction of heat flow is denoted by the arrows on the heat transfer term $\dot{q}_x$ applied at the bounding temperature nodes, and the heat transfer through each of the materials is the same (i.e., $\dot{q}_x = \dot{q}_1 = \dot{q}_2 = \dot{q}_3$). The total thermal resistance for this series circuit is defined as

$$\frac{1}{R_{tot}} = \frac{1}{[R_1 + R_2 + R_3]} \tag{2.16}$$

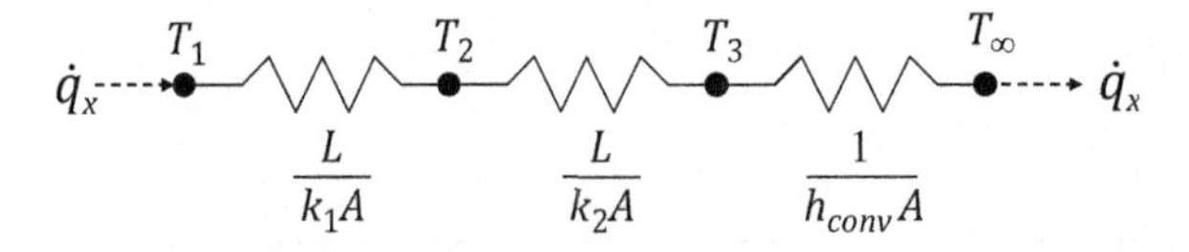

Figure 2.4 Thermal circuit for composite wall structure with a convective boundary condition

where the subscripts on the resistance terms denote the material in question. The total resistance enables us to capture the heat flow between T_1 and T_∞. In the case of the Example 2 problem, application of the series summation for resistances results in equation 2.17.

$$\frac{1}{R_{tot}} = \frac{1}{\left[\frac{L}{k_1 A} + \frac{L}{k_2 A} + \frac{1}{h_{conv} A}\right]} \tag{2.17}$$

Thus, the total heat transfer through this circuit can be recast as

$$\dot{q}_x = \frac{T_1 - T_\infty}{R_{tot}} = \frac{(T_1 - T_\infty)}{\left[\frac{L}{k_1 A} + \frac{L}{k_2 A} + \frac{1}{h_{conv} A}\right]} \tag{2.18}$$

Note that in this equation $1/R_{tot} = C_{tot} = C''_{tot}A$. The heat transfer through the total circuit may thus also be written as

$$\dot{q}_x = C_{tot}(T_1 - T_\infty) \tag{2.19}$$

It is important to properly characterize the thermal conductance when hand calculations are performed or calculations are programmed prior to execution of a software code.

Example 3 Composite Wall with Variable Cross-Sectional Areas As shown in Figure 2.5, the heat flow for this problem will be in the direction of the primary dimension (x). Materials 1, 2 and 3 are all solids with different thermal conductivities. We will assume $A_1 = A_2 = A$ and $A_3 = 0.5A$. Also, we will assume $T_1 > T_2 > T_3 > T_4$. What is the thermal circuit and the heat transfer through it?

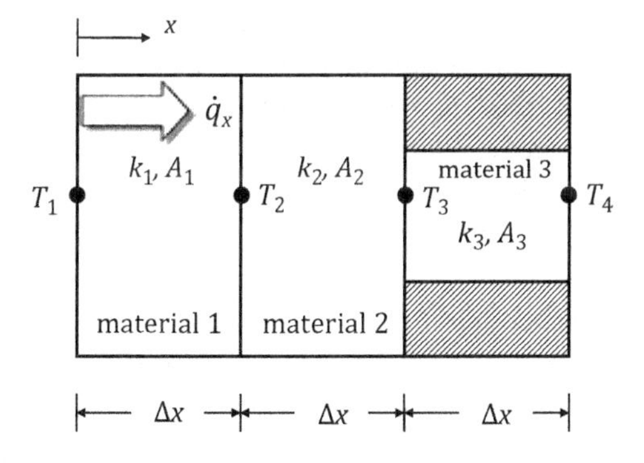

Figure 2.5 Composite wall structure with variable material cross-sectional area

Solution

In the thermal circuit illustrated in Figure 2.6, the common length term of Δx and the area value A have been bookkept for each thermal resistance. Once again, the direction of heat flow is denoted by the arrows on the heat transfer term $\dot{q}_x$ applied at the bounding temperature nodes.

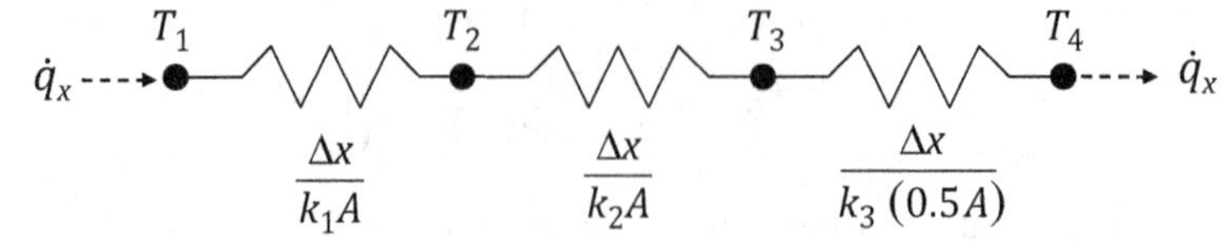

$$\dot{q}_x \rightarrow T_1 \quad \frac{\Delta x}{k_1 A} \quad T_2 \quad \frac{\Delta x}{k_2 A} \quad T_3 \quad \frac{\Delta x}{k_3 (0.5A)} \quad T_4 \rightarrow \dot{q}_x$$

Figure 2.6 Thermal circuit for composite wall structure with variable cross-sectional area

To define the total thermal resistance, we take the thermal resistances as defined in Figure 2.6 and add these together in series format to obtain

$$\frac{1}{R_{tot}} = \frac{1}{\left[\frac{\Delta x}{k_1 A} + \frac{\Delta x}{k_2 A} + \frac{\Delta x}{k_3 (0.5A)}\right]} \tag{2.20}$$

which is also equal to

$$\frac{1}{R_{tot}} = \frac{1}{\left[\frac{\Delta x}{k_1 A} + \frac{\Delta x}{k_2 A} + \frac{2\Delta x}{k_3 A}\right]} \tag{2.21}$$

The total heat transfer then becomes

$$\dot{q}_x = \frac{(T_1 - T_4)}{\left[\frac{\Delta x}{k_1 A} + \frac{\Delta x}{k_2 A} + \frac{2\Delta x}{k_3 A}\right]} \tag{2.22}$$

2.2.2 Interface Resistance

Interface and/or contact resistance between adjoining surfaces can significantly deter heat flow but can be mitigated either using a thermal interface material (TIM) between the surfaces or by applying contact pressure through a bolted joint. Table 2.1 lists some common TIMs used in the aerospace industry. For product specifications on additional TIMs (e.g., GraFoil®, HITHERM™, Stycast® 2850 FT and Eccobond® 285), see references [9–12].

Table 2.2 lists thermal conductance values for bare joint planar surfaces as a function of contact pressure. For a detailed overview of contact pressures as a function of bolted joint configurations, see chapter 8 of the *Spacecraft Thermal Control Handbook* [13].

Table 2.3 lists thermal conductivity values for some common insulating materials at both room temperature and the temperature of liquid helium. For detailed information on the physical properties of the insulators, see the source documents and/or databases referenced.

2.2.3 Limiting Resistance

In a thermal circuit with multiple unequal thermal resistances in series along the heat flow path, the largest thermal resistance will serve as the greatest deterrent to heat flow. This resistance is known as the limiting resistance for the heat flow in the circuit. Let's revisit the problem in Example 3 to determine the limiting resistance. Based on the

Table 2.1 Common aerospace TIMs

Material name	Adhesive	Thermal conductivity (W/m·K)	Temperature range
[1]NuSil™ [2]	Yes	0.609–1.6	158 K–473 K
[2]Cho-Therm® [3]	Optional	1.1–1.4	233 K–473 K
[3]TPLI™ 200 Series [4]	Optional	6	228 K–473 K
[4]Indium [5,6]	No	3,730–81.7	4 K–300 K
[5]Apiezon N (cryogenic) [7]	No	0.095 @2 K, 0.194 @293 K	2 K–303 K
[6]Apiezon L (high vacuum) [8]	No	0.194 @293 K	283 K–303 K

[1] Republished courtesy of NuSil Technology LLC, Carpinteria, CA.
[2] Republished courtesy of Parker-Hannifin Corporation, Chomerics Division, Woburn, MA.
[3] Republished courtesy of LairdTech Corporation, Thermal Materials Division, Cleveland, OH.
[4] Republished with permission of Springer Science from Touloukian, Y.S., Powell, R.W., Ho, C.Y., and Klemens, P.G., "Thermal Conductivity: Metallic Elements and Alloys," *Thermophysical Properties of Matter*, Vol. 1 (1970), pp. 146–151, reference [5]; permission conveyed through Copyright Clearance Center, Inc.
[5] Republished courtesy of Indium Corporation, Clinton, New York.
[6] Republished courtesy of M&I Materials Ltd., Greater Manchester, UK.

Table 2.2 Standard contact resistances for materials in vacuum at room temperature

Surface materials	Contact pressure (MPa)	Surface roughness (μm)	Conductance (W/m·K)
[1]Aluminum 6061-T6	0.5	11.2	625
[2]Aluminum 6061-T6	1	0.3	4,900
[2]Aluminum 6061-T6	5	0.3	12,500
[2]Stainless steel 304	1	0.3	1,200
[2]Stainless steel 304	5	0.3	8,000
[2]OFHC	1	0.2	9,200
[2]OFHC	5	0.2	12,000

[1] Republished with permission of Elsevier from Nishino, K., Yamashita, S., and Torli, K., "Thermal Contact Conductance under Low Applied Load in a Vacuum Environment," *Experimental Thermal and Fluid Science*, Vol. 10 (1995), pp. 258–271, reference [14].
[2] Republished with permission of the American Institute of Aeronautics and Astronautics from Atkins, H.L. and Fried, E., "Interface Thermal Conductance in a Vacuum," *The Journal of Spacecraft and Rockets*, Vol. 2(4) (1965), pp. 591–593, reference [15]; permission conveyed through Copyright Clearance Center, Inc.

individual resistances that comprise R_{tot} (shown in equation 2.21), the one uncommon variable in the fractions in the denominator is the thermal conductivity. The limiting resistance can be determined using the actual value for the thermal conductivity in each material. Alternatively, if we know the relationship between each of the material

Table 2.3 Standard insulators for aerospace applications

		Thermal conductivity (W/m·K)		
Insulator	Material type	@ 293 K	@ 4 K	Material features
[1]G-10 [16]	Fiberglass epoxy laminate	(F) 0.596	(F) 0.072	Difficult to machine, serviceable to ≈140°C
[2]PTFE (i.e., Teflon®)[†] [17]	Fluoropolymer	0.273	0.046	Machinable, CTE sensitive, serviceable to ≈100°C [18]
[3,4]Torlon® 4203 [19,20]	Polyamide-imide	0.260	0.014	Machinable, glass transition @ 277°C
[1,5]Vespel® SP-1 [16,21]	Polyimide	0.35	0.007	Machinable, deflects at 360°C under 2 MPa load
[6]Ultem® 1000 [22]	Polyetherimide	0.388	0.022	See reference [23] for detailed feature info

[†] Polytetrafluoroethylene.
[1] Thermal conductivity data republished with permission of AIP Publishing from Woodcraft, A.L. and Gray, A., "A Low Thermal Conductivity Database," *AIP Conference Proceedings*, Vol. 1185 (2009), pp. 681–684, reference [16].
[2] Republished with permission of Springer Nature from Marquardt, E.D., Lee, J.P., and Radebaugh, R., "Cryogenic Materials Properties Database," *Cryocoolers*, Vol. 11 (2002), pp. 681–687, reference [17].
[3] Republished with permission of Elsevier from Barucci, M., Olivieri, E., Pasca, E., Risegari, L., and Ventura, G., "Thermal Conductivity of Torlon between 4.2K and 300K," *Journal of Cryogenics*, Vol. 45 (2005), pp. 295–299, reference [21].
[4] Room temperature thermal conductivity and material features republished courtesy of Solvay Corporation, Brussels, Belgium.
[5] Material features republished courtesy of DuPont Corporation, Newark, DE.
[6] Thermal Conductivity values provided courtesy of NASA [22].

thermal conductivities relative to a common term, we can still determine the limiting resistance in the circuit. For purposes of the example, let's assume that $k_1 = k_2 = k_3 = k$. If the common term (k) is substituted into the thermal resistances for each material in the example, material 3 is seen to have the largest resistance (i.e., $2\Delta x/kA$), and therefore is the limiting resistance in this thermal circuit. A by-product of this limiting effect might be that temperature T_3 is running warmer than expected. The temperatures in the circuit could be reduced while maintaining the same heat flow in several ways:

i) Selecting a new material 3 that has a higher thermal conductivity;
ii) Altering the geometry (i.e., A/l) to foster a higher thermal conductance; or
iii) Some combination of (i) and (ii).

Identifying the limiting resistances in thermal circuits is a major advantage in the design and analysis of temperature control systems. Silk and Bracken [24] studied spray cooling heat flux performance using POCO high thermal conductivity (HTC) foam attached to an OFHC heater block and compared two TIMs to determine which technique (and/or TIM) resulted in better heat transfer performance. The bonding agents used were Stycast® 2850 FT/Cat-11 and S-Bond® soldering.

Stycast is often used as a bonding agent in the aerospace field as its CTE is similar to that of most metals down to cryogenic temperature ranges. S-Bond® soldering (www.s-bond.com) creates a metallized bond joint. In each case the bonding agent served as the limiting resistance to heat flow in the thermal circuit from the heater block to the liquid droplets impinging the top surface of the foam, therefore the thermal performance of the cooling circuit was a function of the thermophysical properties of the bonding agent. Since the effective thermal conductivity of Stycast is comparable to that of an insulating material and the thermal conductivity of S-Bond® is comparable to that of a moderately conductive metal, the metallized S-Bond® resulted in considerably lower temperature values at the interface of the foam and the copper substrate.

2.2.4 Parallel Resistances

Up to this point we have been assessing thermal circuits with 1-D series resistance configurations where $T = T(x)$. However, real-life problems may contain thermal circuits that include parallel heat flow paths between two elements and/or bounding temperatures. Figure 2.7 shows four parallel resistances.

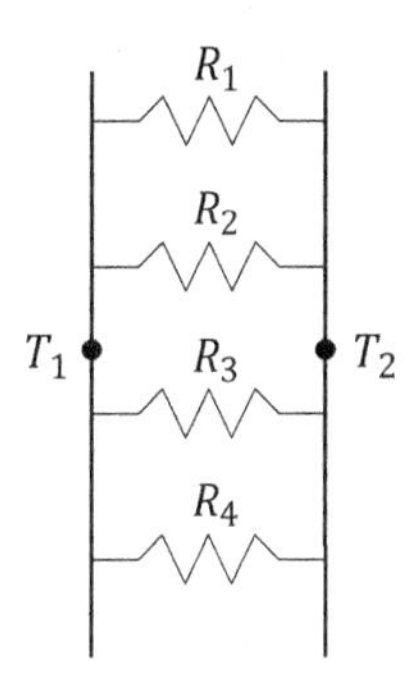

Figure 2.7 Thermal circuit configuration with four parallel resistances

Let's assume we want to define R_{tot} between temperatures T_1 and T_2 for this resistance configuration. Equation 2.23 [25] shows an equality for the overall heat transfer coefficient and R_{tot} as a function of R_1 through R_4.

$$C_{tot} = \frac{1}{R_{tot}} = \frac{1}{R_1} + \frac{1}{R_2} + \frac{1}{R_3} + \frac{1}{R_4} \tag{2.23}$$

Having defined R_{tot} for the parallel resistance configuration, let's now apply this to the analysis of a circuit.

Example 4 Composite Wall with Parallel and Series Heat-Flow Paths Let's assume that in the composite wall configuration shown in Figure 2.8 there is no resistance to heat transfer in the y direction (i.e., temperature does not vary in y) and $T_1 > T_2 > T_3 > T_4$. Materials 1 through 5 are all solids with different thermal conductivities. We will also assume $A_1 = A_5 = A$ and $A_2 = A_3 = A_4 = A/3$. What is the thermal circuit and the heat transfer through it?

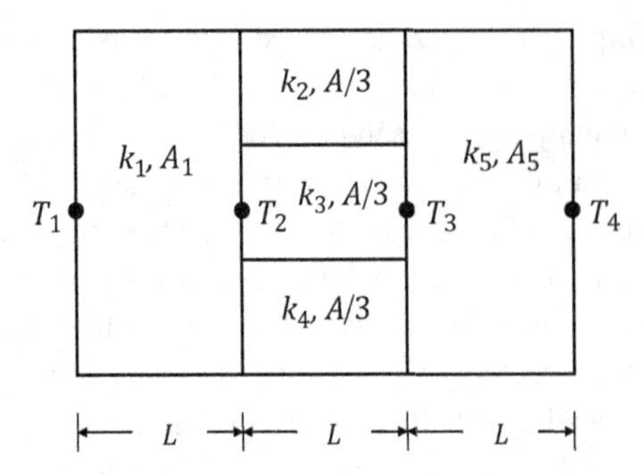

Figure 2.8 Composite wall parallel resistances along heat flow path

Solution

Given that the heat flow is from left to right, materials 2, 3 and 4 represent independent heat flow paths between material 1 and material 5. Two standard approaches are used to capture 1-D heat flows such as those shown in Figure 2.8. These assume that heat flows according to either the isothermal limit or the adiabatic limit [26] within the composite wall. Using an isothermal approach to temperature variation in the y direction at the left and right boundaries with prescribed temperatures of T_1 and T_4 respectively, the thermal circuit will take the form shown in Figure 2.9.

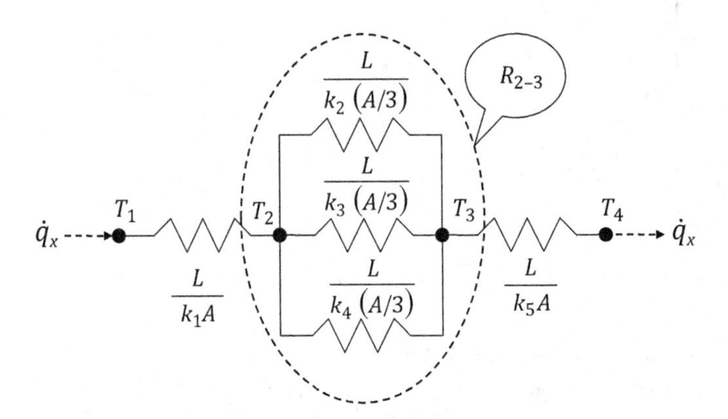

Figure 2.9 Thermal circuit for composite wall with multiple parallel resistances

To capture the individual thermal resistances for elements 1 through 5, elements 2 through 4 can be summed together since they are in parallel, giving the new resistor for this segment of the circuit (R_{2-3}) as equation 2.24.

$$\frac{1}{R_{2-3}} = \frac{k_2 A}{3L} + \frac{k_3 A}{3L} + \frac{k_4 A}{3L} \tag{2.24}$$

If we let $k_2 = k_3 = k_4 = k$, then the thermal resistance for this parallel segment reduces to

$$R_{2-3} = \frac{L}{kA} \tag{2.25}$$

The heat transfer between T_1 and T_4 can now be defined as

$$\dot{q}_x = \frac{(T_1 - T_4)}{\left[\dfrac{L}{k_1 A} + \dfrac{L}{kA} + \dfrac{L}{k_5 A}\right]} \tag{2.26}$$

An alternate approach would be to assume the adiabatic limit in the *y* direction. With adiabatic boundaries along the *y* axis of the composite structure, the thermal circuit is now as shown in Figure 2.10.

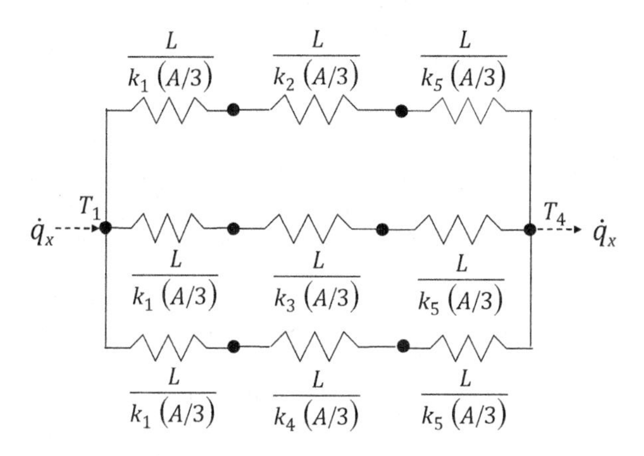

Figure 2.10 Thermal circuit for Example 4 using adiabatic boundaries in the *y* direction

The heat flow is now treated as having three parallel paths between T_1 and T_4. R_{tot} can be expressed using equation 2.27.

$$R_{tot} = \left[\frac{1}{R_{1-2-5}} + \frac{1}{R_{1-3-5}} + \frac{1}{R_{1-4-5}}\right]^{-1} \tag{2.27}$$

Inspection and comparison of the forms for the R_{tot} equations under the adiabatic and isothermal assumptions suggests that these values will be different. However, the full impact of the heat flow assumptions on the total resistance can only be determined through discrete values. Let's calculate R_{tot} with actual thermal conductivity values in both the isothermal and adiabatic limit cases. Table 2.4 provides some examples of thermal conductivity values for standard materials.

Using values from Table 2.4 for the present example, let's assume:

$k_1 = k_{Al} = 177$ W/m · K; $k_2 = k_{Ni} = 90.7$ W/m · K; $k_3 = k_{SS} = 14.9$ W/m · K;
$k_4 = k_{Al} = 177$ W/m · K; $k_5 = k_{CU} = 390$ W/m · K; $A = 11$ cm^2; $L = 0.02$ m

Substituting these values into the two resistance calculations gives an R_{tot} of 0.36 K/W for the isothermal assumption case and 0.42 K/W for the adiabatic assumption case. If we let $T_1 = 40$°C and $T_4 = 10$°C, the heat transfer in the isothermal limit is 83.7 W, whereas in the adiabatic limit it is 71.8 W. In general, the actual resistance in a circuit using a 2-D analysis methodology (which is the most accurate for a 2-D heat transfer problem) will fall somewhere between these values (i.e., $R_{Iso} \leq R \leq R_{Adia}$ [26]). Now let's establish some energy balance constructs for the 2-D problem.

Table 2.4 Thermal conductivity values for some common materials

Material	Thermal conductivity (W/m · K)	Temperature (K)
[1]304 stainless steel [27]	14.9	300
[2]Nickel [28]	90.7	300
[3]2024-T6 aluminum [5]	177	300
[2]6061-T6 aluminum [28]	155.3–167	300
[4]OFHC† [29]	380–400	300
[5]CVD diamond [30]	1,292–1,962	298
[6]Superfluid helium (SHe) [31]	≈85,000	≤2.17‡

† Oxygen-free high thermal conductivity copper.
‡ Lambda point for Helium.
[1] Republished courtesy of CINDAS LLC from *Thermophysical Properties of Selected Aerospace Materials Part II: Thermophysical Properties of Seven Materials*, Touloukian, Y.S. and Ho, C.Y., Editors (1976), pp. 39–46, reference [27], Thermophysical Properties of Matter Database Version 10.0 at http://cindasdata.com/.
[2] Republished courtesy of CINDAS LLC from Touloukian, Y.S., "Recommended values of the thermophysical properties of eight alloys, major constituents and their oxides," reference [28], 1965, Thermophysical Properties of Matter Database Version 10.0 at http://cindasdata.com/.
[3] Republished with permission of Springer Science from Touloukian, Y.S., Powell, R.W., Ho, C.Y., and Klemens, P.G., "Thermal Conductivity: Metallic Elements and Alloys," *Thermophysical Properties of Matter*, Vol. 1 (1970), reference [5]; permission conveyed through Copyright Clearance Center, Inc.
[4] Republished from NIST Monograph 177, Simon, N.J., Drexler, E.S., and Reed, R.P., "Properties of Copper and Copper Alloys at Cryogenic Temperature," (1992), pp. 7–23, reference [29].
[5] Republished with permission of Elsevier from Graebner et al., "Report on a Second Round Robin Measurement of the Thermal Conductivity of CVD Diamond," *Diamond and Related Materials,* Vol. 7 (1998), pp. 1589–1604, reference [30].
[6] Republished with permission of the Licensor (Oxford Publishing Limited) through PLSclear, Barron, R., *Cryogenic Systems*, 2nd edn, 1985, reference [31].

2.2.5 Multiple Energy Sources

Up to this point we have dealt only with examples that include energy input from a single component (and/or location) in the thermal circuit. Typically this has been represented by the dissipation term associated with an E-box (electronics box). The corresponding heat rejection has been to a single heat sink. Let's look at a case where multiple sources for heat input are present. The thermal circuit in Figure 2.11 consists of 4 nodes (each representing an individual element), and Figure 2.12 shows the circuit's thermal resistance network. Heat input to the overall circuit occurs at nodes 1 and 2. Heat rejection occurs at node 4.

To solve the temperature values within this system we need to perform an energy balance at each of the nodes. The energy balance at node 1 is

$$\dot{q}_{in,1} = \dot{q}_{out,1} \tag{2.28}$$

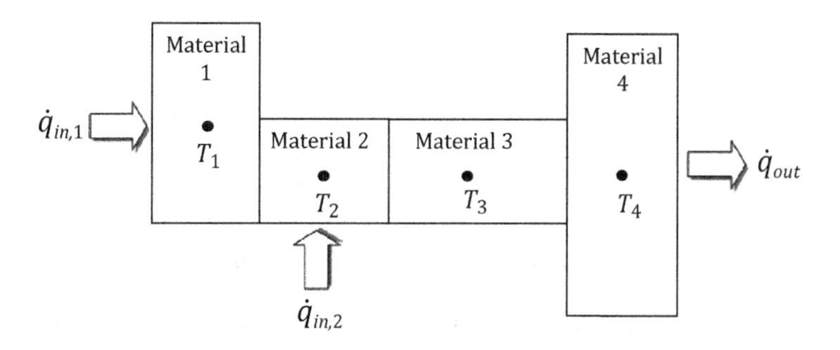

Figure 2.11 Thermal circuit with multiple heat input locations

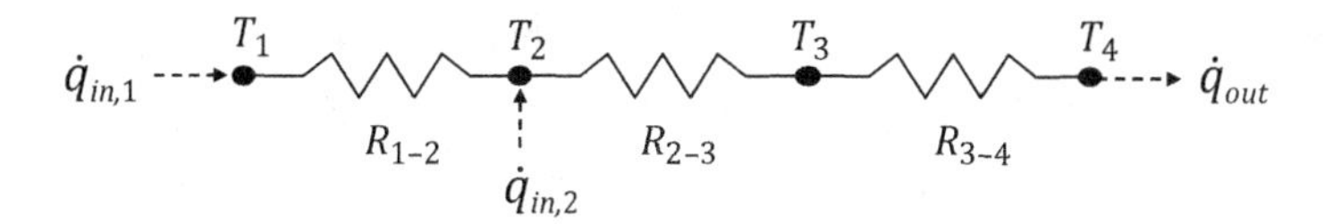

Figure 2.12 Thermal resistance network for circuit with multiple heat input locations

where the energy departing node 1 is transferred to node 2. At node 2, the energy balance is

$$\dot{q}_{in,1} + \dot{q}_{in,2} = \dot{q}_{out,2} = \dot{q}_{out} \tag{2.29}$$

The total energy input to node 2 consists of the thermal energy transferred from node 1 summed with the direct heating to material 2. The summation of these is the amount of thermal energy leaving node 2 and ultimately transferring through the rest of the circuit (in the direction of node 4). The temperature at nodes 3 and 4 is determined by working through the circuit and applying energy balances at nodes 3 and 4.

2.3 Finite Difference, Finite Element and the Energy Balance

Thus far the example problems presented have specified temperatures in the thermal circuit at the boundaries and/or interfaces of the circuit elements. Analysis of thermal problems typically incorporates the use of either a representative temperature for the entire element in question or a temperature function that spatially resolves the temperature across the area of the element. Before advancing any further, let's take a look at these two interpretations.

The way the temperature throughout the material element is treated is a function of whether finite difference or finite element constructs are used to define the elements in the circuit, as well as the solution for the elements' temperatures. In the finite difference approach a single node is assigned at the center location of the element (Figure 2.13). The temperature throughout the entire element is considered uniform and equal to the value at the center node.

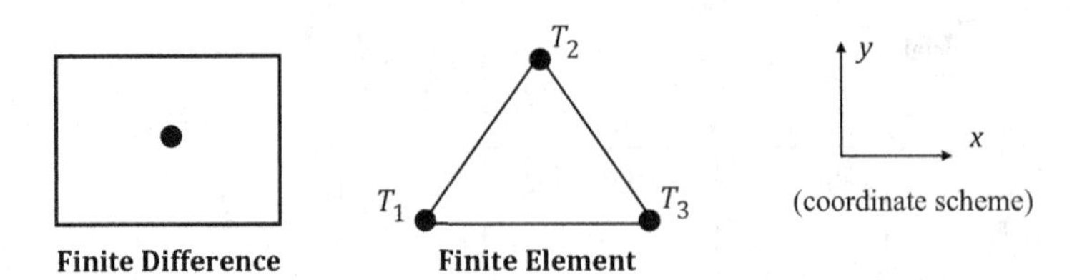

Figure 2.13 Finite difference and finite element nodalization schemes

In the finite element approach to temperature assignment, the boundaries of the element constitute the actual nodes and have temperatures assigned to them. The temperature throughout the element is spatially resolved [32] as a function of the node temperatures and the area the element spans. The temperature function for the triangular finite element shown in Figure 2.13 can be expressed as $T = f(T_1, T_2, T_3, x, y)$. The equation to capture the temperature variation throughout the element would look like

$$T_{FE} = T_1 a + T_2 b + T_3 c \tag{2.30}$$

where a, b and c are area coordinates that provide a relative weight to the temperature nodes of the element [32]. Finite difference methodology applies a differencing scheme to a PDE that serves as the governing equation. The relevant thermal energy equation is:

$$\rho c_p \frac{\partial T}{\partial t} = k\frac{\partial^2 T}{\partial x^2} + k\frac{\partial^2 T}{\partial y^2} + k\frac{\partial^2 T}{\partial z^2} + \dot{g}''' \tag{2.31}$$

Software analysis tools discretize the energy equation by substituting in a differencing scheme for the partial derivatives (both first and second order) and solving it numerically throughout the domain. The technique we have used in our simplified thermal circuits does not incorporate the thermal energy equation given above. Rather, we have applied a differencing scheme to Fourier's Law and incorporated this into an energy balance technique at steady state. Each element in a thermal circuit has been assigned a single node which serves as the representative temperature for the entire element. This approach is an extension of the finite difference method [1,25,26]. Figure 2.14 is a Cartesian-based diagram of a single element surrounded by multiple adjoining elements with which it is in thermal communication. The primary 2-D axes (x and y) are shown in part a) as well as the nodal scheme axes and associated thermal resistances in part b). The primary node of interest is located in element 1 and has nodal coordinates of a, b. If element 1 is part of a larger nodal network representing a material component, then the energy balance can be used to determine heat flows into and out of element 1 (as shown by the heat transfer terms in part b) for each adjoining element). Assuming steady-state heat transfer is taking place without energy generation (i.e., $\dot{E}_{St} = 0$ and $\dot{E}_g = 0$), we know that the energy balance for element 1 (and/or its node) can be expressed as

$$\dot{q}_1 + \dot{q}_2 + \dot{q}_3 + \dot{q}_4 = 0 \tag{2.32}$$

Fourier's Law can be substituted in for the heat transfer terms, representative of the thermal energy exchange experienced with each of the adjoining elements. For heat transfer the relevant thermal energy equation is:

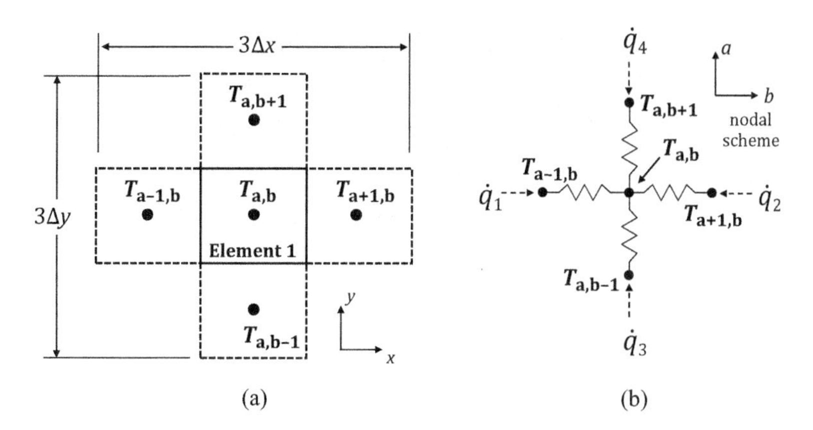

Figure 2.14 Finite difference element configuration in thermal communication with surrounding elements: a) element configuration

Source: Adapted with permission of John Wiley & Sons, Inc. from figure 4.5, *Fundamentals of Heat and Mass Transfer*, Bergman, T., Lavine, A., Incropera, F.P., and Dewitt, D.P., Vol. 7, 2017, reference [1], permission conveyed through Copyright Clearance Center

b) resistance network

$$\frac{k\cdot\Delta y\cdot th}{\Delta x}(T_{a-1,b}-T_{a,b})+\frac{k\cdot\Delta y\cdot th}{\Delta x}(T_{a+1,b}-T_{a,b})+\frac{k\cdot\Delta x\cdot th}{\Delta y}(T_{a,b-1}-T_{a,b})$$
$$+\frac{k\cdot\Delta x\cdot th}{\Delta y}(T_{a,b+1}-T_{a,b})=0 \tag{2.33}$$

where *th* represents the thickness of the material (i.e., into the plane of the page). Also, the thermal conductivity (k) is assumed as isotropic and constant. Based on Figure 2.14a, $\Delta x = \Delta y$. If we incorporate these assumptions into equation 2.33, it reduces to

$$k\cdot th\cdot T_{a-1,\,b}+k\cdot th\cdot T_{a+1,\,b}+k\cdot th\cdot T_{a,\,b-1}+k\cdot th\cdot T_{a,\,b+1}-4k\cdot th\cdot T_{a,\,b}=0 \tag{2.34}$$

This can be further reduced by dividing the equation by $k\cdot th$. The equation now reduces to

$$T_{a-1,\,b}+T_{a+1,\,b}+T_{a,\,b-1}+T_{a,\,b+1}-4T_{a,\,b}=0 \tag{2.35}$$

If an energy balance is applied at each node in the thermal circuit, a series of equations similar to 2.35 is obtained. The comprehensive set of energy balance equations can be used to solve for each temperature in the thermal circuit using EES.

Several of the earlier examples consisted of basic geometries where the length and height were the same. For cases where joined material elements have unequal lengths, the previously established constructs for thermal circuit analysis and the thermal resistance still apply. Borrowing from finite difference methodology, we use a representative node at the center of each element (Figure 2.15). In this case, the centers of the elements are aligned. Assuming materials 1 and 2 are the same and heat flow occurs from element 1 to element 2, the thermal circuit would then have a heat transfer rate of

$$\dot{q}_x = \frac{k \cdot \Delta y \cdot th}{1.5L}(T_1 - T_2) \tag{2.36}$$

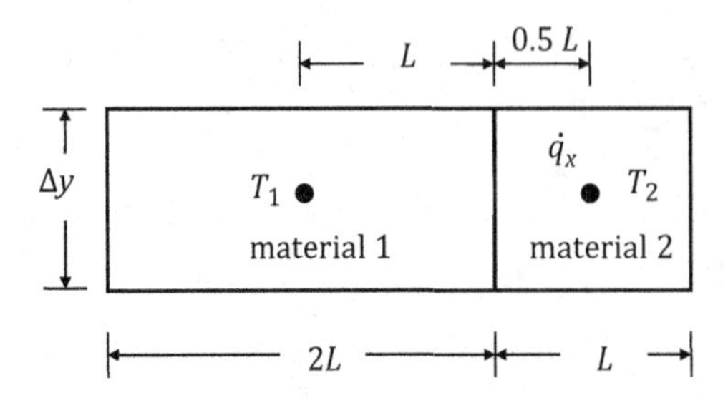

Figure 2.15 Example finite difference nodal scheme

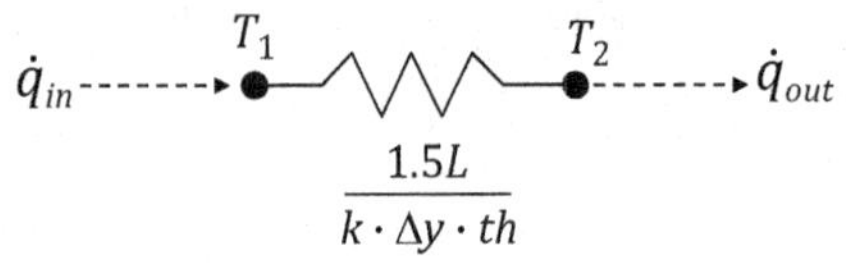

$\dot{q}_{in}$ → T_1 — $\frac{1.5L}{k \cdot \Delta y \cdot th}$ — T_2 → $\dot{q}_{out}$

Figure 2.16 Thermal circuit and resistance for elements of unequal length and aligned nodes

If the material for elements 1 and 2 were dissimilar (see Figure 2.15) then the total resistance for the circuit would be

$$\frac{1}{R_{tot}} = \frac{1}{\left[\frac{L}{k_1 \cdot \Delta y \cdot th} + \frac{0.5L}{k_2 \cdot \Delta y \cdot th}\right]} \tag{2.37}$$

This can be further reduced to equation 2.38.

$$\dot{q}_x = \frac{(T_1 - T_2)}{\left[\frac{L}{k_1 \cdot \Delta y \cdot th} + \frac{0.5L}{k_2 \cdot \Delta y \cdot th}\right]} = \frac{(T_1 - T_2)}{\frac{L}{\Delta y \cdot th}\left[\frac{1}{k_1} + \frac{0.5}{k_2}\right]} \tag{2.38}$$

If the heat flow path is not in a straight line yet the nodes are taken at the center of the elements, we project material 1 into the space of material 2 (Figure 2.17) along the heat flow path from node to node. To determine $\dot{q}_{1-2}$ we must first determine the total resistance.

Let's assume *th* is the thickness of the elements into the page. The total resistance may then be captured as

$$\frac{1}{R_{tot}} = \frac{1}{[R_1 + R_{2a} + R_{2b}]} \tag{2.39}$$

where $R_1 = 0.5L/k_1 \cdot \Delta y \cdot th$, $R_{2a} = 0.5\Delta x/k_{2a} \cdot \Delta y \cdot th$, $R_{2b} = (0.5L - 0.5\Delta y)/k_{2b} \cdot \Delta x \cdot th$

If the materials used are the same, then we can assume $k = k_1 = k_2$. The total resistance becomes

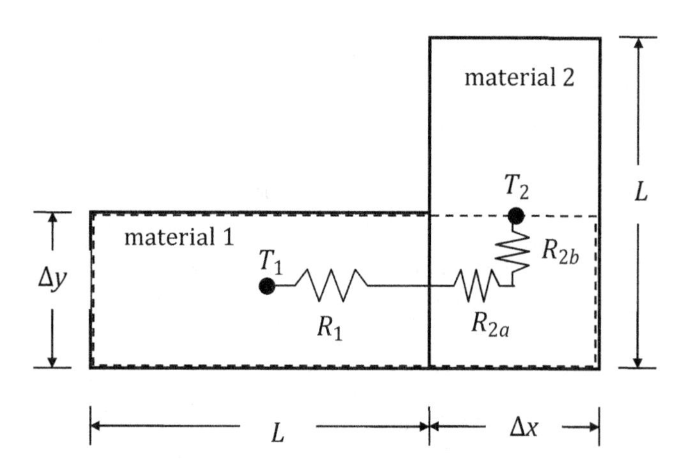

Figure 2.17 Elements with misaligned nodes

$$\frac{1}{R_{tot}} = \frac{1}{\left[\frac{0.5L + 0.5\Delta x}{k \cdot \Delta y \cdot th} + \frac{0.5L - 0.5\Delta y}{k \cdot \Delta x \cdot th}\right]} = \frac{1}{\left(\frac{0.5}{k \cdot th}\right)\left[\frac{L + \Delta x}{\Delta y} + \frac{L - \Delta y}{\Delta x}\right]} \tag{2.40}$$

and the total heat transfer is

$$\dot{q}_{1-2} = \frac{(T_1 - T_2)}{\left(\frac{0.5}{k \cdot th}\right)\left[\frac{L + \Delta x}{\Delta y} + \frac{L - \Delta y}{\Delta x}\right]} \tag{2.41}$$

Figure 2.18 illustrates a circuit composed of three elements, some of which have different cross-sectional areas experiencing heat transfer. The resistance network for the thermal circuit is shown in Figure 2.19.

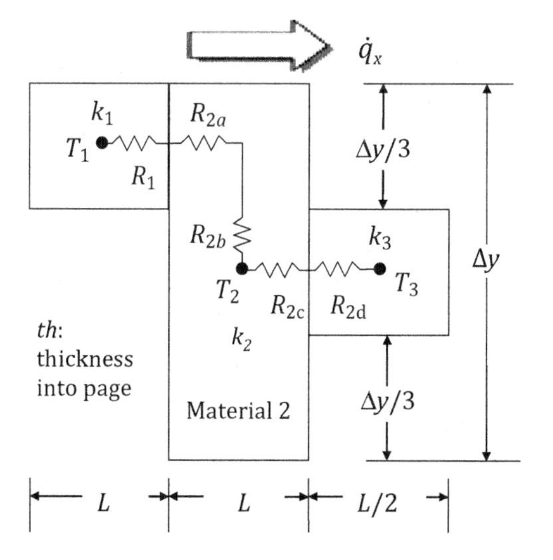

Figure 2.18 Thermal circuit with aligned nodes and different cross-sectional areas

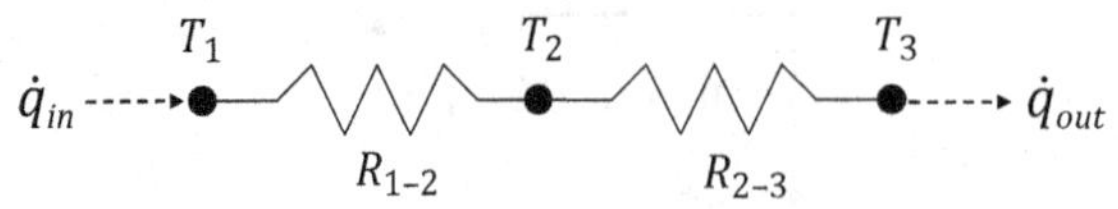

$$\dot{q}_{in} \dashrightarrow T_1 \; R_{1\text{-}2} \; T_2 \; R_{2\text{-}3} \; T_3 \dashrightarrow \dot{q}_{out}$$

Figure 2.19 Resistance network for thermal circuit with aligned nodes and different cross-sectional areas

R_{1-2} may be expressed as the summation of $R_1 + R_{2a} + R_{2b}$. To determine the total thermal resistance, using this approach and assuming a depth of *th* into the page, R_{1-2} can then be defined as

$$R_{1-2} = \frac{0.5L}{k_1\left(\frac{\Delta y}{3}\right)th} + \frac{0.5L}{k_2\left(\frac{\Delta y}{3}\right)th} + \frac{\Delta y/3}{k_2 \cdot L \cdot th} \tag{2.42}$$

Next we must define R_{2-3}, using either the adiabatic or the isothermal assumption. Under the adiabatic assumption, the area used for the R_{2c} calculation would be based on element 3's area (i.e., a projection of element 3's area into element 2). This would give an R_{2-3} relation of

$$R_{2-3} = \frac{0.5L}{k_2\left(\frac{\Delta y}{3}\right)th} + \frac{0.25L}{k_3\left(\frac{\Delta y}{3}\right)th} \tag{2.43}$$

If an isothermal approach is used to define R_{2c}, the R_{2-3} relation would become

$$R_{2-3} = \frac{0.5L}{k_2(\Delta y)th} + \frac{0.25L}{k_3\left(\frac{\Delta y}{3}\right)th} \tag{2.44}$$

The adiabatic approach for determining R_{2-3} (equation 2.43) is consistent with the adiabatic approach used for determining R_{1-2} in the misaligned node scenario. This is reinforced by the fact that if the offset for element 1's node were reduced until it was aligned with the nodes for elements 2 and 3, the third term in equation 2.42 would become zero, thus making the second term in R_{1-2} (i.e., R_{2a}) equal to the first term in R_{2-3} (i.e., R_{2c}) for equation 2.43. Equation 2.44 shows that this does not hold for the isothermal approach, although this does include the entire cross-sectional area in element 2 experiencing heat flow towards element 3. The isothermal resistance will be slightly less than the actual resistance, whereas the adiabatic resistance will be slightly greater. In the present problem, accounting for all of the cross-sectional area experiencing heat flow in resistance R_{2-3} makes the total resistance

$$R_{tot} = \frac{0.5L}{k_1\left(\frac{\Delta y}{3}\right)th} + \frac{0.5L}{k_2\left(\frac{\Delta y}{3}\right)th} + \frac{\Delta y/3}{k_2 \cdot L \cdot th} + \frac{0.5L}{k_2(\Delta y)th} + \frac{0.25L}{k_3\left(\frac{\Delta y}{3}\right)th} \tag{2.45}$$

Assuming that $L = \Delta y/3$ further reduces this relation to

$$R_{tot} = \frac{0.5}{k_1 \cdot th} + \frac{1.66}{k_2 \cdot th} + \frac{0.25}{k_3 \cdot th} \tag{2.46}$$

In some physical cases, as we shall see later, it is more logical to opt for the adiabatic case. Note that resistance approximations for reduced node thermal circuits are subject to estimation error; more accurate results are achieved by increasing the nodal density of the elements in the circuit.

Example 5 The L-bracket Problem An E-box is mounted on an L-bracket as shown in Figure 2.20. What is the temperature of the E-box (i.e., T_e) given the following assumptions?

- Materials 1 and 2 have a depth into the page (relative to the side view) of 0.25 m.
- The electronics box is 0.1 m long and 0.05 m high.
- T_2 is held constant via active cooling at 10°C.
- T_e is generating 30 W of dissipative heat.
- Assume 6061-T6 Al for all materials.
- For analysis purposes, model the electronics box as a solid structure.

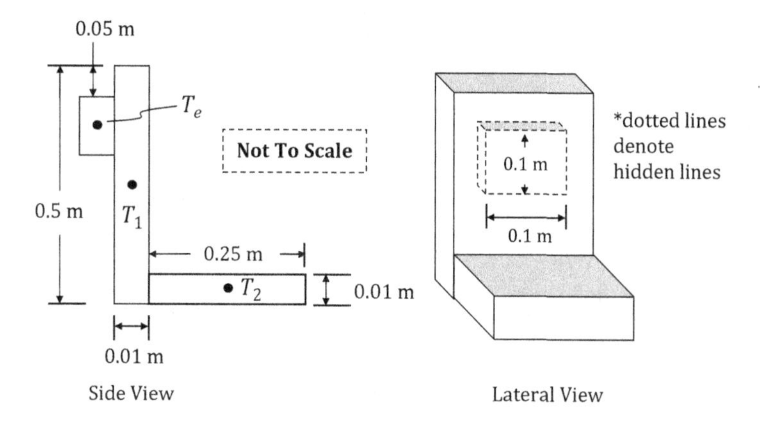

Figure 2.20 Electronics box mounted on an L-bracket

Solution

Figure 2.21 shows the thermal circuit for this configuration.

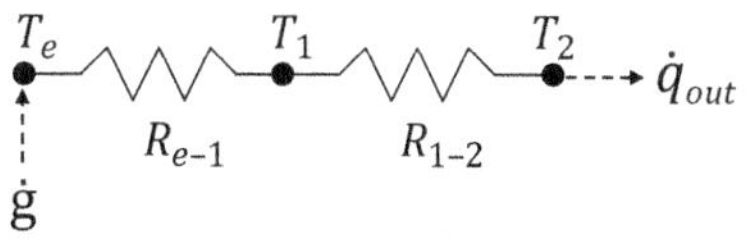

T_e R_{e-1} T_1 R_{1-2} T_2 $\dot{q}_{out}$
$\dot{g}$

Figure 2.21 Resistance network for L-bracket problem

The energy balance can be taken on any of the nodes (i.e., the E-box, element 1 or element 2). A comprehensive energy balance can also be taken on the composite circuit (i.e., from the E-box to node 2). We know $T_2 = 10°C$ and $\dot{g} = 30\,W$. For the heat transfer through this thermal circuit we can take the heat flow from T_e to T_2. Our heat transfer relation would then be

$$\dot{q}_{e-2} = 30\,\mathrm{W} = \frac{(T_e - T_2)}{R_{tot}} \tag{2.47}$$

where

$$\frac{1}{R_{tot}} = \frac{1}{[R_{e-1} + R_{1-2}]} \tag{2.48}$$

In order to determine R_{tot} we must first calculate R_{e-1} and R_{1-2}. R_{e-1} can be determined as being

$$\frac{1}{R_{e-1}} = \frac{1}{[R_e + R_{1a} + R_{1b}]} \tag{2.49}$$

where R_e and R_{1a}/R_{1b} are the resistances from the center of the E-box and element 1 to the interface shared between the two (for the thermal conductivity of 6061-T6 Al, see Table 2.4). Substituting in actual values gives equation 2.50 as

$$\frac{1}{R_{e-1}} = \frac{1}{\left[\frac{0.05\ \mathrm{m}}{(167\ \mathrm{W/m\cdot K})(0.1\ \mathrm{m})^2} + \frac{0.005\ \mathrm{m}}{(167\ \mathrm{W/m\cdot K})(0.1\ \mathrm{m})^2} + \frac{0.15\ \mathrm{m}}{(167\ \mathrm{W/m\cdot K})(0.01\ \mathrm{m})(0.25\ \mathrm{m})}\right]} \tag{2.50}$$

This further reduces to

$$\frac{1}{R_{e-1}} = \frac{1}{[0.03\ \mathrm{K/W} + 0.003\ \mathrm{K/W} + 0.36\ \mathrm{K/W}]} \tag{2.51}$$

Thus $1/R_{e-1} = 2.545$ W/K.

R_{1-2} can now be defined as

$$\frac{1}{R_{1-2}} = \frac{1}{[R_1 + R_2]} \tag{2.52}$$

or

$$\frac{1}{R_{1-2}} = \frac{1}{[R_{1c} + R_{1d} + R_2]} \tag{2.53}$$

Substituting in actual values gives equation 2.54 as

$$\frac{1}{R_{1-2}} = \frac{1}{\left[\frac{0.245\ \mathrm{m}}{(167\ \mathrm{W/m\cdot K})(0.1\mathrm{m})(0.25\ \mathrm{m})} + \frac{0.005\ \mathrm{m}}{(167\ \mathrm{W/m\cdot K})(0.1\ \mathrm{m})(0.25\ \mathrm{m})} + \frac{0.05\ \mathrm{m}}{(167\ \mathrm{W/m\cdot K})(0.01\ \mathrm{m})(0.25\ \mathrm{m})}\right]} \tag{2.54}$$

Reducing each of the thermal resistances will result in equation 2.55.

$$\frac{1}{R_{1-2}} = \frac{1}{[0.587\ \mathrm{K/W} + 0.012\ \mathrm{K/W} + 0.12\ \mathrm{K/W}]} \tag{2.55}$$

Thus $1/R_{1-2} = 1.391$ W/K. Placing the values for R_{e-1} and R_{1-2} into our relation for R_{tot} we get

$$\frac{1}{R_{tot}} = \frac{1}{[0.393\ \mathrm{K/W} + 0.719\ \mathrm{K/W}]} \tag{2.56}$$

This further reduces to

$$\frac{1}{R_{tot}} = \frac{1}{1.112\ \mathrm{K/W}} = 0.9\ \mathrm{W/K} \tag{2.57}$$

Referring back to the heat transfer relation in equation 2.47, we have

$$30\ \mathrm{W} = (T_e - T_2)/R_{tot}$$

If both sides are multiplied by R_{tot}, the new relation becomes

$$R_{tot} \cdot 30\ \mathrm{W} = T_e - T_2 \tag{2.58}$$

Solving for T_e gives

$$T_e = R_{tot} \cdot 30\ \mathrm{W} + T_2 \tag{2.59}$$

Substituting in R_{tot} results in

$$\begin{aligned} T_e &= (1.112\ \mathrm{K/W})(30\ \mathrm{W}) + 283\ \mathrm{K} \\ &= 316.4\ \mathrm{K}\ \text{or}\ 43.4°\mathrm{C} \end{aligned} \tag{2.60}$$

Example 6 The Six-Sided Box with an Energy Source Figure 2.22 shows a box with an electronics box mounted on the top. What is the thermal circuit and the temperature of the E-box given the following assumptions?

- The lower box has a side length of 0.5 m and a wall thickness of 0.005 m.
- The electronics box is dissipating 40 W of thermal energy.
- The electronics box has a side length of 0.01 m.
- Side 5 (i.e., the bottom surface of the lower box) is held at a constant temperature of 10°C.
- Both boxes are made of 2024-T6 Al (i.e., $k_{Al} = 177\ \mathrm{W/m \cdot K}$).

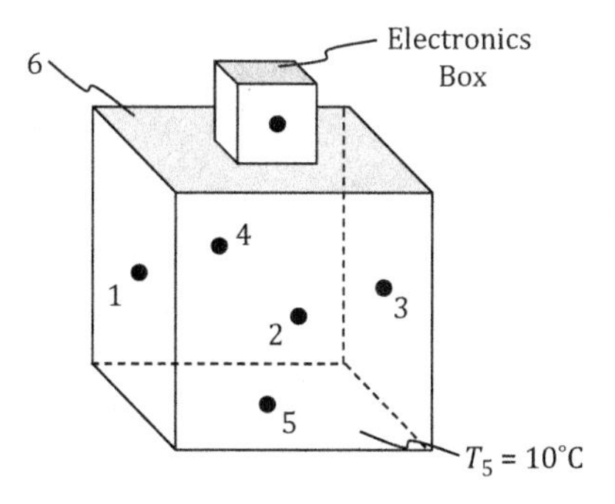

Figure 2.22 Electronics box mounted on top of a six-sided structural box

Solution

The thermal resistances can be determined from the geometry and configuration of the boxes. However, there are six unknown temperatures. In order to solve the entire system of balance equations and temperatures, at least as many equations as unknown temperatures are needed. Since an energy balance can be performed for each node (i.e., on each of the material elements), more than enough equations are available. The geometric symmetry of the configuration confirms that the resistances from the bottom of the lower box (node 5) to the sides of the lower box (nodes 1, 2, 3 and 4) are equal. In other words

$$R_{1-5} = R_{2-5} = R_{3-5} = R_{4-5} \tag{2.61}$$

A similar equality holds true for the top side of the lower box (node 6) and sides (nodes 1, 2, 3 and 4).

$$R_{6-1} = R_{6-2} = R_{6-3} = R_{6-4} \tag{2.62}$$

Due to the symmetry of the lower box all of these resistances are equal.

$$R_{1-5} = R_{2-5} = R_{3-5} = R_{4-5} = R_{6-1} = R_{6-2} = R_{6-3} = R_{6-4} \tag{2.63}$$

For purposes of simplification, these resistances can all be said to be equal to R_A. Regarding the lateral heat flow resistances along the lower box, symmetry gives

$$R_{1-2} = R_{2-3} = R_{3-4} = R_{4-1} \tag{2.64}$$

The remaining resistance is that between the electronics box and the lower box itself (i.e., R_{e-6}). The thermal circuit for this problem is shown in Figure 2.23.

To determine the temperature of the E-box, its energy balance needs to be established using Fourier's Law.

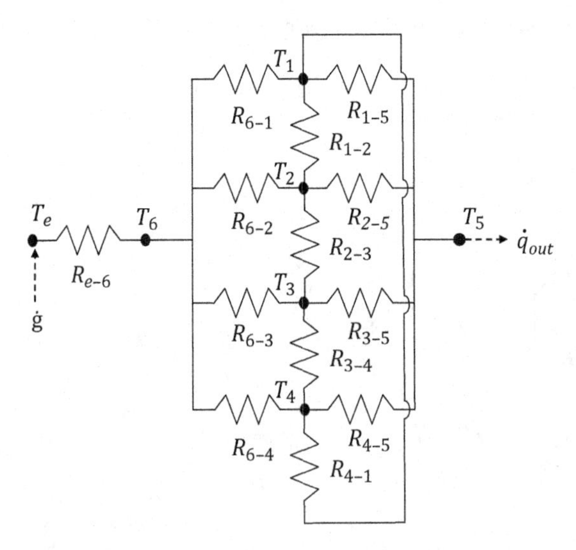

Figure 2.23 Resistance network for E-box on a box problem

$$40\text{ W} = \frac{(T_e - T_6)}{R_{e-6}} \tag{2.65}$$

The long solution to the problem requires that energy balance relations be established for all the nodes associated with the lower box. A simpler approach, however, focuses on nodes 5 and 6, which are termed the boundary nodes for the lower box because they either have a fixed temperature or are where heat flows in and out of the lower box. The energy balance at node 6 can be written out as

$$\dot{q}_{in} = 40\text{ W} = \frac{(T_6 - T_1)}{R_{6-1}} + \frac{(T_6 - T_2)}{R_{6-2}} + \frac{(T_6 - T_3)}{R_{6-3}} + \frac{(T_6 - T_4)}{R_{6-4}} \tag{2.66}$$

Substituting in R_A as the common resistance from the top surface to the side surfaces of the lower box gives

$$\dot{q}_{in} = 40\text{W} = \frac{(T_6 - T_1)}{R_A} + \frac{(T_6 - T_2)}{R_A} + \frac{(T_6 - T_3)}{R_A} + \frac{(T_6 - T_4)}{R_A} \tag{2.67}$$

which reduces to

$$\dot{q}_{in} = \frac{1}{R_A}[4T_6 - T_1 - T_2 - T_3 - T_4] \tag{2.68}$$

Similarly, the energy balance at node 5 can be written out as

$$\dot{q}_{out} = 40\text{W} = \frac{(T_1 - T_5)}{R_{1-5}} + \frac{(T_2 - T_5)}{R_{2-5}} + \frac{(T_3 - T_5)}{R_{3-5}} + \frac{(T_4 - T_5)}{R_{4-5}} \tag{2.69}$$

Once again, substituting in R_A as the common resistance from the bottom surface to the side surfaces of the lower box obtains

$$\dot{q}_{out} = 40\text{W} = \frac{(T_1 - T_5)}{R_A} + \frac{(T_2 - T_5)}{R_A} + \frac{(T_3 - T_5)}{R_A} + \frac{(T_4 - T_5)}{R_A} \tag{2.70}$$

which reduces to

$$\dot{q}_{out} = \frac{1}{R_A}[T_1 + T_2 + T_3 + T_4 - 4T_5] \tag{2.71}$$

The RHS of both energy balance equations for nodes 5 and 6 have common terms of the form

$$T_1 + T_2 + T_3 + T_4.$$

If these terms are isolated, the equations can be solved simultaneously and T_6 determined. For node 5, multiplying both sides of equation 2.71 by R_A and rearranging terms gives

$$\dot{q}_{out} \cdot R_A + 4T_5 = T_1 + T_2 + T_3 + T_4 \tag{2.72}$$

The temperature terms from the RHS can be substituted into the energy balance equation for node 6 once the terms have been isolated.

$$\dot{q}_{in} \cdot R_A = 4T_6 - [T_1 + T_2 + T_3 + T_4] \tag{2.73}$$

$$4T_6 - \dot{q}_{in} \cdot R_A = T_1 + T_2 + T_3 + T_4 \tag{2.74}$$

Substituting in the temperature terms from the final form of the energy balance equation at node 5 gives

$$\dot{q}_{out} \cdot R_A + 4T_5 = 4T_6 - \dot{q}_{in} \cdot R_A \tag{2.75}$$

Letting $\dot{q}_{in} = \dot{q}_{out} = \dot{q}$ and solving for T_6 obtains

$$2\dot{q} \cdot R_A + 4T_5 = 4T_6 \tag{2.76}$$

$$T_6 = \frac{2\dot{q} \cdot R_A + 4T_5}{4} \tag{2.77}$$

which can also be written as

$$T_6 = 0.5\dot{q} \cdot R_A + T_5 \tag{2.78}$$

In order to calculate T_6 we must determine the value for R_A. We know that $R_A = L/kA$. Since we know the geometry and the grade of aluminum, we can determine the value of R_A. With a thermal conductivity value of 177 W/m · K (2024-T6 aluminum), the R_A value becomes

$$R_A = \frac{0.5\ \text{m}}{\left(177 \frac{\text{W}}{\text{m} \cdot \text{K}}\right)(0.5\ \text{m})(0.005\ \text{m})} = 1.13\ \frac{\text{K}}{\text{W}} \tag{2.79}$$

If this value is put back into the equation for T_6 (i.e., equation 2.78), solving the equation gives

$$T_6 = 0.5(40\ \text{W})(1.13\ \text{K/W}) + 10°\text{C} \tag{2.80}$$

$$T_6 = 22.6\ \text{K} + 283\ \text{K} = 305.6\ \text{K} \tag{2.81}$$

T_e in equation 2.65, the energy balance relation for the E-box, can now be determined, giving

$$T_e = T_6 + 40\ \text{W} \cdot R_{e-6} \tag{2.82}$$

Before the E-box temperature is calculated, thermal resistance R_{e-6} is determined as

$$\frac{1}{R_{e-6}} = \frac{1}{[R_e + R_6]} \tag{2.83}$$

The individual resistances for the E-box and side 6 of the lower box can be determined through the following calculations for R_e and R_6.

$$R_e = \frac{0.005\ \text{m}}{\left(177 \frac{\text{W}}{\text{m} \cdot \text{K}}\right)(0.01\ \text{m})^2} = 0.282 \frac{\text{K}}{\text{W}} \tag{2.84}$$

$$R_6 = \frac{(0.005/2)\text{ m}}{\left(177 \frac{\text{W}}{\text{m} \cdot \text{K}}\right)(0.5\text{ m})^2} = 5.65 \times 10^{-5} \frac{\text{K}}{\text{W}} \tag{2.85}$$

Thus $1/R_{e-6}$ becomes

$$\frac{1}{R_{e-6}} = \frac{1}{\left[0.282\text{ K/W} + 5.65 \times 10^{-5}\text{ K/W}\right]} = 3.545\text{ W/K} \tag{2.86}$$

Placing this value in equation 2.82 gives

$$T_e = T_6 + 40\text{ W} \cdot (0.282\text{ K/W}) \tag{2.87}$$

Substituting in the value for T_6 (i.e., 305.6 K) now gives

$$T_e = 305.6\text{ K} + 10\text{ K} \text{ or } T_e = 316.9\text{K}$$

2.4 Radial Geometries

While the prior example problems used Cartesian coordinates to define the thermal resistances, real-life geometries can include curvatures. These type geometries are accommodated by cylindrical and spherical coordinate systems. This section looks at the radial, longitudinal and circumferential resistance relations for thermal conduction in cylindrical coordinates.

2.4.1 The Hollow (or Concentric) Cylinder

A working construct for the radial resistance of a hollow cylinder can be obtained from the thermal energy equation using cylindrical coordinates by taking an energy balance across a differential material volume in the r, θ, and z directions. The final conservative form of the equation is as follows (the thermal conductivity has not been assumed constant and is still inside the partial derivative):

$$\rho c_p \frac{\partial T}{\partial t} = \frac{1}{r}\frac{\partial}{\partial r}\left[kr\frac{\partial T}{\partial r}\right] + \frac{1}{r^2}\frac{\partial}{\partial \theta}\left[k\frac{\partial T}{\partial \theta}\right] + \frac{\partial}{\partial z}\left[k\frac{\partial T}{\partial z}\right] + \dot{g}''' \tag{2.88}$$

Figure 2.25 shows a hollow cylinder of a characteristic geometry. As in the Cartesian development, we need to establish the conductive heat transfer relation in each of the orthogonal directions, beginning with the radial direction. Let's assume that:

i) The direction of heat flow and temperature variation is radial.
ii) The heat transfer rate is steady state.
iii) No energy generation is present.
iv) The thermal conductivity (k) is constant.
v) The thermal conductivity is isotropic (i.e., $k_r = k_\phi = k_z = k$).

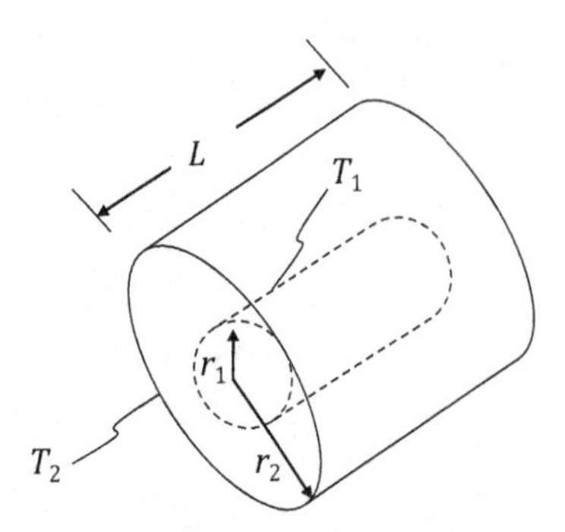

Figure 2.24 Basic hollow cylinder geometry
Source: Adapted with permission of John Wiley & Sons, Inc. from figure 3.7, *Fundamentals of Heat and Mass Transfer*, Bergman, T., Lavine, A., Incropera, F.P., and Dewitt, D.P., Vol. 7, 2017, reference [1]; permission conveyed through Copyright Clearance Center, Inc.

When these assumptions are applied to the energy equation (equation 2.88), it reduces to

$$\frac{k}{r}\frac{d}{dr}\left[r\frac{dT}{dr}\right] = 0 \tag{2.89}$$

The next step is to solve for the temperature distribution. If the differential equation is integrated twice w.r.t. r, the following temperature function is obtained:

$$T(r) = c_1 \ln(r) + c_2 \tag{2.90}$$

The constants of integration (c_1 and c_2) must be determined next. This is accomplished by applying the boundary conditions for $T(r)$ at $r = r_1$ and $r = r_2$ respectively. As shown in Figure 2.24, the temperature values on the boundaries and the corresponding temperature function for each radial location are

$$T = T_1 \; @ \; r = r_1; \; T_1 = c_1 \ln(r_1) + c_2$$

$$T = T_2 \; @ \; r = r_2; \; T_2 = c_1 \ln(r_2) + c_2$$

The equations for the temperature function at the boundaries can be solved simultaneously to determine c_1 and c_2. These constants can then be placed in the equation for the temperature function. The relation for the temperature function now becomes

$$T(r) = (T_1 - T_2)\frac{\ln(r/r_2)}{\ln(r_1/r_2)} + T_2 \tag{2.91}$$

Since the temperature function is now defined, it can be substituted into the differential term in Fourier's Law. In the radial direction this is

$$\dot{q}_r = -k(2\pi rL)\frac{dT}{dr} \tag{2.92}$$

Upon substitution the radial heat transfer relation becomes

$$\dot{q}_r = \frac{2\pi Lk(T_1 - T_2)}{\ln(r_2/r_1)} \tag{2.93}$$

Finally the thermal resistance for this geometry is defined by inspecting the RHS of equation 2.93. Neglecting the temperature difference in the heat transfer relation, the resistance for this geometry is

$$R = \frac{\ln(r_2/r_1)}{2\pi Lk} \tag{2.94}$$

Example 7 Heat Transfer in Concentric Cylindrical Regions A cylindrical cross-section (Figure 2.25) consists of two material elements (A and B). Each element has a length of L (into the page as shown) and a thermal conductivity of k.

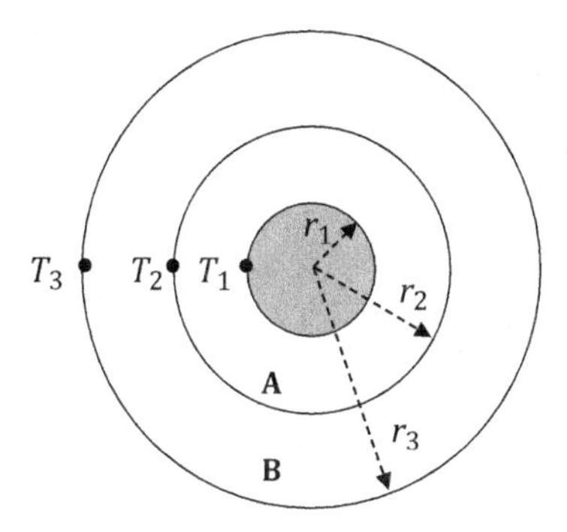

Figure 2.25 Cross-section of concentric cylinders

Assuming that $T_1 > T_2 > T_3$, $r_1 = R$, $r_2 = 1.5R$ and $r_3 = 3.0R$, answer parts (i) through (iii):

i) Draw the thermal circuit.
ii) Define the resistances in material A and material B.
iii) Determine the radial heat transfer through the composite material elements A and B.

Solution

For part i), the temperature gradient suggested by the initial assumption is from inside to outside radially (i.e., T_1 to T_3). The heat flow will be in the direction of the temperature gradient and flow from the inner radius to the outer radius location. The thermal configuration (Figure 2.25) only has two elements (materials A and B). Figure 2.26 shows the thermal circuit.

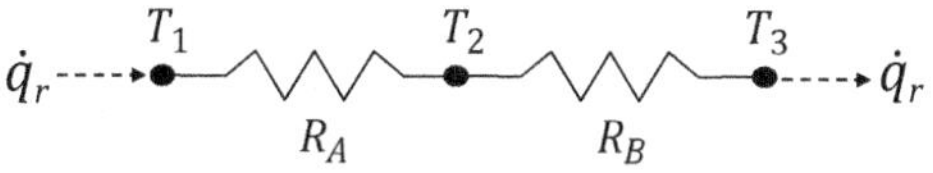

$\dot{q}_r$ T_1 R_A T_2 R_B T_3 $\dot{q}_r$

Figure 2.26 Thermal circuit for concentric cylinder problem

For part ii), elements A and B are both bound on either side by temperatures at different radial distances. Using the resistance definition previously determined and reducing the fractions by substituting in the radial distances recast in radial units of R gives the following resistances for elements A and B:

$$R_A = R_{1-2} = \frac{\ln(r_2/r_1)}{2\pi L k_A} = \frac{\ln(1.5R/R)}{2\pi L k_A} = \frac{\ln(1.5)}{2\pi L k_A} \tag{2.95}$$

$$R_B = R_{2-3} = \frac{\ln(r_3/r_2)}{2\pi L k_B} = \frac{\ln(3R/1.5R)}{2\pi L k_B} = \frac{\ln(2.0)}{2\pi L k_B} \tag{2.96}$$

For part iii) the overall resistance to heat flow through the entire circuit (inner radius to outer radius) must be determined. As the resistances are in series (Figure 2.26), they can be summed to determine R_{tot}. The heat transfer rate is the product of the inverse of R_{tot} and the temperature gradient across the circuit. The resulting equation is

$$\dot{q}_r = \frac{(T_1 - T_3)}{\left[\frac{\ln(1.5)}{2\pi L k_A} + \frac{\ln(2.0)}{2\pi L k_B}\right]} \tag{2.97}$$

which can be rewritten as

$$\dot{q}_r = \frac{(T_1 - T_3)}{\frac{1}{2\pi L}\left[\frac{\ln(1.5)}{k_A} + \frac{\ln(2.0)}{k_B}\right]} \tag{2.98}$$

2.4.2 Longitudinal Resistances in Cylindrical Structures

Figure 2.27 shows two hollow cylinders with a common interface along the z direction, of which the resistance to heat flow is to be determined. For purposes of analysis, each material's longitudinal node is the center of the cylinder sections.

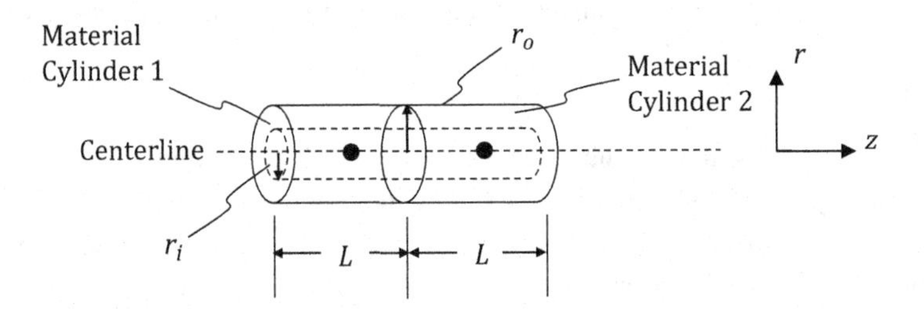

Figure 2.27 Hollow cylinders with a shared interface

The heat transfer in the longitudinal direction is captured using Fourier's Law as

$$\dot{q}_z = \frac{k A_z \Delta T}{\Delta z} \tag{2.99}$$

where A_z is

$$A_z = \pi\left(r_o^2 - r_i^2\right) \tag{2.100}$$

Based on the relation for Fourier's Law in equation 2.99, the resistance can be defined as

$$R = \frac{\Delta z}{kA_z} \tag{2.101}$$

Thus, after substituting in the cross-sectional area definition (equation 2.100) for the hollow cylinder shown in Figure 2.27, R_{1-2} can be defined as

$$R_{1-2} = \frac{L}{k\left[\pi\left(r_o^2 - r_i^2\right)\right]} \tag{2.102}$$

The heat transfer relation from node 1 to node 2 can then be written as

$$\dot{q}_z = \frac{k\left[\pi\left(r_o^2 - r_i^2\right)\right](T_1 - T_2)}{L} \tag{2.103}$$

2.4.3 Circumferential Resistances in Cylindrical Structures

The thermal resistance and heat transfer in the θ direction of a segment for a hollow cylinder needs to be defined. Assuming a non-uniform temperature, heat transfer will occur in the θ direction, and can be modelled using the resistance along the heat flow path. Figure 2.28 shows a hollow cylinder along its longitudinal axis. The cylinder circumference is divided into four sections (each having a single node assigned to it). The temperature associated with each of these nodes is T_1 through T_4 respectively.

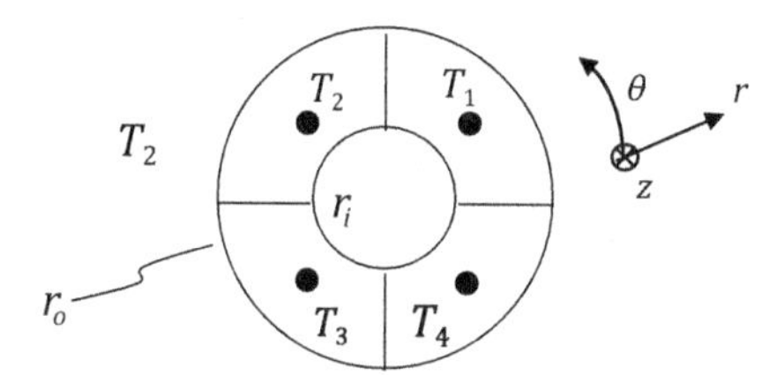

Figure 2.28 Hollow cylinder viewed along longitudinal axis

Assuming each of the nodal elements has the same thermal conductivity (k_θ), the resistance between nodes can be captured using equation 2.104

$$R = \frac{s_{arc}}{k_\theta A_\theta} \tag{2.104}$$

where s_{arc} is the arc length between the two nodes in question at their radial distance and A_θ is the cross-sectional area along the arc. Assuming the nodes are at the radial center of the sectors, the radial distance is the average of the outer and inner radius (i.e., r_{avg}). The area can be recast as

$$A_\theta = (r_o - r_i)th \tag{2.105}$$

where th is the length along the longitude of the sector node into the depth of the page.

Substituting the area, along with the arc length, into the resistance relation (equation 2.104) gives the following relation:

$$R = \frac{r_{avg}\theta}{k_\theta(r_o - r_i)th} \tag{2.106}$$

Note that the θ term in the numerator is in radians. With the resistance between circumferential nodes defined, the heat transfer between any two adjacent nodes can also be defined; for example, between nodes 1 and 2 it will be

$$\dot{q}_{\theta,1-2} = \frac{k_\theta \cdot th \cdot (r_o - r_i)(T_1 - T_2)}{r_{avg}\theta} \tag{2.107}$$

Constructs similar to those used to define the relations for the hollow cylinder in cylindrical coordinates can be applied to determine both resistance and heat transfer as a function of orthogonal direction in a spherical coordinate system. For more on heat transfer relations in spherical coordinates, see references [1] and [26].

2.5 2-D Conduction Shape Factors

The shape factor is the ratio of the cross-sectional area experiencing heat transfer to the conductive path length.

$$S = \frac{A}{L} \tag{2.108}$$

When using the shape factor convention, Fourier's Law can be recast as

$$\dot{Q} = Sk\Delta T \tag{2.109}$$

where the resistance is defined as

$$R = \frac{1}{Sk} \tag{2.110}$$

Shape factor methodology is often used in complex 2-D and 3-D heat transfer geometries. Table 2.5 lists some key shape factors relevant to geometries commonly used in engineering designs.

While physical scenarios that include geometries embedded in infinite media do not apply to spacecraft thermal design, hollow geometries and extrusions may be present on a space structure. The EES software package lists several standard shape factors under [Options > Function Information > Conduction Shape Factors].

Example 8 Shape Factor Problem Figure 2.29 shows a hollow circular cylinder embedded inside a Ni square cylinder of the same length. If there is an energy dissipation of 110 W at the circular/square cylinder interface, and T_1 has a temperature of 20°C, what is T_2?

Table 2.5 Common shape factors applicable to space structures

Geometry	Schematic	Shape factor relation
[1]Square extrusion with a centered circular hole	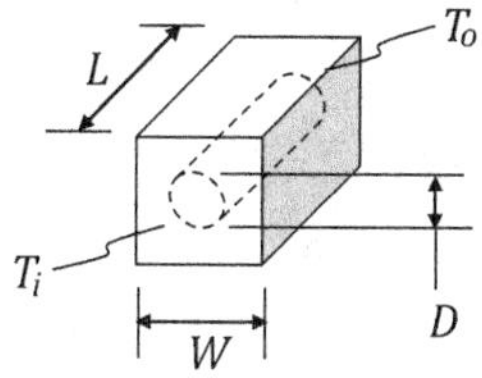 L, T_o, T_i, W, D	$S = \frac{2\pi L}{\ln(1.08W/D)}$ Applicable for $L \gg D$
[1]Square extrusion with a centered square hole	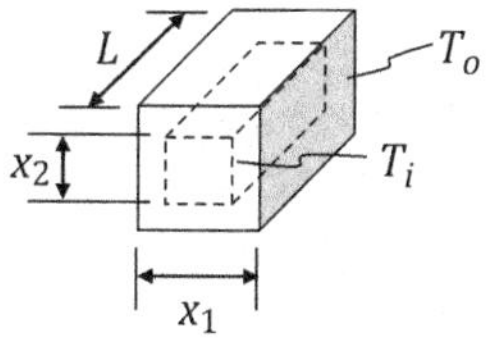 L, T_o, x_2, T_i, x_1	$x_1/x_2 \leq 0.25$ $S = \frac{2\pi L}{0.785 \ln(x_1/x_2)}$ $x_1/x_2 > 0.25$ $S = \frac{2\pi L}{[0.93 \ln(x_1/x_2) - 0.0502]}$
[2]Hollow square of uniform wall thickness	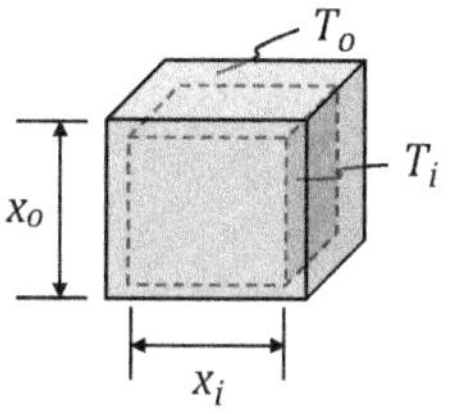 T_o, T_i, x_o, x_i	$Vol = x_o^3 - x_i^3$ $A = 6x_i^2$ $S^* = \frac{2\sqrt{\pi}}{\left[1 + \left(\frac{\sqrt{\pi}}{6}\right)\left[\left(\frac{x_o}{x_i}\right)^3 - 1\right]\right]^{\frac{1}{3}} - 1} + S_\infty^*$ Where $S_\infty^* = 3.391$ and $S = S^* \sqrt{A_i}$
[2]Hollow sphere with uniform wall thickness	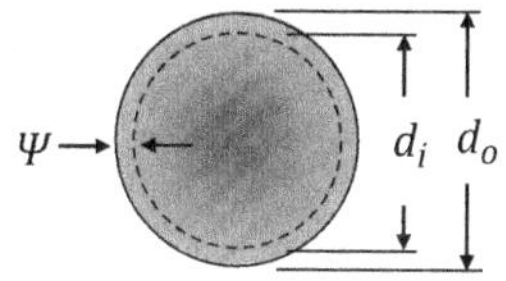 ψ, d_i, d_o	$\psi = \frac{(d_o - d_i)}{2}$ $S^* = \frac{d_i\sqrt{\pi}}{\psi} + S_\infty^*$ Where $S_\infty^* = 2\sqrt{\pi}$

[1] *Source*: Adapted from Table 2-1 Centered Extrusion Figure, *Heat Transfer* by Nellis and Klein [26].
[2] Hollow Square and Hollow Sphere relations republished with permission of the American Institute of Aeronautics and Astronautics from Teerstra, P., Yovanovich, M.M., and Culham, J.R., "Conduction Shape Factors Models for Three Dimensional Enclosures," *The Journal of Thermophysics and Heat Transfer*, Vol. 19(4) (2005), pp. 527–532 [33], conveyed through Copyright Clearance Center, Inc.

Solution

The shape factor (S) is $S = A/L$ (equation 2.108), making Fourier's Law in this case

$$\dot{Q} = Sk(T_1 - T_2)$$

Solving for T_2 gives

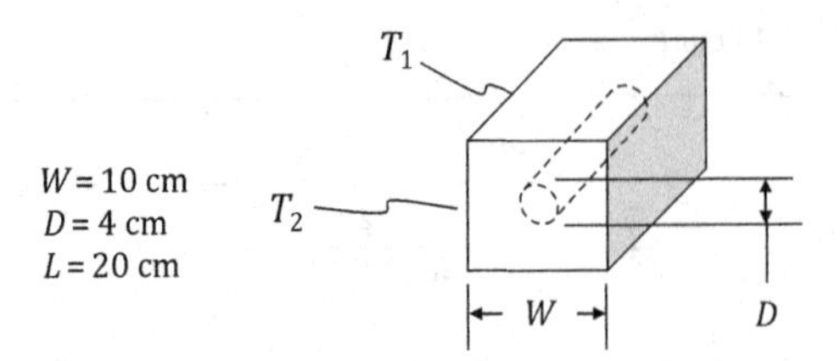

Figure 2.29 Square cylinder with a hollow circular channel along the longitudinal axis

$$T_2 = T_1 - \frac{\dot{Q}}{Sk}$$

Using the shape factor in Table 2.5 for the specified geometry, we can determine S and T_2 using EES. The code for the program is

```
"Example 8: Shape factor problem"
W=0.1              [m]          {Extrusion square length}
D=0.04             [m]          {Diameter of cylinder}
L=0.20             [m]          {Length}
k_Ni=90.7          [W/m-K]      {Thermal conductivity}
q_dot=110          [W]          {Constant heat input}
T_1=293            [K]          {Degrees Celsius}

S=(2*pi*L)/(ln(1.08*(W/D)))     {Shape factor value}
T_2=T_1-(q_dot/(S*k_Ni))     {Temperature at outer surface}
```

When the program is executed, the value of the exterior surface's temperature is found to be $T_2 = 292\,\mathrm{K}$.

2.6 Honeycomb Panel Structures

In spacecraft design, mass is always an issue. The honeycomb panel is a mass-efficient technique used for spacecraft mounting structures and radiators. The honeycomb configuration, borrowed from nature, is not only one of the strongest structural arrangements given the finite thickness of its wall structure but also has significant mass reduction relative to solid materials of a comparable volume. Honeycomb panel structures are created by bonding multiple segments of ribbon material together (Figure 2.30a). The final product is a network of honeycomb cells (Figure 2.30b).

In application, the cells on the honeycomb panel structure are rarely left open; rather, facesheets of contiguous material are applied to both sides (see Figure 2.31). Constructs for the resistance through the facesheets, as well as both planar and through conductance (facesheet to facesheet) can be calculated given knowledge of the material and

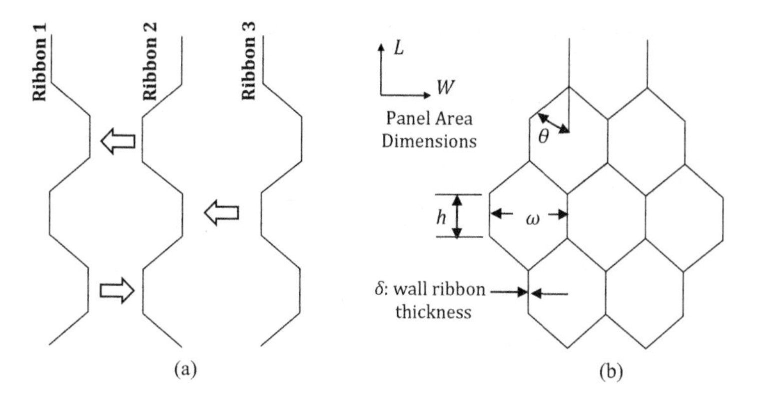

Figure 2.30 Honeycomb structures: a) Ribbon bonding layout; b) Cell-based view
Source: Adapted from *Spacecraft Thermal Control Handbook*, Vol. I, Appendix B, reference [34], with permission of Aerospace Corporation

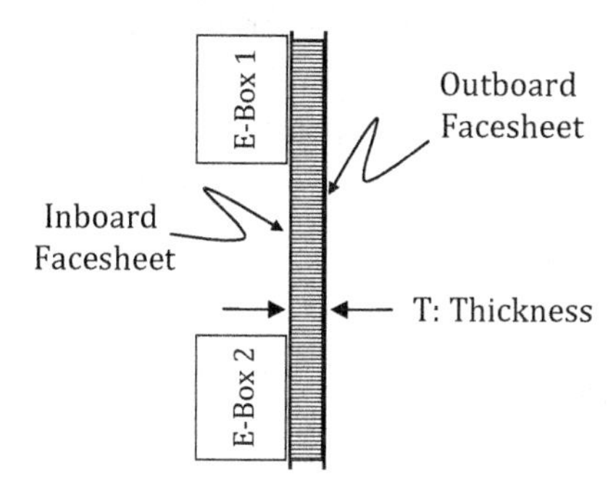

Figure 2.31 Profile of honeycomb panel radiator

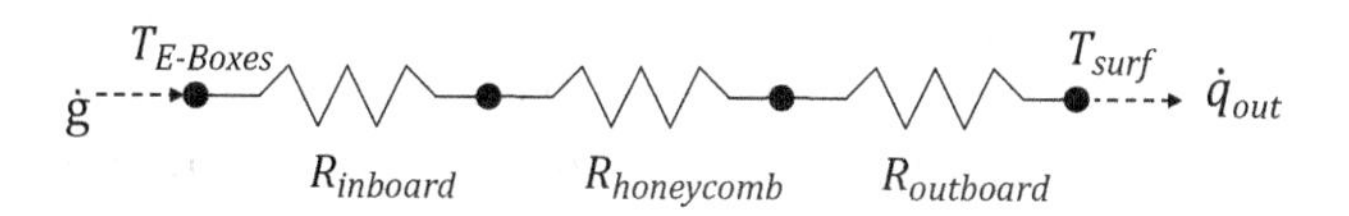

Figure 2.32 Resistance circuit for honeycomb panel configuration

geometry of the honeycomb. Figure 2.32 is a diagram of the heat flow path from the inboard side to the outboard side of the honeycomb panel.

The thermal conductance in the T direction (relative to the Figure 2.31 example) is defined as [34]

$$C_T = \left(\frac{8k\delta}{3\omega}\right)\left(\frac{LW}{\Delta z}\right) \tag{2.111}$$

where

σ: Ribbon-based extension factor ($2h/l$). Equals 4/3 for $\theta = 60°$
δ: Ribbon thickness (m)

k: Ribbon material thermal conductivity (W/m · K)
ω: Cell size (m)
Δz Honeycomb configuration thickness (m).

The corresponding relations for the L and W directions [34] (i.e., planar directions based on Figure 2.30b) are

$$C_L = \left(\frac{3k\delta}{2\omega}\right)\left(\frac{W\Delta z}{L}\right) \tag{2.112}$$

$$C_W = \left(\frac{k\delta}{\omega}\right)\left(\frac{L\Delta z}{W}\right) \tag{2.113}$$

High thermal conductivity materials are usually chosen for radiator honeycomb structures. Since the honeycomb material is sandwiched between two facesheets only about 2.5 cm apart, a low ΔT can be expected from facesheet to facesheet. Thus, any temperature gradients along the facesheets are due to localized heating effects from the dissipation sources on the inboard side, as well as the thermal conductivity of the facesheet material itself (i.e., k_x and k_y). If the heat from the energy sources on the inboard side is not well spread across the outboard facesheet, localized hot spots may adversely affect the heat rejection capability of the radiator surface. This will be explored further in Chapters 3 and 4.

2.7 Lumped Body Heating Methodology

Analysis techniques covered in this chapter have dealt primarily with steady-state phenomena. However, in the design and testing of spacecraft components and ground support equipment, the ability to predict the time required for solid-state objects to transition from one temperature to another is highly advantageous. Imagine a known mass of 2024-T6 aluminum (Figure 2.33) undergoing heating (and/or cooling). Assuming that this body does not experience or have spatial gradients in temperature, if a known amount of heat is added to the body, how can temperature be determined as a function of time?

From thermodynamics, the sensible heating of a fluid undergoing flow is modeled by the equation

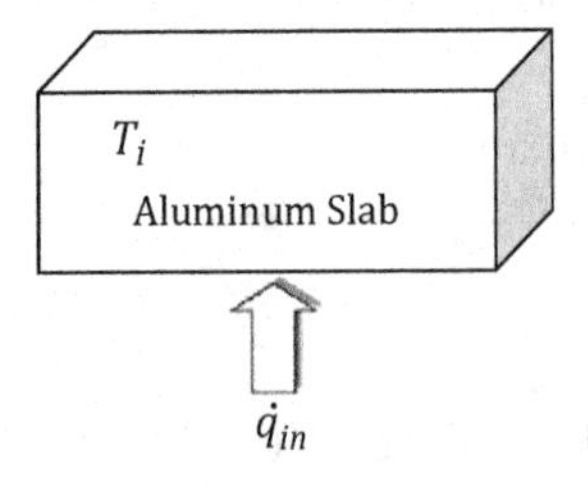

Figure 2.33 Aluminum slab at initial temperature T_i

$$\dot{q} = \dot{m}c_p\Delta T \tag{2.114}$$

where $\dot{m}$ is the mass flow rate of the fluid in kg/s and ΔT is the temperature difference between two discrete times (and/or locations). Equation 2.114 is actually a rate equation that can also be applied to a solid. In fluid flow applications, the time parameter on the RHS is accounted for in the mass flow rate ($\dot{m}$). However, if the system is still, the time parameter can be separated out in the denominator and the sensible heating equation rewritten as

$$\dot{q} = \frac{mc_p\Delta T}{\Delta t} \tag{2.115}$$

which is also equal to

$$\dot{q} = \frac{mc_p\left(T_{final} - T_{init}\right)}{\Delta t} \tag{2.116}$$

A consistency check is provided by alternatively deriving this form from the energy equation, using heat transfer concepts and incorporating the stated assumptions. The standard form of the energy equation is

$$\rho c_p\frac{\partial T}{\partial t} = k\frac{\partial^2 T}{\partial x^2} + k\frac{\partial^2 T}{\partial y^2} + k\frac{\partial^2 T}{\partial z^2} + \Phi_v + \dot{g}''' \tag{2.117}$$

where Φ_v is the viscous dissipation term and $\dot{g}'''$ is the energy generation per unit volume. In the case of a solid there will be no viscous dissipation. Thus, the equation reduces to

$$\rho c_p\frac{\partial T}{\partial t} = k\frac{\partial^2 T}{\partial x^2} + k\frac{\partial^2 T}{\partial y^2} + k\frac{\partial^2 T}{\partial z^2} + \dot{g}''' \tag{2.118}$$

Since no temperature gradients are assumed for this body, the second-order partials in the x, y and z directions also go to zero. The equation now takes the form

$$\rho c_p\frac{dT}{dt} = \dot{g}''' \tag{2.119}$$

The partial derivative representation is dropped because the equation has a functional form of $T = T(t)$. In this relation, the energy generation term can also be interpreted as the heat input to the mass. Both sides of the equal sign have a volume term in the denominator. The product of the density and the volume equals the mass. Multiplying both the LHS and RHS by the volume (*Vol*) gives

$$mc_p\frac{dT}{dt} = \dot{g}'''\cdot Vol \tag{2.120}$$

The equation can be rewritten as

$$mc_p\frac{dT}{dt} = \dot{g} \tag{2.121}$$

If the differential terms are separated and both sides integrated, the result is

$$mc_p \int_{T_1}^{T_2} dT = \dot{g} \int_{t_1}^{t_2} dt \tag{2.122}$$

Evaluating the integrals will result in an equality

$$mc_p(T_2 - T_1) = \dot{g}(t_2 - t_1) \tag{2.123}$$

This can be rewritten as

$$mc_p\Delta T = \dot{g}\Delta t \tag{2.124}$$

where $\Delta t = t_2 - t_1$. The equation can be recast by solving for the heat input

$$\dot{g} = \dot{q} = \frac{mc_p\Delta T}{\Delta t} \tag{2.125}$$

Confidence in the final form is high since the same solution has been achieved using two different methodologies.

Example 9 Sensible Heating of a Metal Slab For a mass of 2 kg with a specific heat of 875 J/kg · K, how long does it take for the temperature to rise by 30 K if 30 W of constant heating is applied?

Solution

The time required can be solved using the sensible heating equation. The form of the equation now becomes

$$\Delta t = \frac{mc_p\Delta T}{\dot{q}} \tag{2.126}$$

where T_2 and T_1 are the final and initial temperatures of the mass respectively. The solution can be determined with the aid of EES. The code for determining the time required for the temperature change is as follows.

```
"Example 9: Sensible heating of a metal slab"

c_p=875             [J/kg-K]           {Specific heat}
m=2                 [kg]               {Mass}
q_dot=30            [W]                {Constant heat input}
DELTAT=30           [K]

delta_t=m*c_p*DELTAT/q_dot            {Time for change in
temperature}
```

The solution in EES is Δt = 1,750 seconds or 29.2 minutes. If the heat load is increased to 60 W, the time for the specified temperature change reduces to 875 seconds or 14.6 minutes.

References

[1] Bergman, T., Lavine, A., Incropera, F.P., and Dewitt, D.P., 2017, *Fundamentals of Heat and Mass Transfer*, John Wiley & Sons, Hoboken, NJ

[2] "NuSil™ Product Guide," 2017, *Space Grade Line*, NuSil Technology LLC, Carpinteria, CA

[3] "Thermal Interface Materials for Electronics Cooling," 2014, *Parker Chomerics Products and Custom Solutions Catalog*, Parker Hannifin Corporation, Woburn, MA

[4] "Tpli 200 Series," 2010, *Gap Filler*, Laird Technologies, London, UK

[5] Touloukian, Y.S., Powell, R.W., Ho, C.Y., and Klemens, P.G., 1970, "Thermal Conductivity Metallic Elements and Alloys," in *Thermophysical Properties of Matter*, Plenum Publishing Corporation, New York, NY

[6] "Metal Thermal Interface Materials," *Properties*, Indium Corporation®, Clinton, NY

[7] "Apiezon Cryogenic High Vacuum Grease," 2012, *N Grease*, M&I Materials Ltd., Manchester, UK

[8] "Apiezon Ultra High and High Vacuum Greases," 2012, *Apiezon L, M & N Greases*, M&I Materials Ltd., Manchester, UK

[9] "GraFoil® Flexible Graphite Engineering Design Manual," 2002, *GraFoil Sheet and Laminate Products*, GrafTech Incorporated, Cleveland, OH

[10] "HITHERM™ Thermal Interface Materials," 2017, *Technical Data Sheet 318*, Advanced Energy Technologies LLC, Cleveland, OH

[11] "Stycast® 2850 FT Thermally Conductive Epoxy Encapsulant," 2001, *Stycast® 2850 FT Technical Data Sheet*, Henkel Loctite Stycast® 2850 FT, Dusseldorf, Germany

[12] "Hysol Eccobond® 285 Thermally Conductive Epoxy Paste Adhesive," 2014, *Loctite Ablestik 285 Technical Data Sheet*, Henkel Corporation, Dusseldorf, Germany

[13] Gluck, D., and Baturkin, V., 2002, "Mountings and Interfaces," *Spacecraft Thermal Control Handbook Vol. I: Fundamental Technologies*, edited by Gilmore, D.G., The Aerospace Press and The American Institute of Aeronautics and Astronautics, Inc., Chap. 8, pp. 247–329.

[14] Nishino, K., Yamashita, S., and Torli, K., 1995 "Thermal Contact Conductance under Low Applied Load in a Vacuum Environment," *Experimental Thermal and Fluid Science*, Vol. 10, pp. 258–271

[15] Atkins, H.L., and Fried, E. 1965 "Interface Thermal Conductance in a Vacuum," *AIAA Journal of Spacecraft and Rockets,* Vol. 2, No. 4, pp. 591–593

[16] Woodcraft, A.L., and Gray, A., 2009, "A low temperature thermal conductivity database," *13th International Workshop on Low Temperature Detectors-LTD13*, Stanford, CA, Vol. 1185, Issue 1, pp. 681–684

[17] Marquadt, E.D., Lee, J.P., and Radebaugh, R., 2002 "Cryogenic Materials Properties Database," *Cryocoolers,* Vol. 11, pp. 681–687

[18] "DuPont™ Teflon® Properties Handbook," 1996, *DuPont™ Teflon® PTFE fluoropolymer resin*, DuPont™ Fluoroproducts, Wilmington, DE

[19] Barucci, M., Olivieri, E., Pasca, E., Risegari, L., and Ventura, G., 2005, "Thermal Conductivity of Torlon between 4.2K and 300K," *Cryogenics*, Vol. 45, pp. 295–299

[20] "Technical Data Sheet: Torlon® 4203L (polyamide-imide)," 2016, *Solvay Specialty Polymers*, Brussels, Belgium

[21] "DuPont™ Vespel® SP-1 Typical ISO Properties," 2014, *DuPont™ Vespel® SP-1 Polyimide Isostatic Shapes*, Wilmington, DE

[22] Whitehouse, P., 2012 "Thermal Conductivity of Unfilled Ultem® 1000 as a function of Temperature," Unpublished

[23] "GE Plastics Ultem® 1000 PEI, Polyetherimide, unfilled, extruded," 2007, *Material Datasheet*, Sabic Innovative Plastics, Pittsfield, MA

[24] Silk, E.A., and Bracken, P., 2010 "Spray Cooling Heat Flux Performance Using HTC Foam," *AIAA Journal of Thermophysics and Heat Transfer*, Vol. 24, No. 1, pp. 157–164

[25] Gilmore, D.G., and Collins, R.L., 2002, "Thermal Design Analysis," *Spacecraft Thermal Control Handbook Vol. I: Fundamental Technologies*, edited by Gilmore, D.G., The Aerospace Press and The American Institute of Aeronautics and Astronautics, Inc., 2002, Chap. 15, pp. 523–598

[26] Nellis, G., and Klein, S., 2009, *Heat Transfer*, Cambridge University Press

[27] *Thermophysical Properties of Selected Aerospace Materials* Part II: *Thermophysical Properties of Seven Materials*, edited by Touloukian, Y.S., and Ho, C.Y. 1977, pp. 39–46, Contract #DSA 900-77-C-37758, CINDAS LLC Thermophysical Properties of Matter Database (TPMD). Version 10.0 at http://cindasdata.com/

[28] "Recommended values of the thermophysical properties of eight alloys, major constituents and their oxides," edited by Touloukian, Y.S., 1965, NBS Sub-Contract No. CST-7590, CINDAS LLC Thermophysical Properties of Matter Database (TPMD). Version 10.0 at http://cindasdata.com/

[29] Simon, N.J., Drexler, E.S., and Reed, R.P., 1972, "Properties of Copper and Copper Alloys at Cryogenic Temperature," NIST Monograph 177, pp. 7–23

[30] Graebner, J., Altmann, H., Balzaretti, N., Campbell, R., Chae, H.-B., Degiovanni, A., Enck, R., Feldman, A., Fournier, D., Fricke, J., Goela, J., Gray, K., Gu, Y., Hatta, I., Hartnett, T., Imhof, R., Kato, R., Koidl, P., Kuo, P., Lee, T.-K., Maillet, D., Remy, B., Roger, J., Seong, D.-J., Tye, R., Verhoeven, H., Worner, E., Yehoda, J., Zachai, R., and Zhang, B., 1998, "Report on a Second Round Robin Measurement of the Thermal Conductivity of CVD Diamond," *Diamond and Related Materials*, Vol. 7, pp. 1589–1604

[31] Barron, R., 1985, *Cryogenic Systems*, 2nd edn, Oxford University Press Inc., New York, NY

[32] Reddy, J.N., 1994, *An Introduction to the Finite Element Method*, McGraw-Hill, Inc., New York, NY

[33] Teertstra, P., Yovanovich, M.M., and Culham, J.R., 2005. "Conduction Shape Factor Models for Three-Dimensional Enclosures," *Journal of Thermophysics and Heat Transfer*, Vol. 19, No. 4, pp. 527–532

[34] Hennis, L., 2002, "Appendix B: Material Thermal Properties – Honeycomb Panel Thermal Properties," *Spacecraft Thermal Control Handbook Vol. I: Fundamental Technologies*, edited by Gilmore, D.G., The Aerospace Press and The American Institute of Aeronautics and Astronautics, Inc., 2002, Appendix B., pp. 813–818

Problems

2.1 For the composite wall shown in Figure P2.1, assume: $k_1 = k_2 = k_3$; $T_1 > T_2 > T_3$.

i) Draw the thermal circuit and label the resistances.
ii) Determine the total composite wall resistance and heat transfer.
iii) Determine the limiting resistance in the system.

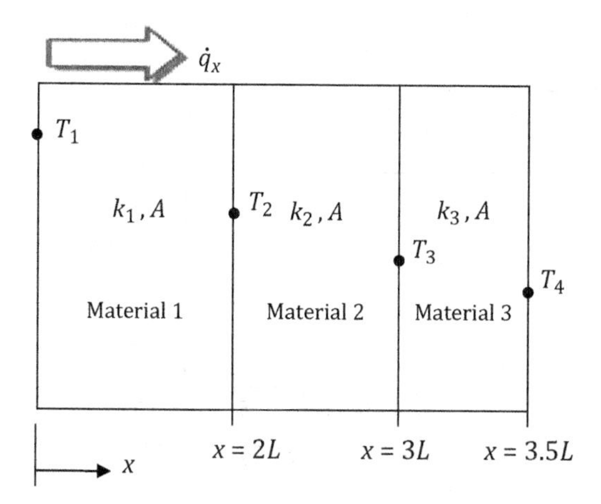

Figure P2.1 Composite wall with common cross-sectional area

2.2 For the composite wall shown in Figure P2.2, assume $T_1 > T_2 > T_3$, *th*: thickness into paper.

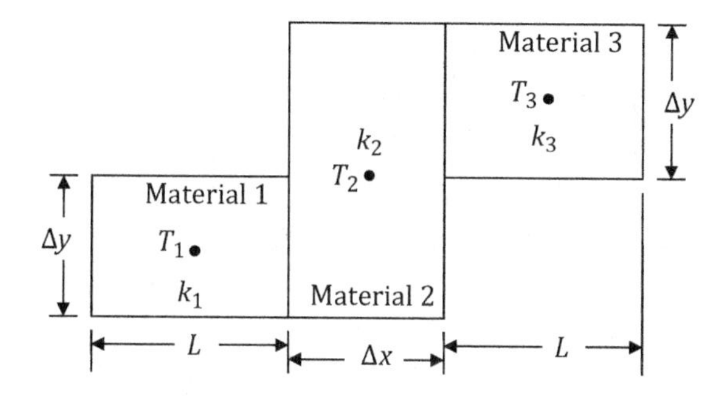

Figure P2.2 Misaligned nodes along heat flow path

i) Write out resistances between materials $1 \leftrightarrow 2$, $2 \leftrightarrow 3$ and $1 \leftrightarrow 3$.
ii) Determine the heat transfer relation for $1 \leftrightarrow 3$.

2.3 For the thermal circuit shown in Figure P2.3,

i) Assuming a depth of *th* into the page, draw the resistance network and label the resistances.

ii) Define $1/R_{tot}$ in terms of the other resistances within the network (Write this out to include resistances R_{1-2}, R_{2-3}, etc. . .).

iii) Assuming $\Delta y = 2L$, material 3 is OFHC, material 4 is 6061 aluminum and materials 1, 2, and 5 are 304 stainless steel, what is the path of least resistance between material 2 and material 5? Show your calculations.

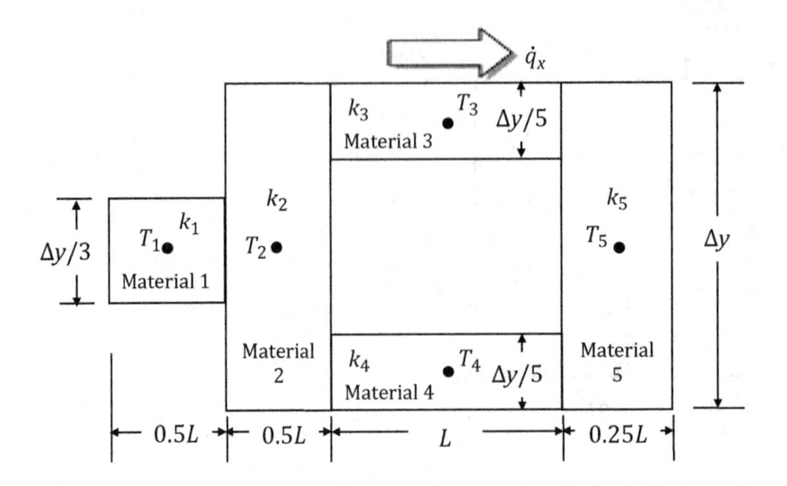

Figure P2.3 Thermal circuit with divergent heat flow path (bi-directional)

2.4 For the thermal circuit shown in Figure P2.4,

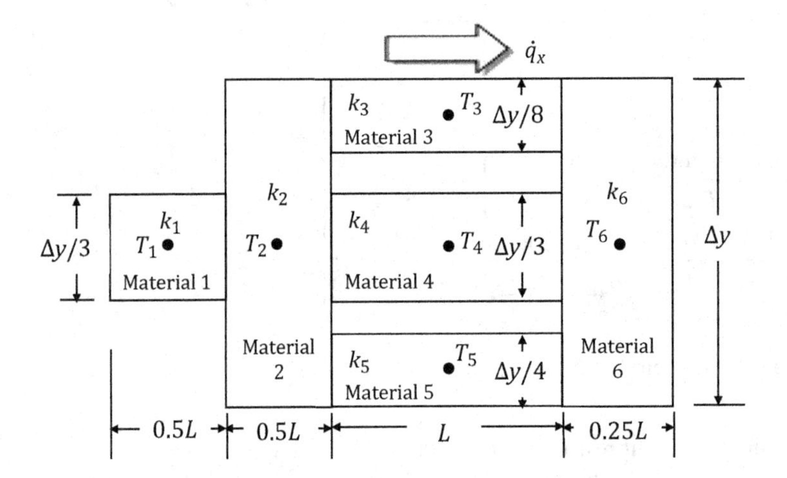

Figure P2.4 Thermal circuit with divergent heat flow path (tri-directional)

i) Assuming a depth of *th* into the page, draw the resistance network and label the resistances.

ii) Define $1/R_{tot}$ in terms of the other resistances within the network (write this out to include resistances R_{1-2}, R_{2-3}, etc.).

iii) Assuming $\Delta y = 2L$, material 3 is OFHC, material 4 is 304 stainless steel and materials 1, 2, 5 and 6 are aluminum 2024-T6, what is the path of least resistance between material 2 and material 6? Show your calculations.

2.5 For the thermal circuit shown in Figure P2.5, assume thickness into paper is *th*.

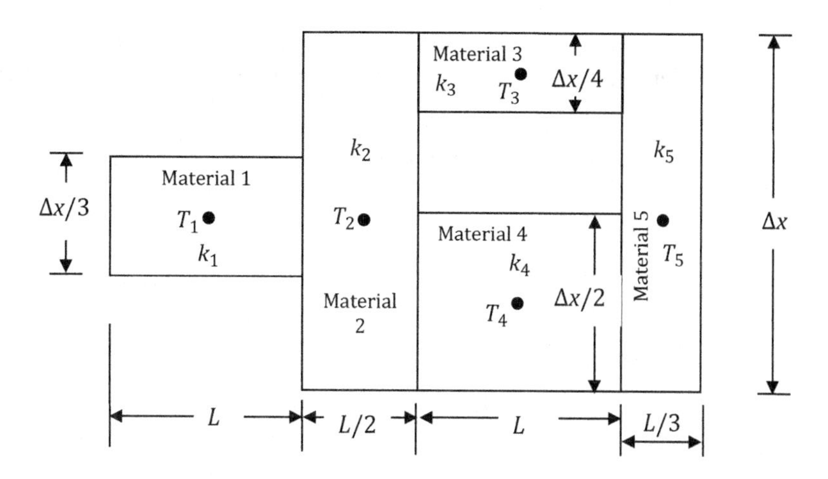

Figure P2.5 Thermal circuit with divergent heat flow path (bi-directional w/different aspect ratios)

i) Write out the resistances between materials 1↔5.

ii) Determine the heat transfer relation for 1↔5.

iii) If $k_1 = k_2 = k_3 = k_4 = k_5$ then what is the limiting resistance in the circuit? (assume $L = 0.5\Delta x$).

iv) Focusing on materials 3 and 4, if $k_2 = k_3 = k_4 = k_5$ then does the path of least resistance to heat flow from material 2 to material 5 pass through material 3 or material 4?

v) Assuming material 2 and material 5 are OFHC (oxygen-free high-conductivity copper), material 3 is 2024-T6 aluminum and material 4 is 304 SS (stainless steel), which material (3 or 4) then serves as the path of least resistance to heat flow between materials 2 and 5?

2.6 For the thermal circuit shown in Figure P2.6,

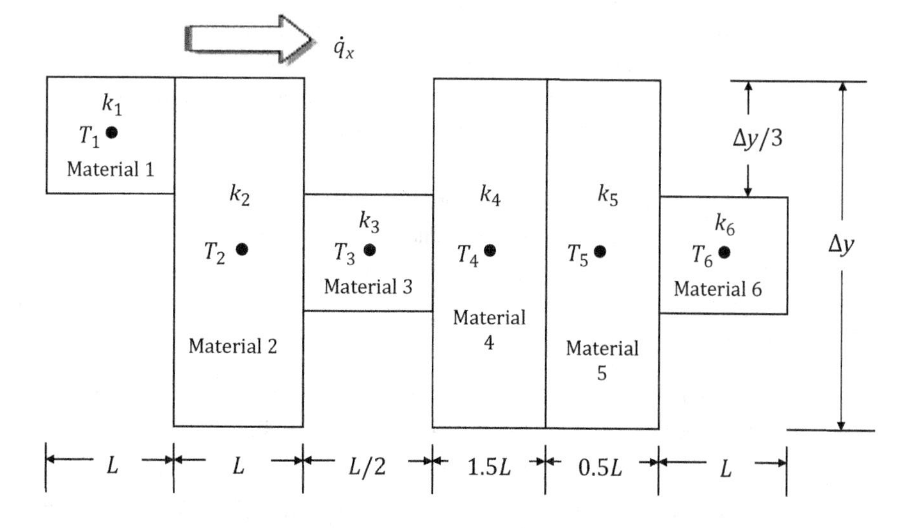

Figure P2.6 Composite wall with variation in element shape factors

i) Assuming a depth of *th* into the page, what is the resistance network?
ii) Assuming $L = \Delta y/3$ and $k_3 = 2k$; $k_6 = 3k$; $k_2 = k_4 = k_5 = k_1 = k$, determine the limiting resistance in the system.

2.7 For the cylindrical cross-section shown in Figure P2.7, assume:

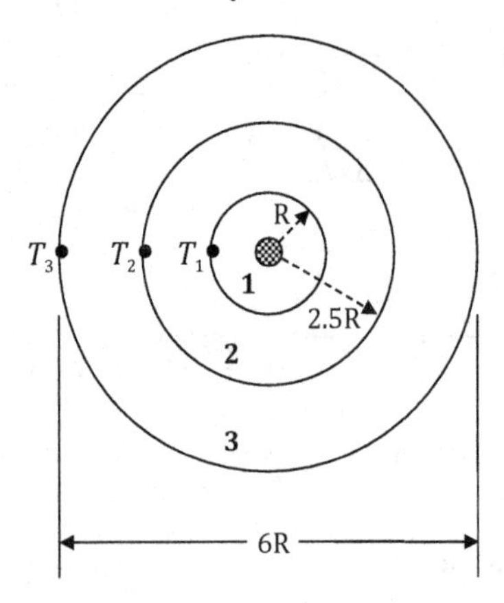

Figure P2.7 Radial configuration of a concentric geometry

$T_1 > T_2 > T_3$;
k_1, k_2, and k_3 are not equal;
R_o is the radius on the inner wall of region 1;
T_o is the temperature along inner wall of region 1;
th: Thickness into paper.

i) Write the thermal circuit for $0 \leftrightarrow 3$.
ii) Determine the heat transfer relation for $0 \leftrightarrow 3$.

2.8 Thermophysical properties for materials in this chapter have been treated as constant. This assumption is permissible for most room-temperature applications. However, in real life properties such as the thermal conductivity have temperature dependence and can vary significantly over large changes in temperature.

i) Provide a single plot of thermal conductivity as a function of temperature between 4 K and 300 K for copper (OFHC)/RRR = 150, 304 stainless steel, aluminum 1100, aluminum 6061-T6 and titanium 15-3-3-3.
ii) Which material experiences the largest variation below 50 K?
iii) Which material has the largest variation between 300 K and 100 K and how would the effective thermal conductance and heat transfer be affected during the material's approach to 100 K?

Note: Thermal conductivities can be determined from NIST's MProps website: http://cryogenics.nist.gov/MPropsMAY/materialproperties.htm

2.9 An aluminum cylinder (2024-T6) has an inner radius of 140 mm and an outer radius of 160 mm. The length of the cylinder is 150 mm. If 50 W of heat input is applied to the cylinder, what is the temperature after 20 minutes of heating? Assume that the initial temperature is 20°C and temperature gradients throughout the material are negligible during the heating process.

2.10 Two cases with similar geometry are shown in Figure P2.10, the primary difference being the level of nodalization.

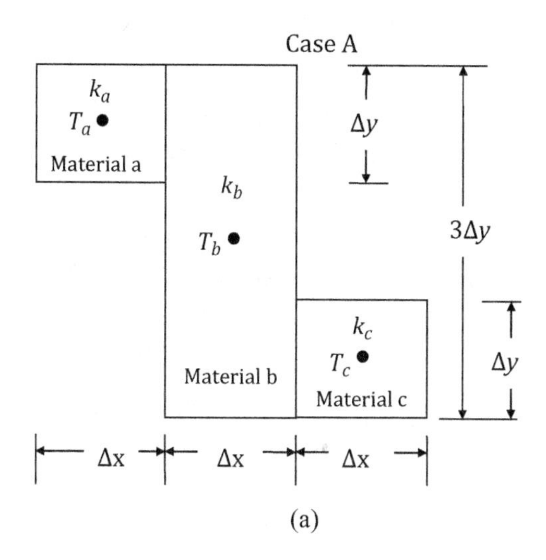

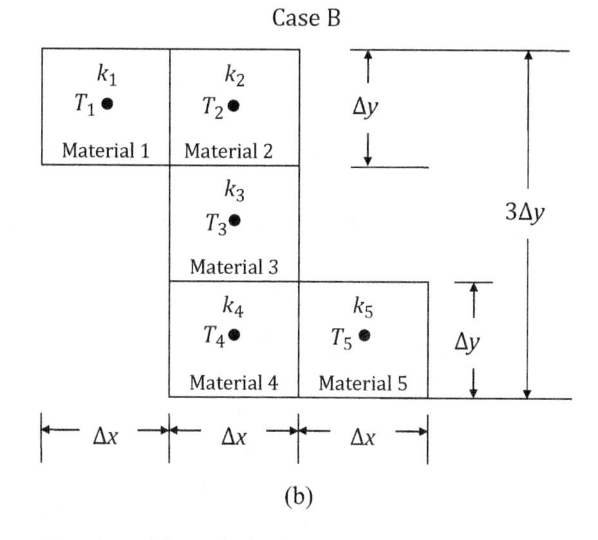

Figure P2.10 Thermal circuit nodalization comparison: a) reduced nodalization for use with projection technique; b) increased nodalization for use without projection technique

For case A

i) Determine the resistances in the thermal circuit.

ii) Assuming 8 W of heat input to material A and a constant $T_c = 5$°C, determine T_a and T_b. (Note: $k_a = k_b = k_c = k_{\text{Al, 2024–T6}}$.)

For case B

iii) Determine the resistances in the thermal circuit.

iv) Assuming 8 W of heat input to material 1 and a constant $T_5 = 5°C$, determine T_1, T_2, T_3 and T_4. (Note: $k_1 = k_2 = k_3 = k_4 = k_{Al,\ 2024\text{–}T6}$.)

Cases A and B

v) Determine the average value for T_2, T_3 and T_4. How does this compare with the value for T_b?

Note: In the problems for both cases A and B you can assume that $\Delta x = \Delta y = 0.05$ m; the thickness into the plane of the paper is 3 mm; and heat flow in each circuit is from left to right.

2.11 Assume you have a bellows tube (shown in Figure P2.11 as partially extended) made of 304 stainless steel with a wall thickness of 0.02 cm.

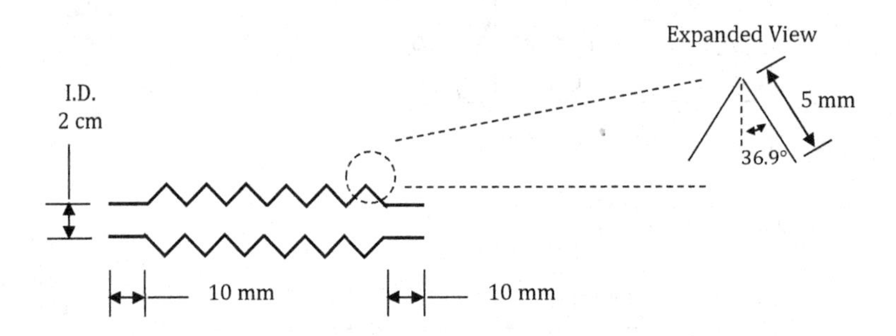

Figure P2.11 Bellows tube

i) What is the thermal conductance of the bellows tube in the longitudinal direction?

ii) What is the ratio of the bellows tube's longitudinal thermal conductance to that of a rigid tube of 304 stainless steel with the same ID and wall thickness as the bellows and a length of 44 mm?

2.12 A C-bracket consisting of nine elements of known material properties and geometry is shown in Figure P2.12. Within the circuit, element 9 is dissipating 10 W of thermal energy and element 5 is dissipating 5 W of thermal energy.

i) Draw the resistance network (label heat inputs and resistances).

ii) Assuming element 1 is a boundary node receiving active cooling, what temperature does element 1 need to operate at in order for element 5 to have a temperature of 20°C?

iii) What is the temperature of element 9 if element 5 is 20°C and element 1 is set to the value determined in part ii)?

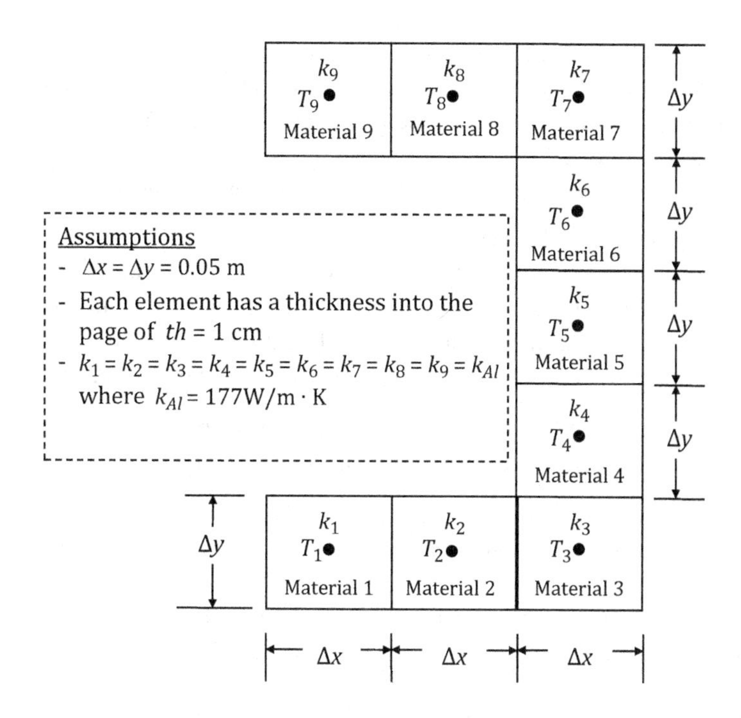

Figure P2.12 Multi-element C-bracket with multiple heat inputs

2.13 A wishbone bracket consisting of eleven elements of known material properties and geometry is shown in Figure P2.13. Within the circuit, element 9 is dissipating 15 W of thermal energy and element 1 is dissipating 10 W of thermal energy.

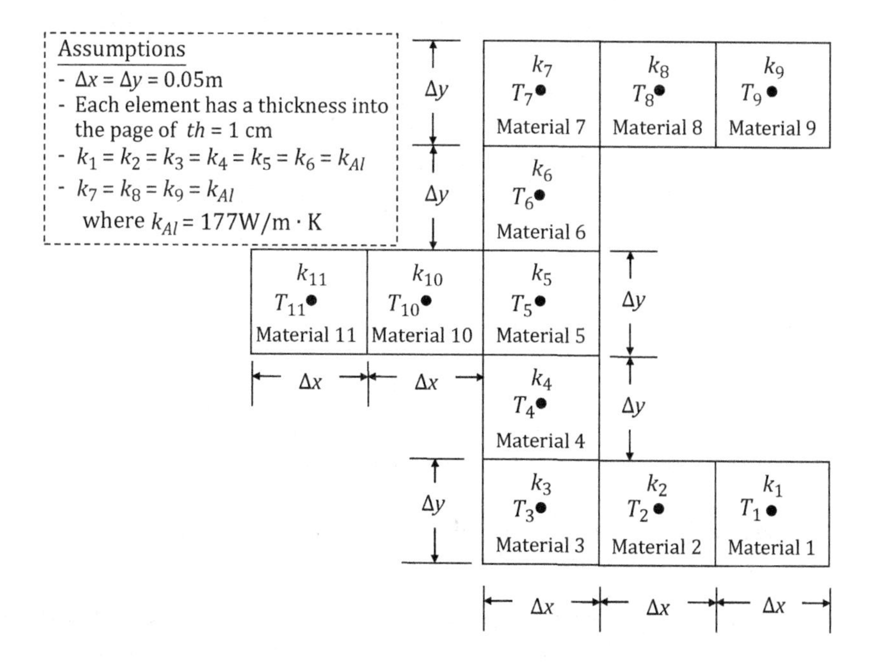

Figure P2.13 Multi-element wishbone bracket with multiple heat inputs

i) Draw the resistance network (label heat inputs and resistances).
ii) Assuming element 11 is a boundary node receiving active cooling, what temperature does element 11 need to operate at in order for element 9 to have a temperature of 20°C?
iii) What is the temperature of element 1 if element 9 is set at a value of 20°C?

2.14 Honeycomb vs. solid material.
i) What is the ratio of the thermal conductance of a piece of aluminum (6061-T6) honeycomb (2.54 cm thick) to that of a solid piece of the same material and thickness. You can assume the length and width are 1 m × 1 m. For the honeycomb, you can assume a ribbon thickness of 1 mm, $\sigma = 4/3$ and $S = 1$ cm.

Note: The ribbon-based area available for conduction from facesheet to facesheet is $A_{ribbon} = 2\sigma LW\delta/S$, making the total volume for the ribbon material $Vol_{ribbon} = 2\sigma\delta LW \cdot th/S$.

ii) What is the conductance-to-mass ratio for both the honeycomb and the solid aluminum in question i)?
iii) For the honeycomb geometry detailed in i), provide a plot of the conductance vs. mass at room temperature for Ni, 304 SS, 6061-T6 Al and OFHC as ribbon materials.
iv) Assuming 300 W is being conducted across the honeycomb panels in iii), provide a plot of ΔT vs. thermal conductivity at room temperature for each material.

2.15 A six-sided cubic enclosure (side length of 50 cm and a wall thickness of 6 mm) has two electronics boxes mounted on the interior wall space (Figure P2.15). The E-box mounted on wall 3 is dissipating 25 W whereas the E-box on wall 4 is dissipating 30 W. Assuming both E-boxes are made of 6061-T6 Al, what temperature is required on the lower box's baseplate (i.e., member five) in order to maintain E-box 4 at 20°C?

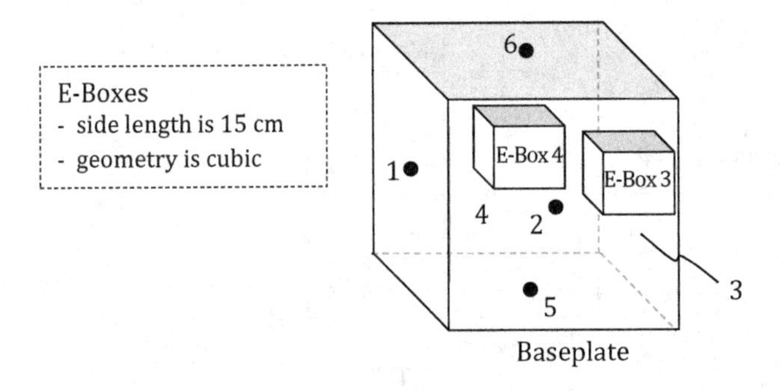

Figure P2.15 Six-sided cubic enclosure with internal E-boxes

Note: Model the E-boxes as solids positioned in the center of the lower box surface they are attached to.

2.16 A multi-tiered plate configuration (Figure P2.16) is populated with two electronics boxes that have the same geometry, yet different dissipations.

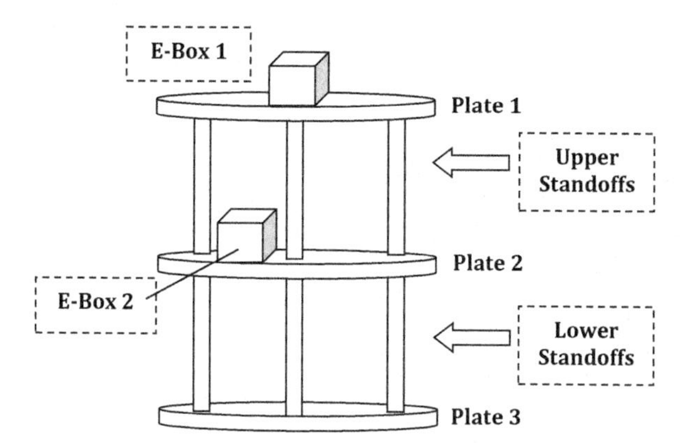

Figure P2.16 Multi-tiered E-box mounting plates

Assumptions:

- Isothermal resistances in the vertical direction

Plates 1, 2 and 3

- 6061-T6 Al
- 0.25″ thick
- Ø 35 cm

Stand-off geometry

- Ø 1.5 cm
- Height of 25 cm

E-box 1 and E-box 2

- Cubic geometry
- Side length of 10 cm
- $\dot{g}_1 = 15$ W
- $\dot{g}_2 = 7.5$ W

i) Draw the thermal circuit and label each resistance.

ii) Assuming that the circuit has nickel stand-offs between each of the plates, what temperature does plate 3 need to operate at in order to maintain a temperature of 25°C on E-box 2?

iii) You are given the additional requirements that plate 3 cannot be below 260 K and E-box 1 must operate between 35°C and 50°C. Given the option of using 304 SS, 6061 Al or OFHC for the stand-off materials, what material selection for the upper and lower stand-offs allows your design to meet temperature requirements?

2.17 For a 6061-T6 aluminum honeycomb panel with a length and width of 50 cm and a thickness of 3 cm:

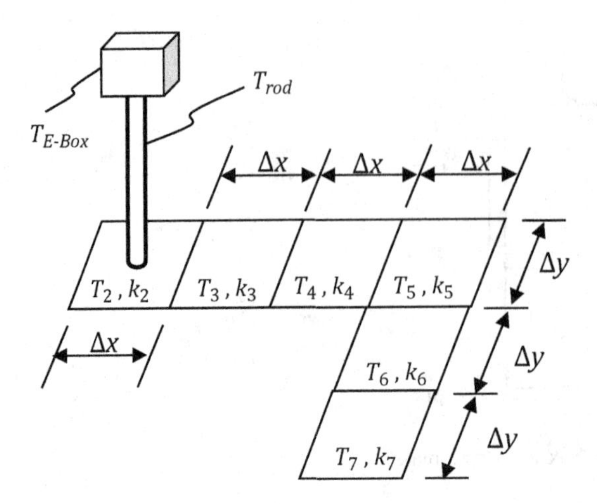

Figure P2.18 E-box with conductive path to a remote temperature sink

i) What is the ratio of the conductance along the ribbon length to the transverse conductance (edge to edge)?
ii) What is the ratio of the through conductance (i.e., in the direction of the thickness) to the conductance along the ribbon length (edge to edge)?

Note: You can assume the ribbon thickness is $\delta = 0.3$ mm and $\sigma = 1.172$ for $\theta = 45°$

2.18 An electronics box sits atop a solid rod of 304 stainless steel. The rod is mounted on the top of one section of a flat L-shaped mounting plate. The E-box is dissipating 10 W.
i) Draw the resistance network for the thermal circuit.
ii) Define each of the resistances between the E-box and element 7 (leave in variable form).
iii) If the E-box is to be operated at a temperature of 35°C, what temperature would element 7 need to be held at? (show your calculations)
iv) If the 304 stainless steel rod is swapped for a rod of identical size made from OFHC, then what temperature does element 7 need to be held at to achieve an E-box temperature of 35°C? (show your calculations)
v) What other measures could be taken to further reduce the temperature difference between the E-box and element 7?

Assumptions:

- Model the E-box as a solid aluminum (6061-T6) cube with a sidewall length of 6 cm.
- The center of the E-box cube is aligned with the centerline axis of the solid rod, which is also aligned with the center of element 2.
- $\Delta x = \Delta y = 9$ cm.
- Members 2 through 7 are aluminum (1100) and have a thickness of 3 mm.
- The rod has a diameter of 3 cm.
- The height of the rod (from material 2 to the E-box) is 18 cm.

2.19 An electronics box is attached to the vertical member of a T-bracket (Figure P2.19). The E-box is dissipating 50 W of energy.

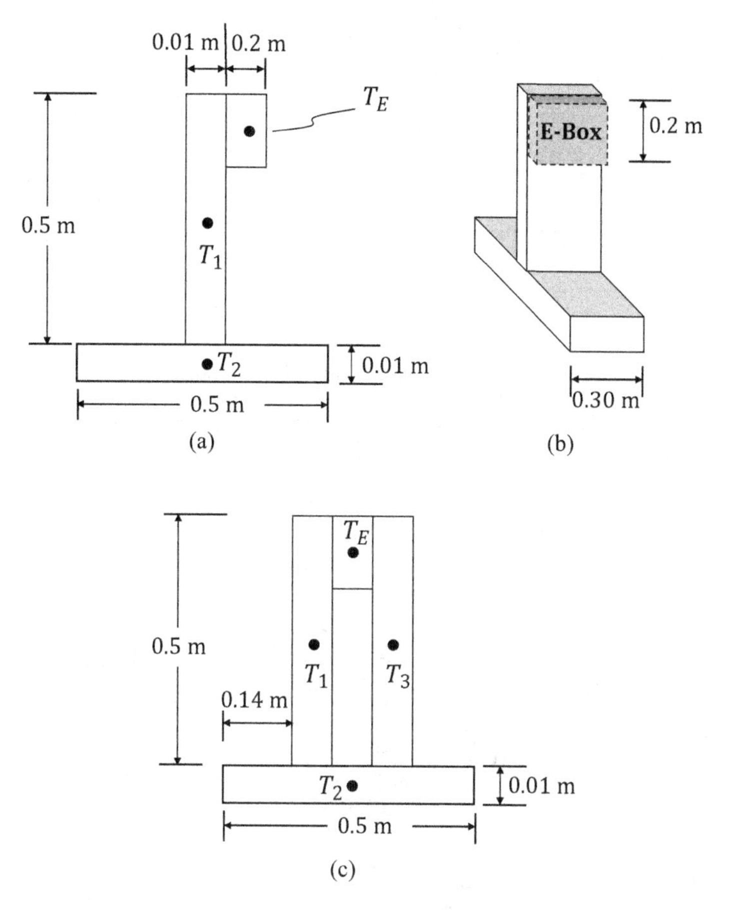

Figure P2.19 T-bracket E-box mount configuration: a) side view; b) lateral view; c) parallel struts for members 1 and 3

i) Assuming all members are made of 2024-T6 aluminum and member 2 is held at a temperature of 10°C via active cooling, what is the temperature of the electronics box?

ii) If a second vertical member (identical to member 1) is added for a new configuration (Figure P2.19c) with member 2 still held constant at 10°C and the same dissipation on the E-box, what is the new temperature for the E-box?

2.20 Let's revisit the hollow cubic enclosure problem with three thermal sources coupled to the interior wall. This time we will model the enclosure as a contiguous piece of metal that experiences uniform heating and cooling (i.e., reminiscent of an oven). The side length of the hollow cube (made of 6061-T6 Al) is 50 cm and it has a wall thickness of 1 cm. E-boxes 1, 2 and 3 mounted on the interior wall (Figure P2.20) have sidewall lengths of 12.5 cm and are dissipating 25 W, 30 W and 40 W respectively. What temperature is required on the exterior of the hollow cube in order to maintain E-box 3 at 20°C?

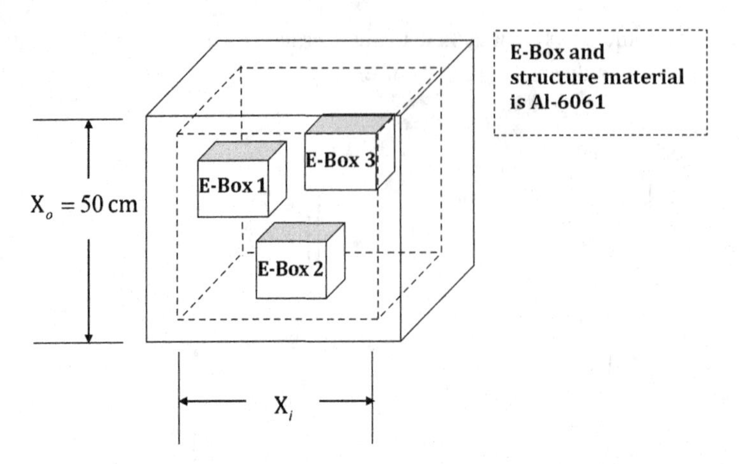

Figure P2.20 Three E-boxes mounted on interior wall of hollow box

2.21 A honeycomb panel (Figure P2.21) consists of two facesheets and a honeycomb matrix, each of which is separated into 9 different nodes. The planar area of the panel is 0.75 m × 0.75 m. Each of the facesheets has a thickness of 0.5 mm whereas the honeycomb matrix is 2.54 cm (i.e., 1.0 in) thick. The honeycomb has a cell width of 1.5 cm, a ribbon thickness of 0.5 mm and is made of 6061-Al.

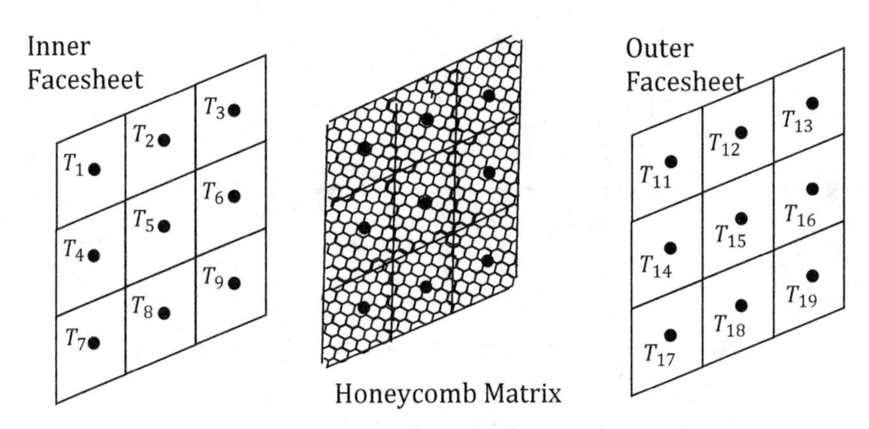

Figure P2.21 Honeycomb panel parts separated

Assume a thermal load of 5 W, 10 W, 30 W, 15 W and 8 W on inner facesheet nodes 1, 3, 5, 7 and 9 respectively. If there is a heat flux rejection of 100 W/m^2 on each of the outer facesheet nodes except node 15:

i) What is the heat flux rejection at node 15 if the inner and outer facesheet material is 6061-T6 aluminum?

ii) For the facesheet material used in i), what is the largest temperature gradient (i.e., ΔT) between any two nodes on the inner facesheet?

iii) If CVD diamond with a thermal conductivity of 1,500 W/m · K is used as the inner and outer facesheet materials, what is the largest temperature gradient across the inner facesheet?

Note: The thermal conduction in the honeycomb in the ribbon and lateral directions can be neglected.

2.22 Two honeycomb panels (Figure P2.22) made from 6061-T6 Al. are joined together in an L-bracket configuration via two corner brackets (80/20® part# 4304). The honeycomb panels consist of a planar area of 0.75 m × 0.75 m. The facesheets are 1 mm thick, and the honeycomb is 2.54 cm thick. On the anti-hinge side of each inner facesheet, the L-bracket configuration is structurally braced by two 80/20® 1515 series T-slot beams (1.06 m in length) attached to the facesheets via a gusseted corner bracket (80/20® part# 4336) at each end of the beams.

Geometry Features

Honeycomb	Corner Brackets (part 4304)	Gusseted Corner Brackets (part 4336)
Material of 6061-T6 Al. Ribbon thickness of 0.5 mm Cell width of 1.5 cm	Material of 6105-T5 Al. Height of 7.62 cm Base length of 7.62 cm Width of 7.137 cm Thickness of 0.635 cm Side Area 52.206 cm^2	Material of 6105-T5 Al. Side length 7.62 cm Diagonal length 10.78 cm Width 3.332 cm Thickness of 0.635 cm Side Area 24.3 cm^2 Diagonal Area 29.07 cm^2

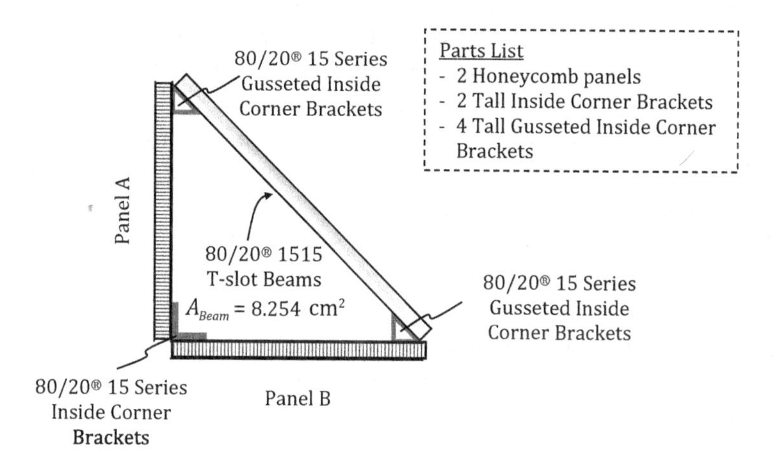

Figure P2.22 Honeycomb L-bracket space platform

i) What is the thermal circuit from the outer facesheet of panel A to the outer facesheet of panel B?

ii) If the outer facesheet of panel B is held at 260 K, what is the temperature of the outer facesheet of panel A if it is receiving a thermal load of 80 W?

Note: Each facesheet and honeycomb material should have its own node; the isothermal assumption applies to heat transfer through each honeycomb panel.

3 Radiative Heat Transfer Analysis

3.1 Fundamentals of Radiation

In Chapter 1 the Stefan–Boltzmann Law (shown in equation 3.1) was presented as the relation used to determine the amount of radiative thermal energy emitted from a surface.

$$\dot{Q} = \varepsilon \sigma A T^4 \tag{3.1}$$

A brief inspection of the Stefan–Boltzmann Law shows that radiation heat transfer has fourth-order temperature dependence. This is fundamentally different from Fourier's law for conduction and Newton's law of cooling, where the thermal energy exchange is dependent upon the first-order temperature difference (i.e., $\dot{Q} = O(T)$). From a physics perspective, the phenomena associated with the propagation of radiant thermal energy can be interpreted according to two different constructs: electromagnetic (EM) wave theory (the classical physics approach) and quantum physics.

From an EM wave perspective, thermal energy is emitted from a surface at select frequencies within the EM spectrum. Quantum theory states that energy is carried away from an emission surface in discrete quantities by photons. Photons are massless and travel at the speed of light in vacuum (c_o = 3.0 × 10^8 m/s). The ability to describe radiative heat transfer using either classical or quantum physics is valid via the principle of wave particle duality [1].

Three wavelength ranges are of interest for radiative heat transfer: the IR (infrared), visible and UV (ultraviolet). The part of each spectrum where thermal energy exchange occurs is listed in Table 3.1.

Table 3.1 Electromagnetic wave scale segments effecting thermal radiation phenomena

Spectral ranges	Wavelength (λ)
Infrared (IR)	1,000 μm ↔ 0.7 μm
Visible	0.7 μm ↔ 0.4 μm
Ultraviolet (UV)	0.4 μm ↔ 0.1 μm

Note that while the UV spectrum of interest spans 0.4 μm to 0.1 μm, the full UV spectrum extends further to 0.06 μm [1]. Thus, a small portion of this spectrum is considered of little significance for thermal energy exchange [2].

3.2 The Blackbody

Emissivity and absorptivity are mathematically defined relative to the radiative performance of a blackbody surface. A number of basic features define a blackbody surface. First, all incident radiation on a blackbody surface is absorbed (i.e., the surface exhibits zero reflection) for all wavelengths. Second, a blackbody surface is an ideal emitter for any given temperature and wavelength. Third, emission (E_b) from a blackbody surface is diffuse regardless of the emission wavelength (λ) and the temperature (T) of the surface [2–5]. As the ideal emitter, a blackbody surface emits the maximum amount of thermal energy that can be emitted from any surface for a particular wavelength and temperature. In real-life applications near room temperature, blackbody performance is unachievable without the application of specialized treatments to enhance the surface's properties. However, near blackbody performance is exemplified in natural objects such as the Sun. Next we define radiation intensity, which can have both spectral and directional dependence and is the basis for determining the rate of thermal energy emitted from and incident upon a surface.

3.2.1 Radiation Intensity

Radiative thermal emission from a surface in any direction can be captured by a receiving surface that is within the originating surface's FOV. This phenomenon is illustrated in Figure 3.1 by two individual area segments located on a virtual hemisphere. The differential surface area (dA) at the base of the virtual hemisphere is the originating surface. The virtual hemisphere surrounding the differential surface (dA) has a radius of R, thus making its surface area $2\pi R^2$.

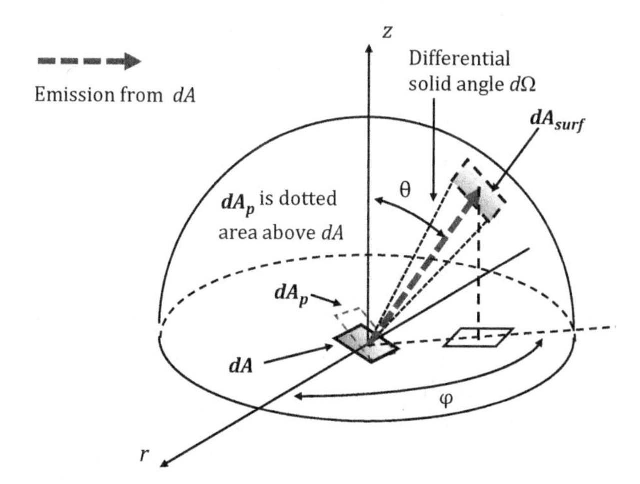

Figure 3.1 Virtual hemisphere for definition of blackbody emissive power
Source: Reproduced with permission of Taylor & Francis Group from *Thermal Radiation Heat Transfer* [4] by Siegel, R. and Howell, J., Copyright 2002; permission conveyed through Copyright Clearance Center, Inc.

For the receiving surface (dA_{surf}) the differential solid angle ($d\Omega$) sweeping out this area on the virtual hemisphere is equal to the differential surface area of interest on the perimeter of the hemisphere divided by the square of the hemisphere's radius. In equation form this is,

$$d\Omega = dA_{surf}/\mathbf{R}^2 \tag{3.2}$$

with units of steradians. Figure 3.2 provides an example of the dimensions for the differential surface area on the virtual hemisphere.

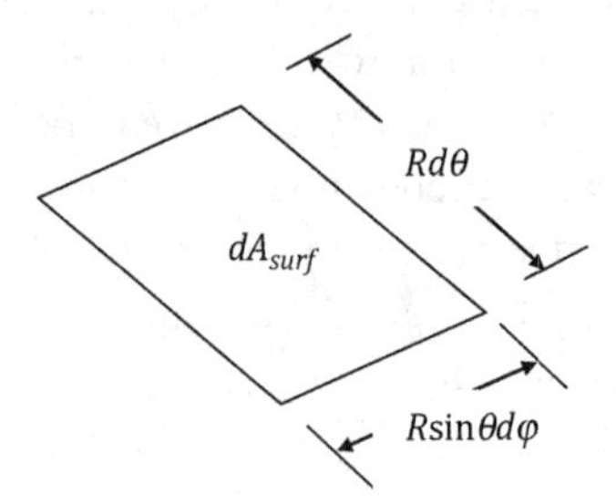

Figure 3.2 Differential surface area on the virtual hemisphere
Source: Reproduced with permission of Taylor & Francis Group from *Thermal Radiation Heat Transfer* [4] by Siegel, R. and Howell, J., Copyright 2002; permission conveyed through Copyright Clearance Center, Inc.

The length and width dimensions for dA_{surf} are defined in spherical coordinates. The differential surface area in Figure 3.2 can be defined as

$$dA_{surf} = \mathbf{R}^2 \sin\theta d\theta d\varphi \tag{3.3}$$

Given the new relation in equation 3.3, the relation for the differential solid angle can be recast as

$$d\Omega = \frac{dA_{surf}}{\mathbf{R}^2} = \sin\theta d\theta d\varphi \tag{3.4}$$

The virtual hemisphere will have a solid angle of 2π steradians (sr) about surface dA at the center of its base [2–4]. In Figure 3.1, the differential surface area dA is the source of the radiative thermal energy received at the virtual hemisphere's surface. The radiation incident upon dA_{surf} is considered as traveling along the surface's normal direction which faces the center of the hemisphere. Correspondingly, energy emitted from the differential surface area at the center of the hemisphere's base is also considered as traveling along the normal for dA_{surf}. Since dA is out of plane with the normal for dA_{surf} we will create a surface at the center of the virtual hemisphere that is aligned with the normal for dA_{surf}. We will name this surface dA_p since it is the projected surface area for dA. Surface dA_p is offset from surface dA by an angle of θ. Since we are interested in the originating surface, we must recast the energy emitted in terms of the unprojected surface area dA. Given that the angle between dA_p and dA is θ, the projected surface area can be rewritten as

$$dA_p = dA\cos\theta \tag{3.5}$$

Radiative thermal energy emission is a function of the intensity of the waves/photons being emitted [2–5]. The intensity may vary with direction or be uniform in all directions. Examples of variation of the intensity over θ and φ for the virtual hemisphere are illustrated in Figure 3.3. For a variation in θ from 0° to 180° (Figure 3.3b), the spectral hemispherical intensity has a constant value, whereas the directional intensity value will vary with the directional angle. To determine the total intensity across all wavelengths emitted from surface dA, we begin by defining the spectral intensity (I_λ), which is the energy emitted per unit time over a specified surface area and wavelength interval ($d\lambda$) that encompasses a solid angle centered relative to the θ and φ directions [2–4]. Thus, the energy associated with the spectral emission of radiation waves passing through the projected surface area (dA_p) and encompassing a solid angle $d\Omega$ can be represented as the product

$$I_\lambda(\lambda, \theta, \varphi)dA\cos\theta \cdot d\Omega$$

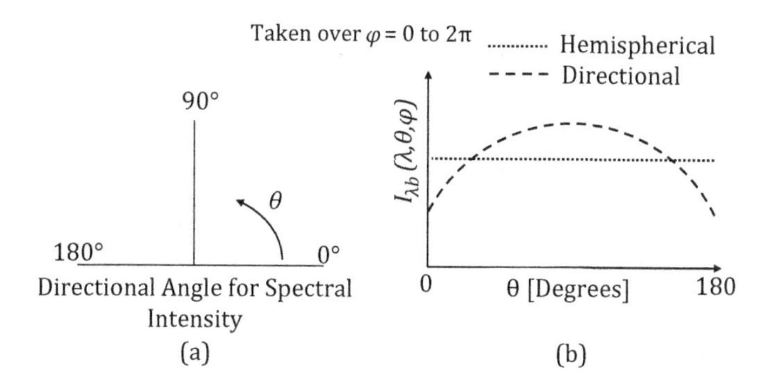

Figure 3.3 Directional dependence for radiation intensity: a) direction of variation; b) spectral hemispherical and directional intensity

This product has units of [W/μm]. If we divide through by dA to make the energy a flux and substitute in for the differential solid angle using equation 3.4, the spectral directional emissive power can then be defined as

$$E_{\lambda b}(\lambda, \theta, \varphi) = I_{\lambda b}(\lambda, \theta, \varphi)\cos\theta\sin\theta d\theta d\varphi \tag{3.6}$$

where $E_{\lambda b}(\lambda, \theta, \varphi)$ has units of [W/μm · m^2]. Given the new flux-based representation for spectral directional emissive power, the blackbody spectral hemispherical emissive power originating at dA can now be determined by integrating the RHS of equation 3.6 over the entire virtual hemisphere for both the θ and φ directions (i.e., all solid angles). Since blackbody emission is diffuse, the intensity term will only be functionally dependent upon the wavelength. As equation 3.7 shows, the blackbody spectral hemispherical emissive power is π times the blackbody spectral intensity.

$$\underbrace{E_{\lambda b}(\lambda)}_{\substack{\textit{Blackbody Spectral}\\ \textit{Hemispherical}\\ \textit{Emissive Power}}} = I_{\lambda b}(\lambda)\int_{\varphi=0}^{2\pi}\int_{\theta=0}^{\pi/2}\cos\theta\sin\theta d\theta d\varphi = \pi I_{\lambda b}(\lambda) \tag{3.7}$$

3.2.2 Planck's Distribution Law

Max Planck experimentally verified that the spectral distributions of hemispherical emissive power and radiant intensity for blackbody surfaces in vacuum are a function of the wavelength and the surface's absolute temperature [4,6,7]. In vacuum this is expressed as

$$E_{\lambda b}(\lambda, T) = \pi I_{\lambda b}(\lambda, T) = \frac{2\pi c_1}{\lambda^5 \left[exp^{\,c_2/\lambda T} - 1 \right]} \tag{3.8}$$

where

$c_1 = hc_o^2$; c_o = speed of light 3.0×10^8 m/s; h = Planck's constant, 6.6256×10^{-34} J · s

$c_2 = hc_o/k_B$; k_B = Boltzmann's constant, 1.3805×10^{-23} J/K

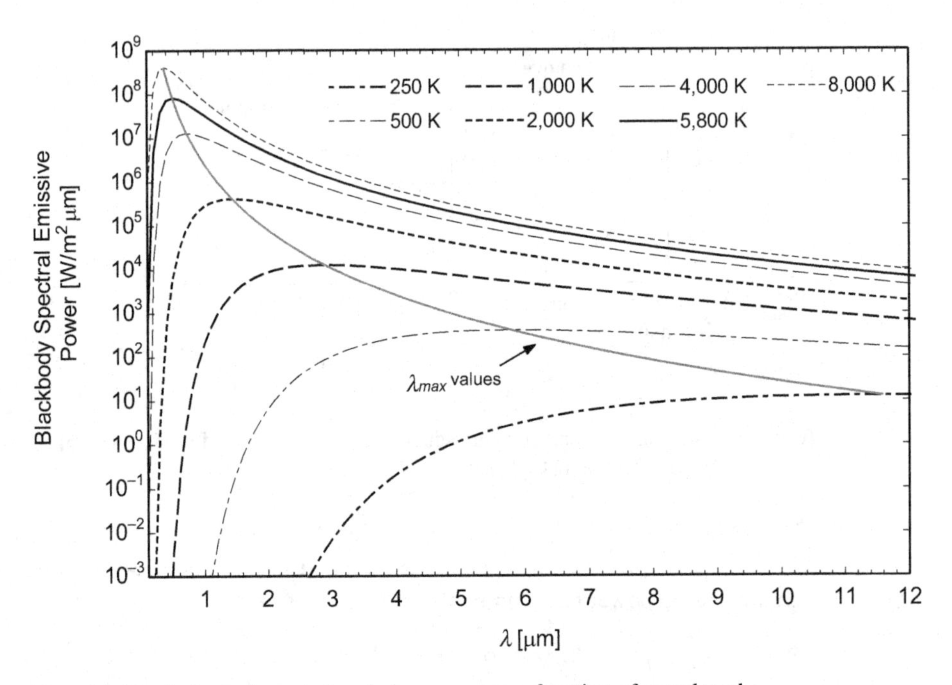

Figure 3.4 Hemispherical spectral emissive power as a function of wavelength

Figure 3.4 is an example of blackbody spectral hemispherical emissive power plotted as a function of wavelength at constant temperature using Planck's spectral distribution. This plot shows that as the temperature of the emission surface increases, the maximum spectral hemispherical emissive power shifts towards smaller wavelengths [4]. This can also be determined through comparison of two curves at different temperatures (e.g., the 250 K curve relative to the 4,000 K curve). Using basic calculus Min/Max theory, the product of the wavelength and temperature at the maximum for each temperature curve in Figure 3.4 can be shown to be a constant. Differentiating Planck's Distribution Law with respect to λ and setting it equal to 0 gives

$$\frac{d}{d\lambda}[E_{\lambda b}(\lambda, T)] = \frac{d}{d\lambda}\left[\frac{2\pi c_1}{\lambda^5[\exp^{c_2/\lambda T} - 1]}\right] = 0 \tag{3.9}$$

Distributing the derivative and then solving for λT would result in

$$\lambda T = \frac{c_2}{5}\frac{1}{[1 - \exp^{-c_2/\lambda T}]} \tag{3.10}$$

where λ actually is a maximum (i.e., λ_{max}). The solution for the product of $\lambda_{max} \cdot T$ is a constant. This relation, known as Wien's displacement law, is

$$\lambda_{max} \cdot T = c_3 \tag{3.11}$$

where $c_3 = 2897.8\ \mu\text{m} \cdot \text{K}$.

Example 1 Blackbody Temperature at Maximum Intensity What is the temperature for a blackbody surface emitting radiative waves with a maximum intensity in the mid-infrared spectrum at $\lambda_{max} = 32\ \mu\text{m}$?

Solution

Using Wien's displacement law to solve for the temperature gives

$$T = \frac{c_3}{\lambda_{max}} = \frac{2897.8\ \mu\text{m} \cdot \text{K}}{32\ \mu\text{m}} \approx 90.5\ \text{K}$$

Thus, for a body emitting its maximum intensity at a wavelength of 32 μm the blackbody temperature falls within the cryogenic temperature regime (i.e., it is <123 K). Alternatively, a body emitting its maximum intensity at a temperature comparable to that of the surface of the Sun (i.e., $\approx 5{,}800$ K) will have a λ_{max} equal to 0.499 μm which is within the lower portion of the visible spectrum.

3.2.3 Blackbody Emissive Power

We can obtain the blackbody total hemispherical emissive power by integrating Planck's Distribution Law (see equation 3.8) over the full spectrum of wavelengths.

$$E_b = \int_0^\infty E_{\lambda b}(\lambda) d\lambda = \pi \int_0^\infty I_{\lambda b}(\lambda) d\lambda \tag{3.12}$$

The blackbody total intensity is defined as the integral of the blackbody spectral intensity over the full spectrum of wavelengths. This is expressed mathematically as

$$I_b = \int_{\lambda=0}^\infty I_{\lambda b}(\lambda) d\lambda \tag{3.13}$$

where $I_{\lambda b}(\lambda)d\lambda$ is the intensity emitted in a small wavelength interval $d\lambda$. Based on the equality in the Planck's law relation (equation 3.8), the blackbody spectral intensity can be defined as

$$I_{\lambda b}(\lambda, T) = \frac{2c_1}{\lambda^5 [exp^{c_2/\lambda T} - 1]} \tag{3.14}$$

In order to determine the blackbody total intensity, equation 3.14 must be integrated over all wavelengths. The blackbody total intensity now becomes

$$I_b = \int_0^\infty \frac{2c_1}{\lambda^5 [exp^{c_2/\lambda T} - 1]} d\lambda \tag{3.15}$$

The integrand in equation 3.15 may be simplified by a change in variable that transforms the integral into a form that can be used to reference its integrated value with the aid of standard definite integral tables [4,5,8]. If we let $\xi = c_2/\lambda T$ for the change in variable, the corresponding solution for the blackbody total intensity becomes

$$I_b = \frac{2c_1 T^4}{c_2^4} \int_0^\infty \frac{\xi^3}{(exp^{\xi} - 1)} d\xi = \frac{\sigma}{\pi} T^4 \tag{3.16}$$

where the constant σ (i.e., the Stefan–Boltzmann constant) is equal to $2c_1\pi^4/15c_2$. Based on equation 3.12, the total emissive power for a blackbody radiating into vacuum would then be

$$E_b = \pi I_b = \sigma T^4 \tag{3.17}$$

This is the Stefan–Boltzmann law for a blackbody surface presented in flux terms [W/m^2].

3.2.4 Irradiation

Spectral irradiation (G_λ) is defined as the radiative energy flux incident on a surface from all directions at a wavelength λ. As with the development for the blackbody total emissive power in Sections 3.2.2 and 3.2.3, the amount of irradiation upon a surface is dependent upon the intensity of the incident radiation. We will define the intensity in spectral directional form as $I_\lambda(\lambda, \theta, \varphi)$. Figure 3.5 is an example of radiation incident upon a differential surface area (dA) swept over a known differential solid angle $d\Omega$.

Assuming that the intensity of the incident radiation is diffuse, the spectral hemispherical irradiation can be defined as

$$G_\lambda(\lambda) = I_\lambda(\lambda) \int_{\varphi=0}^{2\pi} \int_{\theta=0}^{\pi/2} \cos\theta \sin\theta d\theta d\varphi \tag{3.18}$$

Upon evaluating the double integral, the relation reduces to

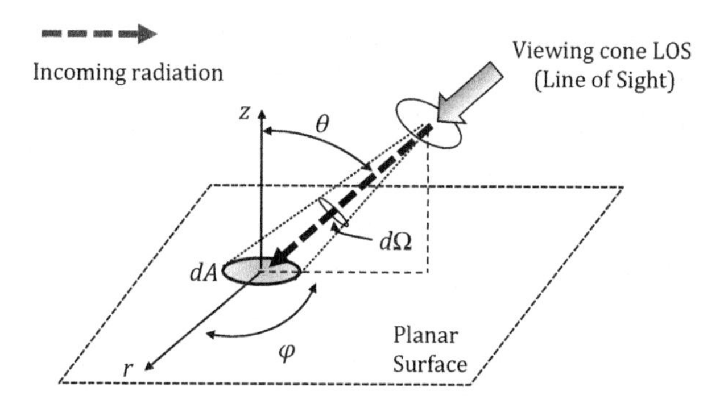

Figure 3.5 Irradiation of a differential surface area
Source: Reproduced with permission of John Wiley & Sons Inc. from *Fundamentals of Heat and Mass Transfer,* Bergman, T.L., Lavine, A.S., Incropera, F.P., and Dewitt, D.P., 7th edn, 2017, reference [2]; permission conveyed through Copyright Clearance Center, Inc.

$$G_\lambda(\lambda) = \pi I_\lambda(\lambda) \tag{3.19}$$

The total hemispherical irradiation may then be obtained by integrating over all wavelengths. The equation for the total hemispherical irradiation then becomes

$$G = \int_0^\infty G_\lambda(\lambda) d\lambda = \pi I \tag{3.20}$$

where it is equal to the product of π times the intensity of the incoming radiation.

3.3 Real Surfaces

The performance of real surfaces can be subject to directional dependence. Thus, the intensity that will be relied upon for the development of the relations capturing thermal emission and incident thermal energy upon a real surface will have a functional dependence of $I_\lambda(\lambda, \theta, \varphi)$. We will begin by defining the radiosity for the real surface.

3.3.1 Radiosity

Radiosity is defined as the sum of all the radiative thermal energy departing a surface. Figure 3.6 provides an example of the departure of thermal energy due to both direct emission and reflection. For reflected thermal energy, the angle of incidence is equal to the angle of reflection (i.e., $\theta_a = \theta_b$) relative to the normal for the energy exchange surface at the point of incidence. Because a blackbody surface absorbs all incident thermal energy, the presence of reflected thermal energy suggests that this phenomenon differs inherently from that occurring with blackbodies.

The spectral directional radiosity defined with respect to the intensity on the incident surface is

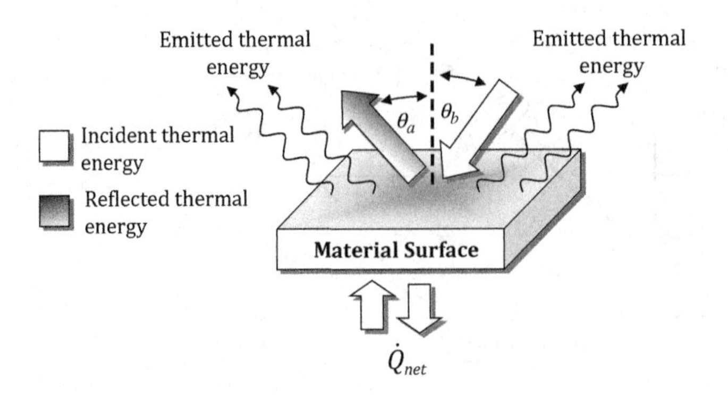

Figure 3.6 Radiosity phenomena on a real surface

$$J_\lambda(\lambda,\theta,\varphi) = \int_{\varphi=0}^{2\pi}\int_{\theta=0}^{\pi/2} I_\lambda(\lambda,\theta,\varphi)\cos\theta\sin\theta d\theta d\varphi \tag{3.21}$$

where I_λ is the summation of the intensity associated with emission and reflection from the surface. In the derivations for both emission and irradiation in Section 3.2 we assumed that the intensity associated with emission and irradiation was diffuse. In this case the radiosity becomes spectral in nature and the spectral hemispherical radiosity can then be defined as

$$J_\lambda(\lambda) = I_\lambda(\lambda)\int_{\varphi=0}^{2\pi}\int_{\theta=0}^{\pi/2} \cos\theta\sin\theta d\theta d\varphi \tag{3.22}$$

Integrating across the virtual hemisphere, the spectral hemispherical radiosity becomes

$$J_\lambda(\lambda) = \pi I_\lambda(\lambda) \tag{3.23}$$

The total hemispherical radiosity is then defined as

$$J = \int_0^\infty J_\lambda(\lambda)d\lambda \tag{3.24}$$

Finally, if the RHS of equation 3.23 is substituted into equation 3.24 and integrated, total radiosity is obtained:

$$J = \pi I \tag{3.25}$$

3.3.2 Emissivity

Since the blackbody is the ideal emitter, the emissive capability of a real surface is measured relative to that of the blackbody surface. The emissivity (ε) is defined as the ratio of the intensity of the radiation emitted by a real surface to that emitted by a blackbody at the same temperature. For the spectral directional case this is written as

$$\varepsilon(\lambda, \theta, \varphi, T) = \frac{I_\lambda(\lambda, \theta, \varphi, T)}{I_{\lambda b}(\lambda, T)} \quad (3.26)$$

The spectral hemispherical emissivity is then defined as

$$\varepsilon_\lambda(\lambda, T) = \frac{E_{\lambda r}(\lambda, T)}{E_{\lambda b}(\lambda, T)} \quad (3.27)$$

where the subscript r in the emissive power term denotes a real surface. The total hemispherical emissivity can then be attained by integrating both sides of equation 3.27 over all wavelengths. The resultant form for the emissivity definition becomes

$$\varepsilon(T) = \frac{E_r(T)}{E_b(T)} \quad (3.28)$$

For additional representations of this fractional relationship, see Bergman et al. [2], Nellis and Klein [3] or Siegel and Howell [4]. While the emissivity relations have been presented here as functions of wavelength and direction, surface conditions can alter emissivity values relative to the smooth, uncontaminated surface analogue (see Modest [5]).

3.3.3 Absorptivity, Reflectivity and Transmissivity

The total hemispherical irradiation on a real surface is equal to the sum of the radiant energy reflected, absorbed and transmitted (for non-opaque materials) (see Figure 3.7). The amount of thermal energy absorbed, reflected and transmitted through the material will therefore be a fraction of the irradiation incident upon the surface. The spectral hemispherical absorptivity may be defined as

$$\alpha_\lambda(\lambda) = \frac{G_{\lambda r, abs}(\lambda)}{G_{\lambda r}(\lambda)} \quad (3.29)$$

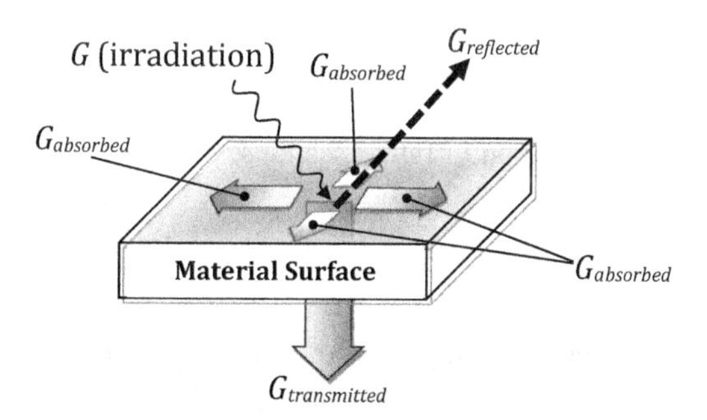

Figure 3.7 Absorption, reflection and transmission on a real opaque surface

where the subscripts r and abs in the numerator denote real surfaces and absorptivity respectively. If both sides of equation 3.29 are integrated over all wavelengths, the total hemispherical absorptivity is obtained:

$$\alpha = \frac{\int_0^\infty \alpha_\lambda(\lambda) \cdot G_{\lambda r}(\lambda) d\lambda}{\int_0^\infty G_{\lambda r}(\lambda) d\lambda} \tag{3.30}$$

which can also be expressed as

$$\alpha = \frac{G_{r,abs}}{G_r} \tag{3.31}$$

Analogous relations can also be developed for the reflectivity and transmissivity. Table 3.2 summarizes the spectral and total hemispherical relations for absorptivity, reflectivity and transmissivity.

Table 3.2 Spectral hemispherical and total hemispherical relations for absorptivity, reflectivity and transmissivity

Representation	Absorptivity (α)	Reflectivity (ρ)	Transmissivity (τ)
Spectral hemispherical	$\alpha_\lambda(\lambda) = \frac{G_{\lambda r,abs}(\lambda)}{G_{\lambda r}(\lambda)}$	$\rho_\lambda(\lambda) = \frac{G_{\lambda r,ref}(\lambda)}{G_{\lambda r}(\lambda)}$ (3.32)	$\tau_\lambda(\lambda) = \frac{G_{\lambda r,trans}(\lambda)}{G_{\lambda r}(\lambda)}$ (3.33)
Total hemispherical	$\alpha = \frac{G_{r,abs}}{G_r}$	$\rho = \frac{G_{r,ref}}{G_r}$ (3.34)	$\tau = \frac{G_{r,trans}}{G_r}$ (3.35)

The equation for total irradiation incorporating absorptivity, reflectivity and transmissivity would be

$$G_r = G_{r,abs} + G_{r,ref} + G_{r,trans} \tag{3.36}$$

Dividing equation 3.36 by the total irradiation would give

$$\frac{G_r}{G_r} = \frac{G_{r,abs}}{G_r} + \frac{G_{r,ref}}{G_r} + \frac{G_{r,trans}}{G_r} \tag{3.37}$$

which can also be expressed as

$$1 = \alpha + \rho + \tau \tag{3.38}$$

We shall return to this summation relation in future sections.

3.3.4 Kirchhoff's Law

Figure 3.8 shows a real body located inside a blackbody enclosure. The real body has a surface area of A. The interior wall of the blackbody enclosure has a uniform temperature of T_b. Assuming the system is in equilibrium, a steady-state energy balance can be applied to the surface of the body located inside the enclosure. Irradiation onto and emission from the body's surface area must be equal. In total hemispherical form the energy balance is

$$G_{r,abs} \cdot A = E_r \cdot A \tag{3.39}$$

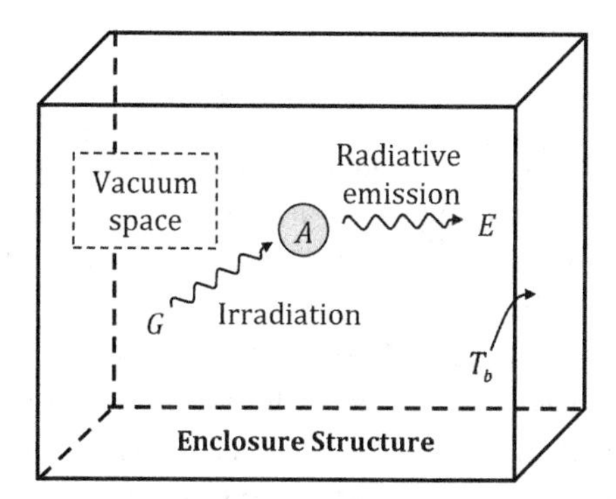

Figure 3.8 Thermal energy exchange for an isothermal enclosure

Irradiation onto surface A, originating along the interior wall of the blackbody, is diffuse and equals the total emission from the blackbody surface.

$$G_r = E_b(T_b) \tag{3.40}$$

Recasting the LHS of equation 3.39 by substituting in $G_{r,abs}$ from equation 3.31 gives

$$\alpha G_r = E_r \tag{3.41}$$

Alternatively, we can solve equation 3.41 for the irradiation as

$$G_r = \frac{E_r}{\alpha} \tag{3.42}$$

Given the previously established assumptions associated with diffuse emission from the interior surface of a blackbody enclosure, we know that the irradiation at surface A is equal to the total emission from the enclosure surface. Using the total hemispherical relation for emissivity (equation 3.28), equation 3.42 can be rewritten as

$$E_b = \frac{\varepsilon E_b}{\alpha} \tag{3.43}$$

If both sides of the equal sign are divided by the blackbody total hemispherical emissive power (E_b), the relation becomes

$$1 = \frac{\varepsilon}{\alpha} \tag{3.44}$$

For any surface interior to the enclosure, therefore, the total hemispherical emissivity and absorptivity are equal (i.e., $\varepsilon = \alpha$). This is Kirchhoff's law for the total hemispherical case. Since both the emissivity and absorptivity are a function of the surface's properties (i.e., independent of spectral and directional distributions) [2], the most general form of Kirchhoff's Law is

$$\varepsilon_\lambda(\lambda, \theta, \varphi) = \alpha_\lambda(\lambda, \theta, \varphi) \tag{3.45}$$

Assuming spectral hemispherical surface properties, the analogous relationship for the spectral case is

$$\varepsilon_\lambda(\lambda) = \alpha_\lambda(\lambda) \tag{3.46}$$

3.3.5 Gray Body Surfaces

In the previous section, the total hemispherical form of Kirchhoff's Law was determined via a derivation that included the assumption of irradiation due to blackbody emission from a surface at a temperature equal to that of the enclosure surface. However, practical engineering applications will not always include surfaces that demonstrate blackbody behavior. A natural extension to these variants of Kirchhoff's Law would be whether or not they also apply to non-blackbody surfaces. The spectral hemispherical form of Kirchhoff's Law (equation 3.46) applies if irradiation is diffuse or the surface fosters diffuse emission. However, additional conditions must be satisfied in the total hemispherical case (equation 3.44). We know that the total hemispherical emissivity and absorptivity in question is associated with real surfaces. Based on the mathematical definition of each, an equality for the emissivity and absorptivity in the total hemispherical case can be recast as

$$\varepsilon = \frac{\int_0^\infty \varepsilon_\lambda(\lambda, T) \cdot E_{\lambda b}(\lambda, T) d\lambda}{E_b} = \frac{\int_0^\infty \alpha_\lambda(\lambda) \cdot G_{\lambda r}(\lambda) d\lambda}{G} = \alpha \tag{3.47}$$

For gray surfaces, the spectral-based absorptivity and emissivity are independent of wavelength [2]. If irradiation resulting from blackbody emission is not present, then in order for this equality to hold, the surface in question must be "gray." Since variation of the spectral values with wavelength is possible, the emissivity and the absorptivity cannot be expected to be equal across all wavelengths in the total hemispherical case. However, if the irradiation and surface emission are concentrated in a spectral bandwidth where properties are approximately equal, the total hemispherical form of Kirchhoff's Law can be invoked. Figure 3.9 illustrates the applicability of Kirchhoff's Law to an arbitrary surface. In this example, the spectral values of emissivity and absorptivity are fairly constant and equal in the visible spectrum, suggesting that Kirchhoff's Law is applicable to the visible wavelength for the surface in question.

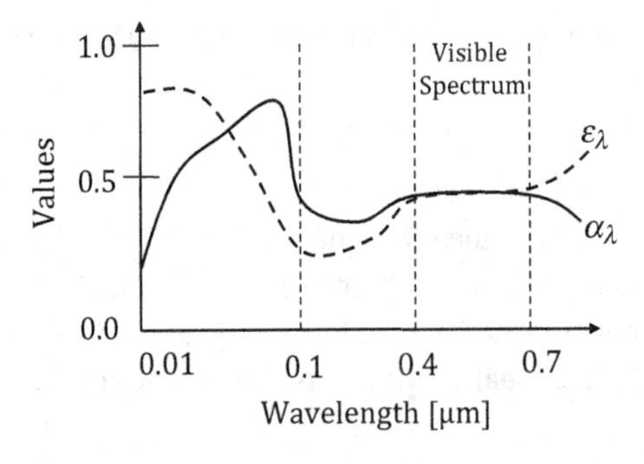

Figure 3.9 Gray body spectral surface properties behavior

How is this relevant to spacecraft thermal analysis, modeling and design? The Sun is much hotter than space platforms and therefore emits at shorter wavelengths in the UV, visible and near IR regions of the EM spectrum. These wavelengths are segregated from those for standard gray body surfaces on spacecraft. We shall use the total hemispherical variant of Kirchhoff's Law for the far IR spectrum (i.e., the long wavelength portion of the IR band) when dealing with radiative heat transfer for gray body surfaces on space platforms.

3.4 Radiative Heat Transfer between Surfaces

This section establishes the constructs (and corresponding energy balance relations) for the net radiation-based thermal energy exchange between two surfaces where no participating media are present. The first step, however, is to determine the amount by which surfaces involved in radiative thermal energy exchange are within each other's FOV (i.e., the view factor).

3.4.1 View Factors

Figure 3.10a shows two surfaces (A_1 and A_2) located inside a virtual enclosure under vacuum conditions (no matter between surface 1 and 2 is participating in their net heat transfer).

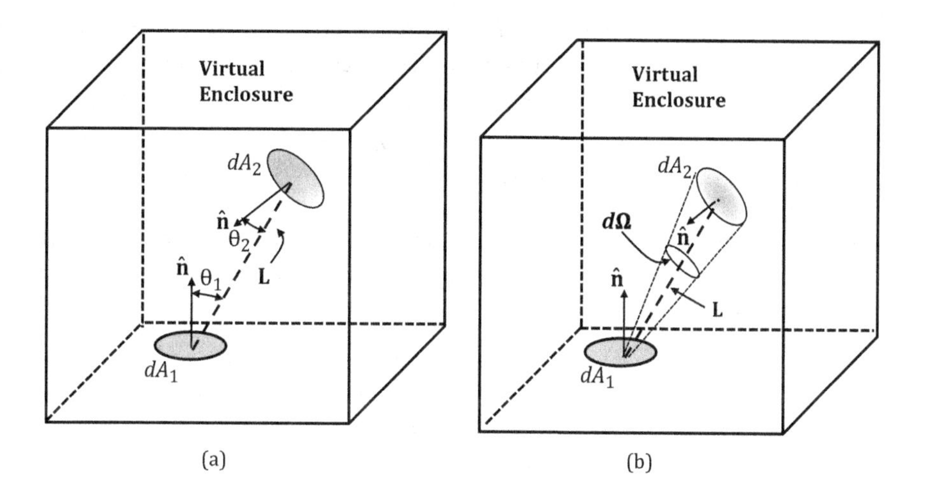

Figure 3.10 Viewing surfaces in the virtual enclosure: a) angular relations between surface normal; b) differential solid angle for view of dA_2 within dA_1's FOV
Source: Reproduced with permission of John Wiley & Sons Inc. from Bergman, T.L., Lavine, A.S., Incropera, F.P., and Dewitt, D.P., *Fundamentals of Heat and Mass Transfer,* 7th edn, 2017, reference [2]; permission conveyed through Copyright Clearance Center, Inc.

The view factor from surface A_1 to surface A_2 is defined as the fraction of thermal radiation leaving surface A_1 that is incident upon surface A_2 [2–5]. Mathematically, this can be expressed as

$$F_{1-2} = \frac{\dot{q}_{1-2}}{\dot{q}_1} \tag{3.48}$$

where F_{1-2} is the view factor for surface 1 to surface 2. Using radiosity constructs, the total radiative thermal energy leaving surface A_1 is

$$\dot{q}_1 = A_1 J_1 \tag{3.49}$$

Both surfaces are composites of differential area elements located in-plane to each surface. A separation distance (**L**) will apply to any combination of differential area elements (e.g., dA_1 and dA_2). In the present case we will assume the differential area elements are located at the center of surfaces 1 and 2. The length vector **L** forms angles with the normal from each differential surface. These angles are θ_1 and θ_2 for surfaces dA_1 and dA_2 respectively. Based on the total intensity of the radiative thermal energy leaving surface 1, the thermal energy incident upon dA_2 can be defined in differential terms as

$$d\dot{q}_{1-2} = I_1 \cos\theta_1 dA_1 d\Omega \tag{3.50}$$

The term $d\Omega$ is the solid angle for the view of surface dA_2 from surface dA_1. Using the solid angle definition for the virtual hemisphere (see Section 3.2.1), the differential solid angle can be recast as

$$d\Omega = \frac{\cos\theta_2 dA_2}{L^2} \tag{3.51}$$

where the cosine term has been included to account for the projection of surface dA_2 onto the line of sight of **L**. Substituting the differential solid angle relation into equation 3.50 expresses the differential heat transfer as

$$d\dot{q}_{1-2} = \frac{I_1 \cos\theta_1 \cos\theta_2}{L^2} dA_1 dA_2 \tag{3.52}$$

If the total intensity is rewritten in terms of the total radiosity and equation 3.52 integrated over surfaces dA_1 and dA_2 to determine the total heat transferred to surface 2, the final relation will be

$$\dot{q}_{1-2} = J_1 \int_{A_2}\int_{A_1} \frac{\cos\theta_1 \cos\theta_2}{\pi L^2} dA_1 dA_2 \tag{3.53}$$

and the final relation for the view factor will be

$$F_{1-2} = \frac{\dot{q}_{1-2}}{\dot{q}_1} = \frac{1}{A_1}\int_{A_2}\int_{A_1} \frac{\cos\theta_1 \cos\theta_2}{\pi L^2} dA_1 dA_2 \tag{3.54}$$

The development used to establish the F_{1-2} view factor relation can also be used to derive a relation for the fraction of radiation leaving A_2 that is incident on A_1 (i.e., F_{2-1}):

$$F_{2-1} = \frac{1}{A_2}\int_{A_2}\int_{A_1} \frac{\cos\theta_1 \cos\theta_2}{\pi L^2} dA_1 dA_2 \tag{3.55}$$

These relations hold for any two surfaces that are diffuse emitters and reflectors. If we multiply the view factor and the double integral term in equation 3.54 by A_1 and multiply both sides in equation 3.55 by A_2, an equality can be established between F_{1-2} and F_{2-1} by way of the double area integral:

$$A_1 F_{1-2} = A_2 F_{2-1} = \int_{A_2} \int_{A_1} \frac{\cos\theta_1 \cos\theta_2}{\pi L^2} dA_1 dA_2 \tag{3.56}$$

This is known as the view factor reciprocity relation for any two surfaces within each other's FOV. In common with most textbooks, we will use the subscripts i and j to denote emitting and receiving surfaces

$$A_i F_{i-j} = A_j F_{j-i} \tag{3.57}$$

3.4.2 Enclosures

For each surface inside an enclosure, the summation of the view factors for each surface equals 1. Thus if the virtual enclosure has multiple surfaces, each surface's view factors to the other surfaces can be expressed as the summation

$$\sum_{j=1}^{N} F_{i-j} = 1 \tag{3.58}$$

In summing view factors, a distinction must be made between contributions from surfaces that view themselves and those that do not. An example is the concave surface. For the surface shown in Figure 3.11, the arrows labeled J represent radiosity (i.e., total radiative energy emitted and reflected) and the possible directions in which thermal energy may be leaving surface i. As the concave surface can view itself, the view factor back upon itself is non-zero (i.e., $F_{i-i} \neq 0$). However, for convex surfaces, $F_{i-i} = 0$!

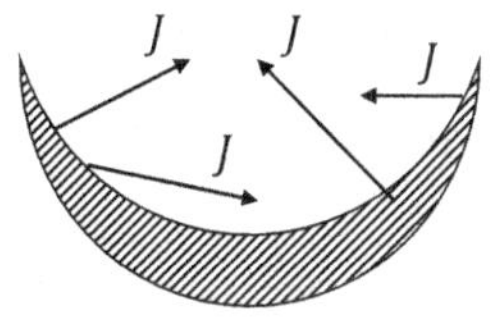

Figure 3.11 Radiosity from a concave surface

Example 2 Flat Disk in an Enclosure A flat disk is located inside an opaque hollow cubic structure (Figure 3.12). The active surface of the disk is surface 1. The interior walls of the hollow cube are surface 2. Using the summation rule, determine each of the view factors in the thermal circuit. Assume the disk has a radius of r and the length(s) of the cube side walls are equal to $6r$.

Solution

Surface A_1, the viewing surface area of the flat disk and A_2, the interior surface area of the hollow cube, are within each other's FOV and will be used in the view factor summation. According to the summation rule (equation 3.58), for surface 1,

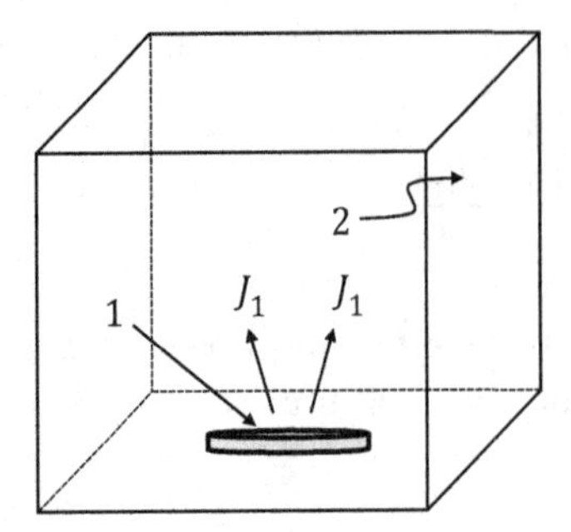

Figure 3.12 Flat disk in a hollow cubic enclosure

$$F_{1-1} + F_{1-2} = 1 \tag{3.59}$$

and for surface 2,

$$F_{2-1} + F_{2-2} = 1 \tag{3.60}$$

The geometry shows that the active surface for the disk does not view itself, therefore $F_{1-1} = 0$. Using equation 3.59, $F_{1-2} = 1$. By reciprocity (from equation 3.57),

$$A_1F_{1-2} = A_2F_{2-1}$$

Algebra is used to determine that

$$F_{2-1} = \left[\frac{A_1}{A_2}\right]F_{1-2} = \left[\frac{A_1}{A_2}\right] \text{ and } F_{2-2} = 1 - \left[\frac{A_1}{A_2}\right]$$

The active surface area for the disk is $A_1 = \pi r^2$ and the active surface area for the inner wall of the enclosure is

$$A_2 = 6s^2 = 6 \cdot (6 \cdot r)^2 = 216 \cdot r^2$$

These values are substituted in to solve for the remaining unknown view factor relations:

$$F_{2-1} = \left[\frac{\pi r^2}{216 \cdot r^2}\right] = 0.0145 \quad \text{and} \quad F_{2-2} = 1 - \left[\frac{\pi r^2}{216 \cdot r^2}\right] = 0.9854$$

Example 3 Subdivision of Surfaces An L-bracket (Figure 3.13) consists of three independent segments. Determine the view factor between surface segment 1 and surface segment 3 (i.e., F_{1-3}).

Solution

Determining F_{1-3} requires reference to Appendix A (common view factors table), as well as the subdivision of surfaces rule. The view factor for two rectangles joined at a 90° angle (VF2) is applicable, but since the geometric case in this example is not

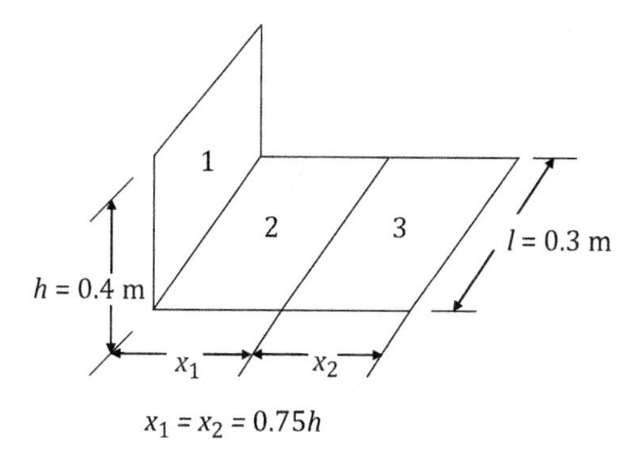

Figure 3.13 L-bracket with three area sections

explicitly listed in the tables, the solution needs to be supported by deriving relationships that can determine the view factor of interest. The geometry in VF2 is matched by taking the view factor from the combined surfaces 2 and 3 to surface 1 (i.e., $F_{2,3-1}$). However, since subdivision only applies to receiving surfaces, surfaces 2 and 3 need to be subdivided (i.e., $F_{1-2,3}$). Mathematically, this is expressed as

$$F_{1-2,3} = F_{1-2} + F_{1-3} \tag{3.61}$$

where the A_1 term has been divided out because it is present on both sides of the equal sign. Solving for F_{1-3} gives

$$F_{1-3} = F_{1-2,3} - F_{1-2} \tag{3.62}$$

EES can be used to solve for $F_{1-2,3}$ and F_{1-2}. The surface dimensions are defined thus:

```
"Right angle view factor"
l=0.3                                   {width of surfaces}
h_lower=0.4                             {height of surface 1}
w_lower=0.75*h_lower+0.75*h_lower       {length of combined
                                          surfaces 1 and 2}
H=h_lower/lW=w_lower/l
```

Each individual surface area and the combined 2/3 surface are defined as:

```
A_1=h_lower*l
A_2=0.75*h_lower*l
A_3=0.75*h_lower*l
A_23=A_2+A_3
```

Each term inside the parentheses of the view factor is individually calculated and then summed:

```
Term1=W*arctan(1/W)+H*arctan(1/H)-((H^2+W^2)^0.5)*arctan(1/&
((H^2+W^2)^0.5))                              {Term 1}
```

```
Term2=(1+W^2)*(1+H^2)/(1+W^2+H^2)                          {Term 2}
Term3=((W^2)*(1+W^2+H^2)/((1+W^2)*(W^2+H^2)))^(W^2)        {Term 3}
Term4=((H^2)*(1+H^2+W^2)/((1+H^2)*(H^2+W^2)))^(H^2)        {Term 4}
F_23_1=(1/(pi*W))*(Term1+0.25*ln(Term2*Term3*Term4)          {View
factor from surface 2/3 to 1}
```

This gives a value of $F_{2,3-1} = 0.1315$. $F_{1-2,3}$ can be determined by reciprocity:

```
F_1_23=A_23*F_23_1/A_1         {Reciprocity rule to determine
                         view factor from surface 1 to 2/3}
```

The value determined for $F_{1-2,3}$ is 0.1972. This calculation can be repeated for F_{1-2} and F_{2-1} to obtain $F_{1-2} = 0.1626$ and $F_{2-1} = 0.2168$. Equation 3.62 can be used to calculate $F_{1-3} = 0.0346$. These values can also be determined in EES using the radiation view factor functions in the pull-down menu under the Heat Transfer & Fluid Flow button.

Example 4 View Factor for a Simple Surface in Orbit The Earth-viewing side of a satellite has a surface area of 1.0 m^2 (Figure 3.14). If the SpaceCube's orbit altitude is 200 km, what is the view factor for its surface to the Earth?

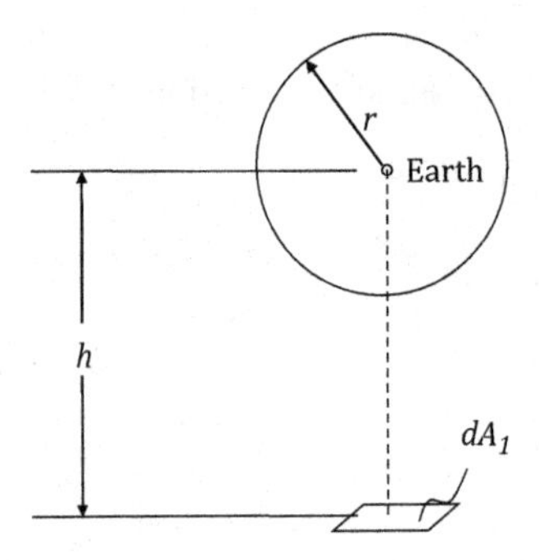

Figure 3.14 Simple surface in Earth orbit

Solution

The view factor table in Appendix A gives the equation for VF9 as

$$F_{dA-2} = \left(\frac{r}{h}\right)^2$$

View factor 9 is applicable because the surface size scale is differential relative to the surface area of the Earth. The radius of the Earth is 6,380 km, the altitude of the SpaceCube is 200 km and the surface area for the Earth-viewing side (i.e., dA_1) is 1.0 m^2, so the value for h in the view factor can be determined as follows.

$$h = r_{earth} + altitude$$

Substituting in discrete values gives

$$h = 6{,}380 \text{ km} + 200 \text{ km}$$

which sums to h = 6,580 km. Placing the values for r and h in view factor 9 and calculating gives

$$F_{d1-2} = \left(\frac{r}{h}\right)^2 = \left(\frac{6{,}380 \text{ km}}{6{,}580 \text{ km}}\right)^2 = 0.969^2$$

$$F_{d1-2} = 0.94$$

3.4.3 Hottel's String Method

An alternative method of determining the value of non-standard view factors is known as Hottel's string method. It is typically applied to complex two-dimensional geometries. In this technique the sum of the length of the crossed strings connecting the outer edges of the viewing surfaces minus the sum of the length of the uncrossed strings for the viewing surfaces must be determined. This value is then multiplied by the common dimension shared by each surface (e.g., *th*) and subsequently divided by twice the area of the receiving surface [9].

Example 5 View Factor between Finite Curved Surfaces For the active surfaces shown in Figure 3.15, determine a mathematical relation for the view to surface 1 from surface 2. The depth of the curved surfaces into the page can be denoted as *th*.

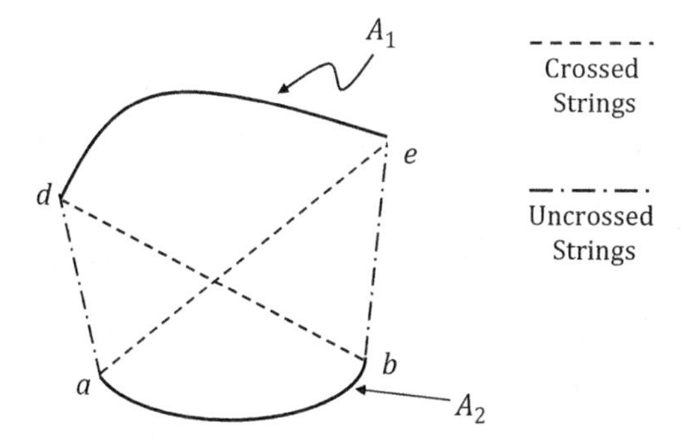

Figure 3.15 Curved surfaces viewing each other

Solution

Using the string technique and the point locations shown in the figure, the view factor can be written in terms of approximate length segments as

$$F_{2-1} = th\frac{db + ea - (eb + da)}{2A_2} \tag{3.63}$$

Hottel's string method is also used to determine the view factors between surfaces with complex geometries and/or complex views to each other. For a detailed treatment of Hottel's string method, including view factor determination with an obstruction present, see Nellis and Klein, Siegel and Howell or Michael Modest [3–5].

The techniques covered here are suitable for RTMs or hand calculations. Most contemporary thermal analysis software packages that perform radiation heat transfer incorporate Monte Carlo techniques to determine the view factors for surfaces within each other's FOV. This statistical approach, which takes into account the probability of a ray emitted from the origin surface actually being intercepted by the receiving surface, is highly advantageous when analyzing detailed thermal models with complex geometries drawn using CAD software. For a detailed overview of the theory behind the Monte Carlo technique, see Siegel and Howell [4] or Modest [5].

3.4.4 Net Radiative Energy Exchange (Blackbody and Gray Surfaces)

Assume two surfaces (surface 1 and surface 2) within each other's FOV are emitting thermal radiation. The rate at which radiation leaves surface 1 and is incident upon surface 2 is defined as

$$\dot{q}_{1-2} = F_{1-2}A_1J_1 \tag{3.64}$$

Equation 3.64 is used to develop the net radiative thermal energy exchange relation between surfaces displaying blackbody and gray body characteristics.

3.4.4.1 Blackbody Surfaces

Assuming the two surfaces have blackbody characteristics (i.e., totally absorbing and emitting with no reflection), by definition the radiosity for both reduces to the blackbody total hemispherical emissive power. As such $J_1 = E_{b1}$ and $J_2 = E_{b2}$. The radiative heat transfer from each surface in either direction (i.e., from 1→2 or from 2→1) then becomes

$$\dot{q}_{1-2} = A_1F_{1-2}E_{b1} \tag{3.65}$$

$$\dot{q}_{2-1} = A_2F_{2-1}E_{b2} \tag{3.66}$$

Thus, the net radiative heat exchange between the two surfaces is

$$\dot{q}_{12} = \dot{q}_{1-2} - \dot{q}_{2-1} \tag{3.67}$$

Equation 3.67 may be rewritten by substituting in the RHS of equations 3.65 and 3.66 for the directional heat transfer terms. This creates a new form of

$$\dot{q}_{12} = A_1 F_{1-2} E_{b1} - A_2 F_{2-1} E_{b2} \quad (3.68)$$

The reciprocity rule for the product of the surface area and the view factor for each surface can be applied to recast equation 3.68 as

$$\dot{q}_{12} = A_1 F_{1-2} [E_{b1} - E_{b2}] \quad (3.69)$$

As a final step, we can substitute the RHS of equation 3.17 for the blackbody total hemispherical emissive power to obtain a form that makes the heat transfer functionally dependent upon the surface temperatures fostering radiative thermal energy exchange. The final form becomes

$$\dot{q}_{12} = A_1 F_{1-2} \sigma [T_1^4 - T_2^4] \quad (3.70)$$

When dealing with multiple surfaces, a general form of this relation applicable to blackbody surfaces undergoing radiative thermal energy exchange would be

$$\dot{q}_{i,net} = \sum_{j=1}^{N} A_i F_{i-j} \sigma \left[T_i^4 - T_j^4\right] \quad (3.71)$$

The following example includes view factor determination and/or radiative thermal energy exchange between two blackbody surfaces.

Example 6: Blackbody Radiative Heat Transfer for a Sphere Inside a Sphere For an enclosed sphere (Figure 3.16) what is the heat transfer from the inner sphere to the outer sphere if the heat transfer surfaces have blackbody characteristics?

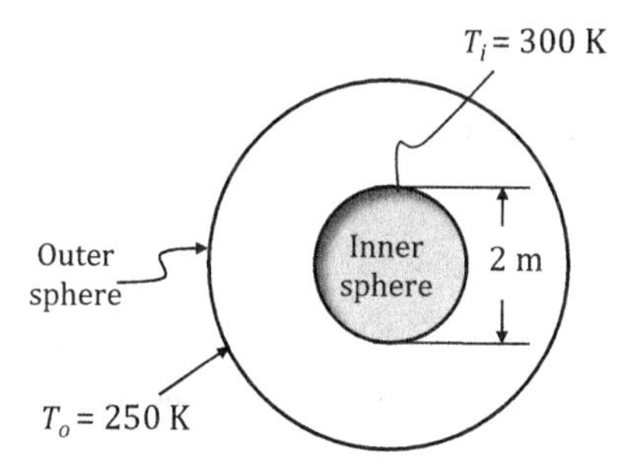

Figure 3.16 A sphere inside a spherical enclosure

Solution

From basic geometry, we know the area of a sphere is $4\pi r_i^2$. If the diameter of the inner sphere is 2.0 m then the radius is r_i = 1.0 m. View factor summation rules for an enclosure give $F_{i-o} = 1$. Thus, an equation for the net heat transfer from the inner sphere to the outer sphere is

$$\dot{q}_{io} = 4\pi\sigma\left[(300\ \text{K})^4 - (250\ \text{K})^4\right] \tag{3.72}$$

where the Stefan–Boltzmann constant is $\sigma = 5.67 \times 10^{-8}$ W/m$^2 \cdot$ K^4. If the value for σ is placed in equation 3.72 and the RHS calculated, the net heat transfer is $\dot{q}_{12} \approx 3.0$ kW.

3.4.4.2 Diffuse Gray Surfaces

Assuming that an active, opaque, radiative surface (see Figure 3.17) is experiencing unequal quantities of irradiation and emission, the difference between the two quantities will be the net radiative heat transfer from the surface ($\dot{q}_{i,net}$). This value is also the rate of heat flow into or out of the subsurface contiguous material.

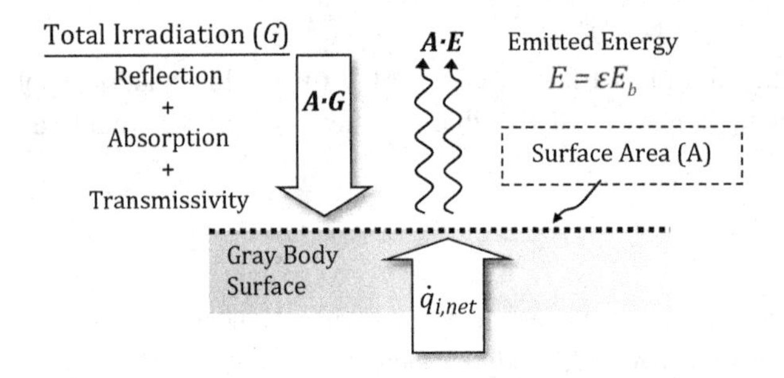

Figure 3.17 Heat flows for an opaque gray body surface

Using the previously defined construct of radiosity (see Section 3.3.1) the net heat transfer from a given surface i can be written as

$$\dot{q}_{i,net} = A_i(J_i - G_i) \tag{3.73}$$

Based on the definition of radiosity, the radiation from surface i can be defined in equation form as

$$J_i = E_i + \rho_i G_i \tag{3.74}$$

where the contributions from the emitted and reflected thermal energy have been summed. The summation relation for absorptivity, reflectivity and transmissivity for an opaque surface (i.e., $\tau = 0$) means that

$$\alpha + \rho = 1 \tag{3.75}$$

Equation 3.75 can be rewritten substituting absorptivity for the reflectivity term.

$$J_i = E_i + (1 - \alpha_i)G_i \tag{3.76}$$

The first term on the RHS can also be rewritten with respect to the blackbody emission with the same temperature as the actual gray surface. Substituting into equation 3.76 results in equation 3.77.

$$J_i = \varepsilon_i E_b + (1 - \alpha_i) G_i \tag{3.77}$$

Using Kirchhoff's Law, equation 3.77 can be rewritten to include the emissivity in the second term on the RHS. The new equation then becomes

$$J_i = \varepsilon_i E_b + (1 - \varepsilon_i) G_i \tag{3.78}$$

The new form for the radiosity relation means the total hemispherical irradiation can be solved as a function of the surface's radiosity, emissivity and corresponding blackbody emission. The new representation is

$$G_i = \frac{J_i - \varepsilon_i E_b}{(1 - \varepsilon_i)} \tag{3.79}$$

Substituting this new representation for G_i into the initial net heat transfer equation for surface i (equation 3.73) gives a new relation

$$\dot{q}_{i,net} = A_i \left[J_i - \left(\frac{J_i - \varepsilon_i E_b}{(1 - \varepsilon_i)} \right) \right] \tag{3.80}$$

which can also be written as

$$\dot{q}_{i,net} = \frac{(E_b - J_i)}{\dfrac{(1 - \varepsilon_i)}{\varepsilon_i A_i}} \tag{3.81}$$

The numerator in this relation expresses the potential for radiative heat transfer from a surface. The denominator is the radiative surface resistance [2,3,5]. When this equation is used to solve for the net radiative heat transfer ($\dot{q}_{i,net}$), the value for the variable J_i must be known. Figure 3.18 shows the heat flow using thermal resistance constructs.

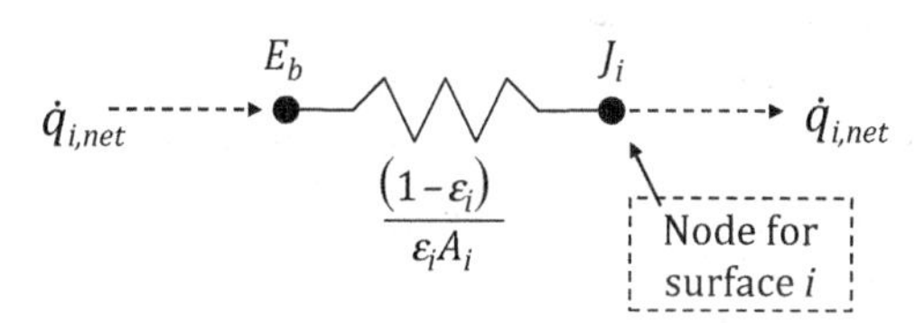

Figure 3.18 Radiative thermal resistance for net heat transfer from a surface

Alternatively, the net radiative heat transfer from a surface can be expressed as a function of its radiosity and that of each surface within its FOV [2,3]. For a surface within the enclosure, the net radiative heat transfer may then be expressed as

$$\dot{q}_{i,net} = \sum_{j=1}^{N} A_i \mathrm{F}_{i-j} J_i - \sum_{j=1}^{N} A_j \mathrm{F}_{j-i} J_j \tag{3.82}$$

Reciprocity enables the product of the area and the view factor in the second term on the RHS to be recast to that shown in the first term. The equation now becomes

$$\dot{q}_{i,net} = \sum_{j=1}^{N} A_i \mathrm{F}_{i-j} J_i - \sum_{j=1}^{N} A_i \mathrm{F}_{i-j} J_j \tag{3.83}$$

The product of the area and the view factor can be factored out from both terms on the RHS while still applying the summation to both. The new form for the net surface radiative heat transfer becomes

$$\dot{q}_{i,net} = \sum_{j=1}^{N} A_i F_{i-j}(J_i - J_j) \tag{3.84}$$

Using resistor/conductor constructs, equation 3.84 can also be written as

$$\dot{q}_{i,net} = \sum_{j=1}^{N} \frac{(J_i - J_j)}{(A_i F_{1-j})^{-1}} \tag{3.85}$$

where the denominator is the geometric resistance due to the FOV for surface i to j. The resistance between surfaces i and j is shown in Figure 3.19.

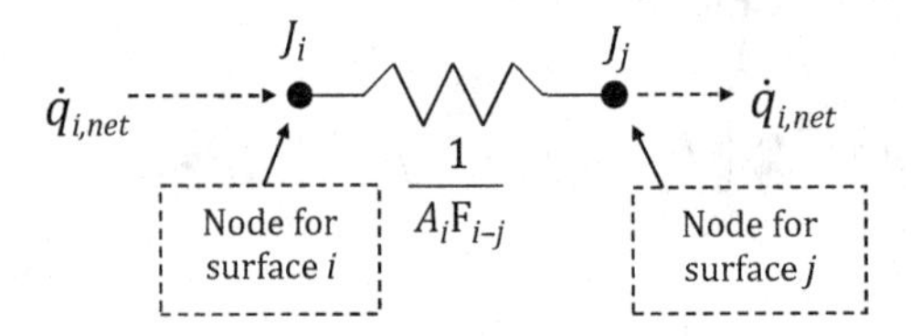

Figure 3.19 Radiative thermal resistance for net heat transfer between two surfaces

An equality can thus be established between the net surface heat transfer relations of equation 3.81 and equation 3.85. This equality is

$$\dot{q}_{i,net} = \frac{(E_{bi} - J_i)}{(1 - \varepsilon_i)/\varepsilon_i A_i} = \sum_{j=1}^{N} \frac{J_i - J_j}{(A_i F_{i-j})^{-1}} \tag{3.86}$$

Example 7 Radiative Circuit for a Network of Three Gray Body Surfaces Assuming that a gray body surface i is in thermal communication with two other gray body surfaces (surfaces 1 and 2), which do not view each other, what does the resistance network for the radiative thermal circuit look like?

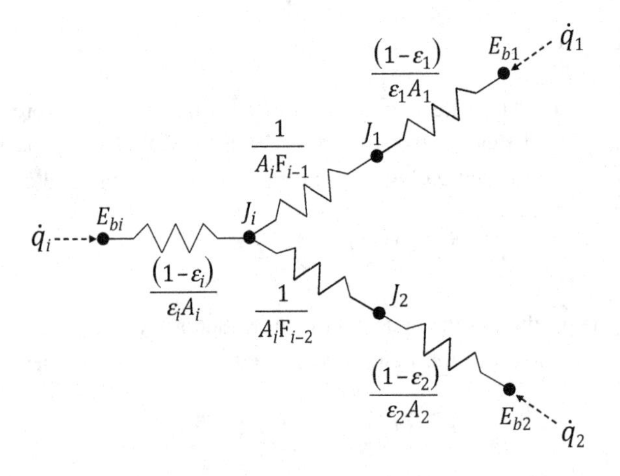

Figure 3.20 Radiative thermal resistance for net heat transfer between surface i and surfaces 1 and 2

Solution

As Figure 3.20 shows, each surface has a blackbody node and a radiosity node. Heat-flow paths and resistances are present between originating and receiving surfaces that are within each other's FOV, but there is no such path between surfaces 1 and 2 because they do not see each other.

Example 8 Radiative Circuit for an Enclosure with Two Gray Body Surfaces Assume an enclosure consists of an L-bracket and a curved surface with two designated surface areas (see Figure 3.21). What is the thermal resistance network and the net heat transfer relation from surface 1 to surface 2?

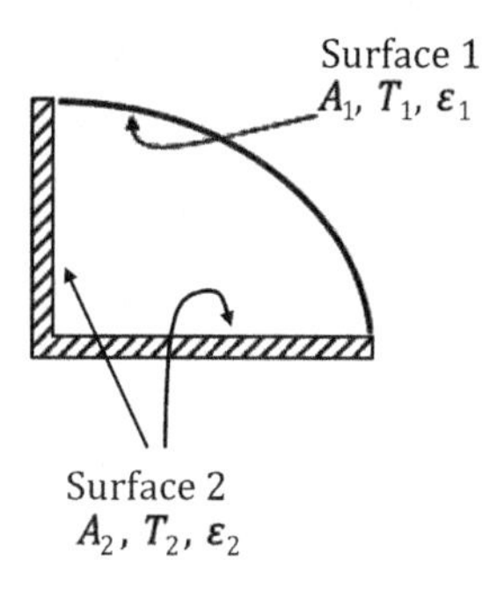

Figure 3.21 Enclosure with two gray body active surfaces

Solution

Since there are only two surfaces in the circuit, the resistance network (Figure 3.22) will contain two nodes (i.e., 1 and 2) with a single heat-flow path between them.

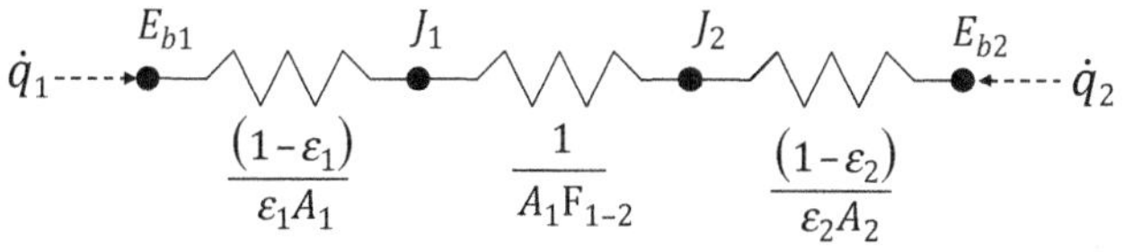

$\dot{q}_1$ → E_{b1} — $\frac{(1-\varepsilon_1)}{\varepsilon_1 A_1}$ — J_1 — $\frac{1}{A_1 F_{1-2}}$ — J_2 — $\frac{(1-\varepsilon_2)}{\varepsilon_2 A_2}$ — E_{b2} ← $\dot{q}_2$

Figure 3.22 Radiative thermal circuit for enclosure with two gray body active surfaces

The net heat transfer between surfaces 1 and 2 can be written as

$$\dot{q}_{12} = \frac{E_{b1} - E_{b2}}{\left[\frac{1-\varepsilon_1}{\varepsilon_1 A_1} + \frac{1}{A_1 F_{1-2}} + \frac{1-\varepsilon_2}{\varepsilon_2 A_2}\right]} \tag{3.87}$$

where the denominator is the summation of the series heat-flow resistances between the two nodes. This can also be rewritten to include the temperature at the nodes as

$$\dot{q}_{12} = \frac{\sigma\left(T_1^4 - T_2^4\right)}{\left[\dfrac{1-\varepsilon_1}{\varepsilon_1 A_1} + \dfrac{1}{A_1 F_{1-2}} + \dfrac{1-\varepsilon_2}{\varepsilon_2 A_2}\right]} \tag{3.88}$$

Example 9 Radiative Circuit for an Enclosure with Three Gray Body Surfaces Figure 3.23 shows a three-surface rectangular enclosure. What is the resistance network and the net heat transfer relation for each surface?

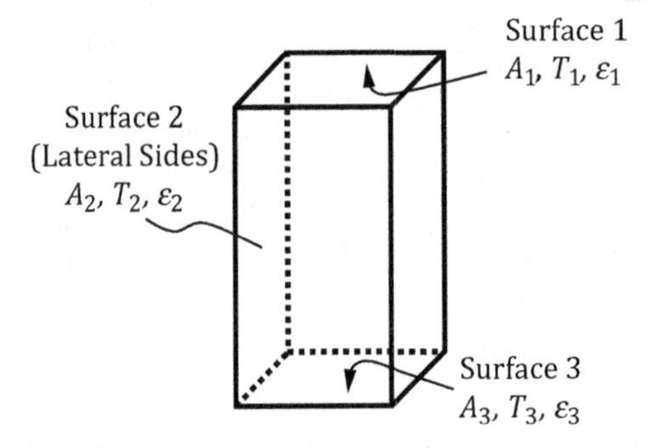

Figure 3.23 Enclosure with three gray body active surfaces

Solution

Since we now have three surfaces in a circuit viewing each other, the resistance network (Figure 3.24) will contain three nodes (1, 2 and 3) and multiple heat-flow paths. As there are more than two surfaces of interest, the temperature at each surface must be obtained by performing an energy balance at the nodes representing each surface.

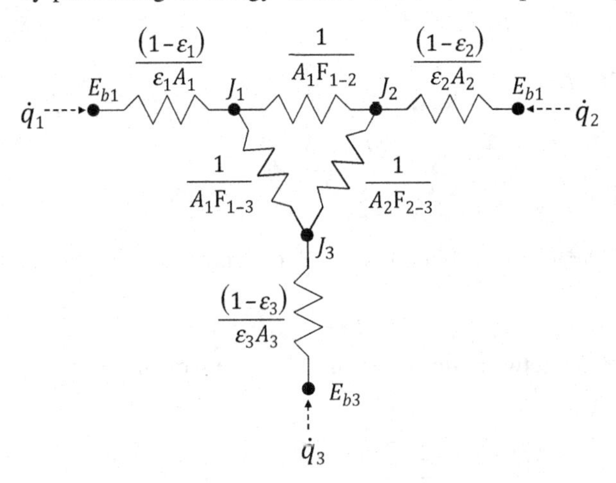

Figure 3.24 Radiative thermal circuit for enclosure with three gray body active surfaces

The energy balances at each of the radiosity nodes for the surfaces would be as follows:

Node J_1 is

$$\frac{E_{b1} - J_1}{\frac{1-\varepsilon_1}{\varepsilon_1 A_1}} = \frac{J_1 - J_2}{(A_1 F_{1-2})^{-1}} + \frac{J_1 - J_3}{(A_1 F_{1-3})^{-1}} \quad (3.89)$$

Node J_2 is

$$\frac{E_{b2} - J_2}{\frac{1-\varepsilon_2}{\varepsilon_2 A_2}} = \frac{J_2 - J_1}{(A_2 F_{2-1})^{-1}} + \frac{J_2 - J_3}{(A_2 F_{2-3})^{-1}} \quad (3.90)$$

Node J_3 is

$$\frac{E_{b3} - J_3}{\frac{1-\varepsilon_3}{\varepsilon_3 A_3}} = \frac{J_3 - J_1}{(A_3 F_{3-1})^{-1}} + \frac{J_3 - J_2}{(A_3 F_{3-2})^{-1}} \quad (3.91)$$

The net heat transfer from each surface can be captured by applying the energy balance relation from equation 3.81 to each of the blackbody nodes (E_{b1}, E_{b2} and E_{b3}). The relations for the net heat transfer from each surface then become

$$\dot{q}_{1,net} = \frac{E_{b1} - J_1}{\frac{1-\varepsilon_1}{\varepsilon_1 A_1}} \quad (3.92),$$

$$\dot{q}_{2,net} = \frac{E_{b2} - J_2}{\frac{1-\varepsilon_2}{\varepsilon_2 A_2}} \quad (3.93),$$

$$\dot{q}_{3,net} = \frac{E_{b3} - J_3}{\frac{1-\varepsilon_3}{\varepsilon_3 A_3}} \quad (3.94)$$

3.4.4.3 Diffuse Gray Body Surface Analysis with Reflection

In the previous section we worked through the energy flux-based approach using the blackbody emission and radiosity to determine the temperatures for elements in a thermal circuit composed of gray body surfaces. That technique included solving the energy balances for the blackbody and radiosity nodes representing each gray body surface in the circuit. Surface temperatures were then solved for using the equation for blackbody energy flux. However, this approach is not able to directly determine the net thermal energy exchange between any two given surfaces within the circuit [3,4]. According to blackbody theory, the net heat transfer between two surfaces (e.g., surface 1 and surface 2) as written in equation 3.70 is

$$\dot{q}_{12} = \sigma A F_{1-2}\left(T_1^4 - T_2^4\right)$$

Since this is in reference to blackbodies, $\varepsilon = 1$. Conversely, for gray body surfaces $\varepsilon < 1$. Thus, an alternate relation must be developed to express net heat transfer occurring between two gray body surfaces.

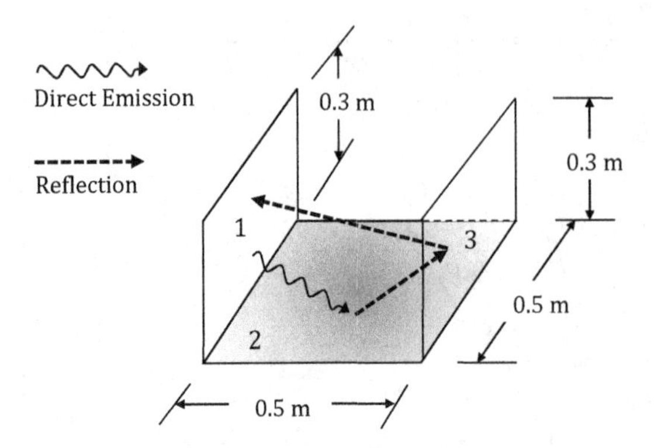

Figure 3.25 Photon reflection from gray body surfaces

Figure 3.25 shows a U-bracket assembly containing three individual surfaces (assume the active surfaces are on the interior sides of the bracket). Photons incident on surfaces are subject to energy absorption at the contact surface, as well as rebound (i.e., reflection). In general, they can undergo numerous rebounds before their energy level diminishes enough to be fully absorbed during contact with a given surface. In Figure 3.25, direct emission from surface 1 to surface 2 is depicted by an oscillating vector. Upon incidence with surface 2, the photon is reflected in the direction of surface 3. In transit from surface 1 to surface 2, a fraction of the photon's energy is absorbed upon initial contact. Energy remaining within the photon after initial contact will be carried along to the next contact surface (i.e., surface 3) via a reflection event. Photons originating at surface 1 may experience several reflections before finally reaching surface 2. If the RHS of the equation for blackbody net heat transfer is multiplied by the emissivity (ε_1), the newly formed relation includes the energy departing surface 1. However, for the diffuse gray body case, it still would not accurately model the energy absorbed at surface 2. Thus, for a gray body

$$\dot{q}_{12} \neq \varepsilon_1 \sigma A F_{1-2} \left(T_1^4 - T_2^4 \right) \tag{3.95}$$

Analytical modeling techniques used to capture the net heat transfer between two gray body surfaces incorporate the product of an energy scaling factor and either the emissivity of the originating surface or the emissivity of both the originating and receiving surfaces multiplied by the RHS of the blackbody net heat transfer relation. The energy scaling factor will have a value between zero and unity and bookkeeps what fraction of the total energy leaving an originating surface either impinges and/or is absorbed by a receiving surface through both direct emission and reflection. Two commonly used conventions are the transform factor method (denoted by F) and Gebhart's energy absorption technique [4] in which the scaling factor term (i.e., B_{i-j}) accounts for the amount of energy departing surface i that reaches and is absorbed by surface j (incorporating both direct emission and reflection) divided by the total thermal energy leaving surface i. For an overview of the Gebhart technique, see Siegel & Howell [4] and NASA's NESC lectures [10].

The $\hat{F}$ technique for bookkeeping net heat transfer in a gray body thermal circuit was introduced by Beckman [11] and is reviewed by Nellis and Klein [3]. In the present text the $\hat{F}$ technique will be used for direct determination of net heat transfer between gray body surfaces.

Beckman's $\hat{F}$ technique uses $\hat{F}_{i-j}$ as the energy scaling factor for net heat transfer between surfaces i and j. This term is defined as the ratio of energy originating from surface i that is transported to surface j divided by the total energy departing surface i. It includes energy contributions from both direct emission and reflection. In generalized form, $\hat{F}_{i-j}$ can be written as

$$\hat{F}_{i-j} = F_{i-j} + \sum_{a=1}^{N} F_{i-a}\rho_a\hat{F}_{a-j} \tag{3.96}$$

where F_{i-j} is the contribution due to direct emission from surface i to surface j and the summation term accounts for energy contributions to the net heat transfer resulting from reflections. The product in the summation term accounts for thermal energy departing surface i that is reflected from surface a and onto surface j. Subscripts i and j are both iterated from 1 to N (i.e., the total number of surfaces). When this convention is applied to the phenomena depicted in Figure 3.16, $\hat{F}_{1-2}$ is

$$\hat{F}_{1-2} = F_{1-2} + F_{1-1}\rho_1\hat{F}_{1-2} + F_{1-2}\rho_2\hat{F}_{2-2} + F_{1-3}\rho_3\hat{F}_{3-2} + F_{1-4}\rho_4\hat{F}_{4-2} \tag{3.97}$$

where the fifth term on the RHS accounts for the surroundings (unspecified in the schematic). Typically, these would be treated as a blackbody, resulting in the term falling out of the equation.

Since $\hat{F}_{i-j}$ only accounts for the thermal energy originating with surface i that is reaching surface j, the fraction of absorption occurring at surface j could be captured by multiplying $\hat{F}_{i-j}$ by the emissivity at surface j (i.e., $\varepsilon_j\hat{F}_{i-j}$). By definition, this product term is equal to the scaling factor (B_{i-j}) used in Gebhart's absorption technique [11]. The relationship between these scaling factors may be written mathematically as

$$\varepsilon_j\hat{F}_{i-j} = B_{i-j} \tag{3.98}$$

As radiation emitted from any surface in the circuit will be absorbed by other surfaces, conservation of energy applies at every surface. The sum of the fractions of the energy absorption occurring at the receiving surfaces must equal unity for any originating surface. In $\hat{F}_{i-j}$ notation this can be written as

$$\sum_{j=1}^{N} \varepsilon_i\hat{F}_{i-j} = 1 \tag{3.99}$$

The radiative thermal energy that leaves surface 1, reaches surface 2 and gets absorbed is

$$\dot{q}_{1-2} = \varepsilon_1\alpha_2A_1\sigma T_1^4\hat{F}_{1-2} \tag{3.100}$$

Kirchhoff's Law can be invoked to recast this equation as

$$\dot{q}_{1-2} = \varepsilon_1\varepsilon_2 A_1 \sigma T_1^4 \hat{F}_{1-2} \tag{3.101}$$

Using a similar construct, the heat transfer from surface 2 to surface 1 can be written as

$$\dot{q}_{2-1} = \varepsilon_1\varepsilon_2 A_2 \sigma T_2^4 \hat{F}_{2-1} \tag{3.102}$$

The net heat transfer would then become

$$\dot{q}_{12} = \varepsilon_1\varepsilon_2 A_1 \sigma T_1^4 \hat{F}_{1-2} - \varepsilon_1\varepsilon_2 A_2 \sigma T_2^4 \hat{F}_{2-1} \tag{3.103}$$

As with view factors, reciprocity rules are applicable to the $\hat{F}$ term, so

$$A_i \hat{F}_{i-j} = A_j \hat{F}_{j-i} \tag{3.104}$$

Equation 3.103 can then be rewritten as

$$\dot{q}_{12} = \varepsilon_1\varepsilon_2 A_1 \sigma \hat{F}_{1-2}\left(T_1^4 - T_2^4\right) \tag{3.105}$$

Figure 3.26 illustrates the net heat transfer between two gray body surfaces using $\hat{F}_{i-j}$ notation. The radiative resistance between the two surfaces is $1/\varepsilon_1\varepsilon_2 A_1 \hat{F}_{1-2}$.

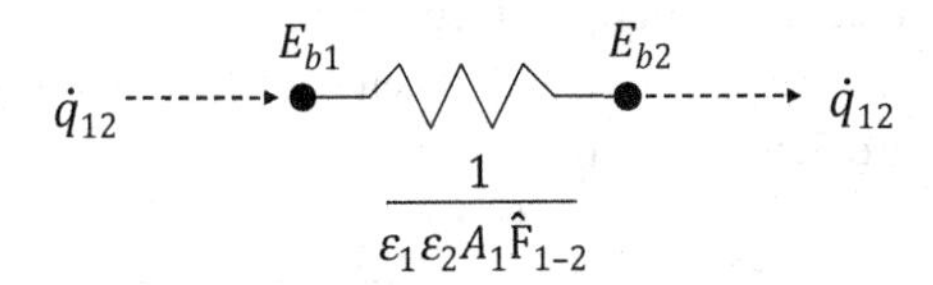

Figure 3.26 Resistance for net heat transfer between gray body surfaces

---

Example 10 Net Radiative Heat Transfer on Members of a U-Bracket Referring to Figure 3.25, determine the net heat transfer from surface 1 to surface 2 if T_1 = 300 K, T_2 = 250 K and each surface is coated with Sperex white paint (i.e., ε = 0.85).

Solution

When the following EES code is used to determine the view factors and $\hat{F}$ values, the heat transfer from surface 1 to surface 2 is 7.09 W.

```
{Gray body net heat transfer between surface 1 and 2}
{ Code adapted from solution to example problem 10.5-3 in
Heat Transfer by Nellis and Klein [3] }

{Geometry}
a=0.5     [m]
b=0.3     [m]
c=0.5     [m]

A_1=a*b
A_2=a*c
```

```
A_3=a*b
A_4=2*a*b+a*c

{View factors}
F[1,1]=0
F[2,2]=0
F[3,3]=0

F[2,1]=f3d_2(a,b,c)
F[1,2]=A_2*F[2,1]/A_1

F[2,3]=F[2,1]
F[3,2]=F[1,2]

F[1,3]=f3d_1(a,b,c)
F[3,1]=F[1,3]

F[1,4]=1-(F[1,1]+F[1,2]+F[1,3])
F[2,4]=1-(F[2,1]+F[2,2]+F[2,3])
F[3,4]=1-(F[3,1]+F[3,2]+F[3,3])

F[4,1]=A_1*F[1,4]/A_4
F[4,2]=A_2*F[2,4]/A_4
F[4,3]=A_3*F[3,4]/A_4
F[4,4]=1-(F[4,1]+F[4,2]+F[4,3])

{emissivities}
eps_1=0.85; eps_2=0.85; eps_3=0.85; eps_4=1.0

{reflectivities}
rho[1]=1-eps_1; rho[2]=1-eps_2; rho[3]=1-eps_3; rho[4]=0

{temperatures}
T_1=300
T_2=250

{F-hat values}
Duplicate M=1,4
  Duplicate N=1,4
    F_hat[M,N]=F[M,N]+SUM(rho[K]*F[M,K]*F_hat[K,N],K=1,4)
  End
End

Q_dot_12=eps_1*eps_2*sigma#*A_1*F_hat[1,2]*(T_1^4-T_2^4)
```

Computer analysis tools used for radiative heat transfer analysis of gray body surfaces often rely on this technique, which will be revisited later on in the text.

3.5 Multilayer Insulation

Multilayer insulation (MLI) consists of a blanket-like layered stack of materials that limit the heat transfer through conduction and convection mechanisms in a vacuum environment. Heat transfer through this layered construction is dominated by radiation. The theory behind MLI heat transfer performance is based on fundamental radiation heat transfer concepts. The following subsections review the fundamental theory and performance of MLI and consider practical implications of implementing MLI on space platforms.

3.5.1 Theory and Background

The radiative heat transfer performance associated with individual MLI sheets facing one another is analogous to that of the infinite parallel plane scenario (shown in Figure 3.27 in 3-D) and its associated view factor.

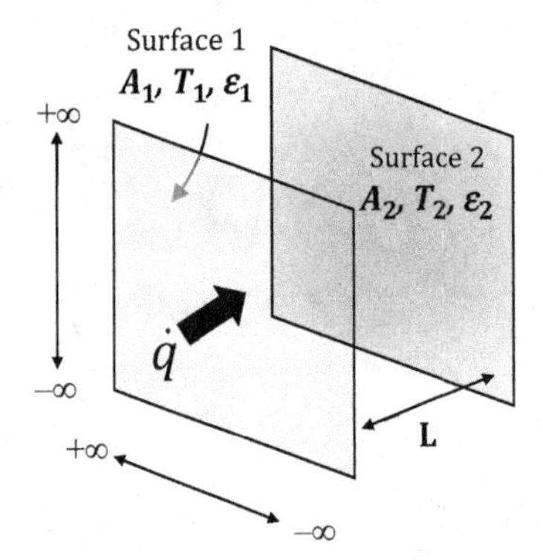

Figure 3.27 Infinite parallel planes viewing each other

Conceptually, given that the dimensions for each plane are positive and negative infinity, it can be concluded that they both have the same surface area. As such, $A_1 = A_2 = A$. Given a finite separation distance between the planes (i.e., L), it can also be assumed that the view factor between the two surfaces is ideal (i.e., $F_{1-2} = 1$). Under these assumptions, the heat transfer between surface 1 and surface 2 may be written as

$$\dot{q}_{12} = \frac{A\sigma\left(T_1^4 - T_2^4\right)}{\left[\dfrac{1}{\varepsilon_1} + \dfrac{1}{\varepsilon_2} - 1\right]} \tag{3.106}$$

Having determined the heat transfer between a single pair of infinite parallel planes, let's now look at the case for a series of parallel plane surfaces of finite area (i.e., radiation shields). Radiation shields are typically made from low-emissivity surfaces [2,5,12]. Kirchhoff's Law states that $\varepsilon + \rho = 1$. Thus, the standard radiation shield surface will be highly reflective, reducing the net radiative heat transfer through multiple radiation shield surfaces when placed together in series. Figure 3.28 shows a new configuration in which radiation shields are placed between the two parallel surfaces. The bounding surfaces are 1 and 4, and 2A, 2B, 3A and 3B are the double-sided shield surfaces. Focusing on shield material 2, side 2A faces surface 1 and side 2B faces surface 3A. In practical engineering applications, surfaces will not have infinite dimensions. Thus, in the present case we will assume that each of the shields has finite dimensions in x and y . All the shields have the same planar dimensions in x and y, so $A_1 = A_2 = A_3 = A_4 = A$.

In the infinite case, the determination that the view factor between two surfaces was equal to unity was predicated on the surfaces being dimensionally infinite. We cannot automatically make that assumption in the present case. However, if the ratio of each individual plane's primary dimension to the separation distance between successive planes is much greater than unity (i.e., $\Delta x \gg \Delta z$ and $\Delta y \gg \Delta z$), we can assume that the FOV between the two surfaces approximates to the infinite case (i.e., $F_{1-2} = F_{2-3} = F_{3-4} = 1$), giving a view factor of unity. The validity of this assumption can be checked using VF1 from Appendix A. For a surface area 1.0 m × 1.0 m with a separation distance of 1.5 mm VF 1 will approximate to a value of 1.0. For the shields shown in Figure 3.28, the net heat transfer between bounding surfaces 1 and 4 is

$$\dot{q}_{14} = \frac{\sigma\left(T_1^4 - T_4^4\right)}{\left[\frac{1-\varepsilon_1}{\varepsilon_1 A_1} + \frac{1}{A_1 F_{1-2}} + \frac{1-\varepsilon_{2A}}{\varepsilon_{2A} A_2} + \frac{1-\varepsilon_{2B}}{\varepsilon_{2B} A_2} + \frac{1}{A_3 F_{2-3}} + \frac{1-\varepsilon_{3A}}{\varepsilon_{3A} A_3} + \frac{1-\varepsilon_{3B}}{\varepsilon_{3B}} + \frac{1}{A_3 F_{3-4}} + \frac{1-\varepsilon_4}{\varepsilon_4 A_4}\right]} \tag{3.107}$$

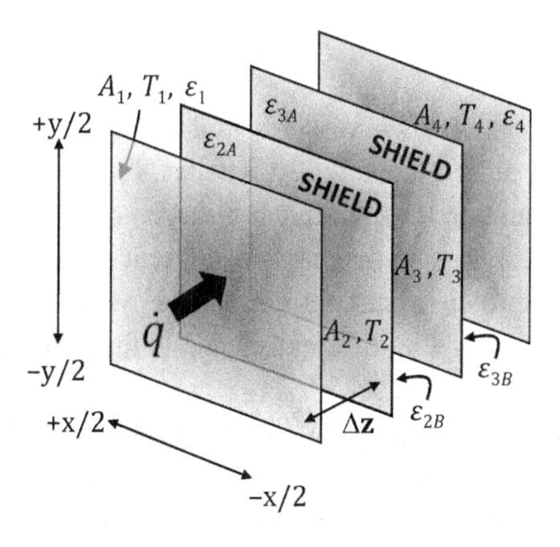

Figure 3.28 Radiation shield between two parallel planes

If the variable A is substituted for the common areas and unity for the view factors, the heat transfer relation becomes

$$\dot{q}_{14} = \frac{A\sigma\left(T_1^4 - T_4^4\right)}{\left[3 + \frac{1-\varepsilon_1}{\varepsilon_1} + \frac{1-\varepsilon_{2A}}{\varepsilon_{2A}} + \frac{1-\varepsilon_{2B}}{\varepsilon_{2B}} + \frac{1-\varepsilon_{3A}}{\varepsilon_{3A}} + \frac{1-\varepsilon_{3B}}{\varepsilon_{3B}} + \frac{1-\varepsilon_4}{\varepsilon_4}\right]} \tag{3.108}$$

If the emissivity values on either side of the shields are chosen to match the emissivity for the same side on the bounding surfaces (i.e., let $\varepsilon_{2B} = \varepsilon_{3B} = \varepsilon_1$; $\varepsilon_{2A} = \varepsilon_{3A} = \varepsilon_4$) then substituted back into the heat transfer relation and reduced, the new relation is

$$\dot{q}_{14} = \frac{A\sigma\left(T_1^4 - T_4^4\right)}{3\left[\frac{1}{\varepsilon_1} + \frac{1}{\varepsilon_4} - 1\right]} \tag{3.109}$$

The factor of 3 in the denominator is equal to the number of shields plus one (i.e., N+1). This equation captures the phenomena associated with MLI. One term often referred to when discussing multiple layers of insulation is the effective emittance (ε^*). The effective emittance for a blanket of N non-contacting layers of emissivities ε_A and ε_B on opposite sides is calculated using the following equation.

$$\varepsilon^* = \frac{1}{(\mathrm{N}+1)\left[\frac{1}{\varepsilon_A} + \frac{1}{\varepsilon_B} - 1\right]} \tag{3.110}$$

With this relationship established, the general heat transfer relation for N layers of an MLI blanket can be rewritten in terms of the effective emissivity between bounding layers A and B as

$$\dot{q}_{AB} = \varepsilon^* A\sigma\left(T_A^4 - T_B^4\right) \tag{3.111}$$

3.5.2 MLI Blanket Constructions

MLI blankets comprise multiple layers of low-emissivity material with low thermal conductivity (i.e., insulating) material placed interstitially between the successive radiative layers. The reduced thermal conduction (i.e., conductive shorts) occurring between the successive layers helps limit the heat transfer mechanisms present to radiation alone. A four-layer MLI blanket and a photo of a sample coupon of MLI are shown in Figure 3.29. Figure 3.30 is an example of the different elements that can be used in the application of an MLI blanket.

Common blanket materials used include Kapton, Mylar and Beta cloth [12]. However, Mylar has been known to disintegrate when exposed to UV radiation so it is not used for exteriors with visibility to sunlight, or as an innermost blanket layer due to flammability concerns. Common insulating materials used include dacron, nomex, nylon and silk [12]. MLI techniques for attachment to spaceflight components include taping, buttons and Velcro.

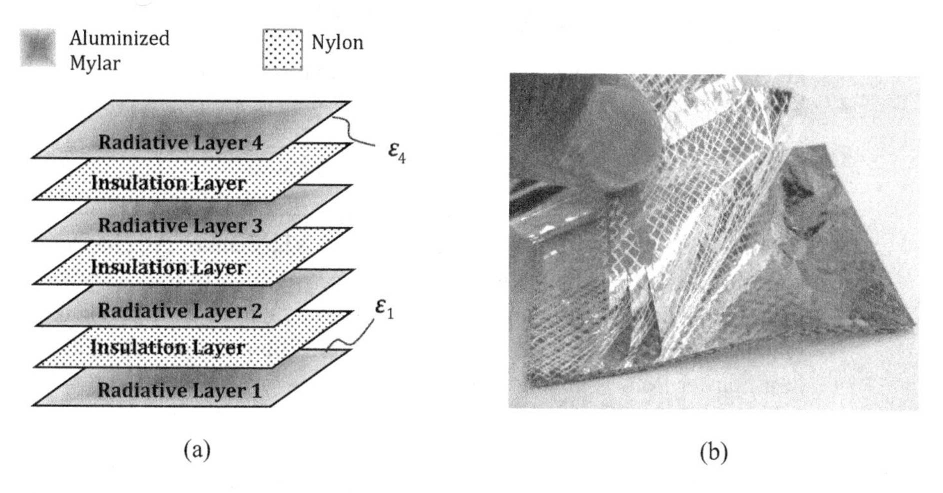

Figure 3.29 MLI stack-up: a) radiative and insulating MLI layers; b) sample MLI coupon

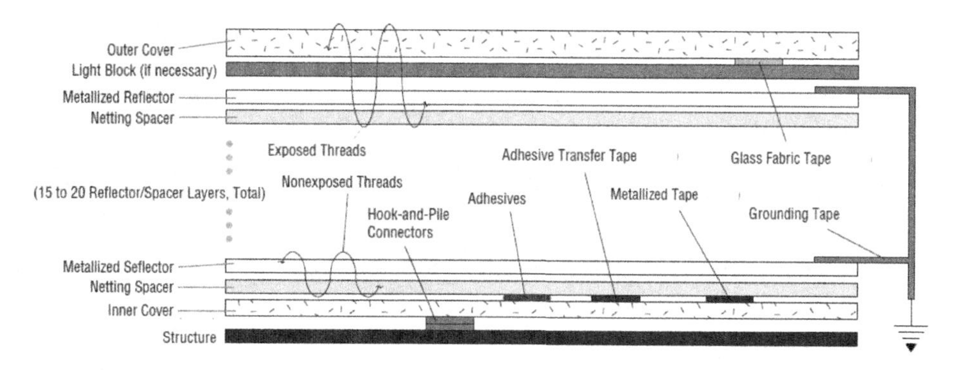

Figure 3.30 MLI blanket elements
Source: NASA [13]

Space is much colder than items on the space platform, so items insulated with MLI typically lose heat. For modeling purposes, either the inner layer of the blanket and the adjacent surface are treated as having the same temperature or the blanketed surface (and its emissivity) is treated as a radiative resistance in series with the MLI blanket. Building upon equation 3.111, and assuming that heat loss in a space environment is from inside to out with the blanket's inner side having the same temperature as the adjacent surface, the heat transfer can be expressed as

$$\dot{q}_{MLI} = \varepsilon^* A\sigma\left(T_i^4 - T_o^4\right) \tag{3.112}$$

where T_i is the inner-layer temperature and T_o is the outer-layer temperature. The effective thermal conductivity can be determined by establishing an equality between the Fourier's law for conduction representation and the net radiative heat transfer relation for the blanket. Solving for the thermal conductivity using this equality will result in equation 3.113. In this relation Δx is the thickness of the blanket, and as the boundary temperatures and the effective emittance for the radiative layers are also known, the effective thermal conductivity can be determined as

$$k_{eff} = \frac{\varepsilon^{*} \cdot \Delta x \cdot \sigma \left(T_i^4 - T_o^4\right)}{(T_i - T_o)} \tag{3.113}$$

Example 11 Heat Transfer Performance of an MLI Blanket Assume an MLI blanket consists of 15 layers of double-sided aluminized Mylar (0.15 mil). The blanket is known to have bounding temperatures of $T_1 = 300$ K and $T_2 = 150$ K. If the blanket has a cross-sectional area of 0.6 m^2, what is (i) the effective emittance for the blankets and (ii) the heat transfer from bounding surface 1 to 2?

Solution

Appendix B shows that 0.15 mil aluminized Mylar has an emissivity of 0.28. Equation 3.110 is used to determine effective emissivity; the corresponding EES code is

```
"Heat transfer performance of an MLI blanket"

A=0.6               [m^2]                      {Shield area}
N=15                                           {Number of blankets}
eps_1=0.28                                     {Shield side one}
eps_2=0.28                                     {Shield side two}

eps_star=1/((N+1)*(1/eps_1+1/eps_2-1)) {Effective emissivity}
```

Upon execution, the value for eps_star is found to be 0.01017, which can be placed in equation 3.112 to determine the heat transfer. In EES this would be

```
T_1=300                   [K]
T_2=150                   [K]
Q_dot_12=eps_star*A*sigma#*(T_1^4-T_2^4)              {sigma# is
  Boltzmann's Constant}
```

Execution of the code for part (ii) results in *Q_dot_12* = 2.629 W.

3.5.3 Practical Considerations for Application

Table 3.3 lists issues that can degrade MLI heat transfer performance in real-life applications. These effects become magnified at cryogenic temperatures. For detailed spaceflight information on these degradation impacts, see chapter five of the *Spacecraft Thermal Control Handbook* [12]. For a detailed overview of MLI design and application, see NASA's report, "Multilayer Insulation Material Guidelines" [13], an electronic version of which can be found on the NASA Technical Reports Server website.

Many investigators have sought to develop predictive methods that more accurately account for the blanket heat transfer taking place [14].

Table 3.3 Potential causes of MLI performance degradation

Environmental and/or blanket feature	Effect
Emissivity	Scaling parameter for fourth-order temperature difference affecting the net radiation heat transfer between layers. The lower the emissivity, the lower the contribution to radiative heat transfer through the MLI.
Number of blankets	The effective emittance decreases with increasing blanket layers down to a threshold value beyond which there are diminishing returns for using additional blanket layers [12].
Compression/thickness	Conductive thermal shorts become pronounced with increasing blanket thickness [12]. Compressed blankets tend to have less residual gas internal to the layered construction.
Planar size	Small surface areas have greater impacts on heat leak due to imperfections, edge effects and penetrations. For large surface areas, these features have a reduced effect on insulation performance [12].
Insulation layer thermal conductivity	Insulation layers with low thermal conductivity leads to a higher resistance for conduction-based heat leak through the blanket materials.
Pressure and/or trapped gasses	The inability to eliminate gas from the blanket construction while under vacuum conditions can promote gas conduction-based thermal shorts. This reduces the insulating performance of the MLI. Ideally, the MLI is most effective under vacuum conditions (i.e., $\leq 10^{-5}$ Torr).

These methods take into account the heat transfer contribution from each of the mechanisms involved in the overall heat transfer through the blanket. In equation form, the MLI heat transfer relationship now becomes

$$\dot{q}_{MLI} = \dot{q}_{gas,conduction} + \dot{q}_{layer,conduction} + \dot{q}_{radiation} \tag{3.114}$$

In relation 3.114, the gas conduction term accounts for molecular conduction through the blanket due to the presence of non-vented gasses. Figure 3.31 illustrates the magnitude of the gas thermal conductivity that can occur between confined surfaces (as is the case with MLI) at different levels of vacuum, showing that the thermal conductivity does not settle out at its minimum until pressures $\leq 10^{-5}$ Torr are attained.

The layer conduction term accounts for thermal conduction through the blanket. The radiation term is the net radiation heat transfer through the blanket (which is actually the RHS of equation 3.111). One of the key empirical relationships relied upon in the industry is the Lockheed Equation [16] (equation 3.114):

$$\dot{q}''_{MLI} = \frac{C_s \bar{N}^{3.56} T_m}{N_S + 1}(T_H - T_C) + \frac{C_r \varepsilon_{tr}}{N_S}\left(T_H^{4.67} - T_C^{4.67}\right) \tag{3.114}$$

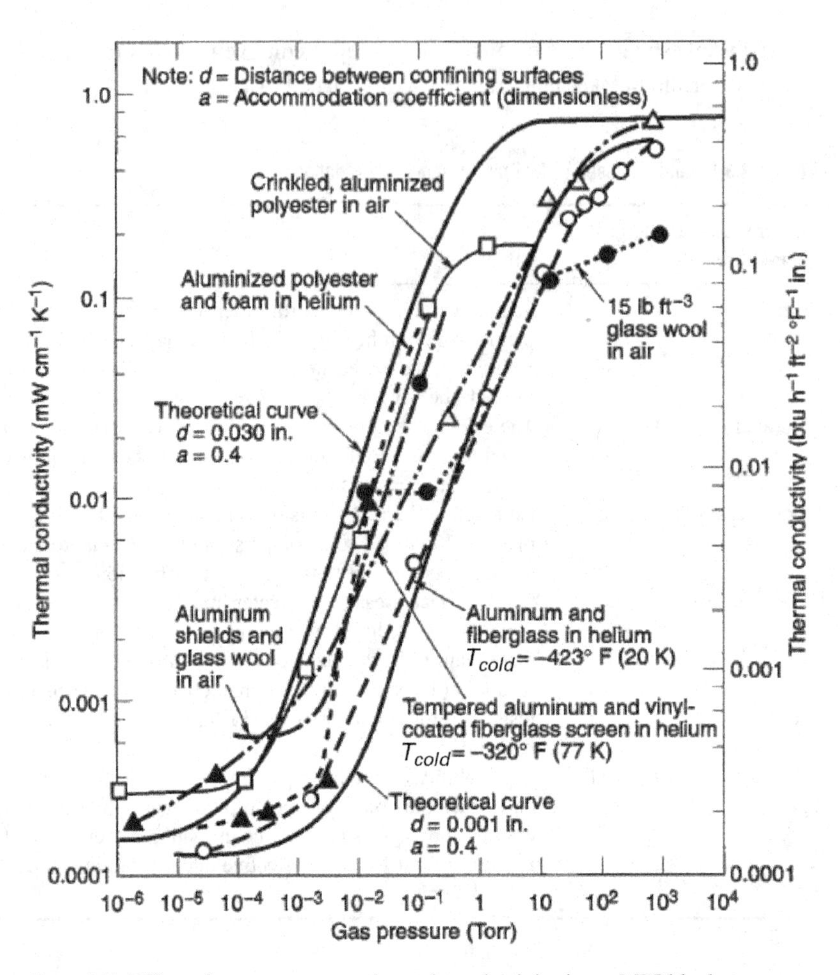

Figure 3.31 Effect of gas pressure on thermal conductivity in an MLI blanket
Source: *Spacecraft Thermal Control Handbook*, Vol. I, reference [12], courtesy of Aerospace Corporation

where

$\bar{N}$: Layers/cm	T_c: Cold-side temperature
N_S: Number of shields	T_h: Hot-side temperature
$C_S = 2.11 \times 10^{-9}$	$C_r = 5.39 \times 10^{-10}$
$T_m = (T_H + T_C)/2$	$\varepsilon_{tr} = 0.031$

The Lockheed Equation, which was developed from heat transfer performance data for double-sided aluminum blankets, is functionally dependent on several variables: layer density of the blanket ($\bar{N}$), total number of layers (N_S), and hot- and cold-side temperatures (T_h and T_c). For purposes of the analysis techniques covered, when invoking the non-ideal MLI assumption this text will predominantly rely on the analytically determined MLI relationship from Section 3.5.1 unless otherwise stated. The Lockheed Equation will be revisited in connection with the space environment and cryogenics in Chapters 4 and 7 respectively.

References

[1] Serway, R.A., 1986, *Physics for Scientists and Engineers*, 2nd edn, Saunders College Publishing
[2] Bergman, T., Lavine, A., Incropera, F.P., and Dewitt, D.P., 2017, *Fundamentals of Heat and Mass Transfer*, John Wiley & Sons, Hoboken, NJ
[3] Nellis, G., and Klein, S., 2009, *Heat Transfer*, Cambridge University Press, Cambridge
[4] Siegel, R., and Howell, J., 2002, *Thermal Radiation Heat Transfer*, 4th edn, Taylor & Francis, New York, NY
[5] Modest, M., 2013, *Radiative Heat Transfer*, 3rd edn, Academic Press, Oxford
[6] Planck, M., 1959, *The Theory of Heat Radiation*, Dover Publications, New York, NY
[7] Planck, M., 1901, "Distribution of Energy in the Spectrum," *Annals of Physics*, Vol. 4, No. 3, pp. 553–563
[8] Spiegel, M., 1997, *Mathematical Handbook of Formulas and Table*, McGraw-Hill Corporation, New York, NY
[9] Hottel, H.C., and Sarofim, A.F., 1967, *Radiative Transfer*, McGraw-Hill Corporation, New York, NY
[10] Rickman, S.L., "Form Factors, Grey Bodies and Radiation Conductances: Part 3," NASA, 31 Oct. 2012, http://nescacademy.nasa.gov/category/3/sub/3.
[11] Beckman, W.A., 1968, "Temperature Uncertainties in Systems with Combined Radiation and Conduction Heat Transfer," Proceedings of ASME Annual Aviation and Space Conference, Beverly Hills, CA
[12] Donabedian, M., Gilmore, D.G., Stultz, J.W., Tsuyuki, G.T., and Lin, E.I., 2002, "Insulation," *Spacecraft Thermal Control Handbook Vol. I: Fundamental Technologies,* edited by Gilmore, D.G., The Aerospace Press and The American Institute of Aeronautics and Astronautics, Inc., Chap. 5, pp. 161–206
[13] Finckener, M.M., and Dooling, D., 1999, "Multilayer Insulation Material Guidelines," NASA/TP-1999-209263, Marshall Space Flight Center and D^2 Associates
[14] McIntosh, G., 1994, "Layer by Layer MLI Calculation Using a Separated Mode Equation," *Advances in Cryogenic Engineering*, Vol. 39, pp. 1683–1690
[15] Glaser, P., Black, I., Lindstrom, R., Ruccia, F., and Wechsler, A., 1999, "Thermal Insulation Systems, A Survey," NASA SP-5027, NASA
[16] Keller, C.W., Cunnington, G.R., and Glassford, A.P., 1974, "Thermal Performance of Multi-Layer Insulations." Final Report, Contract NAS3–14377, Lockheed Missiles & Space Company

Problems

3.1 Determine:

i) The spectral emissive power for a blackbody surface at a temperature of 800 K and a wavelength of 100 μm (i.e., within the IR spectrum).

ii) The total emissive power for a blackbody at the same temperature as in part i).

iii) The maximum intensity of a radiating blackbody is at a wavelength of 100 μm. What is its temperature?

3.2 Revisit Section 3.2.2. Derive a proof showing that the first derivative of Planck's Distribution Law with respect to wavelength (λ) results in equation 3.10. Show all work in your derivation.

3.3 For the example shown in Figure P3.3

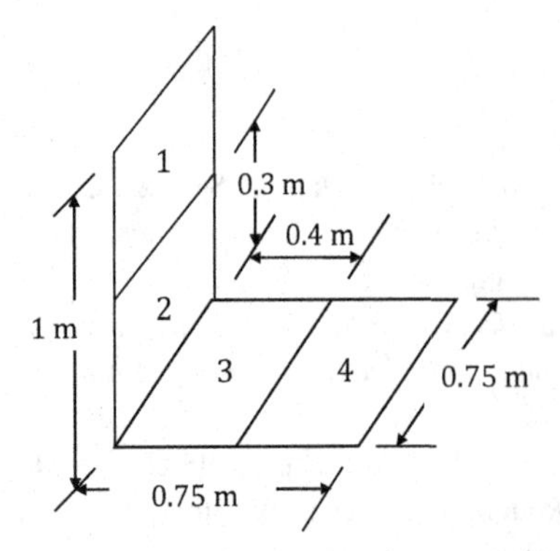

Figure P3.3 Subdivided elements perpendicular to one another joined at a common edge

i) Determine the view factor for F_{1-4}

ii) Assuming each of the surfaces in the diagram are blackbody surfaces, determine the heat transfer between surface 1 and 4 if surface 1 is at 300 K and surface 4 is at 250 K.

3.4 For the example in Figure P3.4,

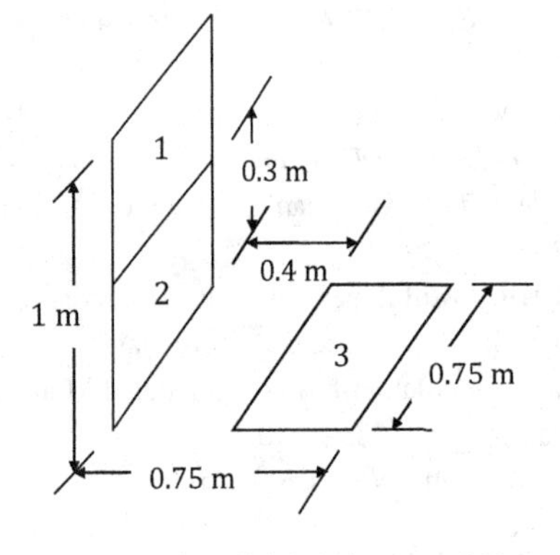

Figure P3.4 Subdivided elements 90° out of plane with a remote element

i) Determine the view factor for F_{1-3}

ii) Assuming each of the surfaces in the diagram are blackbody surfaces, determine the heat transfer between surface 1 and surface 3 if surface 1 is at 330 K and surface 3 is at 263 K.

3.5 For the arrangement of surfaces shown in Figure P3.5,

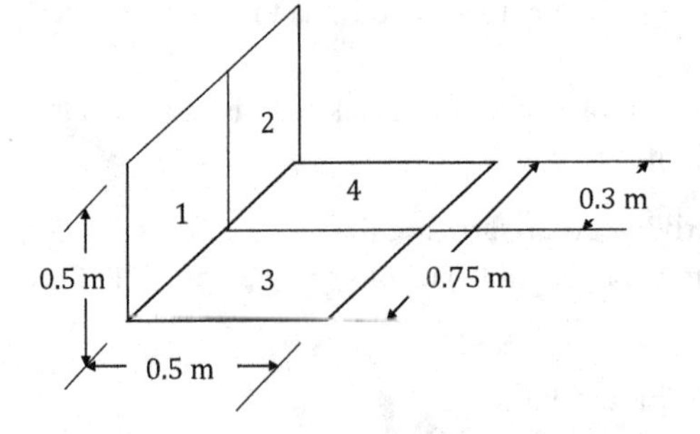

Figure P3.5 Laterally subdivided elements perpendicular to one another joined at a common edge

i) What is the value for F_{1-4}?

ii) Assuming blackbody radiative heat exchange, with $T_1 = 320$ K and $T_4 = 280$ K, what is $\dot{q}_{14}$?

iii) Assuming diffuse gray body radiative heat exchange, what is the thermal resistance network for $\dot{q}_{14}$?

iv) Assuming diffuse gray body radiative heat exchange, determine $\dot{q}_{14}$ for $T_1 = 320$ K and $T_4 = 280$ K if both surfaces are coated with Catalac black paint ($\varepsilon = 0.88$).

Note: you can assume that $A_1F_{1-4} = A_2F_{2-3}$

3.6 A gray body surface has an area approximately the size of a sheet of A4 paper (8.5 inches × 11.0 inches).

i) Assuming ideal emissivity (i.e., $\varepsilon = 1.0$), plot the total emissive power for the gray surface for temperatures varying from 3 K to 300 K.

ii) At what temperature does the surface begin to emit less than 50 μW of thermal energy?

iii) Provide plots for the total emissive power for the gray surface between the temperatures of 3 K and 300 K, assuming emissivity values of 0.8, 0.6, 0.4 and 0.2. Overlay these on top of the plot from part i).

Note: Calculations and plots should be performed in metric.

3.7 For a group of radiation shields (assuming the infinite parallel plate case with equal area), calculate and plot the effective emittance as a function of the number of shields with N ranging from 1 to 30. Overlay the plots for the following case

i) ε_1 = 0.25 mil aluminized Kapton; ε_2 = Beta cloth

ii) ε_1 = VDA (vapor-deposited aluminum); ε_2 = 0.25 mil aluminized Kapton

iii) $\varepsilon_1 = \varepsilon_2$ = Black Anodize Sample 1.

3.8 An enclosure consists of nested radiation shields as shown in Figure P3.8. Assuming diffuse gray body radiation for the heat transfer occurring,

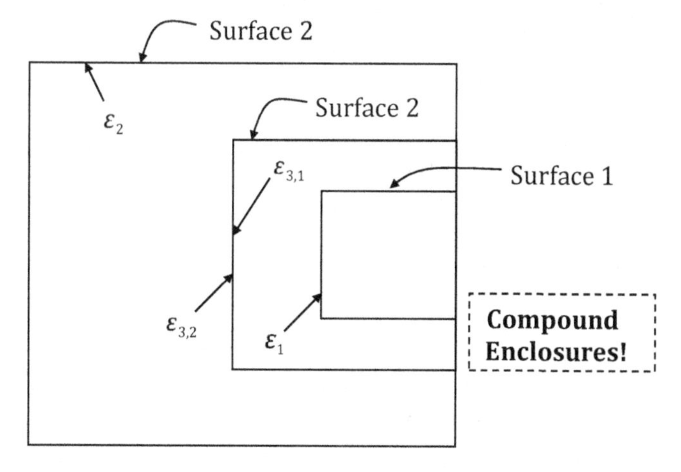

Figure P3.8 Compound enclosures in a nested oven configuration

i) Diagram the resistance network circuit for heat transfer between surface 1 and surface 2.
ii) Assuming $A_1 = 0.5A_3$ and $A_3 = 0.3A_2$, ε_1 = polished gold (BOL), ε_2 = polished aluminum (BOL) and $\varepsilon_1 = \varepsilon_2$ = VDA (vapor-deposited aluminum), what is the total and the limiting resistance in the network?
iii) If $T_1 = 200$ K and $T_2 = 20$ K, then what is $\dot{q}_{12}$? Assume $A_2 = 0.75\ \text{m}^2$.

3.9 You are designing an insulator with a series of 5 radiation shields (i.e., N = 5). Each radiation shield has a surface area of 0.75 m^2 with a reflectivity of 0.3 on one side and an absorptivity of 0.4 on the opposite side. It can be assumed that the shields are opaque and gray body assumptions are applicable.

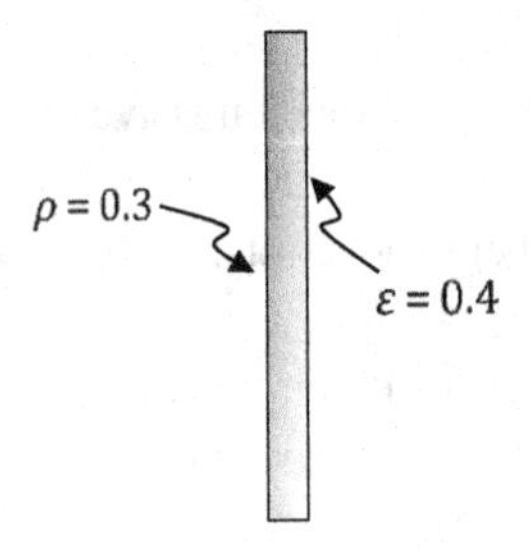

Figure P3.9 Example of single radiation shield

i) What is the effective emittance through the series of shields? Show the calculation and your assumptions.
ii) Assuming the shields are bound on one side by a surface at 300 K and on the opposite of the shield network there is a surface at 150 K, what is the heat transfer between these two surfaces?

3.10 A metal sleeve with a square cross-section has a plan view of its four sides (i.e., 1, 2, 3 and 4) as shown in Figure P3.10. The interior sides of the sleeve are all active. Surfaces 1, 2 and 4 are all gray body surfaces. Surface 3 is a blackbody surface.

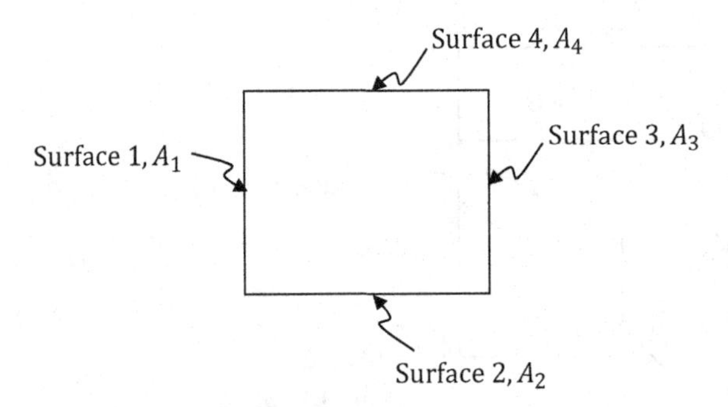

Figure P3.10 Metal sleeve with four surfaces

For the geometry shown:

i) Draw the thermal resistance network.

ii) What is the energy balance relation at J_1, J_2, J_3 and J_4 (leave balance equations in variable form)?

iii) Assuming surface three has an area of 0.25 m^2 and a temperature of 320 K, what is the radiosity for surface 3?

3.11 A multi-enclosure box cross-section consists of six separate surfaces (Figure P3.11). All interior surfaces of the box are active gray body surfaces.

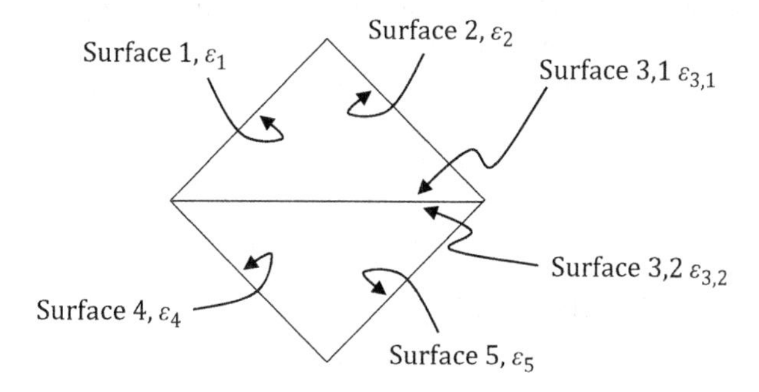

Figure P3.11 Opposing triangular enclosures with a common base

For the geometry shown:

i) Draw the thermal resistance network.

ii) What is the energy balance relation at node 3? Please leave the balance equation in variable form.

iii) Find the total emissive power of surface 2 if $\rho_2 = 0.3$, $A_2 = 0.5$ m^2 and $T_2 = 300$ K.

3.12 An enclosure contains four separate surfaces (i.e., 1, 2, 3 and 4) as shown in Figure P3.12. The interior sides are all active. Surfaces 1, 2 and 4 are all gray body surfaces whereas surface 3 is a blackbody surface. For the geometry shown:

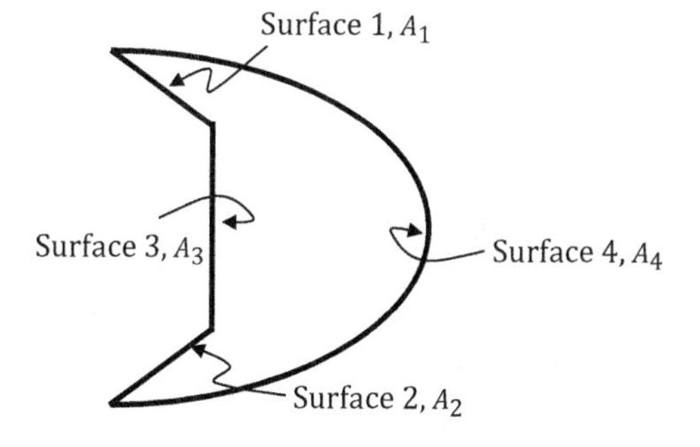

Figure P3.12 Surface enclosure with blocked view to interior from corner areas

i) Draw the thermal resistance network.
ii) Write out the energy balance at node J_4 (in variable form).
iii) Assuming surface 3 has an area of 0.5 m^2 and a temperature of 295 K, what is the radiosity for surface 3?

3.13 Figure P3.13 shows a circular surface positioned in the center of the base of a box with a side length of 1 m. A spherical ball is suspended (via 1/1,000th-inch Teflon wire) from the top of the box's interior. The ball is aligned with the center of the circular surface at the base of the box. For the configuration shown determine the view factor from the base surface to the interior of the enclosure box. The figure shows required dimensions and arrows to all active surfaces.

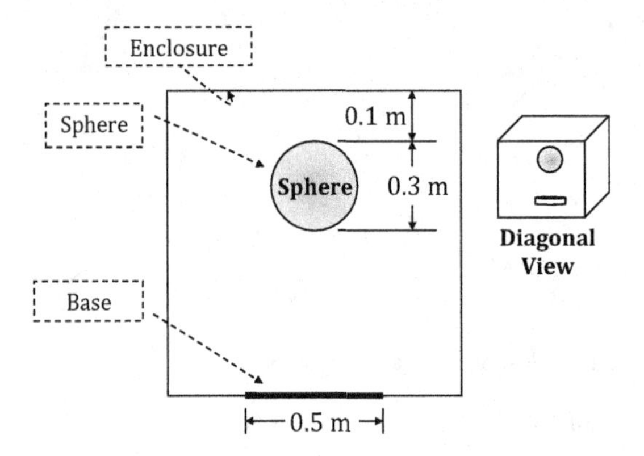

Figure P3.13 Sphere suspended in a cubic enclosure

3.14 A sphere (surface 1) is suspended from the top side of an enclosed cylinder via 1/1,000-inch diameter Teflon string (Figure P3.14). Treating the enclosure network as consisting of four surfaces, determine the view factor for the inner wall of the cylinder to the sphere, including any summation and/or reciprocity rules applicable in the solution.

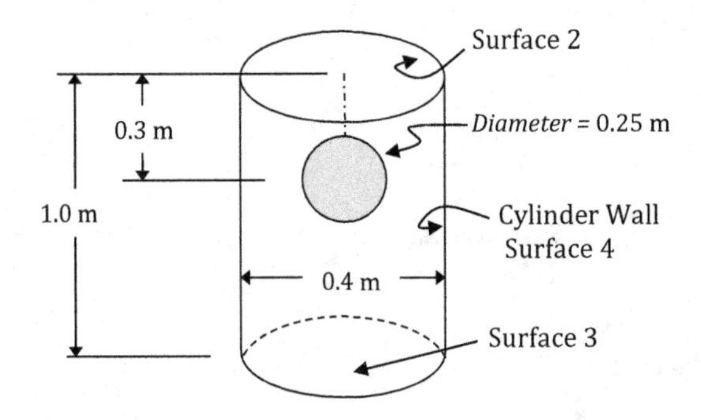

Figure P3.14 Sphere suspended in a cylindrical enclosure

3.15 Determine the view factor to Earth for a satellite surface (on a SpaceCube) facing the Earth edgewise from an orbit altitude of 350 km (Figure P3.15).

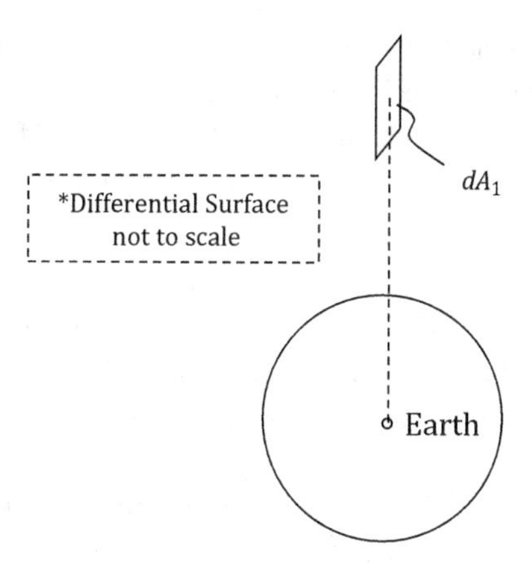

Figure P3.15 Flat surface with normal perpendicular-to-nadir direction

3.16 Determine the view factor to Earth for a SmallSat with a cubic geometry (side length of 1.0 m) positioned at an orbit altitude of 250 km (Figure P3.16).

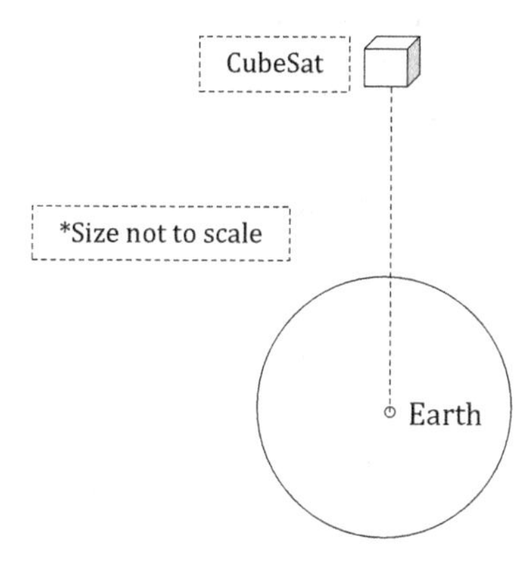

Figure P3.16 Cubic structure with one side facing nadir direction

3.17 An MLI blanket has an effective emittance (ε*) of 0.015 and a cross-sectional area of 0.5 m^2 with boundary temperatures of 348 K and 248 K. Calculate the effective conductance (using conduction theory) of the MLI blanket.

3.18 A spherical probe with a diameter of 0.5 m is perfectly encapsulated in MLI while in transit to a distant planet (Figure P3.18).

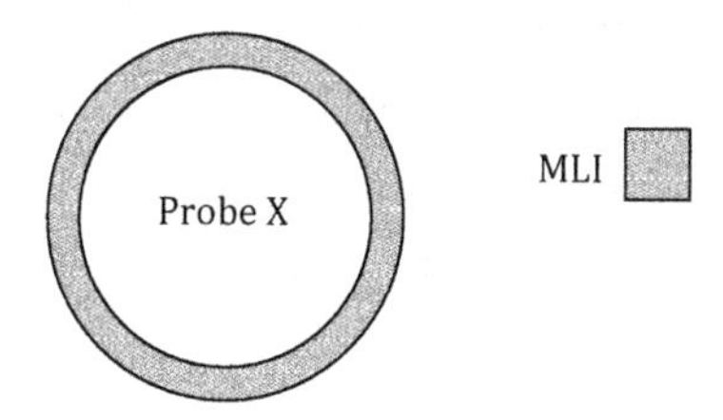

Figure P3.18 Probe exterior surface covered with MLI

i) Assuming values of $\varepsilon_1 = 0.26$ and $\varepsilon_2 = 0.35$ for the surfaces of the recurring MLI shields, how many shields are required to achieve an effective emittance of 0.01753?
ii) Assuming a heat loss of 6 W from the probe through the MLI, and a temperature of 142 K on the outermost shield, what is the temperature of the innermost shield (and/or the sphere's surface)?

3.19 A cryogen transfer line, approximately 1 m in length, is circumscribed by an outer sleeve. The diameter of the inner tube itself is 1.27 cm, whereas the outer sleeve has a diameter of 3.81 cm. Both have a wall thickness of 1 mm and are made of 304 stainless steel.
i) Assuming that the gap between the outer wall of the transfer line and the inner wall of the sleeve is filled with polystyrene insulation ($k_{eff} = 300\ \mu\text{W/cm} \cdot \text{K}$), what is the radial thermal conductance through the sleeve/polystyrene/tube composite lay-up?
ii) Alternatively, if there is a vacuum between the inner tube and the outer sleeve, what is the effective thermal conductance (radially) if the inner tube has a wall temperature of 77 K and the outer sleeve has a wall temperature of 300 K? Assume polished stainless steel as the coating for the surfaces of the inner tube and the outer sleeve facing each other. Disregard endpoint effects.

3.20 An L-bracket is located inside a temperature-controlled enclosure (Figure P3.20). The bracket has an electronics box sitting on the top flange that is dissipating 15 W of

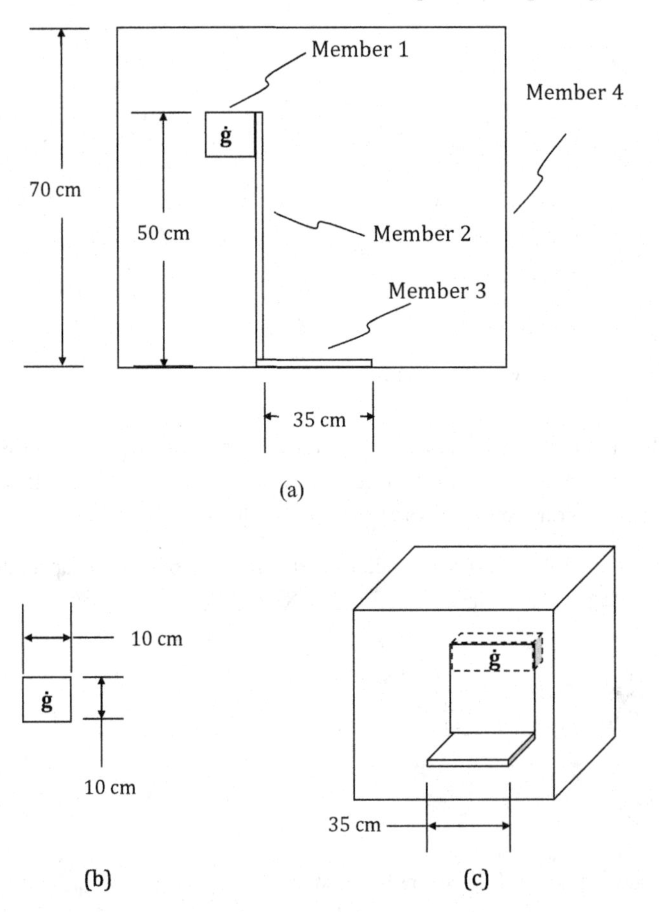

Figure P3.20 Enclosure with E-box mounted on an L-bracket: a) side profile; b) E-box dimensions; c) L-bracket base dimension

thermal energy. All interior surfaces of the box are gray bodies. The member surfaces are 1) Black Kapton BOL; 2) aluminized Kapton (0.15); 3) aluminized Kapton (0.15); and 4) black anodized.

i) Draw the thermal circuit with identification of all resistances and heat flows.
ii) Determine the view factors for each member of the circuit to all the others.
iii) Write out the energy balance relations for the circuit.
iv) Determine the temperature of the electronics box (i.e., member 1) given the following assumptions:

- Member 4 is a hollow cubic that is temperature controlled at 5°C
- The L-bracket is positioned in the center of the base of member 4 (i.e., the center of member 3 and the bottom side of member 4 are aligned).
- Members 1, 2 and 3 have a depth into the page of 35 cm.
- Member 3 is square in footprint.
- Model the electronics box (member 1) as a solid.
- The thickness for members 2, 3 and 4 is 5 mm.
- All members are made of aluminum 1100 ($k_{\text{Al, 1100}} \approx 210$ W/m · K) with thermal conductivity at room temperature.
- Steady-state conditions.

3.21 A calorimetric sensor is seated along the interior wall of a spherical cavity with a diameter of 1.0 m opposite a circular hole (i.e., aperture) providing an FOV to space. The center of the square sensor and the circular aperture are aligned.

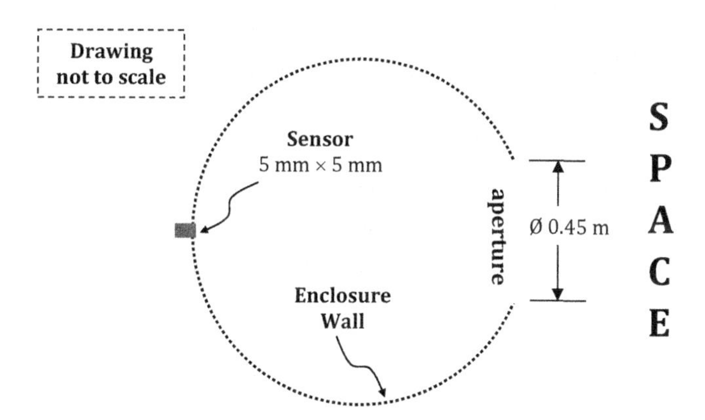

Figure P3.21 Calorimetric sensor in a spherical cavity

For the sensor and aperture dimensions shown in Figure P3.21, determine:

i) The view factor from the sensor surface to the cavity wall.
ii) The view factor from the aperture to the cavity wall.
iii) The heat transfer from the sensor surface through the aperture to space, assuming the sensor is a blackbody surface operating at a temperature of 220 K.

3.22 Three E-boxes are positioned inside a stainless-steel cubic enclosure. The cubic enclosure has a side length of 1.0 m and a wall thickness of 7.5 mm. Each E-box is cubic (side length of 25 cm) and made of Al-6061. The top of E-box 3 is directly opposite E-box 2, and E-box 2 is directly opposite E-box 1.

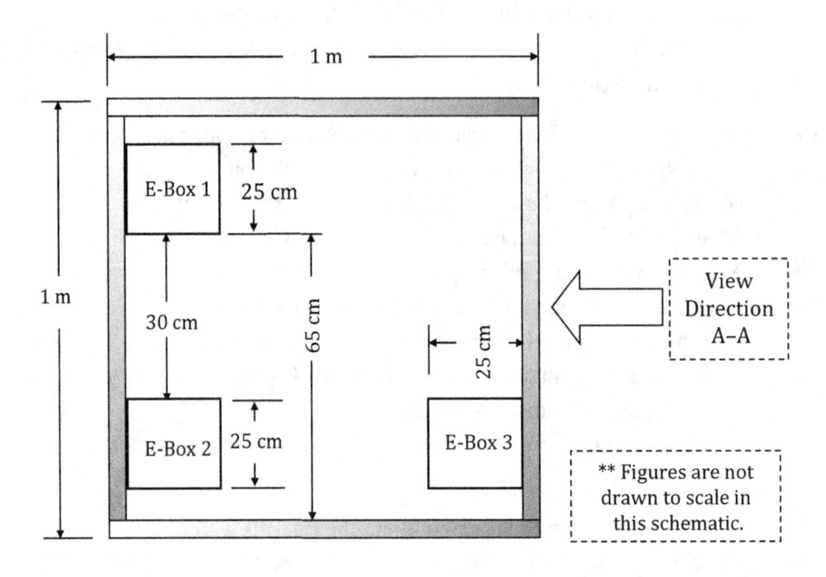

Figure P3.22a Internal geometry of SpaceCube

i) For the configuration shown in Figure 3.22a and b, determine the view factor between each of the E-boxes internal to the structure, as well as the view factors to the interior of the structure itself.

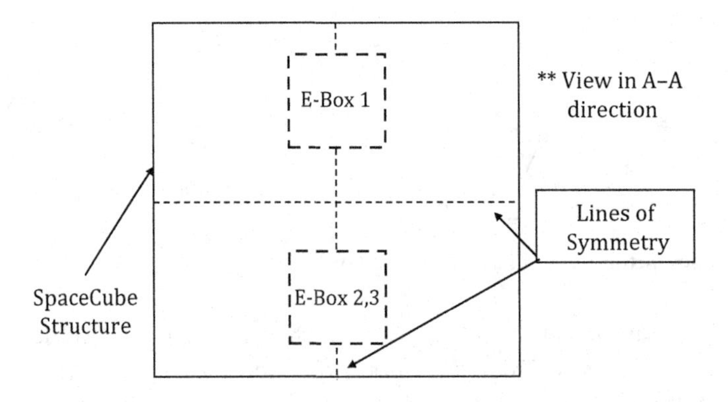

Figure P3.22b SpaceCube interior components, view in A–A direction

ii) Determine the temperature of E-boxes 1, 2 and 3 assuming energy generations of 55 W, 60 W and 70 W respectively and a fixed temperature for the structure of 310 K. Use the adiabatic assumption for the conductive coupling between the structure inner wall and the E-boxes. Coatings for the interior surfaces are:

E-box 1: NASA/GSFC NS-37 white paint
E-box 2: Paladin black lacquer
E-box 3: Vapor-deposited aluminum
Structure: S13GLO white paint

4 The Space Environment

4.1 Design Considerations

Spacecraft, which we will also refer to as satellites in this text, comprise one or multiple instruments and subsystems that enable measurements and/or observations, platform orientation, health monitoring and uplink/downlink communications. Generally, instruments are comprised of subsystems and subsystems are comprised of components. Thermal control system (TCS) design considerations are applied to all items onboard a spacecraft, and the most stringent requirements typically apply to items associated with instruments. This is because instruments typically carry specially designed hardware built to perform a specific type of measurement (or observation) in space. Adherence of spaceflight system operations to properly defined thermal parameters (i.e., requirements) enables the success of the thermal control system applied.

The most obvious operational thermal requirement is the operating temperature. Table 4.1 lists standard spaceflight items and their operational and survival temperatures. The operational temperatures denote temperature ranges over which the items can operate without risking temperature-induced failure (note that this does not take into account potential CTE mismatches for mated parts). Survival temperatures indicate the extremes that should not be exceeded.

Table 4.1 Operational and survival temperature ranges for standard spacecraft components

Component	Operating temperatures (°C)	Survival temperatures (°C)
[1]**Batteries**	0 ↔ 15	−10 ↔ 25
[2]**Electronics boxes**	−10 ↔ 50	−30 ↔ 60
[1]**Reaction wheels**	−10 ↔ 40	−20 ↔ 50
[1]**Star trackers**	0 ↔ 30	−10 ↔ 40
[1]**Gyros/IMUs**	0 ↔ 40	−10 ↔ 50
[1]**Antennas**	−100 ↔ 100	−120 ↔ 120
[1]**Solar panels**	−150 ↔ 110	−200 ↔ 130
Detectors	−272↔ −173	−269 ↔ 35

[1] Republished with permission of Springer Science from Gilmore, D.G., Hardt, B.E., Prager, R.C., Grob, E.W., and Ousley, W., *Space Mission Analysis and Design: Spacecraft Subsystems-Thermal* [1], Edited by Larson, W.J., and Wertz, J.R., 3rd edn, 1999; permission conveyed through Copyright Clearance Center, Inc.
[2] Republished with permission of Oxford University Press from Mehoke, D., *Fundamentals of Space Systems: Spacecraft Thermal Control* [2], Edited by Pisacane, V., 2nd ed., 2005; permission conveyed through Copyright Clearance Center, Inc.

Table 4.2 Standard design requirements and/or specifications

Type	Description	Unit of measure
Temperature	Usually a given setpoint with desired high and low operating values	°C or K
Stability	Degrees change per unit time	°C/minute
Spatial	Degree change per unit length (i.e., surface gradients)	°C/m
Rejection capability	Heat removal from a surface or a component	W
Heat flux	Amount of thermal energy acquisition required at a surface	W/cm^2 or W/m^2

However, temperature is just one type of requirement that may be imposed. Table 4.2 summarizes the various other types of requirements that may be imposed. The thermal subsystem works with others to create viable operating conditions on the space platform. For example, the temperature control architecture used will depend not only on the choice of conductive materials but also waste heat from electronic components, thermal energy input from resistance-based heaters, environmental heat input and MLI. Table 4.3 lists various subsystems connected with the thermal system. Mechanical, electrical, optical, guidance navigation and control, detector and laser systems are all coupled to the TCS. A satisfactory design for each subsystem is typically reached through design iterations and negotiations. Subsystems with predefined temperature boundaries, such as detectors and lasers, require a TCS architecture that takes these into account. For the optical subsystem, the TCS needs to maintain a given temperature gradient on an optical bench or nominal temperature for a space telescope assembly.

A key interface is between thermal and mechanical subsystems. All components onboard a space platform must survive vibrational (and/or acceleration) loads experienced during launch; components of all other susbsystems must clear this design/performance hurdle as well. In the thermal subsystem, both materials and mechanical configurations chosen will directly affect conductive heat flow paths. Temperature control hardware that experiences wear (e.g., mechanical heat pumps) is subject to weigh-in on design and integration and test practices. Successful operation of TCS hardware relies on access to power, as well as the operational capability of circuitry on the space platform.

Redundancy of spaceflight subsystems is typically a function of the Mission Class designation (e.g., Class A through E). One area where redundancy is routinely implemented is in the electrical subsystem, which determines (through negotiation) the amount of in-flight power allocated to the TCS, as well as the total power available to the space platform. The use of solar arrays reinforces the critical role that environmental heating plays in mission design and TCS architecture. Thermal energy capture in flight is subject to orbit parameters defined and maintained by guidance, navigation and control (GN&C). Ultimately, power available to a space platform will depend on whether it is an earth-based or a deep space mission.

Table 4.3 Spaceflight subsystems coupled to thermal

Subsystem	Responsibility	Interface with thermal
Detectors	• Detector design and in-flight energy-based photonic measurements	• Regulation of detector temperature
Lasers	• Laser system design and in-flight operation	• Regulation of laser temperature
Optical	• Space telescope design and in-flight operation • Optical bench design and in-flight operation	• Regulation of space telescope or optical bench temperature and gradients
Mechanical	• Structural integrity of system components through launch into orbit and during flight • Mechanisms	• Thermophysical properties of materials • Conductive paths • Mechanically operated thermal control hardware
Electrical	• Circuit architecture and redundancy levels • Power system architecture	• Power budget • Power delivery system • Electrical circuitry • Temperature readouts
GN&C	• Orbit selection and in-flight maneuvers	• Environmental heating • Orbit statistics • FOV to environmental sources/sinks

4.2 Earth-Based vs. Deep Space Missions

Space platforms that fly Earth-based missions typically use a combination of solar arrays (photovoltaics), batteries and environmental heating from the Sun and Earth for thermal heating and electrical power to insure temperatures for flight components are maintained within desired ranges. Solar heating and Earth-emitted thermal loading are therefore important in the design of a TCS. Heat flux values for the Solar constant and Earth emitted thermal energy are determined via measurements taken immediately outside the Earth's atmosphere. One instrument that measures solar intensity is the total irradiance monitor onboard the SORCE (Solar Radiation and Climate Experiment) mission. Based on total solar irradiance (TSI) measurements made with this instrument, average values for the solar constant (i.e., the intensity of the power supplied by the Sun) are determined annually and later released to the global community by the World Radiation Center in Davos, Switzerland [3,4]. The most recent values measured have shown a solar constant value of approximately 1,361 W/m^2 [5]. Power available to a satellite is a function of solar array area and cell thermal-to-electric conversion efficiency. State-of-the-art PV (photovoltaic) cells flying on modern spaceflight missions have conversion efficiencies on the order of 30% $\leftrightarrow$ 35% (laboratory-based measurements have shown efficiencies as high as 45% [6]). In eclipse (when transiting through the Earth's shadow) the space platform typically relies heavily on battery power. Traditional batteries flown on space platforms include nickel hydride (NiH_2), nickel cadmium (NiCd) and lithium ion (Li-ion). Li-ion batteries are the most common type used today. SOH (state of health) monitoring and maintenance is key to the successful operation of the space platform after launch. Depending on the location of available downlink stations, power may be required during eclipse for updates on a satellite's SOH.

Imagine a deep space mission to Jupiter using a similar architecture to that of Earth-based missions where in-flight power is based on a combination of battery power and thermal-to-electric energy conversion (i.e., batteries and solar arrays). The Sun is a spherical electromagnetic emitter (i.e., it emits at wavelengths across the electromagnetic wave scale). Assuming that energy emitted from the Sun is uniform across its surface area, we can extrapolate that the Sun emits thermal energy as spherical waves (Figure 4.1). For an electromagnetic energy source that emits waves in a spherical format, the intensity of the energy experienced at a known distance (r) from the source is defined as the average emissive power of the source divided by the surface area swept out by a sphere having a radius equal to the distance between the source and the receiver. The intensity relation for an emitter at a distance r is given by equation 4.1. In the equality, P_{avg} is the average emissive power of the source. In the case of the Earth, the solar constant (1,361 W/m^2) serves as the intensity (I) [3].

$$I = \frac{P_{avg}}{A_{sphere}} = \frac{P_{avg}}{4\pi \cdot \bar{r}^2} \tag{4.1}$$

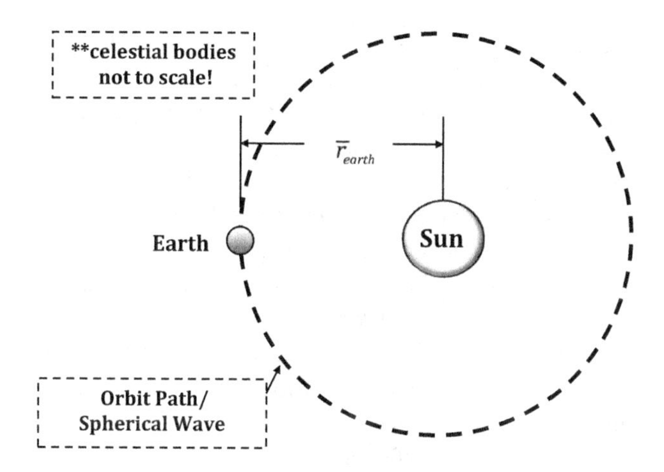

Figure 4.1 Radiative waves experienced by Earth at distance (r) from the Sun

While it is known that the Earth's orbit path around the Sun is somewhat elliptical, for purposes of interpretation and analysis we shall approximate it as circular. The average distance between the Earth and the Sun is 93 million miles. Given knowledge of the solar constant for Earth, we can calculate the average emissive power of the Sun. In addition, with knowledge of the average orbit distance for Jupiter around the Sun (483 million miles), we can determine the intensity of the Sun's rays at that distant planet. Another approach might be to solve the intensity calculations for the Earth and Jupiter simultaneously, considering that the average emissive power of the Sun is the same in both calculations. This equality would be of the following form.

$$I_{earth} \cdot 4\pi \cdot \bar{r}^2_{earth} = P_{avg} = I_{jupiter} \cdot 4\pi \cdot \bar{r}^2_{jupiter} \tag{4.2}$$

Solving for the intensity at the planet Jupiter gives the following relationship

$$I_{jupiter} = I_{earth} \frac{\bar{r}^2_{earth}}{\bar{r}^2_{jupiter}} \tag{4.3}$$

Given the assumptions for radial distances made previously for both Earth and Jupiter, $I_{jupiter}$ approximates to

$$I_{jupiter} \approx 0.04 \cdot I_{earth} \tag{4.4}$$

Thus the intensity of the Sun's rays at the planet Jupiter is approximately 1/25th of the intensity of the Sun's rays at planet Earth. The solar array area needed to perform a comparable amount of thermal energy capture for conversion to electricity at Jupiter would therefore need to be 25 times larger than the solar array area used at the radial distance of the Earth. Traditionally this has been considered highly impractical from a design perspective, so alternate energy sources have routinely been opted for when designing deep space probes to destinations such as Jupiter. Among the most frequently used alternate power system approaches are radio-isotope thermoelectric generators (RTGs) [7].

RTGs use thermal energy from the radioactive decay of nuclear materials such as ^{238}Pu (plutonium 238) for conversion to electricity using thermoelectric devices. Thermoelectrics have two operating modes: Peltier mode and Seebeck mode (for detailed discussion on thermoelectric devices see Section 5.5). In Peltier mode a thermoelectric that is exposed to a drive current creates a temperature gradient. In Seebeck mode, a thermoelectric subjected to a temperature gradient produces an electric current. RTGs, which use thermoelectrics in Seebeck mode, are capable of generating upwards of 1 kWe. In application, these are coupled with Brayton cycles for electricity production [4]. This presents an interesting dynamic. Nuclear power systems for space application have been the subject of much scrutiny for several decades. This has primarily been due to concerns of a potentially catastrophic event taking place during launch and ascent, as well as during departure from the Earth's atmosphere. Nonetheless, the Sun is known to be the largest nuclear power source in our solar system. The thermal energy provided for solar array/battery/environmental heating architectures on contemporary Earth-based mission platforms is fueled by a continuous H_2 fission reaction at the Sun. Thus, the use and benefit of nuclear power is already present (albeit remote) on Earth-based missions.

With continual improvements in thermal-to-electric conversion efficiency for PV cells, at the beginning of the new millenium mission planners and engineers decided to reassess the space community's previous policy of not using PV technology on deep space missions. In 2011 NASA launched the Juno Spacecraft which flew solar arrays rather than a nuclear energy source. Juno arrived at Jupiter in 2016 and has been performing its mission successfully without power supply interruptions. While there is limited information regarding the PV cells flown on Juno, the mission fact sheet states that Juno has 420 W power available while in orbit around Jupiter. This mission is the first of its kind and reinforces the potential for successful use of advanced PV cell technology on exploration missions as far out as Jupiter. In the decades to come it is expected that an increasing number of deep space missions will opt for power generation using PV cells as opposed to RTGs.

4.3 Astrodynamics Fundamentals

4.3.1 General Definitions

A key aspect of spacecraft thermal system design is environmental heating, the amount of which is a function of the space platform's orbit statistics. We begin by defining the ecliptic and the equatorial planes. Over the course of a calendar year, the Earth's path of travel around the Sun is elliptical in form. The locus of day-to-day stationary points for the travel path of the Earth may also be used to establish a plane of its motion around the Sun. This conceptual plane, which includes the Sun, the Earth and the Earth's travel path, is known as the ecliptic plane. Figure 4.2 shows the relationship between the ecliptic plane, the equatorial plane, the solar vector and the solar cone at the winter solstice. The solar vector, which denotes the direction of the Sun's rays towards the planet at any point along its travel path, lies within the ecliptic plane. The Sun is much bigger than the Earth. However, given the distance between the two (an average of 93 million miles over the course of a calendar year), for solar rays that lie outside the ecliptic plane the angle of approach is very shallow by the time they reach Earth. They are considered to have a parallel travel path and are modeled as parallel to the standard representation of the solar vector and the ecliptic plane in general. The equatorial plane, which is perpendicular to the Earth's spin axis and is inclined 23.4° relative to the ecliptic plane, lies along the Equator. An object in orbit around the Earth has an orbit inclination, a prescribed beta angle (β) and an orbit period (OP). The inclination is the angle between the orbit plane and the equatorial plane. In this text, if the inclination angle references the ecliptic plane it will be stated explicitly. The beta angle is the minimum angle between the orbit plane and the solar vector. The OP is the time it takes for one revolution around the planet [1,2]. Figure 4.3 shows example beta angles for an orbit inclination of 66.6° relative to the equatorial plane (i.e., 90° relative to ecliptic plane).

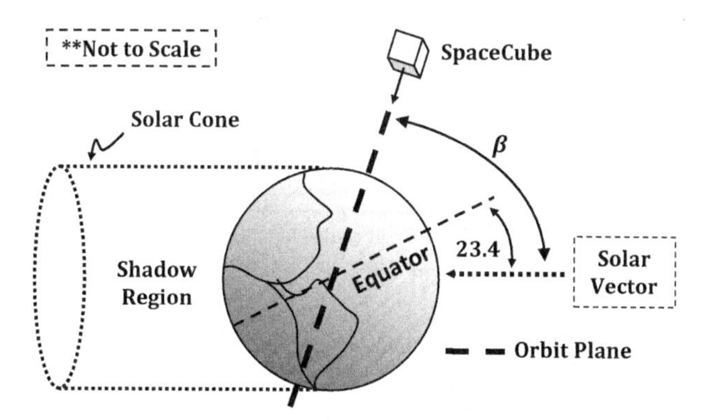

Figure 4.2 Earth and beta angle, side profile at winter solstice location
Source: Reproduced from figure 2.12a, *Spacecraft Thermal Control Handbook Vol. I: Fundamental Technologies*, reference [4] with permission from Aerospace Corporation

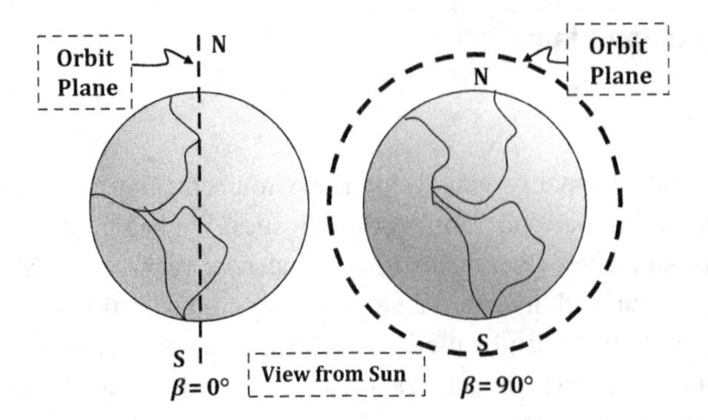

Figure 4.3 Beta angles at 66.6° inclination relative to the equatorial plane as viewed from the Sun Source: Reproduced from figure 2.12b, *Spacecraft Thermal Control Handbook Vol. I: Fundamental Technologies*, reference [4] with permission from Aerospace Corporation

4.3.2 Types of Orbits

Earth-based orbits can be either circular or elliptical and are categorized according to their altitude. There are three basic categories: LEO (low Earth orbit), MEO (medium Earth orbit) and HEO (high Earth orbit) [8] (Figure 4.4). LEOs fall within an altitude range of 180–2,000 km. MEOs span an altitude range of 2,000–35,760 km and HEOs have altitudes greater than 35,760 km. While Figure 4.4 shows the MEO and HEO paths as elliptical and the LEO path as circular, each of these orbit types can be either circular or elliptical. For an elliptical orbit the altitude can be determined given knowledge of the semi-major axis.

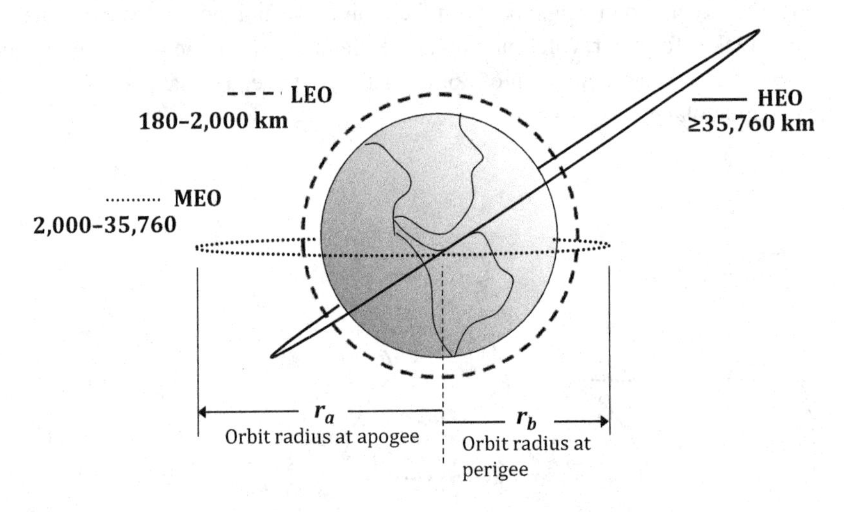

Figure 4.4 Circular and elliptical Earth-based orbits

The semi-major axis is defined as

$$a = \frac{r_a + r_b}{2} \tag{4.5}$$

where r_a and r_b are the radius of apogee and perigee respectively (as shown for the example MEO orbit case in Figure 4.4). In a circular orbit path, these two radii are equal and the semi-major axis simply reduces to the orbit radius. Since the semi-major axis is measured relative to the center of the Earth, the orbit altitude can be determined by subtracting the Earth's radius from the semi-major axis value. The semi-major axis can also be used to determine the orbit period. The orbit period is defined as

$$OP = 2\pi \left(\frac{a^3}{\mu} \right)^{1/2} \tag{4.6}$$

where

OP: Orbit period [s]
a: Semi-major axis (for elliptical cases) of the orbit (or the radius for circular orbits)
μ: The product of the universal gravitational constant and the mass of the planet. This value for Earth is 3.98603×10^{14} m^3/s^2

In the absence of the declaration of a particular orbit type, the orbit category the spacecraft is in can be determined from the orbit period. Equation 4.6 suggests that increases in orbit period will correspond with increases in altitude. HEOs have the largest orbit periods and LEOs the smallest. Most LEO-based satellites have orbit periods on the order of 90 minutes, whereas HEOs can span upwards of 24 hours [8].

Selection of a particular orbit for use is a function of the spaceflight mission and its overall goals (e.g., measurements and/or observations to be performed). Under each orbit category there are specialized orbits. The Sun-synchronous orbit is a special case of LEO in which the satellite has a continuous view of the planet at a fixed Sun angle. With repeated passes, this orbit allows different planetary locations to be viewed at the same local time. This orbit is often chosen for reconnaissance satellites and weather observation platforms. The Molniya orbit, prevalent under MEO, is highly elliptical with large inclination (upwards of 62°) and an orbit period on the order of 12 hours [4]. These orbits provide a good view of the North Pole region. In the HEO category, a very common specialized orbit is GEO (geo-synchronous orbit), which has an altitude of approximately 35,786 km. GEOs are typically circular with low inclinations relative to the equatorial plane. Their orbit period is approximately 24 hours which enables the satellite to fly over a fixed location [1,2,8]. This orbit is used on weather observation, communication and surveillance satellites.

4.4 Lagrange Points

The Lagrange points are named after the mathematician Joseph-Louis Lagrange who initially proposed them as locations for static positioning of small celestial bodies in his work *Essai sur le Problème des Trois Corps* [9]. From an astrodynamics perspective, the five Lagrange points are physical locations in a two-body Keplerian system in which a third body of much smaller size scale can be placed and remain relatively fixed. Figure 4.5 shows the relative positioning of Lagrange points L1 through L5 in the

Table 4.4 Past, present and future missions staged at Sun–Earth Lagrange points

Mission	Mission science	Observing location/ orbit	Launch date
SOHO Solar and Heliospheric Observatory [11]	Study internal structure of the Sun, its outer atmosphere and source of solar winds	L1 Halo	1995
WIND [12]	Plasma and energetic particle-based solar wind measurements	L1	2004
DSCOVR Deep Space Climate Observatory [13]	Measure solar wind velocity distribution and irradiance upon the Earth	L1	2015
WMAP Wilkinson Microwave Anisotropy Probe [14]	Investigate temperature differences in the CMB	L2 Halo	2001
Herschel Space Telescope [15]	IR wavelength observation of matter and phenomena in the early universe	L2	2009
Planck [16]	Investigate cosmic history and the origins of the universe	L2	2009
JWST James Webb Space Telescope [17]	IR observation of early galaxies and phenomena in our solar system	L2 Halo	2021
WFIRST Wide Field IR Survey Telescope [18]	Exoplanet detection, study of dark energy and measure of cosmic acceleration	L2	TBD[†]

[†] To be determined.

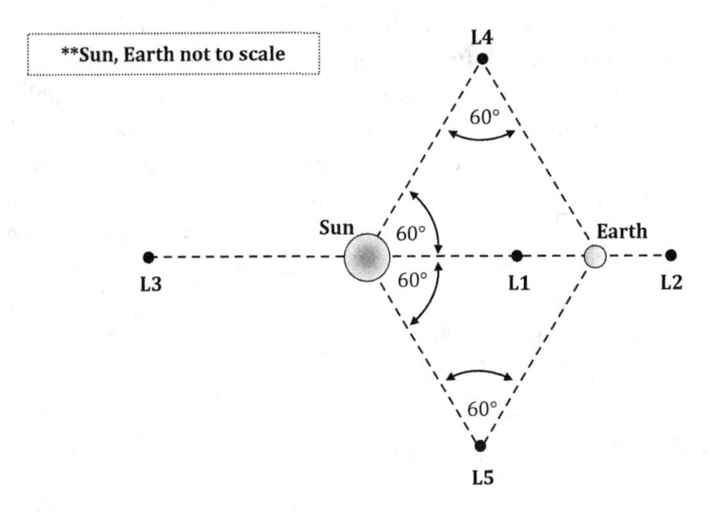

Figure 4.5 Sun–Earth system Lagrange points

Sun–Earth system. Points L1, L2 and L3 are known to be slightly unstable, requiring periodic orbit and/or attitude adjustment of the space platform, whereas L4 and L5 are not. The distance from Earth to L1 and L2 is approximately 1/100th the average distance between the Earth and the Sun [8,10]. The distance from L3 to the Sun is approximately the same as the average distance from the Earth to the Sun. The positioning of a space platform at one of the five Lagrange points presents a unique opportunity for continuous observations of celestial bodies and/or the celestial sky. Lagrange point missions are listed in Table 4.4. Previous L1 missions have focused on observations of either the Sun or the sunlit Earth. Missions at L2 have traditionally been space-based observatories viewing the celestial body. From a thermal perspective, since the Lagrange points are fairly fixed positions in the Sun–Earth system, unlike the standard Earth-based orbit they foster a constant thermal environment. As such, the primary design point for consideration in the TCS architecture will be the continuous thermal loading of the space platform from sources in the FOV at the relevant Lagrange point.

4.5 Environmental Thermal Heating

In the space environment there are four potential sources of thermal energy input to onboard components. These are:

1) Solar energy (from the Sun)
2) Albedo (solar energy reflected off of the Earth)
3) Earth-IR (emitted IR energy from the Earth)
4) Flight-system energy dissipations.

Solar energy, albedo and Earth-IR are all inherent to the near-Earth space environment. Energy dissipations will be based on artificial heating provided to the component of interest either from the operation of a component producing resistive heating or from heater elements flown onboard. As an example of a simple surface in the space environment, Figure 4.6 shows a single flat plate in space with one active side.

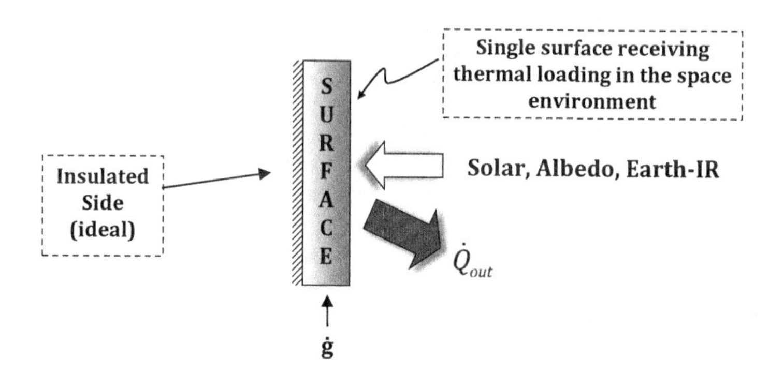

Figure 4.6 Thermal equilibrium case for a bare surface in space

A basic energy balance, assuming ideal insulation on the LHS of the plate, is shown in equation 4.7. When expanded to include individual environmental contributions the energy balance relation will take the form of equation 4.8. The $\dot{g}$ term in equation 4.8 is for energy generation (and/or dissipation) associated with the object.

$$\dot{Q}_{in} = \dot{Q}_{out} \tag{4.7}$$

$$\dot{g} + \dot{Q}_{Solar} + \dot{Q}_{Albedo} + \dot{Q}_{Earth\text{-}IR} = \dot{Q}_{out} \tag{4.8}$$

$\dot{Q}_{out}$ is the heat rejection term. Since we are rejecting heat in a vacuum environment, the Stefan–Boltzmann law for gray body surfaces applies. This is captured in equation 4.9 (see also Figure 4.7).

$$\dot{Q}_{out} = \varepsilon A_{surf} \sigma T^4_{surf} \tag{4.9}$$

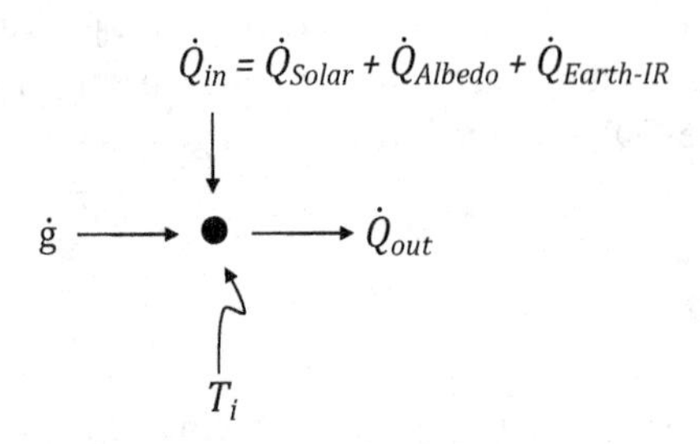

Figure 4.7 Node i with energy inputs and outputs

Figure 4.8 illustrates the case where another component with energy generation is embedded within the spacecraft yet in thermal communication with the spacecraft boundary surface (exposed to space).

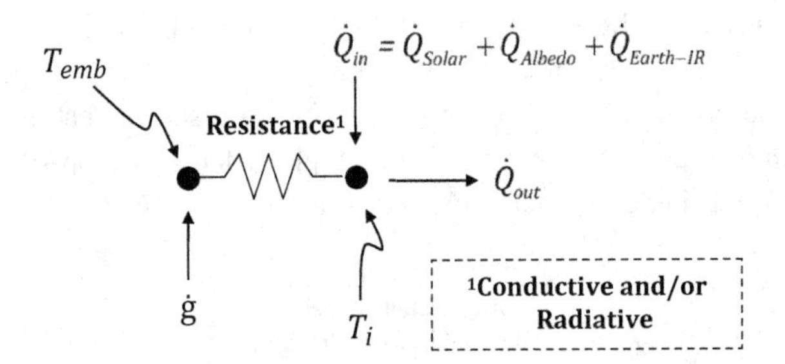

Figure 4.8 Embedded node coupled to spacecraft exterior structure node

Equation 4.8 is also applicable for finding the energy balance at node i.

$$\dot{g} + \dot{Q}_{Solar} + \dot{Q}_{Albedo} + \dot{Q}_{Earth\text{-}IR} = \dot{Q}_{out}$$

This can also be written as

$$\frac{(T_{emb} - T_i)}{R_{emb\text{-}i}} + \dot{Q}_{Solar} + \dot{Q}_{Albedo} + \dot{Q}_{Earth\text{-}IR} = \dot{Q}_{out} \tag{4.10}$$

where R_{emb-i} is the resistance between the embedded node and node i (at the boundary surface). It should be noted that we have only assumed a conductive thermal coupling exists between the two nodes in this case. If the energy balance in equation 4.10 were to have the $\dot{Q}$ terms expanded out to explicitly include environmental energy contributions, the new form would be

$$\dot{g} + \alpha_S SA_i + Alb \cdot \alpha_S SA_i \mathrm{F}_{i\text{-}Earth} + \varepsilon EIR_{flux} A_i \mathrm{F}_{i\text{-}Earth} = \varepsilon A_i \sigma T_i^4 \tag{4.11}$$

where

A_i: Area of exterior exposed surface
S: Solar constant (assumed to be 1,361 W/m^2)
α_S: Solar absorptivity of the incident surface
ε: Emissivity of exterior exposed surface i
Alb: Fraction of Earth-reflected solar energy (ranges $0 \leftrightarrow 1$)
EIR_{flux}: Earth-emitted IR radiation (W/m^2)
$F_{i\text{-}Earth}$: Surface-to-Earth view factor.
σ: Stefan–Boltzmann constant (5.67×10^{-8} W/m$^2 \cdot$ K^4)

4.5.1 Flat Plate Analysis

We now return to the single-node basic surface and focus on the heat rejection capability of the plate. Taking a special case for environmental loading that neglects energy generation, albedo and Earth-IR, the energy balance equation becomes

$$\dot{Q}_{Solar} = \dot{Q}_{out} \tag{4.12}$$

which can also be written as

$$\alpha_S S A_{surf} = \varepsilon \sigma A_{surf} T_i^4 \tag{4.13}$$

or alternatively as

$$\alpha_S S = \varepsilon \sigma T_i^4 \tag{4.14}$$

Rearranging equation 4.14 to solve for the temperature gives the following

$$T_i = \left[\underbrace{\left(\frac{S}{\sigma}\right)}_{Constant} \cdot \left(\frac{\alpha_S}{\varepsilon}\right) \right]^{1/4} \tag{4.15}$$

On the RHS of this relation, the ratio of the solar constant to the Stefan–Boltzmann constant is a constant value regardless of the surface in question. The temperature (T_i) is functionally dependent upon the surface coating which determines the α_S/ε ratio. The α_S/ε ratio is a measure of the absorbing–emitting capabilities of a surface receiving solar thermal energy while also emitting in the IR spectrum. This is used as a metric in selecting coatings for radiators and radiator design. As a rule of thumb, if $\alpha_S/\varepsilon > 1$ then the thermal energy capture surface is primarily absorbing solar thermal energy. Alternatively, if $\alpha_S/\varepsilon < 1$ then the surface is primarily emitting thermal energy (in the IR).

Example 1 Flat Plate Facing Deep Space Assuming a surface facing deep space that is rejecting 50 W, what is the temperature for a 1.0 m^2 surface area coated with polyurethane white paint (referencing Appendix B, $\alpha_S = 0.27$, $\varepsilon = 0.84$)?

Solution

Neglecting environmental loading, we know that $\dot{Q}_{in} = \dot{Q}_{out}$. Thus

$$50\,\text{W} = \varepsilon\sigma A_{surf} T_i^4$$

Substituting in for actual values on the RHS of the balance equation gives

$$50\ \text{W} = (0.84)\left(5.67 \times 10^{-8}\ \text{W/m}^2 \cdot \text{K}^4\right)\left(1.0\ \text{m}^2\right) \cdot T_i^4$$

If we then solve for T_i we will attain

$$T_i = 180\,\text{K or } -93°\text{C}$$

If this calculation is repeated for a dissipation of 500 W then $T_i = 320$ K $\approx$ 47°C. For a thermal dissipation of 1 kW, $T_i = 380$ K $\approx$ 107°C. If a coating for the surface is selected that provides an emissivity of 0.92 (i.e., $\varepsilon = 0.92$), then with a 1 kW thermal dissipation the temperature prediction becomes $T_i = 372$ K $\approx$ 99°C. If a sensitive flight component (e.g., electronics box) were thermally coupled to this plate, it would be projected to operate near this elevated temperature as well. Exposure of electronics to such high temperatures is not desirable so the temperature would need to be lowered. However, considering that the emissivity is already very high, the options are limited. One solution is to increase the surface area for heat rejection, which will cause the temperature to go down.

While ideal insulation is assumed in the example problems, MLI does not perform as an ideal insulator in real life, and therefore heat loss through the MLI will occur. In order to model this loss, we will treat the plate as having a uniform temperature and assume that the MLI consists of 15 layers of double-sided .15 aluminized kapton $\varepsilon = 0.34$. The heat transfer between the bounding layers of the MLI can be modeled as

$$\dot{Q}_{i\text{-}o} = \varepsilon^* A_{surf}\sigma\left(T_i^4 - T_o^4\right) \tag{4.16}$$

where ε^* is the effective emissivity for the 15 layers of insulation. We shall treat the surface temperature of the MLI immediately adjacent to the plate as the same as that of the plate, the outer surface of the MLI being taken to be a single massless node characterized by its own temperature (e.g., T_o). Figure 4.9 shows the coupling of the MLI node to the structure surface node. This outer surface will be in thermal communication with environmental sources and sinks within its FOV. Given that the thermal circuit now has two unknowns (i.e., T_o and T_i) two equations are needed to solve for these temperature values. The energy balance on the plate now becomes

$$50\ \text{W} = \dot{Q}_{i-o} + \dot{Q}_{i,out} \tag{4.17}$$

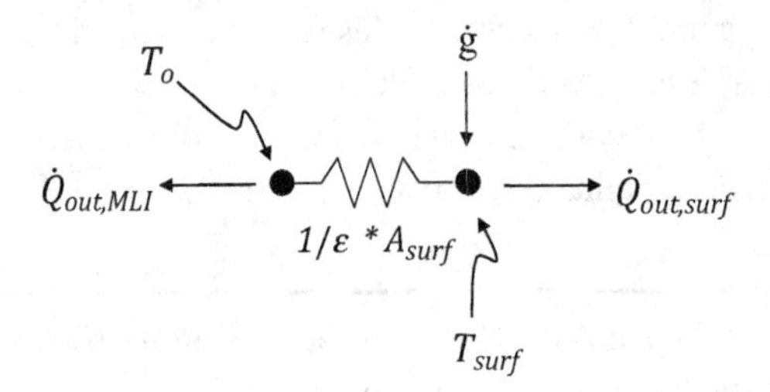

Figure 4.9 MLI node coupled to spacecraft exterior structure node

or

$$50\ \text{W} = \varepsilon^* A_{surf}\sigma\left(T_i^4 - T_o^4\right) + \varepsilon\sigma A_{surf} T_i^4$$

The surface energy balance on the outer portion of the MLI becomes

$$\dot{Q}_{MLI,out} = \dot{Q}_{i-o} \tag{4.18}$$

or

$$\varepsilon\sigma A_{surf} T_o^4 = \varepsilon^* A_{surf}\sigma\left(T_i^4 - T_o^4\right) \tag{4.19}$$

A numerical analysis tool is required for solution of the unknown temperature values. Using EES as the analysis tool, the code for the solution of this problem is

```
{Flat plate with 15 layers of MLI on one side}

sigma=5.67e-8          {Boltzmann's constant: W/m^2-K^4}
g_dot=50               {Energy dissipation}
A_surf=1.0             {Surface area}
N=15                   {Number of MLI layers}
eps=0.34               {0.15 Aluminized kapton}
eps_i=0.92             {Bare plate surface}
eps_1=0.34             {MLI layers side 1}
eps_2=0.34             {MLI layers side 2}

eps_eff=1/((N+1)*(1/eps_1+1/eps_2-1))

{Energy balance on plate}
g_dot=eps_eff*A_surf*sigma*(T_i^4-T_o^4)+eps_i*A_surf*sigma*T_i^4

{Surface energy balance on outer part of MLI}
eps_1*A_surf*sigma*T_o^4=eps_eff*A_surf*sigma*(T_i^4-T_o^4)
```

The solution values are

```
A_surf=1 [m^2]    eps=0.34            eps_1=0.34
eps_2=0.34        eps_eff=0.0128      eps_i=0.92
g_dot=50 [W]      N=15                sigma=5.670E-08 [W/m^2-K]
T_i=175.4 [K]     T_o=76.54 [K]
```

Example 2 Solar Loading with No Dissipation In this case a surface with an area of 1.0 m^2 is receiving full frontal loading from the Sun and has no thermal energy dissipation. Assume the surface is coated with 3 mils of Z306 black polyurethane paint ($\alpha_S = 0.95$, $\varepsilon = 0.87$).

Solution

The energy balance would be

$$\dot{Q}_{in} = \dot{Q}_{solar} = \dot{Q}_{out}$$

If the $\dot{Q}$ terms are expanded out (recall equation 4.13), we have

$$\alpha_S S A_{surf} = \varepsilon \sigma A_{surf} T_i^4$$

$$(0.95)(1,361 \text{ W/m}^2)(1.0 \text{ m}^2) = (0.87)(5.67 \times 10^{-8} \text{ W/m}^2 \cdot \text{K}^4)(1.0 \text{ m}^2) \cdot T_i^4$$

Solving for T_i gives T_i = 402.4 K ≈ 129.4°C. If the solar absorptance value is lowered by opting to use Z93 white paint ($\alpha_S = 0.19$, $\varepsilon = 0.89$), the energy balance becomes

$$(0.19)(1,361\text{W/m}^2)(1.0 \text{ m}^2) = (0.89)(5.67 \times 10^{-8} \text{ W/m}^2 \cdot \text{K}^4)(1.0 \text{ m}^2) \cdot T_i^4$$

This modification of the surface coating obtains T_i = 267.6 K ≈ −5.4°C. Modeling this in EES with 15 layers of double-sided .15 aluminized kapton (ε = 0.34) covering the anti-Sun side of the flat plate gives a corresponding EES code of

```
{Flat plate receiving solar loading with 15 layers of MLI on anti-Sun side}

sigma=5.67e-8                {Boltzmann's Constant: W/m^2-K^4}
S=1361                       {Solar constant}
A_surf=1.0                   {Surface area}
N=15                         {Number of MLI layers}
eps=0.34                     {0.15 Aluminized kapton}
alpha_S=0.19                 {Solar absorptivity for Z93 white paint}
eps_i=0.89                   {Emissivity for Z93 white paint}
eps_1=0.34                   {MLI layers side 1}
eps_2=0.34                   {MLI layers side 2}

eps_eff=1/((N+1)*(1/eps_1+1/eps_2-1))

{Energy balance on plate}
S*A_surf*alpha_S=eps_eff*A_surf*sigma*(T_i^4-T_o^4)+eps_i*A_surf*&
  sigma*T_i^4

{Surface energy balance on outer part of MLI}
eps_1*A_surf*sigma*T_o^4=eps_eff*A_surf*sigma*(T_i^4-T_o^4)
```

The solution values are

alpha_S=0.19	A_surf=1 [m^2]	eps=0.34
eps_1=0.34	eps_2=0.34	eps_eff=0.0128
eps_i=0.89	N=15	S=1361 [W/m^2]
sigma=5.670E-08 [W/m^2-K^4]	T_i=266.6 [K]	T_o=116.4 [K]

4.5.2 Surface Area Projections

For space platforms, environmental heating due to incident solar thermal energy must take into consideration the projected surface area that is in Sun view. This is determined via the γ angle between the normal for the surface in question and the solar constant (see Figure 4.10).

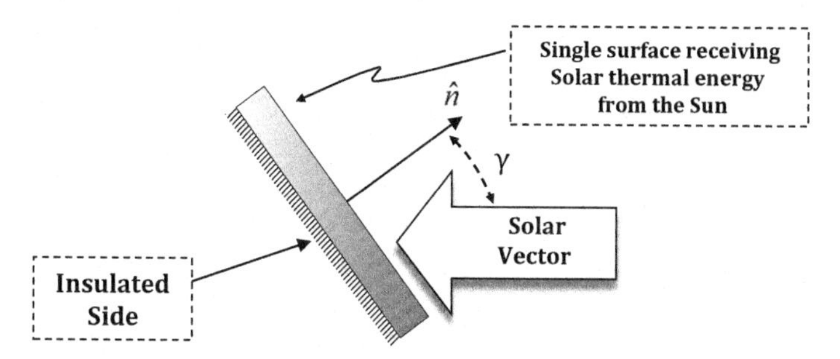

Figure 4.10 Relationship between the solar constant and the projected surface area

The projected surface area (A_p) may be calculated as

$$A_p = A_{surf}\cos\gamma \tag{4.20}$$

In light of this technique, we now revisit the calculation for the Z306 black 1.0 m^2 surface with ideal MLI on one side.

Example 3 Solar Loading of a Projected Surface Area with No Dissipation Assuming zero thermal dissipation, a γ angle of 60° and 20 layers of double-sided .15 aluminized kapton ($\varepsilon = 0.34$) covering the anti-Sun side of the flat plate, determine the plate temperature.

Solution

The energy balance would now take the form of the following

$$\alpha_S S A_p = \varepsilon\sigma A_{surf}\left(T_i^4 - T_o^4\right) + \varepsilon\sigma A_{surf} T_i^4$$

or

$$\alpha_S S A_{surf}\cdot\cos 60^\circ = \varepsilon\sigma A_{surf}\left(T_i^4 - T_o^4\right) + \varepsilon\sigma A_{surf} T_i^4$$

The energy balance on the outer surface of the MLI would be

$$\varepsilon_{MLI}\sigma A_{surf} T_o^4 = \varepsilon\sigma A_{surf}\left(T_i^4 - T_o^4\right)$$

Modeling this in EES would give this code for the analysis

```
{Flat Plate receiving solar loading with 20 layers of MLI on anti-Sun side}

sigma=5.67e-8                {Boltzmann's Constant}
S=1361                       {Solar constant}
Beta=60                      {Beta angle in degrees}
A_proj=A_surf*cos(Beta)      {Projected absorbing surface area}
```

```
A_surf=1.0                  {Surface area}
N=20                        {Number of MLI layers}
eps=0.34                    {0.15 Aluminized kapton}
alpha_S=0.19                {Solar absorptivity for Z93 white paint}
eps_i=0.89                  {Emissivity for Z93 white paint}
eps_1=0.34                  {MLI layers side 1}
eps_2=0.34                  {MLI layers side 2}

eps_eff=1/((N+1)*(1/eps_1+1/eps_2-1))

{Energy balance on plate}
S*A_proj*alpha_S=eps_eff*A_surf*sigma*(T_i^4-T_o^4)+eps_i*&
  A_surf*sigma*T_i^4

{Surface energy balance on outer part of MLI}
eps_1*A_surf*sigma*T_o^4=eps_eff*A_surf*sigma*(T_i^4-T_o^4)
```

On execution of the code, the solution set in EES would be

```
alpha_S=0.19            A_proj=0.5 [m^2]    A_surf=1 [m^2]
Beta=60 [Degrees]       eps=0.34            eps_1=0.34
eps_2=0.34              eps_eff=0.009753    eps_i=0.89
N=20                    S=1361 [W/m^2]      sigma=5.670E-08 [W/m^2-K^4]
T_i=224.4 [K]           T_o=91.7 [K]
```

Example 4 Spinning Plate with Solar Loading If the temperature for a plate that is spinning in and out of Sunview is to be calculated, the average surface area over a full revolution (i.e., 0 to 2π radians) needs to be determined. This would include integrating the projected surface area over a full revolution

$$A_{p,average} = \frac{1}{2\pi}\int_0^{2\pi} A_p d\gamma \tag{4.21}$$

Substituting in the relation for the projected surface area (i.e., equation 4.20) obtains

$$A_{p,average} = \frac{1}{2\pi}\int_0^{2\pi} A\cos\gamma d\gamma \tag{4.22}$$

The zero radians position is with the surface normal facing away from the solar vector. The integrand then goes to zero when the surface normal rotates between $\pi/2$ and $3\pi/2$ (i.e., the active surface is out of Sunview). Thus the integral limits can be reduced and the $A_{p,average}$ relation rewritten as

$$A_{p,average} = \frac{A}{2\pi} \int_{3\pi/2}^{\pi/2} \cos\gamma d\gamma = \frac{A}{\pi} \quad (4.23)$$

Using this new average area for the flat plate in Example 4.3, the plate temperature can be determined using the following EES code.

```
{Spinning flat plate with ideal insulation receiving solar loading on
  one side}

sigma=5.67e-8                  {Boltzmann's constant: W/m^2-K^4}
S=1361                         {Solar constant}
A_surf=1.0                     {m}
A_proj=A_surf/pi               {Projected absorbing surface area}
alpha_S=0.19                   {Solar absorptance for Z93 white paint}
eps_i=0.87                     {Emissivity of Z93 white paint}

{Energy balance on plate}
S*A_proj*alpha_S=eps_i*A_surf*sigma*T_i^4
```

When solved for the temperature, the plate has now reduced to T_i = 202.1 K.

4.6 Analysis Methodologies

Both transient and steady-state analysis methodologies are used for satellite on-orbit temperature predictions. Transient modeling is performed when knowledge of the time-based temperature variation of an item between two discrete locations during a single orbit is required. Steady-state modeling is chosen when determining the maximum and minimum (i.e., Max and Min) temperatures on orbit. In the steady-state case two approaches are used to predict on-orbit temperatures for spaceflight components:

1) Determine the Max and Min temperatures using orbit average environmental loads and component Max and Min thermal dissipations at orbit hot and cold case extremes.
2) Determine temperatures at worst-case hot (Max) and worst-case cold (Min) locations experiencing respectively maximum and minimum environmental thermal loading and component thermal dissipation.

Approach 1) is typically chosen with computer analysis. Approach 2) is often used with basic hand calculations for RTM designs. A key aspect of both approaches is the determination of the Max and Min locations. For the simple surface example problems, the flat plate was placed in LEO with an inclination of 90° relative to the ecliptic plane (i.e., 66.6° inclination relative to equatorial plane) and β = 0°. When viewed from the Sun the orbit path is that of the dotted line shown in Figure 4.11. Figure 4.12 shows the orbit path along directional view A–A.

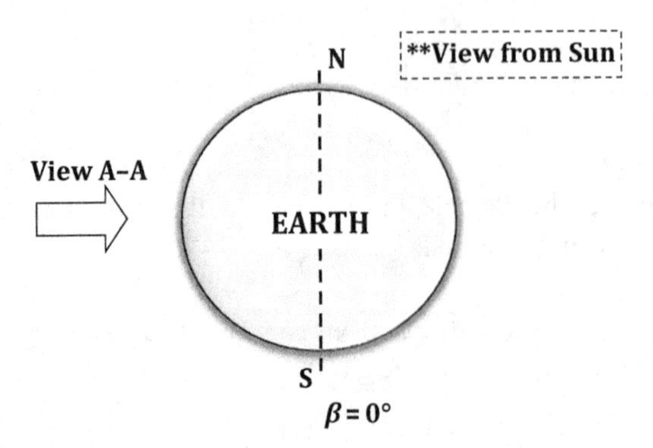

Figure 4.11 Simple surface orbit path viewed from the Sun

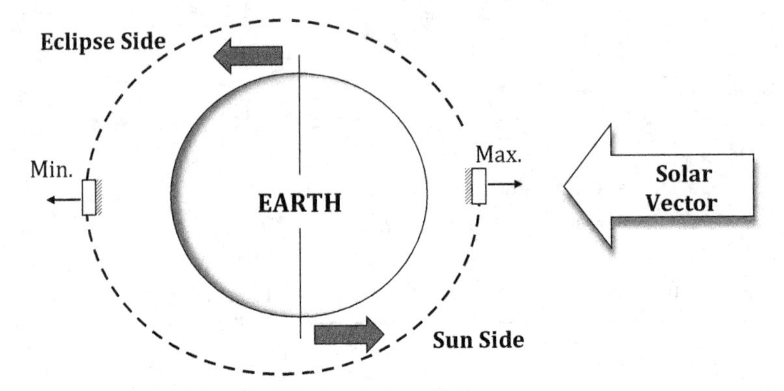

Figure 4.12 A–A view of maximum and minimum orbit positions (shown with orbit inclination of 66.6° relative to the equatorial plane and $\beta = 0°$)

The Max and Min locations may be considered as being in full Sunview and in full eclipse. Note that for an orbit like that shown in Figure 4.12, with an active surface facing the Sun, the active surface would be facing deep space while in eclipse. This would produce very cold temperature conditions onboard in the absence of thermal dissipation. We now return to the flat plate case and look at Max and Min performance while in a given orbit.

Example 5 Max and Min On-Orbit Temperature Analysis for the Anti-Nadir-Facing Flat Plate The Max and Min temperatures of a 1.0 m^2 plate are to be calculated for the orbit case in Figure 4.12. The orbit period is 90 minutes. The plate is coated with Z93 white paint ($\alpha_S = 0.19$, $\varepsilon = 0.89$) on its active side. The non-active side of the plate is covered with non-ideal insulation and the plate is dissipating 30 W of thermal energy. The MLI is assumed to comprise 10 layers of 0.15 mil double-sided aluminized kapton ($\alpha_S = 0.25$, $\varepsilon = 0.34$). What are the Max and Min temperatures for the plate in the given orbit?

Solution

Max Location (Full Sunview)

Given that the normal for the plate's active surface is aligned with the solar vector, we know that $A_p = A_{surf} \cos\gamma$ which also is $A_p = A_{surf} \cos 0°$. As such, we can conclude that

$A_p = A_{surf}$. In order to determine the temperature of the plate we must perform an energy balance on it. Using general constructs, we know our energy balance will be of the form

$$\dot{g} + \dot{Q}_{in} = \dot{Q}_{out,Total}$$

where the $\dot{Q}_{out,Total}$ term will include both rejection away from the plate's active surface and heat loss through the MLI. There will be two energy balances: one for the plate and one for the MLI. The balance equation for the plate is

$$\dot{g} + \dot{Q}_{Solar} = \dot{Q}_{out,rej} + \dot{Q}_{plate-MLI}$$

Expanding the $\dot{Q}$ terms out, the equation can be rewritten as

$$\dot{g} + \alpha_S S A_{surf} = \varepsilon_{surf} A_{surf} \sigma T^4_{surf} + \varepsilon^* A_{surf} \sigma \left(T^4_{surf} - T^4_o\right)$$

Here, using a lumped capacitance approach, the surface temperature is taken as equal to that of the entire plate. T_o is the temperature of the outer surface of the MLI. Next we define the energy balance for the outer surface of the MLI facing the Earth. This relationship is

$$\dot{Q}_{in} + \dot{Q}_{plate-MLI} = \dot{Q}_{out,rej}$$

which can also be written as

$$\dot{Q}_{EIR} + \dot{Q}_{Alb} + \varepsilon^* A_{surf} \sigma \left(T^4_{surf} - T^4_o\right) = \varepsilon_{MLI} A_{surf} \sigma T^4_o$$

Expanding out the $\dot{Q}$ terms gives

$$\varepsilon_{MLI} EIR_{flux} A_{surf} \mathrm{F}_{surf-Earth} + Alb \cdot \alpha_S S A_{surf} \mathrm{F}_{surf-Earth} + \varepsilon^* A_{surf} \sigma \left(T^4_{surf} - T^4_o\right) = \varepsilon_{MLI} A_{surf} \sigma T^4_o$$

With the exception of the temperatures EIR_{flux}, Alb and $\mathrm{F}_{surf-Earth}$, all the inputs to these equations are known. Since the plate in question is in a LEO orbit with a known orbit period of 90 minutes, the altitude can be determined using equation 4.6. Given knowledge of the altitude, the view factor can be calculated using VF9 from Appendix A. This results in a value of $\mathrm{F}_{surf-Earth} = 0.9196$. However, to determine the albedo factor (Alb) and the Earth-IR flux, we must reference the empirical data located in the Earth-IR tables in Appendix C. For the active MLI surface facing the Earth, our α_S/ε ratio is less than 1, therefore the IR-sensitive portion of Table C.2 applies for an inclination of 66.6° relative to the equatorial plane. For the inclination angle and orbit period specified, the $EIR_{flux} = 250\ \mathrm{W/m^2}$. The corresponding albedo factor is 0.26. Placing these in the energy balance equations for the plate and the MLI surface gives

Plate

$$30\ \mathrm{W} + (0.19)\left(1{,}361 \frac{\mathrm{W}}{\mathrm{m}^2}\right)(1.0\ \mathrm{m}^2) = (0.89)(1.0\ \mathrm{m}^2)\sigma T^4_{surf} + \varepsilon^* (1.0\ \mathrm{m}^2)\sigma \left(T^4_{surf} - T^4_o\right)$$

MLI Surface

$$(0.92)\left(1.0\ \mathrm{m}^2\right)\left[(0.34)\left(250\frac{\mathrm{W}}{\mathrm{m}^2}\right)+(0.26)(0.25)\left(1,361\frac{\mathrm{W}}{\mathrm{m}^2}\right)\right]$$
$$+\varepsilon^*\left(1.0\ \mathrm{m}^2\right)\sigma\left(T_{surf}^4-T_o^4\right)=(0.34)\left(1.0\ \mathrm{m}^2\right)\sigma T_o^4$$

where $\sigma = 5.67 \times 10^{-8}\ \mathrm{W/m^2 \cdot K^4}$ and $\varepsilon^* = 0.01862$. Solving these equations simultaneously in EES we have

```
{Anti-Nadir-facing flat plate w/MLI on non-active side, max case}
sigma=5.67e-8 [W/m^2-K^4]
g_dot=30                    [W]
Solar=1361                  [W/m^2]
EIR_Hot=250                 {Hot case Earth-IR flux}
EIR_Cold=193                {Cold case Earth-IR flux}
A=1.0 [m^2]
Alb_H=0.26
F_surf_earth=0.92
N=10
eps_1=0.34                  {Aluminized kapton}
eps_2=0.34                  {Aluminized kapton}
alpha_S_MLI=0.25            {Absorptivity of aluminized kapton}
eps_surf1=0.89              {Z93 white paint}
alpha_surf1=0.19            {Z93 white paint}
eps_surf2=0.84              {Polyurethane white paint}

estar=1/((N+1)*(1/eps_1+1/eps_2-1))

{Energy balance at the plate}
alpha_surf1*Solar*A+g_dot=estar*A*sigma*(T_surf^4-T_o^4)+&
  eps_surf1*A*sigma*T_surf^4

{Energy balance on the outer MLI surface}
eps_1*A*EIR_Hot*F_surf_earth+Alb_H*alpha_S_MLI*Solar*A*F_surf_&
  earth+estar*A*sigma*(T_surf^4-T_o^4)=eps_1*A*sigma*T_o^4
```

Executing the program gives $T_{surf} = 275.6\ \mathrm{K} \approx 2.6°\mathrm{C}$ and $T_o = 300.4\ \mathrm{K} \approx 27.4°\mathrm{C}$.

Min Location (Full Eclipse with No Solar Heating or Albedo)

In full eclipse where the active surface is facing deep space, there is no direct environmental heat input to the active surface. However, the energy generation previously designated will be present, as well as some Earth-IR thermal loading that will be incident on the outer surface of the MLI. The energy balance for the plate is

$$\dot{g}=\varepsilon^* A_{surf}\sigma\left(T_{surf}^4-T_o^4\right)+\varepsilon_{surf}A_{surf}\sigma T_{surf}^4$$

The corresponding energy balance relation for the MLI will be

$$\dot{Q}_{EIR} + \varepsilon^* A_{surf}\sigma\left(T_{surf}^4 - T_o^4\right) = \varepsilon_{MLI}A_{surf}\sigma T_o^4$$

Expanding out the $\dot{Q}$ terms, the MLI energy balance relation becomes

$$\varepsilon_{MLI}EIR_{flux}A_{surf}F_{surf-Earth} + \varepsilon^* A_{surf}\sigma\left(T_{surf}^4 - T_o^4\right) = \varepsilon_{MLI}A_{surf}\sigma T_o^4$$

As for the hot case, we can substitute in values that are known to solve for the temperatures. The values for σ and ε^* are the same. However, EIR_{flux} for the eclipse location needs to be determined, using Appendix Table C.1. For an IR-sensitive surface in a 90-minute orbit with an approximate inclination angle of 66.6°, the data shows a flux of 193 W/m^2. Substituting these values into the plate and MLI energy balance relations will give the following relations as the final form

Plate

$$30\text{ W} = \varepsilon^* \cdot \left(1.0\text{ m}^2\right)\sigma\left(T_{surf}^4 - T_o^4\right) + (0.89)\left(1.0\text{ m}^2\right)\sigma T_{surf}^4$$

MLI Surface

$$(0.34)\left(193\frac{\text{W}}{\text{m}^2}\right)(0.92)\left(1.0\text{ m}^2\right) + \varepsilon^*\left(1.0\text{ m}^2\right)\sigma\left(T_{surf}^4 - T_o^4\right) = (0.34)\left(1.0\text{ m}^2\right)\sigma T_o^4$$

The temperature of the plate and the outer surface of the MLI can be determined by substituting in these energy balance equations for the plate and the MLI in the EES code for the Max case. The corresponding syntax is

```
{Energy balance at the plate}
g_dot=estar*A*sigma*(T_surf^4-T_o^4)+eps_surf1*A*sigma*T_surf^4

{Energy balance on the outer MLI surface}
eps_1*A*EIR_Cold*F_surf_earth+estar*A*sigma*(T_surf^4-&

T_o^4)= eps_1*A*sigma*T_o^4
```

On execution of the code we get T_{surf} = 159.3 K ≈ −113.7°C, T_o = 234.1 K ≈ −38.9°C. One of the non-intuitive constructs in this approach was the incorporation of heat transfer from the Earth to the spacecraft. The Earth is typically taken to have an equivalent sink temperature between −15°C and −23°C, making the spacecraft typically warmer than the Earth. However, the approach used here implies net heat transfer from the Earth to the spacecraft. Why does this occur? The view factor to Earth was ignored in the heat rejection portion of the calculation. Another way to interpret this is that we assumed the Earth does not block the simple surface's view to space and added the difference in surface IR energy back in as heating from the Earth.

Example 6 Transient On-Orbit Temperature Analysis for the Anti-Nadir-Facing Flat Plate In this example, temperature as a function of time is to be calculated and plotted for the 1.0 m^2 flat plate in orbit shown in Figure 4.12. The orbit period is 90 minutes and the plate has an active surface coating of Z93 white paint ($\alpha_S = 0.19$, $\varepsilon = 0.89$) and ideal insulation on the non-active side.

Solution

We will use EES to calculate the temperature of the plate as a function of time during orbit. We can define the constants in the main program, the geometry and thermophysical properties of the plate with the following

```
{Transient analysis of anti-nadir-facing flat plate in orbit
  with ideal MLI on back side}

{Main program}
sigma=5.68e-8                    {Boltzmann's Constant}
Solar=1361                       {Solar constant}

{Plate/radiator features}
l_rad=1.0
w_rad=1.0
thick_rad=0.0254
A_surf=l_rad*w_rad
Vol_rad=l_rad*w_rad*thick_rad
rho=2700.0
c_p=896.0
```

The orbit statistics are defined by dividing the circular orbit into 72 positions.

```
{Orbit statistics}
Orbit_Periodm=90          {Orbit period in minutes}
Orbit_Periods=90*60       {Orbit period in seconds}
Orbit_Degrees=360         {Degrees in orbit}
Delta_Degrees=5
Del_T=Delta_Degrees*Orbit_Periods/Orbit_Degrees
INK=Orbit_Degrees/5
Revs=10.0                 {#Revolutions in transient analysis}
Full=Revs*INK
```

The optical properties for the Z93 white paint surface coating and thermal dissipation are

```
{Surface coatings}
alpha_S=0.19; eps=0.89;

{Plate dissipation}
g_dot=0.0
```

Since thermal loads and temperatures at each orbit position will be bookkept, arrays will be used to capture values as a function of the orbit time. The initial values for each of the primary variables are defined using the following syntax

```
{Initial array settings}
Ang[1]=0
T[1]=273
Time[1]=0.0
Time_min[1]=0.0

Q_in[1]=alpha_S*A_surf*F_surf_earth*Solar
```

To determine the temperature of the plate at each orbit position a transient analysis is required. An iterative loop is set up to perform calculations at each timestep using the "duplicate" command. The iterative loop will consist of the following code

```
Duplicate K=1,FULL
  Ang[K+1]=Ang[K]+5
  Time[K+1]=Time[K]+75.0
  Time_min[K+1]=Time[K+1]/60

  z[K]=floor(K/INK)
  iter[K]=K-z[K]*INK

  Q_in[K+1]=test(iter[K],Solar,g_dot)

  T[K+1]=(Q_in[K]-eps*sigma*A_surf*((T[K])^4))*(Del_T/&
    (rho*Vol_rad*c_p))+T[K]
End
```

The temperature equation is derived from the energy balance on the plate, which can be modeled as

$$\rho \cdot Vol \cdot c_p \frac{dT}{dt} = \dot{Q}_{in} - \varepsilon\sigma A T^4 \tag{4.24}$$

where $\dot{Q}_{in}$ is the sum of energy dissipations and environmental thermal loads on the plate. For the derivative of the temperature with respect to time a forward difference scheme is applied

$$\frac{dT}{dt} = \frac{T^{p+1} - T^p}{\Delta t} \tag{4.25}$$

where p denotes the present timestep and $p+1$ denotes the next value moving forward in time. Using an implicit approach to the definition of the temperature values on the RHS of equation 4.24 and solving for T^{p+1} gives the temperature relation shown in the EES code. To solve this problem we need to incorporate the environmental heating profile as a function of position (and/or time) during orbit by calculating the thermal load at each orbit position based on the normal direction of the plate's active surface. This requires conditional statements for the appropriate amount of environmental loading. Since EES does not allow conditional statements in the main body of program syntax, a separate function containing the conditional statements will be created and called from within the iterative loop. The function, which will be placed at the top of the code and before the main body of the program, is

```
{A is iter, B is Solar, C is g_dot and Y is Q_in}
Function test(A,B,C)
        If (A<18) Then
        Y:=B* cos(A* 5)+C
        Endif

        If (A>54) Then
        Y:=B* cos((72-A)*5)+C
        Endif

        If (A>17) and (A<55) Then
        Y:=C
        Endif

        test:=Y
End
```

Figure 4.13 is a plot of the plate temperature profile spanning 10 revolutions. The plate temperature reaches harmonic equilibrium after 800 minutes with a maximum value of 315 K and a minimum of 292 K (a temperature swing of 23 K). Increasing the thermal mass of the object will dampen the temperature swing.

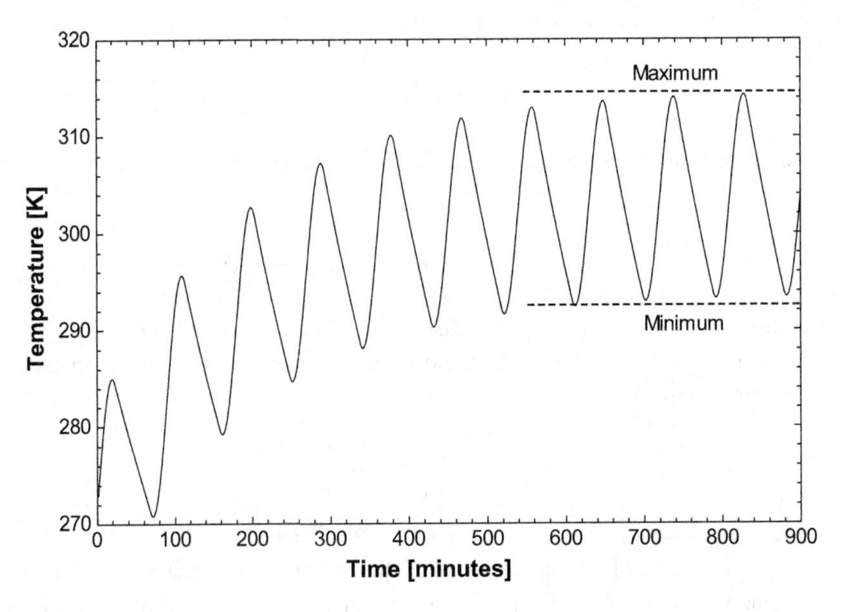

Figure 4.13 Transient vs. orbit time for an anti-nadir-facing flat plate in orbit

4.7 Reduced Node SpaceCube Analysis

An additional level of complexity arises if the simple surface in orbit is a hollow SpaceCube with a wall thickness of $th = 0.015$ m and one E-box inside (Figure 4.14).

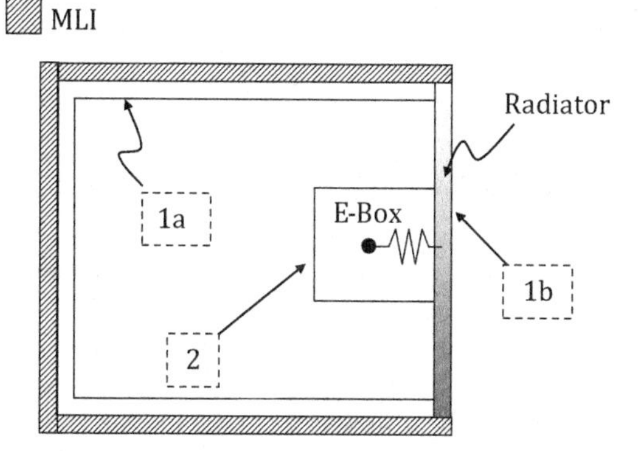

Figure 4.14 SpaceCube configuration

The following assumptions will be applied to the SpaceCube:

- The E-box will be treated as if it were a solid with a side length of 0.4 m.
- The E-box has a thermal energy dissipation of 60 W.
- Surfaces 1a, 1b and 2 are all gray body surfaces.
- All the MLI covered surfaces have ideal insulation (i.e., $\varepsilon^* = 0.0$).
- All materials are made of 6061 aluminum.
- The effective radiator surface (1b) is part of the larger enclosure structure.

For radiative heat exchange, heat transfer is dependent upon surface properties. The convention of designating surfaces a and b (i.e., 1a and 1b) allows the surface properties to be distinguished from one another. In addition, surface 1b which is serving as the radiator is the only part of the SpaceCube in thermal communication with the boundary environment (i.e., space). Regardless of the surface property designations for the structure, from a conduction standpoint surfaces 1a and 1b are part of an assembled metal structure. In this simple model, the structure itself will have a single node and a single temperature.

In this example, the SpaceCube is in an orbit similar to the one in Example 5, but this time the active surface (i.e., the radiator) is Nadir facing. The SpaceCube and the viewing direction of the active radiator surface at the Max and Min locations are shown in Figure 4.15 (MLI covering the exterior of the SpaceCube is not shown).

To solve this problem we need to determine the thermal couplings inherent to the thermal circuit, which will aid in later establishing the energy balance relations at each of the nodes. Figure 4.16 shows the thermal circuit diagram, with the radiative path for heat flow from the embedded node (i.e., node 2) to the radiator highlighted. The conductive couplings are not included explicitly because the units for the radiative and conductive resistances are different. Conductive heat flows are therefore denoted by

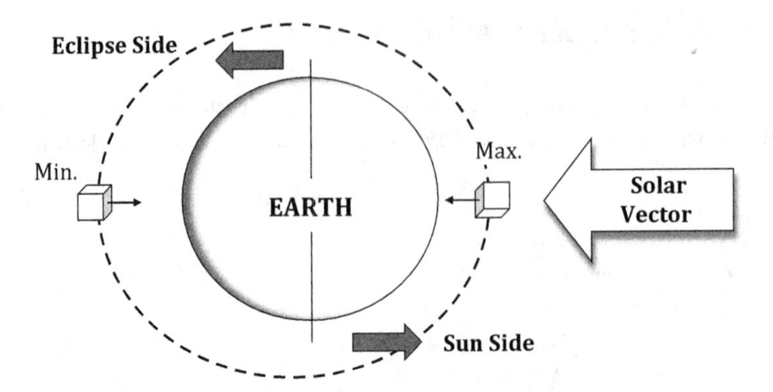

Figure 4.15 Maximum and minimum orbit positions for SpaceCube (shown with orbit inclination of 66.6° relative to the ecliptic plane and $\beta = 0°$)

directional arrows. It is assumed that the heat flow is from the interior to the exterior of the SpaceCube. The conduction from the E-box to node 1 (i.e., the spacecraft structure) may be captured as

$$\dot{q}_{21,cond} = \frac{(T_2 - T_1)}{R_{2-1}} \tag{4.26}$$

where R_{2-1} is the resistance to heat flow between node 2 and node 1. At steady state, the thermal energy entering and dissipated on the SpaceCube structure has to equal that which is rejected to ambient by way of the radiator. Thus, the heat dissipating at node 2 (the E-box) eventually has to work its way to the rejection surface 1b (the radiator). Surface 1b has been designated as the only surface in thermal communication with the ambient environment (recall that the other 5 exterior surfaces on the SpaceCube have all been covered with ideal MLI).

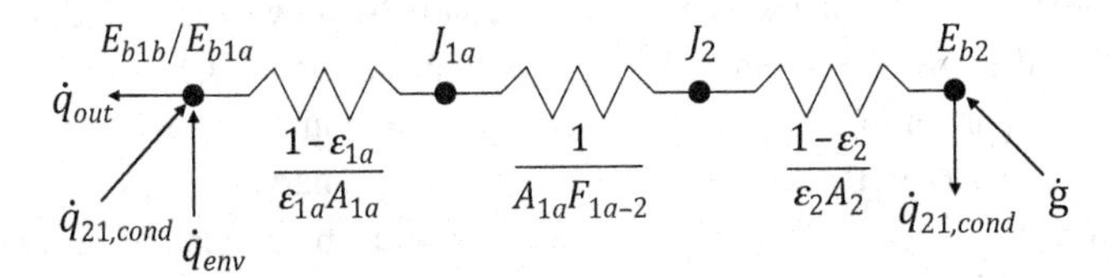

Figure 4.16 SpaceCube thermal couplings for 60 W energy dissipation case

To determine the temperatures throughout the thermal circuit, an energy balance is performed at the heat rejection surface, working through the circuit from outside to in. The goal is to determine temperatures at both nodes 1 and 2. Since the radiator is part of the SpaceCube structure, the structure temperature is determined first. The environmental energy balance (i.e., the balance at node 1, Figure 4.17) is

$$\dot{Q}_{env} + \dot{q}_{21,total} = \dot{Q}_{out} \tag{4.27}$$

where $\dot{q}_{21,total} = \dot{q}_{21,rad} + \dot{q}_{21,cond}$ (i.e., the total summation of radiative and conductive heat transfer between nodes 2 and 1). Both radiative and conductive heat transfer arriving at node 1 is leaving node 2. Given the present assumption of 60 W thermal energy dissipation at node 2, the summation of the thermal energy leaving node 2 must be equal to 60 W (Figure 4.18). Thus, the previous equation (and/or energy balance at node 1) may be rewritten as

$$\dot{Q}_{env} + \dot{g} = \dot{Q}_{out} \tag{4.28}$$

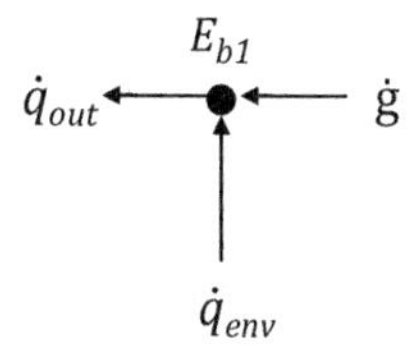

Figure 4.17 Energy balance at node 1

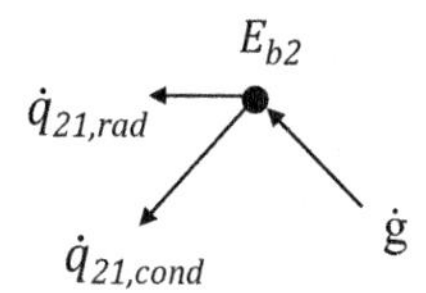

Figure 4.18 Energy balance at node 2

The environmental heat load term will either be due to Earth-IR or solar inputs (specifically albedo). Equation 4.28 can also be written as

$$\dot{Q}_{env} + \dot{g} = \varepsilon_{1b} A_{1b} \sigma T_1^4 \tag{4.29}$$

Using this equation we can find T_1 and place it into the equation for the energy balance at node 2 to determine T_2. Similar to the flat plate cases from Section 4.5, we will assume the radiator surface area on the structure (i.e. surface 1b) is 1.0 m^2. Assuming that surface 1b is coated with Z93 white paint, EES code can be used to determine the temperature of the SpaceCube structure.

```
{Max temperature determination for nadir-facing SpaceCube
  dissipating 60 W}

sigma=5.67e-8               {Boltzmann's constant: W/m^2-K^4}
S=1361                      {Solar constant}
Alb_H=0.36                  {Including inclination correction}
EIR_Hot=250
EIR_Cold=193

th=0.015                    {SpaceCube wall thickness}
g_dot=60                    {E-box dissipation}

{Radiator surface area}
A_1b=1.0
```

```
{View factor to Earth}
F_surf_earth=0.9196

{Coatings}
eps_1a=0.02 {VDA}
eps_2=0.87              {Z306 Black polyurethane paint}
eps_1b=0.89             {Z93 White paint}
alpha_1b=0.19           {Z93 White paint solar absorptance}

{Energy balance on SpaceCube}
g_dot+Alb_H*alpha_1b*S*A_1b*F_surf_earth+eps_1b*A_1b*&
  EIR_Hot*F_surf_earth=eps_1b*A_1b*sigma*T_1^4
```

The temperature values for the structure in the hot and cold cases will then be T_1 = 288.5 K = 15.5°C and T_1 = 260.3 K = −12.7°C respectively. Next an energy balance must be performed at node 2. The equation for the energy balance at node 2 is

$$\dot{g} = \dot{q}_{21,total} = \dot{q}_{21,rad} + \dot{q}_{21,cond} \tag{4.30}$$

The $\dot{Q}$ terms can be expanded out and the equation rewritten as

$$\dot{g} = \frac{\sigma\left(T_2^4 - T_1^4\right)}{\left[\frac{1-\varepsilon_{1a}}{\varepsilon_{1a}A_{1a}} + \frac{1}{A_{1a}F_{1a-2}} + \frac{1-\varepsilon_2}{\varepsilon_2 A_2}\right]} + \frac{(T_2 - T_1)}{R_{2-1}} \tag{4.31}$$

In EES the corresponding syntax is

```
{E-Box (Node 2) Energy balance}
g_dot=(T_2-T_1)/R_2_1+sigma*(T_2^4-T_1^4)/Rad_Res

Rad_Res=(1-eps_1a)/(eps_1a*A_1a)+1/(A_2*F_2_1a)+(1-eps_2)/(eps_2*A_2)
```

However, a cursory inspection of equation 4.31 reveals that there are three unknowns that must be defined prior to solving for T_2. These are the active surface area for the SpaceCube interior wall (A_{1a}), the active surface area of the E-Box (A_2) and the conductive thermal resistance between the E-box and the SpaceCube structure (R_{2-1}). These unknown variables can be defined in EES using the following.

```
E_Box_length=0.4       [m]
E_Box_side=E_Box_length^2
A_2=5*(E_Box_side)
s_length=1.0            [m]    {SpaceCube exterior wall length}
in_length=s_length-2*th
A_1a=(in_length^2)-E_Box_side+5*(in_length^2)
```

For the E-box's active surface area only five sides of the cubic structure are included because one side is intimately coupled to the interior surface of the enclosure. To determine the SpaceCube interior's active surface area, the wall thickness is subtracted from the exterior wall length on either side, giving the interior wall length. The square of this value is multiplied by a factor of five to account for the surfaces not in contact with the E-box. This value is summed with the surface area for the interior wall coupled to the E-box that remains unobstructed by the mating surface in contact with the E-box. Using the methodologies for conduction-based resistance calculations (see Chapter 2) R_{2-1} can be determined with the following EES code

```
k_6061=155.3                              [ W/m-K]

R_2_1=0.5*E_Box_length/(k_6061*E_Box_side)+0.5*th/(k_6061*6*in_length^2)
```

Finally the view factor between surfaces 1a and 2 must be determined (using the rules pertaining to summation of surfaces in an enclosure summarized in Section 3.4.2). We know that

$$F_{2-1a} + F_{2-2} = 1.0 \tag{4.32}$$

However, since $F_{2-2} = 0.0$, we can conclude that $F_{2-1a} = 1.0$. Also, by reciprocity, we know that

$$A_2 F_{2-1a} = A_{1a} F_{1a-2} \tag{4.33}$$

Rearranging terms, we can establish a relation for F_{1a-2} as

$$F_{1a-2} = \frac{A_2}{A_{1a}} F_{2-1a} \tag{4.34}$$

The corresponding EES code will be

```
{View factors}
F_2_1a+F_2_2=1.0
F_2_2=0.0
F_1a_2=(A_2/A_1a)*F_2_1a
```

These relations will result in F_{1a-2} being 0.1458. Having determined all the unknown variables, we can now determine T_2 using equation 4.31. Substituting the known values into equation 4.31 gives

$$60\ \text{W} = \dot{g} = \frac{\left(5.67 \times 10^{-8}\ \text{W/m}^2 \cdot \text{K}^4\right)\left[T_2^4 - T_1^4\right]}{\left[\dfrac{1-0.02}{0.02 \cdot (5.0\ \text{m}^2)} + \dfrac{1}{5.0\ \text{m}^2 \cdot (0.16)} + \dfrac{1-0.87}{0.87 \cdot (0.8\ \text{m}^2)}\right]} + \frac{(T_2 - T_1)}{R_{2-1}} \tag{4.35}$$

Since T_2 has both first-order and fourth-order terms in the relation, the T_2 term cannot be isolated, so a numerical solver or a trial-and-error technique must be used to find it. Using the code we have developed for this problem, we can solve for the max (hot) case to determine that $T_1 = 288.5$ K and $T_2 = 289.0$ K. Changing the environmental heat load term in the code for the cold case to only include Earth-IR for the surface sensitivity of the Z93 white paint, the calculation can be repeated for the min (cold) case. Temperature values for the cold case are $T_1 = 260.3$ K and $T_2 = 260.8$ K. Due to their geometry and structure material, the E-box and structure temperatures for the two-node SpaceCube model are fairly similar. The thermal resistance between node 2 and node 1 was approximately 0.008057 K/W, which corresponds to an extremely high thermal conductance (approximately 125 W/K). The temperature of node 2 is therefore likely to follow the temperature at node 1. In contemporary spacecraft thermal design, the temperature-controlled item is not always in intimate contact with the inboard side of the platform's radiator. The conductive coupling between the radiator and the temperature-controlled item will also not always be as strong as the one modeled in the SpaceCube example.

The thermal coupling between node 1 and node 2 can be reduced by the addition of a spacer made of insulating material. From the list of spacer candidate materials given in Table 2.3, three stand out as potential options: G10, Vespel and PTFE. Each material has a different thermal conductivity, so the total resistance will vary from one spacer material to the next. The spacer thickness is assumed to be 1.0 cm. The new resistance calculation can be captured by replacing the *R_2_1* equation with the following EES syntax

```
l_space=0.010              [m]          {Spacer length}

R_1=0.5*th/(k_6061*6*in_length^2)
R_2=0.5*E_Box_length/(k_6061*E_Box_side)
R_spacer=l_space/(k_ins*E_Box_side){k_ins is thermal
                                  conductivity of spacer}

R_2_1=R_1+R_2+R_spacer
```

In the new syntax for determining the total resistance between nodes 2 and 1, *R_1* is the resistance from the mid-plane of node 1 to the interface with the spacer, and *R_2* is the resistance from the center of the E-box to the interface with the spacer. The thermal conductivity for each of the materials is substituted into the variable *k_ins*. Table 4.5 summarizes the temperatures for nodes 1 and 2 in both the hot and cold cases for each candidate spacer material. The temperature gradient shown between node 2 and node 1 in the presence of a spacer implies that as the resistance to heat flow towards the sink (i.e., radiator) increases, so does the localized temperature of the source (in this case the E-box). The insulator with the largest thermal conductivity (G10) has the smallest temperature gradient in both the hot and cold cases. The material with the lowest thermal conductivity (PTFE) has the largest temperature gradient between the structure and the E-box. The G10 spacer had an effective resistance of 0.1129 K/W, whereas the Vespel and PTFE spacers had effective resistances of 0.1866 K/W and 0.237 K/W respectively. Options for reducing the temperatures in the hot case include re-orienting the active surface (the radiator) to a different FOV that results in lesser environmental loading, and/or changing the surface coating to one with a lower absorptivity and emissivity.

Table 4.5 Hot- and cold-case SpaceCube temperatures for E-box-to-radiator spacer

	Max (hot) case		Min (cold) case	
Insulator	T_1 [K]	T_2 [K]	T_1 [K]	T_2 [K]
G10	283.6	290.0	256.4	262.9
Vespel	283.6	293.8	256.4	266.8
PTFE	283.6	296.2	256.4	269.4

4.7.1 Non-ideal Space Structure MLI Performance

In real life, MLI is not a perfect insulator. When using MLI there will be some measure of thermal communication with the space environment. There are two interpretations for heat loss from a surface and/or structure insulated with MLI. The first assumes that heat loss through the MLI is uniform. Figure 4.19 illustrates uniform heat loss through the MLI used to insulate the example SpaceCube. Recall the energy balance relation for node 1, written as

$$\dot{Q}_{env} + \dot{g} = \dot{Q}_{out}$$

However, now the heat loss from the SpaceCube structure will include both that occurring through the MLI and that which is rejected from the radiator surface. Accounting for each of these explicitly, the energy balance relation is now expressed as

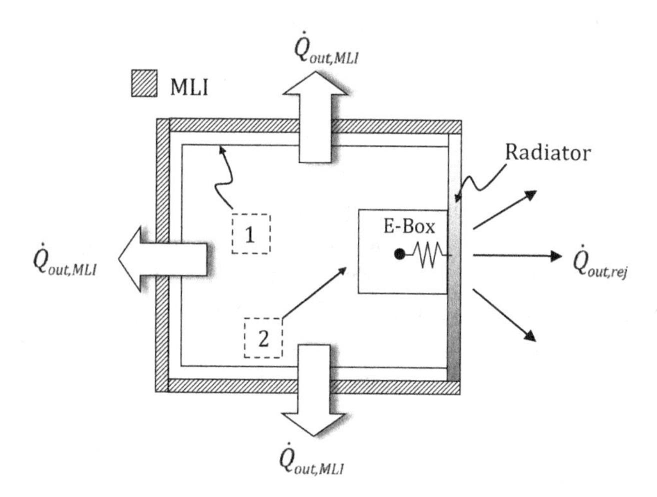

Figure 4.19 SpaceCube with uniform heat loss through MLI

$$\dot{Q}_{env} + \dot{g} = \dot{Q}_{out,MLI} + \dot{Q}_{out,rej} \tag{4.36}$$

Expanding out the radiator's heat rejection term to include the temperature at node 1 gives

$$\dot{Q}_{env} + \dot{g} = 5 \cdot \dot{Q}_{out,MLI} + \varepsilon_{1b} A_{1b} \sigma T_1^4 \tag{4.37}$$

The MLI heat loss term has a factor of 5 in front of it to account for the fact that there are five insulated surfaces. The solution for temperature T_1 would then be

$$T_1 = \left[\frac{\dot{Q}_{env} + \dot{g} - 5 \cdot \dot{Q}_{out,MLI}}{\varepsilon_{1b} A_{1b} \sigma}\right]^{1/4} \tag{4.38}$$

This approach assumes that any environmental loading on the exterior of the MLI surfaces will also be uniform, so equations 4.37 and 4.38 apply. If thermal loading is not uniform, temperature predictions obtained using this approach may be inaccurate.The second interpretation does not assume uniform heat loss through the MLI, nor does it assume uniform environmental thermal loading on the exterior of the MLI surfaces. A general balance relation for node 1 under this interpretation would be

$$\dot{Q}_{env} + \dot{g} = \sum_{n=1}^{N} \dot{Q}_{out,MLI,n} + \varepsilon_{1b} A_{1b} \sigma T_1^4 \tag{4.39}$$

where the MLI heat rejection is the sum of the MLI heat rejections from N MLI surfaces.

Example 7 SpaceCube On-Orbit Analysis with Non-ideal MLI Having Non-uniform Heat Loss For this problem, we can update the earlier analysis to include non-ideal MLI on the exterior surface. The MLI is assumed to consist of 10 layers of 0.15 mil double-sided aluminized kapton (α_S = 0.25, ε = 0.34). If a 1.0 cm G10 spacer is positioned between the E-box and the enclosure structure, what are the Max and Min temperatures in the given orbit?

Solution

Given the orientation of the SpaceCube, which has a Nadir-facing radiator, one MLI surface (the one opposite the radiator on the SpaceCube) will be receiving solar thermal loading and the other four surfaces will be receiving Earth-IR loading edgewise. To determine the Earth-IR loading that is occurring on the surfaces with an edgewise view of the planet, we need to know the view factor. Given the diameter of the planet and the orbit period for the SpaceCube (90 minutes), we can determine the altitude of the SpaceCube using equation 4.6 and incorporate it into the relation for VF10 in Appendix A to determine the view factor. This results in a value of 0.322 for the view factor. Turning to the EES code for the Max and Min cases, each case must be updated to account for heat loss and/or loading through the MLI to and from the structure, as well as thermal loading on the exterior (i.e., outboard) surface. The thermal loads for the MLI surfaces in the Max case are

```
{MLI Exterior Surface Thermal Loads}
A_MLI=1.0 [m^2]
alpha_S_MLI=0.25          {Absorptivity of aluminized kapton}
F_MLI_Earth=0.322         {Edgewise MLI to Earth view factor}
```

```
Q_MLI1=alpha_S_MLI*S*A_MLI                {Sun-viewing surface}

Q_MLI_side=eps_MLI*A_MLI*F_MLI_Earth*EIR_Hot+Alb_H*&
  alpha_S_MLI*S*F_MLI_Earth*A_MLI

Q_MLI2=Q_MLI_side
Q_MLI3=Q_MLI_side
Q_MLI4=Q_MLI_side
Q_MLI5=Q_MLI_side
```

The code for the MLI effective emissivity will be

```
{Non-Ideal MLI Definition}
N=10                                        {Number of MLI shields}
eps_MLI=0.34                                {Aluminized kapton}
estar=1/((N+1)*(1/eps_MLI+1/eps_MLI-1))     {Effective emissivity}
```

The energy balance equations for determination of the MLI surface temperatures will be

```
{MLI Surface Energy Balances}
T_MLI1^4=Q_MLI1+estar*A_MLI*sigma*(T_1^4-T_MLI1^4)/(eps_MLI*A_MLI*sigma)
T_MLI2^4=Q_MLI2+estar*A_MLI*sigma*(T_1^4-T_MLI2^4)/(eps_MLI*A_MLI*sigma)
T_MLI3^4=Q_MLI3+estar*A_MLI*sigma*(T_1^4-T_MLI3^4)/(eps_MLI*A_MLI*sigma)
T_MLI4^4=Q_MLI4+estar*A_MLI*sigma*(T_1^4-T_MLI4^4)/(eps_MLI*A_MLI*sigma)
T_MLI5^4=Q_MLI5+estar*A_MLI*sigma*(T_1^4-T_MLI5^4)/(eps_MLI*A_MLI*sigma)
```

where the MLI temperature in each equation is for its exterior surface. The energy balance on the SpaceCube structure will now become

```
{Energy balance on SpaceCube}
g_dot+Alb_H*alpha_1b*S*A_1b*F_surf_earth+eps_1b*A_1b*EIR_Hot*&
F_surf_earth=estar*A_MLI*sigma*(T_1^4-T_MLI1^4)+estar*A_MLI*&
sigma*(T_1^4-T_MLI2^4)+estar*A_MLI*sigma*(T_1^4&
-T_MLI3^4)+estar*A_MLI*sigma*(T_1^4-T_MLI4^4)+estar*&
A_MLI*sigma*(T_1^4-T_MLI5^4)+eps_1b*A_1b*sigma*T_1^4
```

where the "&" denotes the line wrap feature. For the min (cold) case, there is no longer solar or albedo loading on the MLI surfaces and the radiator. The thermal load on the MLI covering the side of the SpaceCube opposite the radiator and on the lateral sides becomes

```
Q_MLI1=0.0                                  {Space-viewing surface}

Q_MLI_side=eps_MLI*A_MLI*F_MLI_Earth*EIR_Cold
```

The corresponding energy balance for the SpaceCube structure is

```
{Energy balance on SpaceCube}
g_dot+eps_1b*A_1b*EIR_Cold*F_surf_earth=estar*A_MLI*sigma*&
(T_1^4-T_MLI1^4)+estar*A_MLI*sigma*(T_1^4-T_MLI2^4)+estar*&
A_MLI*sigma*(T_1^4-T_MLI3^4)+estar*A_MLI*sigma*(T_1^4-&
T_MLI4^4)+estar*A_MLI*sigma*(T_1^4-T_MLI5^4)&
+eps_1b*A_1b*sigma*T_1^4
```

Table 4.6 shows the temperature results for the SpaceCube structure and the E-box in the hot and cold cases. As shown in the table, the temperature in the non-ideal MLI model is colder than in the ideal model, and the temperature difference between the ideal and non-ideal models is greater (≈6 K) in the cold case. This suggests that the use of non-ideal MLI allowed for greater heat loss from the structure than that which occurred in the ideal MLI model.

Table 4.6 SpaceCube temperatures for G10 E-box–radiator spacer and non-ideal MLI

	Max (hot) case		Min (cold) case	
Insulator	T_1 [K]	T_2 [K]	T_1 [K]	T_2 [K]
Ideal MLI	283.6	290.0	256.4	262.9
Non-ideal MLI	277.0	283.4	250.4	256.9

However, the thermal loading associated with being in Sunview in the hot case made up for much of the temperature reduction observed in eclipse in the cold case. Furthermore, we could approximate closer to the temperature performance of the ideal MLI scenario by increasing the number of MLI layers, decreasing the emissivity of the MLI surfaces to a value lower than that for aluminized kapton (i.e., <0.34) or a combination of both. Note that the theoretical performance of non-ideal MLI does not take into account shortcomings in heat transfer performance due to penetrations and poor workmanship.

4.7.2 SpaceCube with Radiator Thermally Decoupled from Structure

In the design of spacecraft systems the radiator is typically separated from the spacecraft structure (and associated temperature-controlled items) by low thermal-conductivity spacers and/or truss structures, as well as low-emissivity coatings and MLI. The thermal break between the spacecraft structure and the radiator reduces errant thermal paths and radiative backloads, enabling better temperature control of flight components. In a more realistic example, where the radiator is actually separated from the SpaceCube structure, the radiator side of the spacecraft structure becomes its own separate node (node 3). The radiator can be thermally decoupled by using four 304 SS L-brackets to mount it to the SpaceCube (one in each corner of the radiator as shown by the dotted circles in Figure 4.20).

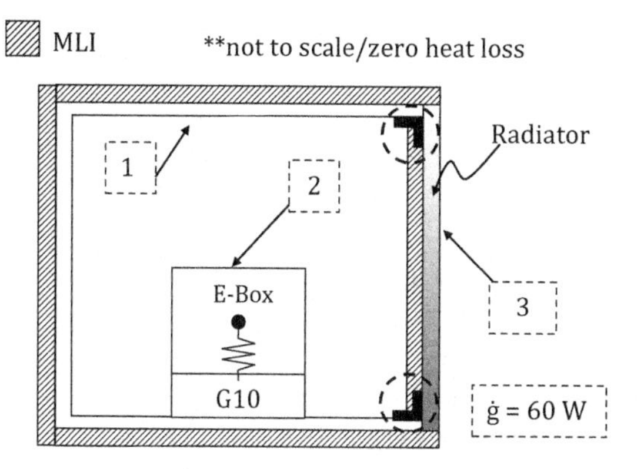

Figure 4.20 SpaceCube with E-box separated from radiator

The L-brackets will have an area of 3.5 cm × 3.5 cm on each mounting surface and a thickness of 1 cm. The effective resistance between the structure and radiator now becomes

$$\frac{1}{R_{1-3,eff}} = \frac{1}{R_{3-Brackets} + R_{Brackets-1}} \tag{4.40}$$

which results in a value of $R_{1-3,\,eff} = 1.116$ K/W. We will also add ideal insulation to the inner spacecraft side of the radiator to limit the heat transfer mechanism to conduction for heat flow from the structure to the radiator. The E-box will be treated as remote from the radiator (which is often the case for temperature-controlled items in contemporary spacecraft thermal design).

Turning to the network, since node 1 is conductively coupled to node 3 more than two diagrams are needed. The first (Figure 4.21) includes thermal couplings for nodes 1 and 2. As with the previous example, the 60 W heat input in the energy balance at the radiator has to be included. Even though the E-box now has an effective insulator (G10) between it and the SpaceCube structure, the heat dissipated from the E-box will still make its way to the radiator (provided the E-box temperature is higher) and be rejected. A heat flow diagram for the radiator is shown in Figure 4.22.

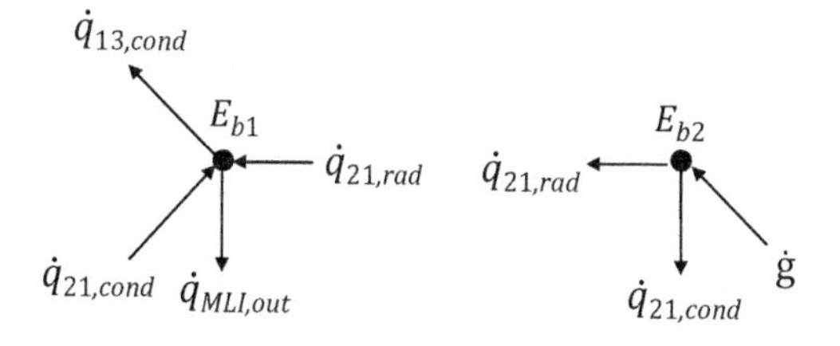

Figure 4.21 SpaceCube heat flows for nodes 1 and 2 with 60 W energy dissipation from remote E-box

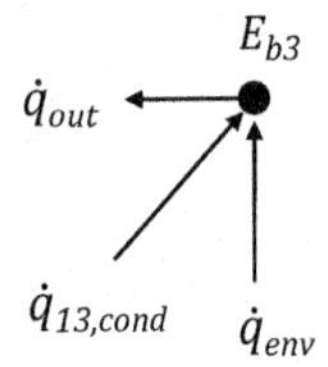

Figure 4.22 SpaceCube thermal couplings at node 3 with 60 W energy dissipation from remote E-box

Applying the energy balance at each of the nodes gives three equations. At node 3 (the radiator), the energy balance relation is

$$\dot{Q}_{env} + \dot{Q}_{1-3,cond} = \varepsilon_3 A_3 \sigma T_3^4 \tag{4.41}$$

At node 2 (the E-box) we have

$$\dot{g} = \dot{Q}_{2-1,rad} + \dot{Q}_{2-1,cond} \tag{4.42}$$

The energy balance at node 1 (the SpaceCube structure) is

$$\dot{Q}_{2-1,rad} + \dot{Q}_{2-1,cond} = \dot{Q}_{1-3,cond} + \dot{Q}_{MLI,out} \tag{4.43}$$

Before the energy balance relations for the new 3-node configuration are placed into the EES program, the view factor section of the code is defined as

```
{View factors}
F_2_1a+F_2_2=1.0
F_2_2=0.0
F_1a_2=(A_2/A_1a)*F_2_1a

F_surf_earth=0.9196
```

The effective thermal resistance between nodes 1 and 3 (i.e., $R_{1-3,eff}$) can be defined as

```
{SS L-Bracket resistance}
R_Bracket=2*(0.005/(k_SS*0.035*0.035))+(1.75/100)/(k_SS*0.03*0.01)

{SS L-Brackets coupling SpaceCube to radiator}
R_1_3_eff=1/(4/(R_Bracket+R_1))
```

For the Max case, the energy balance relation for the SpaceCube structure (including heat loss through the MLI) is replaced by a new relation that solves the energy balance for the temperature at node 1. This will be

```
{Energy balance on SpaceCube}
T_1=T_2-((Eb_1-J_1a)/((1-eps_1a)/(eps_1a*A_1a))+(T_1-T_3)/(R_1_3_eff)&
+estar*A_MLI*sigma*(T_1^4-T_MLI1^4)+estar*A_MLI*sigma*(T_1^4-T_MLI2^4)&
+estar*A_MLI*sigma*(T_1^4-T_MLI3^4)+estar*A_MLI*sigma*(T_1^4-T_MLI4^4)&
+estar*A_MLI*sigma*(T_1^4-T_MLI5^4))*R_2_1
```

We will also add an energy balance at surface 1a to the code

```
{Energy balance on Surface 1a}
(Eb_1-J_1a)/((1-eps_1a)/(eps_1a*A_1a))=(J_1a-J_2)/(1/(A_1a*F_1a_2))
```

The previous code used for the energy balance on node 2 also applies in the present example

```
{E-box (node 2) energy balance}
g_dot=(T_2-T_1)/R_2_1+(Eb_2-J_2)/((1-eps_2)/(eps_2*A_2))
```

The energy balance at the surface 2 radiosity node is

```
{Energy balance on Surface 2}
(Eb_2-J_2)/((1-eps_2)/(eps_2*A_2))=(J_2-J_1a)/(1/(A_2*F_2_1a))
```

In order to solve for the radiator temperature, the energy balance at node 3 is accounted for using the following code

```
{Energy balance on radiator}
A_3a=in_length^2
A_3=1.0                    [m^2]
eps_3a=0.02                {VDA}
eps_3=0.89                 {Z93 white paint}
alpha_3=0.19               {Z93 white paint solar absorptance}

T_3=(((eps_3*A_3*EIR_Hot*F_surf_earth+Alb_H*alpha_3*S*A_3&
  *F_surf_earth+(T_1-T_3)/(R_1_3_eff)))/(eps_3*A_3*&
  sigma))^0.25
```

The blackbody heat fluxes for nodes 1 and 2 must also be defined in the code.

```
{Blackbody emission definitions}
Eb_1=sigma*T_1^4
Eb_2=sigma*T_2^4
```

A code can also be created for the cold case by modifying the thermal loads incident on the radiator and the outer surfaces of the MLI to account for the thermal environment experienced while in eclipse.

The E-box dissipation of 60 W assumed in the previous examples is rather low given that most free-flying spacecraft in LEO have dissipations on the order of 500 W. Table 4.7 lists the nodal temperature results for both the hot and cold cases given thermal dissipations of 60 W and 120 W from the E-box.

Table 4.7 Hot- and cold-case temperatures for 3-node SpaceCube with separated radiator

	Max (hot) case			Min (cold) case		
E-box dissipations	T_1 **[K]**	T_2 **[K]**	T_3 **[K]**	T_1 **[K]**	T_2 **[K]**	T_3 **[K]**
60 W	297.7	304.1	274.6	279.6	286.1	246.8
120 W	340.7	353.1	282.0	327.1	339.7	257.2

In both the Max and Min cases, attention should be paid to the temperature gradient. In each case the E-box has the warmest temperature while the radiator has the coldest. The SpaceCube structure temperature is between these two values. Since the E-box is the dissipation source and the radiator is in direct thermal communication with the ambient of space, this gradient is consistent with expected temperature performance. At a thermal dissipation level of 60 W, the temperatures calculated are well within acceptable ranges for yellow and red limits. However, at 120 W dissipation, the E-box temperature is in excess of operational values for standard electronics boxes.

The thermal coupling between the E-box and the SpaceCube structure was reduced in the previous example by placing the finite-thickness G10 between them. In the present example, decoupling the structure and the radiator further increases the total effective thermal resistance between the E-box and the radiator, resulting in significantly warmer temperatures as the E-box dissipation increases. In the case of large thermal dissipations, another conductor could be added to create a configuration similar to that in the previous SpaceCube examples where the E-Box was directly coupled to the radiator (Figure 4.23). This will decrease the total resistance to heat flow away from the E-box and into the radiator, reducing the E-box temperatures to a value closer to that of the radiator.

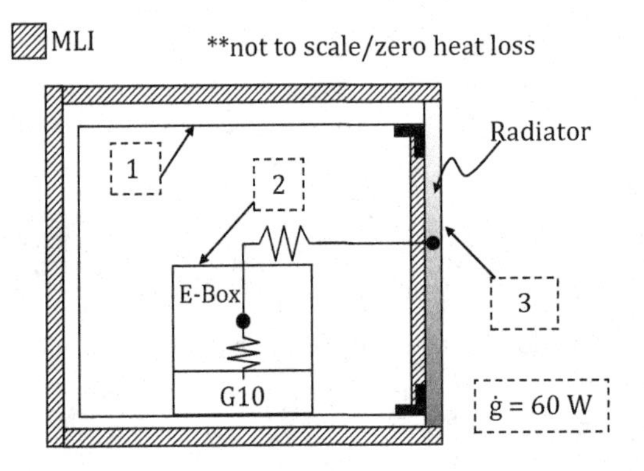

Figure 4.23 Box (or SpaceCube) configuration with remote E-box and added conductor coupling to radiator

Given the extreme temperatures obtained in the calculations for the hot and cold cases under large dissipations, simply reducing the effective thermal resistance between the E-box and the radiator may not suffice to bring the E-box temperature values back to within prescribed operational limits. Options to be considered would include alternate radiator coatings, β angle variation and possibly changing the FOV of the radiator to limit direct thermal loading. Standard approaches for increasing the thermal coupling between the E-box and the radiator include thermal straps, heat pipes, loop heat pipes, thermo-electric coolers (TECs), traditional heat pumps and single-phase mechanical pumps. These technologies are reviewed in Chapter 5.

4.7.3 PV Cells and/or Solar Array Analysis

Figure 4.24 is a generic illustration of a PV cell stack-up sitting on top of a honeycomb solar array panel structure made of aluminum. The facesheets of the honeycomb panel

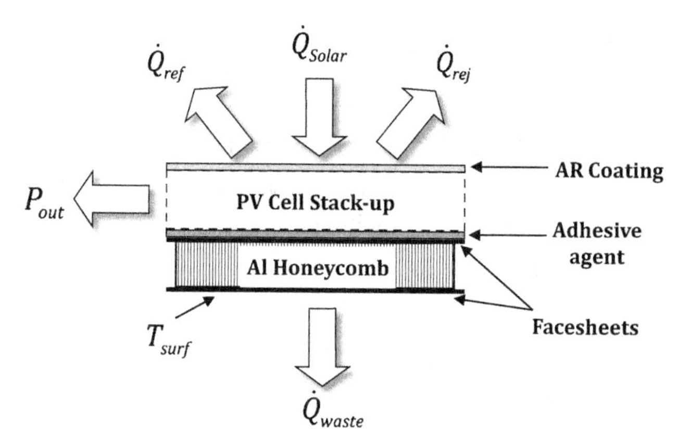

Figure 4.24 Heat flow for PV cell stack mounted on a honeycomb substrate

are typically made of M55J (graphite fiber) or K1100 (carbon fiber). Given that solar loading is occurring on the Sun-viewing side of the PV stack-up, some solar energy capture will occur. The power output for the PV cells can be defined mathematically as

$$P_{out} = \eta_{PV} \cdot \dot{Q}_{Solar} \tag{4.44}$$

where η_{PV} is the thermal-to-electric conversion efficiency for the cells. Assuming the cells shown in Figure 4.24 have a conversion efficiency of 30%, the electrical power produced will be

$$P_{out} = (0.3) \cdot \dot{Q}_{Solar}$$

The top surface of the stack-up is rejecting heat to space. Although the stack-up will usually have an anti-reflective coating, some thermal energy reflection may be taking place. Thermal energy that is neither reflected, rejected nor converted to electrical power will ultimately be transferred to the opposite side of the honeycomb panel and rejected to ambient as waste heat. The energy balance on the PV cell stack-up circuit will then be

$$\dot{Q}_{solar} = P_{out} + \dot{Q}_{rej} + \dot{Q}_{ref} + \dot{Q}_{waste} \tag{4.45}$$

which can be rewritten as

$$\dot{Q}_{solar} = (0.3) \cdot \dot{Q}_{solar} + \dot{Q}_{rej} + \dot{Q}_{ref} + \varepsilon\sigma A_{surf} T_{surf}^4 \tag{4.46}$$

GaAs PV cells are predominantly used for space-based applications. The standard emissivity and absorptivity quoted for GaAs PV cells is ε_{IR} = 0.92 and α_{solar} = 0.82. Depending on the number of junctions used, these coating values may increase a little. Figure 4.25 shows solar array panels on a spacecraft, viewed from the direction of the Sun. The panels are mounted on the spacecraft's cylindrical housing. The bottom of the mounting structure will be in the FOV of the spacecraft structure (Figure 4.26).

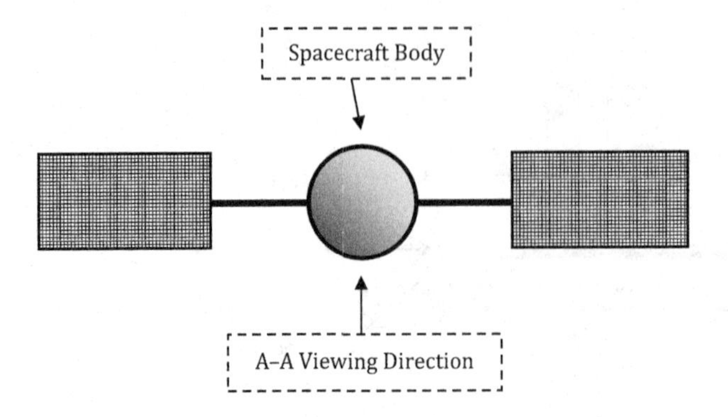

Figure 4.25 Spacecraft and solar array PV cell thermal energy capture area

A portion of the waste heat from the PV cells that is rejected from the bottom of the honeycomb panel structure will be incident on the spacecraft structure. Radiative thermal loading from the spacecraft to the solar array heat rejection surface may also occur. The net thermal energy exchange between these two surfaces may be non-negligible and should be taken into account in the overall thermal analysis of the spacecraft.

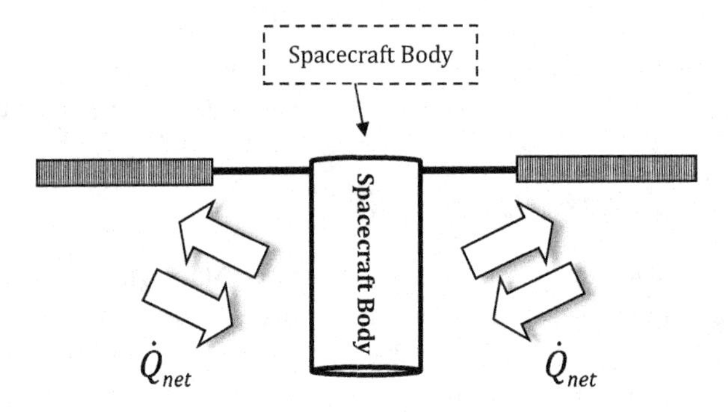

Figure 4.26 Spacecraft and solar arrays, A–A view

References

[1] Gilmore, D.G., Prager, R.C., Grob, E.W., and Ousley, W., 2002, "Spacecraft Subsystems: Thermal," in *Space Mission Analysis and Design*, edited by Larson, W.J., and Wertz, J.R., Microcosm Press and Springer, Portland, Chapter 11.5, pp. 428–458

[2] Mehoke, D., 2005, "Spacecraft Thermal Control," in *Fundamentals of Space Systems*, edited by Pisacane, V.L., Oxford University Press, Inc., Chap. 7, pp. 423–464

[3] Serway, R.A., 1986, *Physics for Scientists & Engineers,* Saunders College Publishing, Philadelphia

[4] Clawson, J.F., Tsuyuki, G.T., Anderson, B.J., Justus, C.G., Batts, W., Ferguson, D., and Gilmore, D.G., 2002, "Spacecraft Thermal Environments," in *Spacecraft Thermal Control Handbook Vol. I: Fundamental Technologies*, edited by Gilmore, D.G., The Aerospace Press and The American Institute of Aeronautics and Astronautics, Inc., Chap. 2, pp. 21–69

[5] Koller, S., 2018, "Solar Constant: Construction of a Composite Total Solar Irradiance (TSI) Time Series from 1978 to the Present," World Radiation Center, www.pmodwrc.ch/en/?s=solar+constant#1512395072314-75289b48-d22b
[6] "Best Research-Cell Efficiencies," National Renewable Energy Laboratory, 2018, www.nrel.gov/pv/assets/images/efficiency-chart-20180716.jpg
[7] Dakermanji, G., and Sullivan, R., 2005, "Space Power Systems," in *Fundamentals of Space Systems*, edited by Pisacane, V.L., Oxford University Press, Inc., Chap. 6, pp. 326–422
[8] Riebeek, H., "Catalog of Earth Satellite Orbits," National Aeronautics and Space Administration, 2009, https://earthobservatory.nasa.gov/Features/OrbitsCatalog
[9] Cornish, N., "WMAP: The Lagrange Points," National Aeronautics and Space Administration, 2012, https://map.gsfc.nasa.gov/mission/observatory_l2.html
[10] Pisacane, V.L., 2005, "Astrodynamics," in *Fundamentals of Space Systems*, Oxford University Press, Inc., 2005, Chap. 3, pp. 102–170
[11] "SOHO Mission Fact Sheet," National Aeronautics and Space Administration, 2015, www.nasa.gov/pdf/156578main_SOHO_Fact_Sheet.pdf
[12] "The WIND Spacecraft," National Aeronautics and Space Administration, https://wind.nasa.gov/
[13] "DSCOVR: Deep Space Climate Observatory," National Oceanic and Atmospheric Administration, 2016, www.nesdis.noaa.gov/content/dscovr-deep-space-climate-observatory
[14] "WMAP Mission Overview," National Aeronautics and Space Administration, 2017, https://map.gsfc.nasa.gov/mission/
[15] "Herschel Mission Fact Sheet," European Space Agency, April 29, 2013, http://sci.esa.int/herschel/47356-fact-sheet/
[16] "Planck Mission Fact Sheet," European Space Agency, May 13, 2015, http://sci.esa.int/planck/47365-fact-sheet/
[17] "The James Webb Space Telescope: Webb Vital Facts," 2018, National Aeronautics and Space Administration, https://jwst.nasa.gov/facts.html
[18] "WFIRST: Observatory," 2018, National Aeronautics and Space Administration, https://wfirst.gsfc.nasa.gov/observatory.html

Problems

4.1 Assume you have an active surface in Sunview receiving solar loading. The surface has an area of 1.0 m^2. Calculate and plot the temperature of the surface as a function of the α/ε ratio when varied from 0.1 to 1.5.

4.2 A surface coated with Z306 black polyurethane paint has its normal oriented relative to the solar constant by an angle of γ. Assuming ideal insulation on the anti-Sun side and an active surface area of 0.75 m^2, determine the temperature for angles of 0°, 15°, 30°, 45°, 60° and 75°.

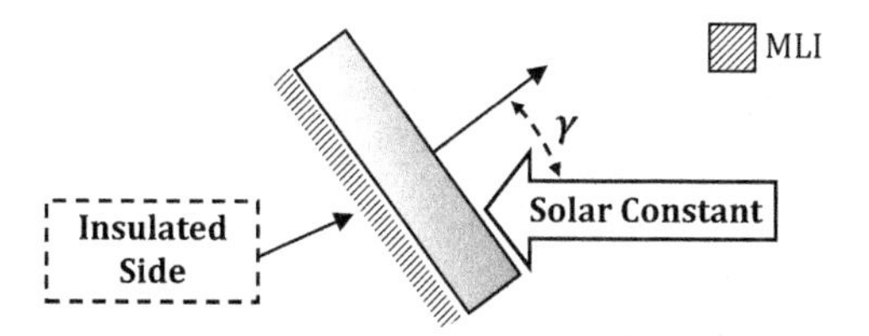

Figure P4.2 Thermal performance of a flat plate at varying angles of the surface normal relative to the solar constant

4.3 Provide a plot of the view factor to Earth as a function of altitude (ranging 200–35,000 km) for a 1.0 m^2 surface under the following two orientation conditions.

i) Nadir-facing flat plate with a surface normal directed towards the center of the planet;

ii) Flat plate viewing deep space with a surface normal offset by 90° from the Earth's nadir direction.

4.4 A 1.0 m^2 flat plate has one nadir-facing surface that is coated with Z93 white paint ($\alpha_S = 0.19$, $\varepsilon = 0.89$). The opposite surface is covered with 10 layers of 0.15 mil double-sided aluminized kapton ($\alpha_S = 0.25$, $\varepsilon = 0.34$). Assuming an energy dissipation of 30 W on the plate, what are the Max and Min temperatures for the plate if it has an orbit period of 90 minutes? (Note: Assume $F_{surf\text{-}earth} = 1.0$)

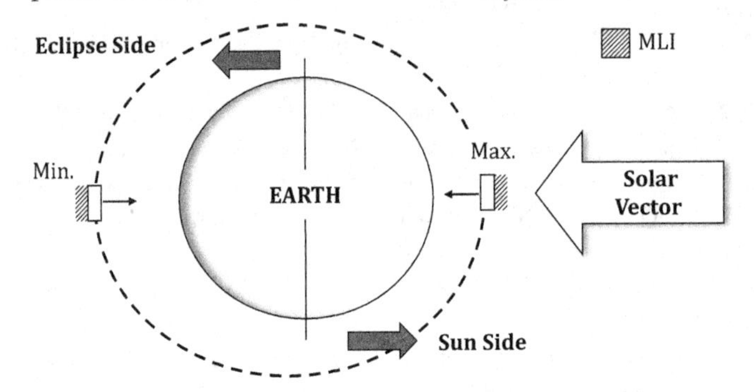

Figure P4.4 Maximum and minimum orbit positions with orbit inclination of 66.6° relative to the equatorial plane and $\beta = 0°$

4.5 A flat plate is in an orbit with an inclination of 90° (relative to the ecliptic plane) and a $\beta = 0°$. The plate has one active surface which is facing Earth during orbit (as shown in Figure P4.5). The active surface area is 1.0 m^2. The anti-nadir side of the plate is covered with ideal MLI.

i) Assuming an orbit period of 90 minutes, calculate the Max and Min temperatures for the active surface given a surface coating of Martin Black Velvet paint.

ii) Assuming an orbit period of 90 minutes, calculate the Max and Min temperatures for the active surface given a surface coating of Polyurethane white paint.

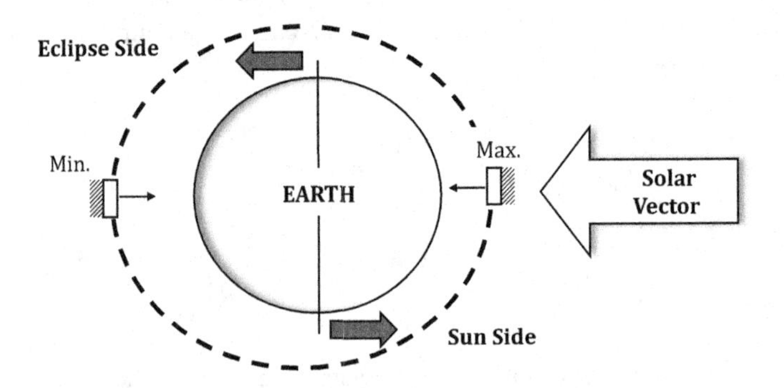

Figure P4.5 Flat plate in orbit receiving thermal loading

4.6 A cubic structure with a side length of 0.5 m resides in space. The outer surface of the structure is coated with Paladin black lacquer ($\alpha_S = 0.95$; $\varepsilon = 0.75$) and the only environmental thermal energy source within view of the cubic structure is the Sun (and/or the solar constant). If the cubic structure is spinning (spin axis perpendicular to the solar constant, as shown in Figure P4.6), what is the surface temperature of the cubic structure?

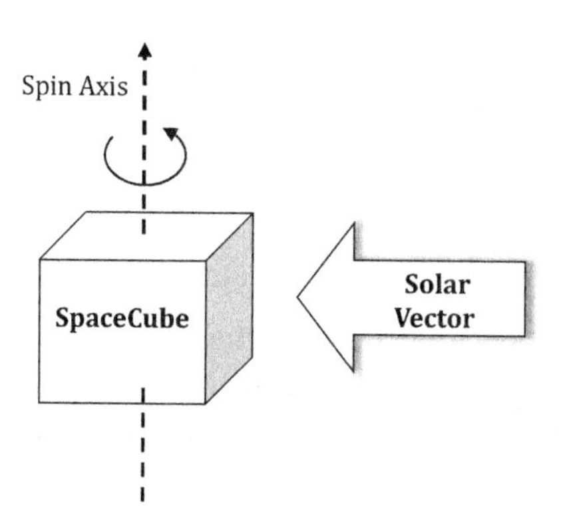

Figure P4.6 Spinning cubic structure illuminated by the Sun

4.7 A flat plat with an orbit inclination of 90° (relative to the ecliptic plane) and $\beta = 0°$ is in orbit around the Earth. The plate is made of aluminum (6061-T6) and has one active side (denoted by the arrow in Figure P4.7) that is continually facing Earth while in orbit. It has a surface coating of Z93 white paint. All other surfaces are covered in ideal MLI. The orbit period for the plate is 90 minutes.

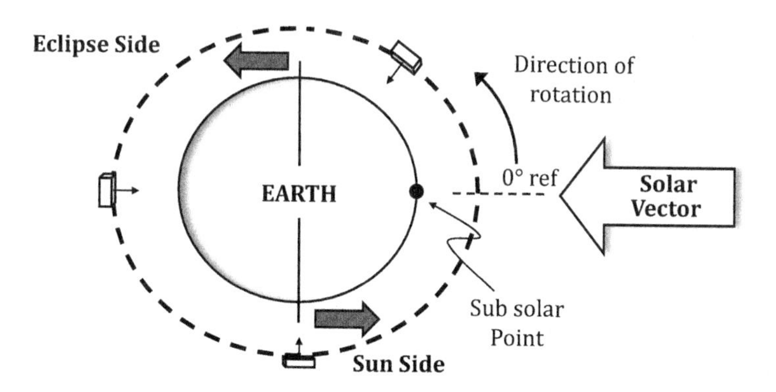

Figure P4.7 Example orbit positions for a flat plate transient analysis

i) Assuming a starting temperature of 273 K for the plate at the 0° reference point (i.e., sub-solar position) at time $t = 0$ and 5° increments along the revolution path, provide a plot of the temperature as a function of time for the first 5 revolutions of the orbit.

ii) How many watts of additional heater power are required for the plate to reach harmonic equilibrium during the first five revolutions? Plot the temperature as a function of time for purposes of validation.

Note: You can assume the following

- the plate has dimensions of 1.0 m × 1.0 m × 0.0254 m.

Hint: For the partial derivative of the temperature with respect to time you can apply a forward difference (i.e., explicit) approach.

$$\frac{\Delta T}{\Delta t} = \frac{T^{p+1} - T^{p}}{\Delta t}$$

where p denotes a specific timestep and $p + 1$ denotes the next timestep.

4.8 Three thermal shields made from 0.25 mil aluminized kapton have their centers aligned and are positioned with a separation distance of 0.5 m (Figure P4.8). Each shield is square in shape and has a side length of 1.5 m. Assuming that shield 1 is receiving full solar loading on its outward-facing side and shield 3 is facing deep space on the side opposite that facing shield 2:

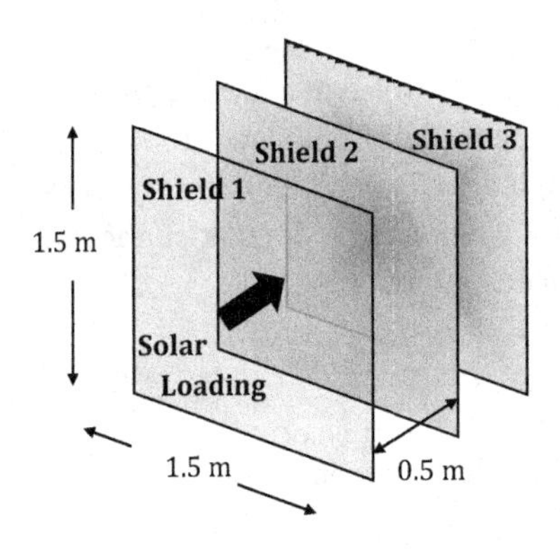

Figure P4.8 Sunshields mitigating solar loading

i) Determine the net heat transfer from shield 2 to shield 3 and the temperature of shield 3 if the separation distance between successive shields is 0.5 m.
ii) Determine and plot the temperatures for shields 1, 2 and 3 if the separation distance is varied between 0.5 m and 1.5 m.

4.9 An electronics box is mounted on the exterior of a spacecraft (Figure P4.9). The E-box has one active side ($A_s = 0.5$ m^2) which faces deep space while in eclipse. The E-box itself has a constant energy generation of 40 W through the entire orbit. The advisory "do not exceed" low operating temperature limit for the box is −20°C.

Note: Assume heat loss through the MLI in this case is 0 W. The orbit inclination is 36.6° and the period is 6 hours.

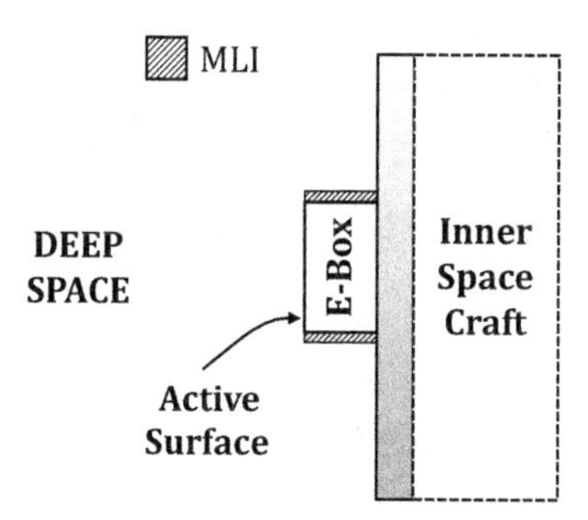

Figure P4.9 Spacecraft profile

i) Assuming a coating of S13 GLO white paint on the active surface, can the box stay above the cold temperature limit specified with no environmental loading on the surface? Show work.

ii) How much additional heater power is needed to reach −20°C (i.e., 253 K)?

iii) If the power budget only allows 20 W to the E-box while in eclipse, what is an alternate approach to reaching the temperature requirement?

4.10 A SpaceCube is in orbit around the Sun at a distance of 1 AU (Figure P4.10). The only thermal energy source in near proximity to the SpaceCube is the Sun. Assuming the entire exterior of the SpaceCube is coated with Dow Corning white paint, determine:

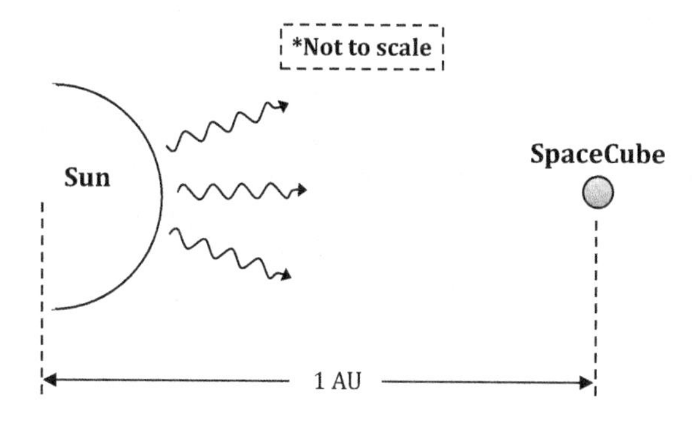

Figure P4.10 SpaceCube in orbit around the Sun at 1 AU

i) The outer surface temperature of the SpaceCube if it is spherical in shape.

ii) The outer surface temperature of the SpaceCube if it is cubic in shape with one side's surface normal aligned with the solar constant at all times of flight.

4.11 Let's assume a SpaceCube has an inclination of 0° relative to the equatorial plane, and an altitude of 250 km. The SpaceCube has a side length of 1.0 m and no insulation. If the SpaceCube is in full eclipse, what is the total environmental thermal load upon the SpaceCube surfaces if they are coated with Martin Black Velvet paint?

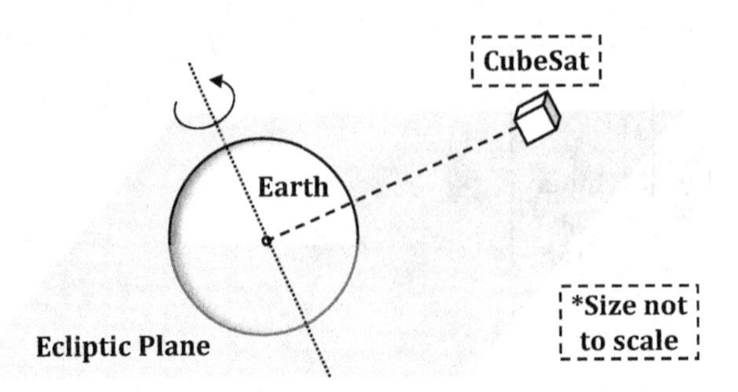

Figure P4.11 Total thermal environmental load for a nadir-facing CubeSat in orbit

4.12 In the flat plate analysis in Example 5, the MLI relation used was that for the theoretical net heat transfer through MLI. This relation idealizes workmanship imperfections and perforations through the blanket structure. An alternate relation (in heat flux form) which does account for quality of workmanship and heat leaks is the Lockheed Equation for Double-Sided Aluminum (shown in equation 3.114 as well as below).

$$\dot{q}'' = \frac{C_s \bar{N}^{3.56} T_m}{N_S + 1}(T_H - T_C) + \frac{C_r \varepsilon_{tr}}{N_S}\left(T_H^{4.67} - T_C^{4.67}\right)$$

Assuming the same orbit as in Example 5 (p. 132) and Z93 white paint as the active surface's coating:

i) If there are 10 layers of MLI (0.15 mil aluminized Mylar) on the non-active side of the plate with an outboard MLI layer made of Beta cloth, calculate the plate temperatures at the Max and Min orbit locations when the active surface is nadir facing and the Lockheed Equation is used for the MLI heat transfer.

ii) How does the predicted temperature performance using the Lockheed Equation compare to that determined using the theoretically based MLI heat transfer relation?

4.13 A hollow spherical probe that is 1.0 m in diameter contains multiple E-boxes that are comprehensively dissipating 50 W into the interior wall of the sphere. The entire exterior surface of the sphere is unshielded by MLI and thereby in thermal communication with the ambient environment. Four potential coatings are being considered for use on the exterior surface: Catalac black paint, NASA/GSFC NS-53-B green paint, Hughson A-276 white paint and Magnesium Oxide white paint.

i) Determine the temperature of the spherical structure for each of the candidate paints if the sphere is positioned at L1.

ii) Determine the temperature of the spherical structure for each of the candidate paints if it was positioned at L3.

4.14 A hollow cubic SpaceCube with a sidewall length of 0.8 m and a wall thickness of th = 0.015 m has one E-box located inside (Figure P4.14). The effective radiator surface (i.e., 1b) is one full side on the exterior of the SpaceCube enclosure that is nadir facing. The other five sides are covered in non-ideal MLI consisting of 12 layers of aluminized kapton. The inner surface of the SpaceCube is coated with VDA, and the exterior radiator surface is coated with Z93 white paint. The E-box has a side length of 0.4 m and is decoupled from the SpaceCube inner wall via a G10 spacer that is 1 cm thick. The outer surface of the E-box is coated with Z306 black polyurethane paint. Assuming an orbit inclination of 66.6°, $\beta = 0°$ and an orbit period of 90 minutes, determine the Max and Min temperature predictions for the structure and the E-box if it has a dissipation of 60 W. Radiative thermal exchange to and from the spacer can be neglected.

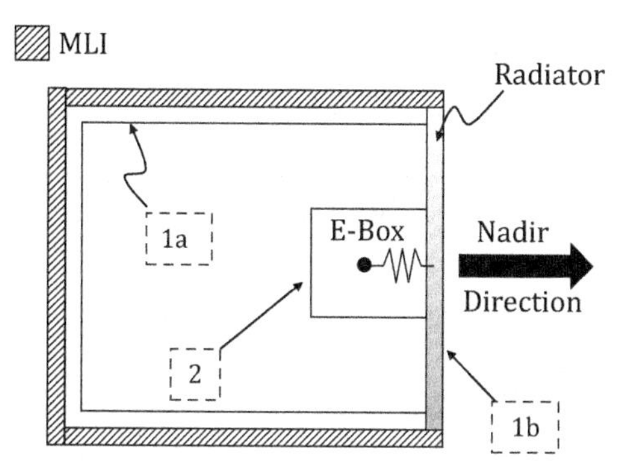

Figure P4.14 Two-node SpaceCube with active surface viewing Earth

4.15 A hollow cubic SpaceCube with a sidewall length of 0.75 m and a wall thickness of th = 0.015 m has one E-box located inside (Figure P4.14). The effective radiator surface (i.e., 1b) is one full side on the exterior of the SpaceCube enclosure. The other five sides are covered in non-ideal MLI consisting of 15 layers of aluminized kapton. The inner surface of the SpaceCube is coated with VDA and the exterior radiator surface is coated with Z93 white paint. The E-box has a side length of 0.3 m and is decoupled from the SpaceCube inner wall via a G10 spacer that is 1 cm thick. The outer surface of the E-box is coated with Z306 Black Polyurethane paint. The SpaceCube is in a circular orbit with an altitude of 10,385 km. Assuming an orbit inclination of 0°, determine the Max and Min temperature predictions for the structure and the E-box if the E-box has a dissipation of 75 W. Radiative thermal exchange to and from the spacer can be neglected.

4.16 A hollow cubic SpaceCube (Figure P4.16) with a sidewall length of 0.75 m and a wall thickness of th = 0.015 m has one E-box located inside. The effective radiator surface (i.e., 1b) is one full side on the exterior of the SpaceCube enclosure. The other five sides are covered in non-ideal MLI consisting of 11 layers of aluminized kapton. The inner surface of the SpaceCube is coated with VDA, and the exterior radiator surface is coated with Z93 white paint. The E-box has a side length of 0.4 m and is decoupled from the center of the SpaceCube inner wall via a G10 spacer that is 1 cm thick. The outer surface of the E-box is coated with 3M Black Velvet paint (BOL). The SpaceCube is in a circular LEO orbit with an altitude of 273 km. Assuming an E-box dissipation of 65 W and an orbit inclination of 6.6°

i) Determine the temperature predictions for the structure and the E-box in the Max case.

ii) Determine the temperature predictions for the structure and the E-box in the Min case.

iii) Determine the required amount of heat input on the structure if the E-box were to be brought up to a temperature of 273 K in the Min case.

Note: Radiative thermal exchange to and from the spacer can be neglected.

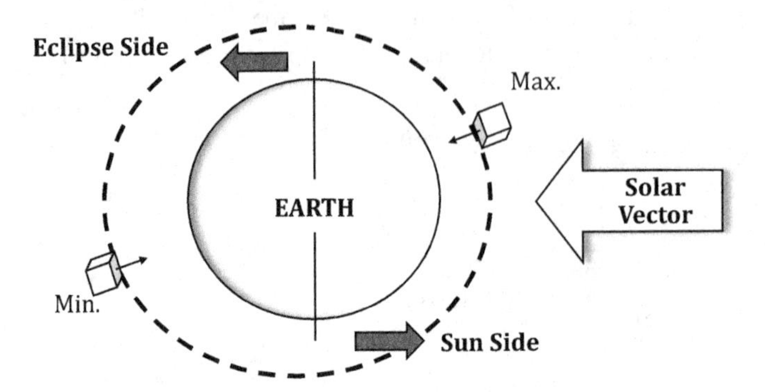

Figure P4.16 SpaceCube in LEO at inclination of 6.6° relative to the equatorial plane

4.17 A hollow cubic SpaceCube with a sidewall length of 0.8 m and a wall thickness of th = 0.015 m has one E-box located inside (Figure P4.17). The effective radiator surface (i.e., 1b) is one full side on the exterior of the SpaceCube enclosure and is nadir facing. The other five sides are covered in non-ideal MLI consisting of 20 layers of aluminized kapton. The inner surface of the SpaceCube is coated with black anodize (sample 3), and the exterior radiator surface is coated with Z93 white paint. The E-box has a side length of 0.4 m and is decoupled from the SpaceCube inner wall via a Vespel spacer that is 1 cm thick. The outer surface of the E-box is coated with black anodize (sample 3). Assuming an E-box dissipation of 60 W, determine:

i) The temperature predictions for the structure and the E-box while at L1 with the active surface viewing the Earth.

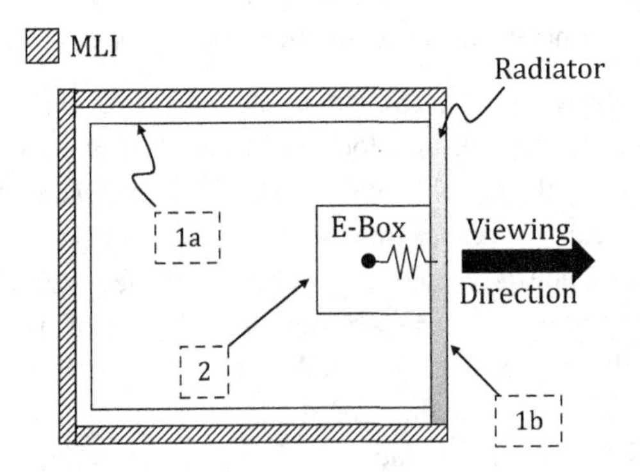

Figure P4.17 Two-node SpaceCube with active surface viewing celestial bodies

ii) The temperature predictions for the structure and the E-box while at L3 with the active surface viewing the Sun.

Note: Radiative thermal exchange to and from the spacer can be neglected.

4.18 A hollow cubic SpaceCube (Figure P4.18a) with a sidewall length of 0.7 m and a wall thickness of $th = 0.015$ m has an L-bracket mounted on one side (recall Problem 3.20 from Chapter 3, p. 110). The base of the L-bracket has a footprint of 35 cm × 35 cm and a thickness of 5 mm. The vertical segment of the L-bracket has a width (into the page as shown) of 35 cm and a thickness of 5 mm. At the top of the L-bracket is an electronics box with the geometry shown in Figure P4.18b. The E-box has a length dimension (i.e., into the page) of 35 cm. The effective radiator surface (i.e., 4b) is one full side on the exterior of the SpaceCube enclosure. The other five sides are covered in non-ideal MLI consisting of 12 layers of aluminized kapton. The inner surface of the SpaceCube is coated with black anodize (sample 3), and the exterior radiator surface is coated with Z93 white paint. The L-bracket and the outer surface of the E-box are also coated with black anodize (sample 3). The SpaceCube has a circular LEO orbit with an altitude of 273 km. Assuming an E-box dissipation of 35 W and an orbit inclination of 6.6°, determine the Max and Min case temperature predictions for the E-box.

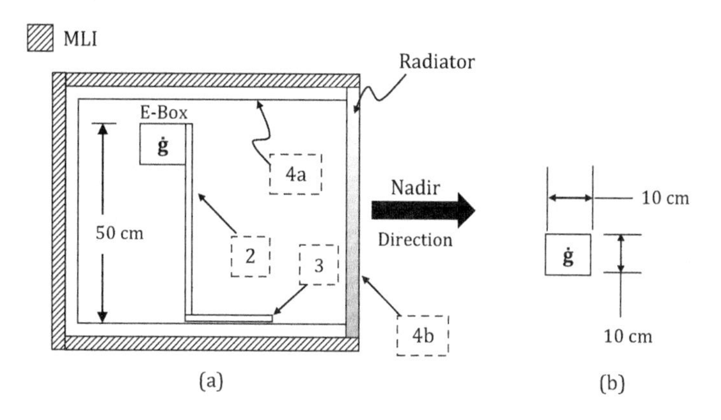

Figure P4.18 SpaceCube enclosure with an E-box mounted on an L-bracket: a) internal configuration; b) E-box dimensions

4.19 A honeycomb panel is populated with solar arrays covering an area of 4.0 m^2. The multi-junction PV cells for the array have a solar absorptivity of 0.92 and an emissivity of 0.85. Assuming $\beta = 15°$ with the solar constant:

i) Determine the power produced by the array for a thermal-to-electric conversion efficiency of 35%.

ii) Determine the total heat rejected to space from both the array panel and the honeycomb substrate (i.e., waste heat).

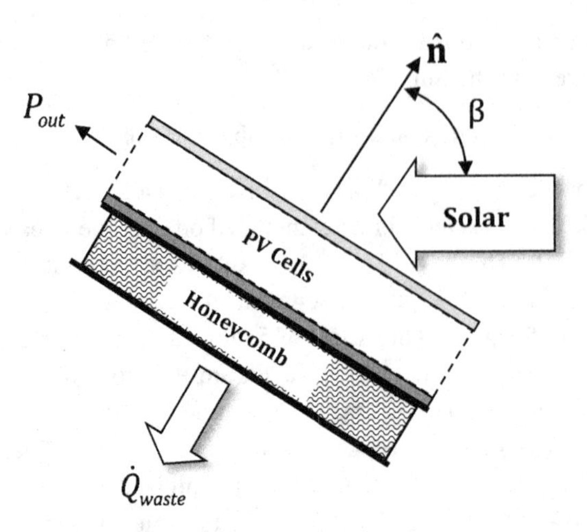

Figure P4.19 Solar array thermal loading in space

iii) Assuming the panel array temperature is 60°C, what is the temperature of the space-viewing facesheet (anti-Sun side) on the honeycomb substrate if it is coated with Magnesium Oxide white paint?

5 Space-Based Advanced Thermal Conductance and Storage Technologies

5.1 Space-Based Technologies

At the end of Chapter 4 we had begun to perform temperature analysis of components internal to a SpaceCube in LEO. In the analysis procedure we determined that it would be necessary to implement a thermal coupling between the E-box and the radiator in order to remove heat from the E-box and reduce its operational temperature. How might that additional thermal coupling between the E-box and the radiator be implemented? Standard approaches used for the implementation of thermal couplings between spaceflight components include:

- Thermal straps
- Single-phase mechanical pumps
- Heat pipes
- Loop heat pipes (LHPs)
- Thermoelectric coolers (TECs)
- Heat pumps (standard terrestrial cycles typically not used on space-based platforms).

While many of these components (and/or techniques) are used throughout the terrestrial engineering community, some technologies, such as LHPs, may be regarded as exclusive to space applications. Standard heat pump and/or refrigeration systems have typically not been chosen for application in unmanned Earth-based satellites. However, their performance features do serve as a baseline for alternate heat pump system and/or thermal control technologies (e.g., TECs and Cryocoolers). This chapter reviews contemporary spacecraft thermal control technologies that enable heat transfer between a temperature source and a temperature sink, as well as energy storage. Operational aspects of each of these technologies specific to the space environment are also addressed.

5.2 Transfer Processes

Transfer processes in nature that occur near room temperature (i.e., considerably above the zero point) abide by three fundamental rules. For a given measurable quantity that is being transferred between two discrete locations:

i) The transfer of the quantity of interest is resultant from a potential difference (i.e., a driving force) between the two discrete locations.

ii) The transfer of the quantity of interest will occur along a designated path.
iii) During the transfer process along the designated path, the quantity of interest will experience a resistance to the transfer of the quantity.

The manifestation of these rules can be identified in multiple engineering disciplines. We see these rules in electrical circuits via the Ohmic behavior of electrically conductive materials. In equation form this is expressed through Ohm's law

$$V = I \cdot R_e \tag{5.1}$$

where I (i.e., current) is the flow of electrons, R_e is the resistance to current in the electrically conductive material along the flow path and V actually represents a potential that fosters the flow between two discrete locations. This equation can be recast [1] as

$$I = \frac{(V^+ - V^-)}{R_e} \tag{5.2}$$

In regard to heat transfer (specifically thermal conduction and convection), the process for thermal energy transfer can be can be captured in equation form as

$$\dot{Q} = \frac{(T_1 - T_2)}{R_{1-2}} \tag{5.3}$$

where the thermal resistance to heat flow (R_{1-2}) is L/kA for conductive processes and $1/h_{conv}A$ for convective processes. In both of these cases, the quantity being transferred is thermal energy, whereas the potential fostering the heat flow is the temperature gradient between the two locations. In thermal conduction, the path consists of the thermally conductive material through which the thermal energy flows. For convection, the path is through the fluid/solid interface and the fluid which is experiencing flow. The resistance to thermal energy flow is along both the fluid/solid interface and the bulk fluid. Last but not least, this dynamic is also present in fluid mechanics. It is most easily observed in internal flow processes. A common example is that of Poiseuille flow in a circular, horizontal channel. Assuming a constant cross-section with a diameter of D over a channel of length L, Bernoulli's equation can be applied over the length of the channel to determine Darcy's pressure drop formula [2] (also known as the Darcy–Weisbach equation)

$$\frac{\Delta P}{\rho g} = f \cdot \frac{L}{D} \cdot \frac{Vel_{avg}^2}{2g} \tag{5.4}$$

where ΔP is the pressure gradient, ρ is the fluid density, f is the friction factor, g is gravitational acceleration and Vel_{avg} is the average velocity of the fluid over the channel cross-sectional area. This relation also is an identity for the head loss (h_f) in the channel. Substituting in for the density using the mass flow rate relation $\dot{m} = \rho \cdot A \cdot Vel_{avg}$, we can isolate the mass flow rate variable as

$$\dot{m} = \frac{\pi \Delta P D^3}{2 Vel_{avg} L f} \tag{5.5}$$

This can also be recast as

$$\dot{m} = \frac{(P_1 - P_2)}{R_{flow}} \tag{5.6}$$

where

$$R_{flow} = \frac{2Vel_{avg}Lf}{\pi D^3} \tag{5.7}$$

Based on the form presented in equation 5.6, the quantity being transferred is the fluid mass, whereas the potential promoting the flow between the two locations is the pressure gradient. The path is the channel itself. Given the no-slip condition present at the wall, the channel has an effective resistance to the fluid flow. The addition of flow elements in the channel's cross-section will result in additional flow resistances. However, this is often accounted for by pressure drop.

The rules for transfer processes are directly applicable to the functionality of thermal energy transfer devices such as heat pumps, heat pipes, loop heat pipes and thermoelectric coolers. In each of these devices, the effect of thermal energy transfer is dependent upon compound transfer processes (sometimes competing with one another) taking place concurrently. This will be explored further in subsequent sections dedicated to each of these devices.

5.3 Fundamental Technologies

Of the thermal conductance hardware and/or methods highlighted at the beginning of the chapter, with the exception of single-phase mechanical pumps the performance of the devices in each of the suggested approaches is often referenced against the performance features associated with heat pump systems. This is due to the fact that heat pump systems are regarded as having performance features that are appealing for application on space platforms. An assessment of the basic features associated with the operation of heat pump systems is therefore worthy of review.

We know that heat pumps:

- **acquire heat at a source location**
- **transport heat from the source location to the rejection location**
- **reject heat at the heat rejection location**
- use a working fluid
- provide temperature lift of the working fluid at the rejection location relative to the source location's temperature
- use evaporation and condensation for thermal energy exchange with the environment at the source and sink locations respectively
- have a mechanism (compressor) which institutes flow
- require electricity.

The primary features to be considered for analogy with other approaches/devices are highlighted in bold at the top of the list. A cursory review of the suggested thermal coupling approaches reveals some distinctions in features and performance compared to those in the list of heat pump features. Some devices incorporate the use of a working fluid, whereas others do not. Some require a mechanical pump to operate, whereas others do not. Table 5.1 groups each of the aforementioned thermal coupling methods by feature. These categorizations are important because they help define the general

Table 5.1 Categorization of standard spaceflight thermal conductance devices

Single phase	Multiphase
• Single-phase pump loop • TEC • Thermal straps	• Heat pipe • Loop heat pipe • PCMs (phase change materials) • Heat pumps
Solid state	**Mechanical pump driven**
• TEC • Thermal straps	• Heat pumps • Single-phase pump loop

characteristics of the approaches under consideration. Devices that cause vibration are generally undesirable for most unmanned space science platforms. Mechanical pumps foster vibration/jitter and their components can wear over time. The wear and degradation of the pump mechanism implies that the cooling capabilities of the system may be compromised with extended run time. This could impact the overall mission lifetime. Vibration isolation can be implemented to mitigate jitter. Whilst mission designers opt to simply avoid sources of vibration if at all possible, mechanical pumps were used on the space shuttle fleet and are presently in use on the space station. The space station pumps are large capacity (i.e., multi-kW) single-phase systems. These multi-kW systems are beyond the dissipation levels (and power budgets) observed on most modern Earth-based satellites. However, mechanical heat pump systems that provide cooling down to cryogenic temperatures (i.e., Cryocoolers) are increasingly being used on space-based missions [3].

Since the heat pump is the ideal thermal energy transport device for use on the space platform, due to its abundance of appealing performance features, let's take a closer look at heat pump systems as a thermal conductance technology.

5.3.1 Heat Pumps

Figure 5.1 shows a schematic of a heat pump system with its associated thermo-machinery. In this figure the state points are labeled 1 through 4. Starting at state point 4, heat is absorbed into the working fluid (which is primarily in the liquid state) via the evaporator on the low-pressure side of the loop. By state point 1 the working fluid is all (if not predominantly) vapor. That vapor is compressed via work input from a compressor to a higher-pressure level (as denoted by the high-pressure side of the loop). That high-pressure vapor condenses back into liquid by heat rejection via the condenser. By state point 3 the fluid is all (or predominantly) liquid. However, it is still at high pressure. The working fluid is then expanded through a valve to the low-side pressure. This is only feasible through manipulation of the Joule–Thomson effect. The Joule–Thomson effect fosters either an increase or decrease in temperature of a fluid when it undergoes expansion [4]. Within the refrigeration and/or thermodynamics community it is usually expressed as the Joule–Thomson coefficient (μ_{J-T}):

$$\mu_{J-T} = \left(\frac{\partial T}{\partial P}\right)_h \tag{5.8}$$

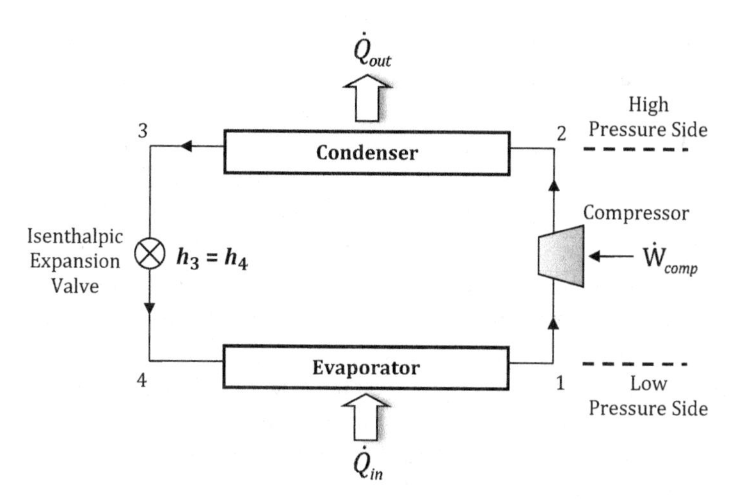

Figure 5.1 Heat pump system with associated thermomachinery

It is the rate of change of the temperature for the fluid as a function of the change in pressure applied at constant enthalpy. In the study by Hendricks et al. [5], the Joule–Thomson inversion curves for several common fluids (both cryogenic and non-cryogenic) were determined experimentally. In addition, the researchers provided polynomial relations for the data collected for each fluid. The Joule–Thomson inversion curve for neon, based on a polynomial, is reproduced in Figure 5.2. The Joule–Thomson value is applicable at any point on the curve.

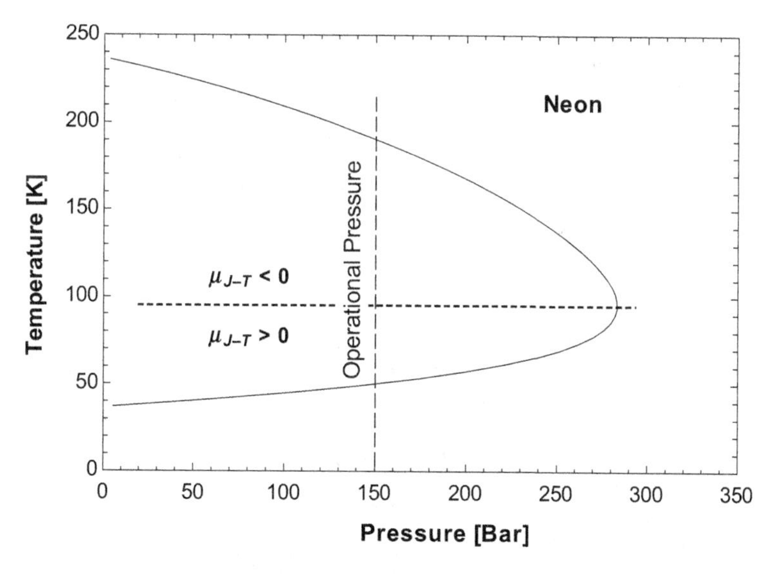

Figure 5.2 Plot for Neon Joule–Thomson coefficient curve

Source: Adapted with permission of John Wiley & Sons, Inc., *Advanced Thermodynamics*, Bejan, A., 1988, reference [6]; permission conveyed through Copyright Clearance Center, Inc.

If we draw a horizontal line through the inversion point (i.e., the inflection point where the sign value changes on the curve) and focus on the upper region, we can determine by inspecting the slope of the curve that a decrease in pressure would result in a negative Joule–Thomson value (and an associated increase in the temperature of the fluid). Alternatively, below the inversion point, a decrease in pressure will result in a positive Joule–Thomson value (and ultimately a decrease in fluid temperature). In a heat pump system, the thermodynamic cycle can only be completed if there is a reduction in the temperature of the working fluid (i.e., the refrigerant) corresponding with the reduction in pressure from state point 3 to state point 4. Thus, for given operating conditions (i.e., temperature and pressure) μ_{J-T} must be positive in order for the heat pump cycle to work. Upon reaching state point 4, the thermodynamic cycle has been completed and is poised to repeat for continuous thermal energy exchange with the environment. Last but not least, while the plot and the phenomena discussed here are specific to neon, the general shape of the inversion curve and the temperature variation associated with the Joule–Thomson effect due to fluid expansion applies to all pure fluids. Joule–Thomson expansion valves are used heavily in the cryogenics community for cryogenic tanks, thermodynamic vent systems and cryocooler systems (see Chapter 7).

Figure 5.3 is an example schematic of the vapor compression cycle on a *P–h* (i.e., pressure-enthalpy) diagram. Note that phase change occurs at a constant temperature and pressure for a pure fluid. However, this does not hold true for fluid mixtures. As mentioned previously, the heat pump provides heat acquisition at the evaporator and heat rejection at the condenser.

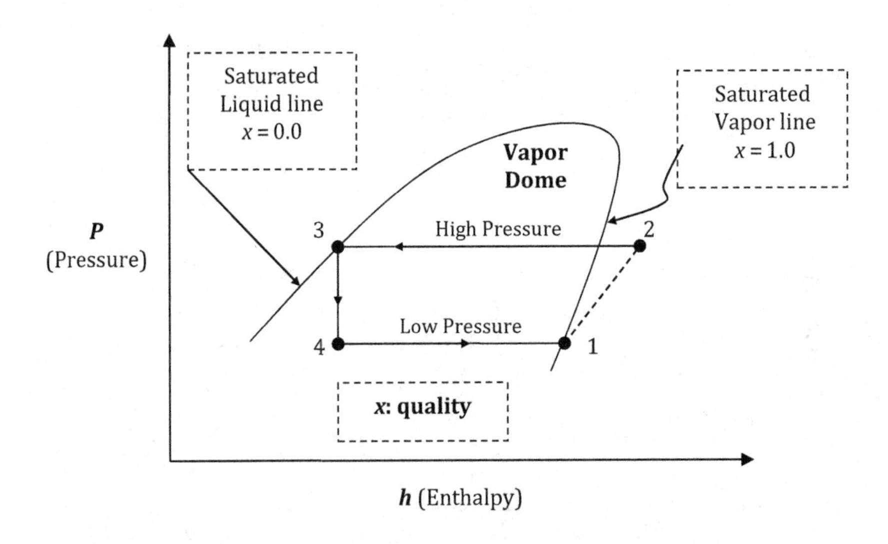

Figure 5.3 Vapor compression cycle on P–h diagram
Source: Adapted with permission of John Wiley & Sons, Inc. from *Fundamentals of Engineering Thermodynamics*, Moran, M.J., Shapiro, H.N., Boettner, D.D., and Bailey, M.B., 8th edn., 2014, reference [4]; permission conveyed through Copyright Clearance Center, Inc.

As such it transports thermal energy from a source location to a sink location. The liquid-to-vapor (in the evaporator) and vapor-to-liquid (in the condenser) phase change processes are occurring at constant temperature and pressure. For any given pure fluid, a given saturation pressure has a corresponding saturation temperature. Since the saturation pressure for condensation is greater than that for evaporation, the corresponding saturation temperature for condensation is also greater. The heat pump rejects heat at a higher temperature from the condenser than that absorbed in the evaporator. Thus, a temperature lift is associated with a heat pump's thermal energy transport process. How might this have an impact on a space platform? The radiative emissive power of a surface is strongly dependent upon its temperature. The higher the temperature, the more heat can be rejected. The potential to provide a temperature increase at the heat rejection location (which is what heat pumps accomplish) means more waste heat can be rejected. A device that could perform heat acquisition and transport from one location to another while also providing temperature lift without using mechanical components that wear would be very appealing for use on a space platform.

Example 1 Basic Heat Pump Cycle An R134a heat pump cycle (schematic shown in Figure 5.1) with a mass flow rate of 0.01 kg/s is operating with an evaporator temperature of 10°C and a condenser temperature of 35°C. Assuming the fluid is saturated vapor upon exit from the evaporator and saturated liquid when exiting the condenser, what is the cooling capacity (on the low-pressure side), heat rejection (from the condenser on the high-pressure side) and cycle efficiency for a compressor performing isentropically? Please include the Carnot efficiency in the calculations as well.

Solution

The solution of this problem requires knowledge of the thermodynamic conditions at each of the state points in the cycle. We will use EES which has equations of state for multiple pure fluids (including R134a) built into the software package. We will begin by establishing the known quantities of mass flow rate and temperature within the evaporator and the condenser.

```
{R134a Vapor Compression Cycle}
m_dot=0.01          [kg/s]
T_low=10              {Evaporator-side temperature, Celsius}
T_high=35             {Condenser-side temperature, Celsius}
```

In this problem we are performing calculations with the temperatures in degrees Celsius. However, it should be noted that this unit convention needs to be set under [Unit System > Options] located under the main menu. At state point 1 we know the quality of the fluid is saturated vapor. We can use both the quality and the temperature in the evaporator to determine the pressure and enthalpy at state point 1 by referencing the properties tables.

```
{State Point 1}
x_1=1.0
P_low=pressure(R134a,T=T_low,x=x_1)
h_1=enthalpy(R134a,T=T_low,x=x_1)
s_1=entropy(R134a,P=P_low,x=x_1)
```

The equation for the compressor work between state points 1 and 2s is

$$\dot{W}_{comp} = \dot{m} \cdot (h_{2s} - h_1) \tag{5.9}$$

where the subscript 2s denotes the entropy is unchanged relative to the entropy at the previous state point in the cycle. If the compression process was assumed to be irreversible, then state point 2 would not be under isentropic conditions and the subscript would simply be 2. Given that the mass flow rate, compressor work input and enthalpy at state point 1 are known, equation 5.9 can be written in EES as

```
{Compressor work input}
W_dot_comp=m_dot*(h_2s-h_1)          {units of kW}
```

The values at state point 2 may be written is EES as

```
{State point 2}
s_2=s_1
h_2s=enthalpy(R134a,P=P_high,s=s_2)
```

At state point 3, the quality is known to be saturated liquid. In addition, the temperature is known. Thus, the pressure and enthalpy can be determined by referencing the properties tables.

```
{State point 3}
x_3=0.0
P_high=pressure(R134a,T=T_high,x=x_3)
h_3=enthalpy(R134a,T=T_high,x=x_3)
```

The fluid expansion process occurring across the expansion valve (from state point 3 to state point 4) takes place at constant enthalpy. Thus

```
{State point 4}
h_4=h_3
```

Given knowledge of the enthalpy at each of the state points and the mass flow rate (which was previously defined), energy balances can be applied to both the evaporator and the condenser.

```
{Evaporator cooling capacity}
Q_dot_c=m_dot* (h_1-h_4)

{ Condenser heat rejection}
Q_dot_h=m_dot* (h_2s-h_3)
```

From basic thermodynamics, we know the Carnot efficiency for refrigeration is defined as

$$COP_{carnot} = \frac{T_{low}}{T_{high} - T_{low}} \tag{5.10}$$

where T_{low} and T_{high} are the temperature of the evaporator on the low-pressure side of the cycle and the temperature of the condenser on the high-pressure side of the cycle respectively. The actual coefficient of performance for the refrigeration effect associated with a heat pump cycle is defined as the cooling capacity divided by the compressor work input.

$$COP_{actual} = \frac{\dot{Q}_c}{\dot{W}_{comp}} \tag{5.11}$$

In EES this is

```
{COPs}
T_low_K=T_low+273
T_high_K=T_high+273

COP_Carnot=T_low_K/(T_high_K-T_low_K)

COP_actual=Q_dot_c/W_dot_comp
```

Execution of the code will result in a cooling capacity $\left(\dot{Q}_c\right)$ of 1.553 kW and a heat reject $\left(\dot{Q}_h\right)$ of 1.71 kW. The Carnot efficiency is 11.32, whereas the actual COP (coefficient of performance) for refrigeration is 9.86.

A common technique used in refrigeration technology to improve the efficiency of vapor compression cycles is the incorporation of an LLSL-HX (liquid line/suction-line heat exchanger) [7]. Figure 5.4 provides a schematic of a standard vapor compression cycle that includes an LLSL-HX. As shown in the figure, the LLSL-HX provides a thermal coupling between the condenser exit liquid and the evaporator exit vapor along the compressor suction line. This coupling allows for preheating of the saturated vapor exiting the evaporator, prior to entering the compressor. This technique is known as "recuperative" heat exchange. The transfer of thermal energy to the compressor suction line results in a loss of thermal energy in the fluid entering the expansion valve. This fosters lower temperatures in the evaporator and produces a net result of improved efficiency in the cycle [7]. Let's analyze a heat pump cycle that includes an LLSL-HX.

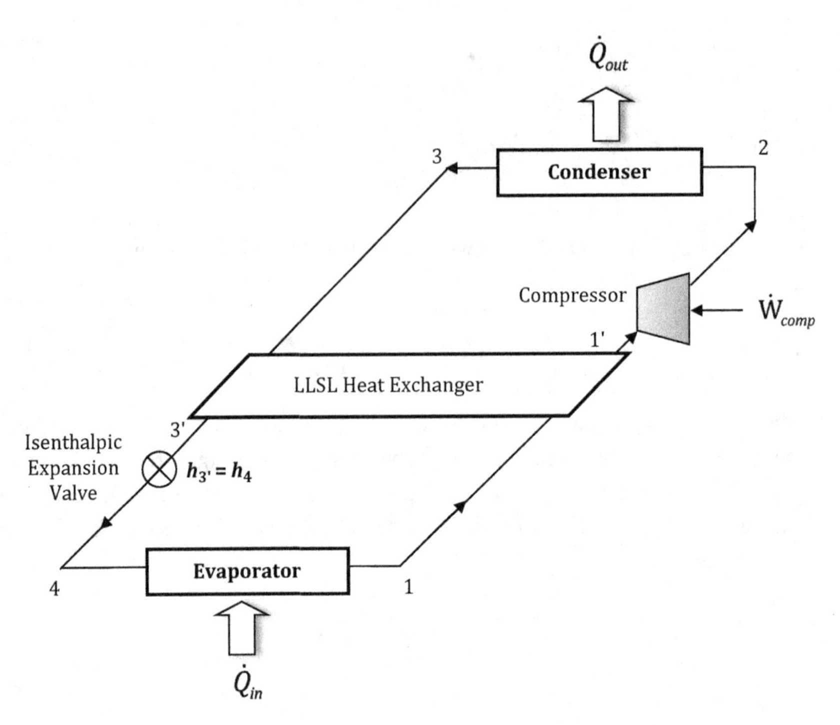

Figure 5.4 Heat pump system with an LLSL-HX

Example 2 Heat Pump Cycle with an LLSL Heat Exchanger Let's return to the R134a heat pump cycle described in Example 1 and incorporate an LLSL-HX. The new heat pump cycle is depicted in Figure 5.4. As in the isentropic heat pump cycle in Example 1, the work input is 157.5 W and the evaporator and condenser temperatures are 10°C and 35°C respectively. The fluid is saturated vapor upon exit from the evaporator and saturated liquid when exiting the condenser. Assuming an LLSL-HX effectiveness of 0.99, determine the cycle efficiency relative to that determined for the standard case in Example 1.

Solution

The solution of the problem requires knowledge of the thermodynamic conditions at each of the cycle's state points. Thus, once again we will use EES to assist in determining the solution. The environmental boundary conditions are the same as those for the cycle in Example 1. Thus, we can capture those in EES as

```
{R134a Vapor compression cycle w/LLSL heat exchanger}

T_low=10            {Evaporator-side temperature, Celsius}
T_high=35           {Condenser-side temperature, Celsius}
```

```
{Compressor work input}
W_dot_comp=0.158        {units of kW}
```

The conditions at state point 1 are the same also.

```
{State point 1}
x_1=1.0
T_1=T_low
P_low=pressure(R134a,T=T_low,x=x_1)
h_1=enthalpy(R134a,T=T_low,x=x_1)
```

However, this time we have added an additional line in this block of code to assign an explicit temperature of T_1 to state point 1. In Figure 5.4, state point 1′ is at the exit of the LLSL-HX on the suction-line side of the compressor. This suggests that the pressure at 1′ really is the low-side pressure of the cycle. However, this is the only independent variable for state point 1′ known at this time. Ideally, both the temperature and the enthalpy should also be known. Since the effectiveness of the LLSL-HX was defined in the problem statement, we can use the heat exchanger effectiveness relation to determine the temperature at 1′. As regards the heat exchanger in the cycle schematic, that relation is

$$Eff_{HX} = \frac{T_{1p} - T_1}{T_3 - T_1} \tag{5.12}$$

where subscript 1p denotes 1′. The equation is the ratio of the actual temperature glide to the maximum temperature glide possible on the heat exchanger suction-line side. Inherent to equation 5.12 is the assumption that at 100% effectiveness the heat transfer into the suction-line flow stream will raise the temperature at state point 1′ to the same value as that at state point 3. On determining the temperature, we can then find the enthalpy by referencing the properties tables. In EES the corresponding code will be

```
{State point 1p; LLSL heat exchanger relations}
P_1p=P_low

HX_Eff=0.99
HX_Eff=(T_1p-T_1)/(T_3-T_1)

h_1p=enthalpy(R134a, P=P_1p, T=T_1p)
s_1p=entropy(R134a, P=P_1p, T=T_1p)
```

At state point 2s we will have

```
{State point 2s}
P_2=P_high
s_2=s_1p
h_2s=enthalpy(R134a,P=P_high,s=s_2)
```

In the present problem we were able to determine the enthalpy at state point 2s given the compressor work value provided and the enthalpy determined at state point 1′. If the compressor work input was not provided, the isentropic work input could be determined from the difference between the enthalpy at state point 2s and the enthalpy at state point 1′. Given knowledge of the compressor efficiency (η_{comp}) we could then use equation 5.13 to determine the actual work

$$\eta_{comp} = \frac{\dot{W}_{isen}}{\dot{W}_{actual}} = \frac{h_{2s} - h_{1p}}{h_2 - h_{1p}} \tag{5.13}$$

At state point 3 we have

```
{State point 3}
x_3=0.0
T_3=T_high
P_high=pressure(R134a,T=T_high,x=x_3)
h_3=enthalpy(R134a,T=T_high,x=x_3)
```

The enthalpy conditions at all the inflow/outflow locations of the LLSL-HX are known, with the exception of the enthalpy at state point 3′. An energy balance can be applied to the heat exchanger to determine this one unknown. The equation for the energy balance will be

$$h_3 - h_{3p} = h_{1p} - h_1 \tag{5.14}$$

where the subscript 3p denotes 3′ and the mass flow rate has been dropped from both sides of the equal sign because it is constant. The corresponding EES code will be

```
{State point 3p}
h_3-h_3p=h_1p-h_1
T_3p=temperature(R134a, P=P_high,h=h_3p)
```

The isenthalpic expansion from state point 3 to state point 4 is captured as

```
{State point 4}
h_4=h_3p
```

The evaporator cooling capacity, condenser heat rejection and the actual coefficient of performance is

```
{Evaporator cooling capacity}
Q_dot_c=m_dot*(h_1-h_4)

{Condenser heat rejection}
Q_dot_h=m_dot*(h_2s-h_3)
```

```
{Compressor energy balance}
W_dot_comp=m_dot* (h_2s-h_1p)                    { Units of kW}

{ COPs}
COP_actual=Q_dot_c/W_dot_comp
```

Execution of the EES code will result in an actual COP of 9.98. This is a modest increase relative to the baseline case for the heat pump cycle from Example 1 that does not include an LLSL-HX. Additional performance improvement can be achieved by varying the pressure ratio in the cycle. The topic of recuperative heat exchange will be revisited in Chapter 7.

5.3.2 Thermal Straps

One basic thermal coupling that is often used is the thermal strap. A thermal strap may also be referred to as an FCL (flexible conductive link) [8]. Thermal straps are used in the electronics industry (in laptops and desktop computers) to conduct heat away from processors, as well as extensively in the aerospace industry. In aerospace applications they are typically made of copper and aluminum, whereas the computer/electronics industry has been known to use graphite thermal conductors as well. Figure 5.5 provides an example schematic for the general configuration of a thermal strap (TS). Thermal strap construction consists of a conductor section and mounting flanges to be attached to a temperature source and sink. The conductor cross-sections are typically made of wires or thin film layers (i.e., foils). Figure 5.6 shows example endpoint cross-sections at the header/footer mounting flanges for braided wire and foils. In addition to heat flow passing through the header and/or footer to and from the mounting surfaces, heat flow will also pass through the contact joints between the header and footer and the conductor materials. For high thermal conductivity materials, this most often results in the limiting resistance being located at either the mounting interfaces or the conductor to header/footer interfaces.

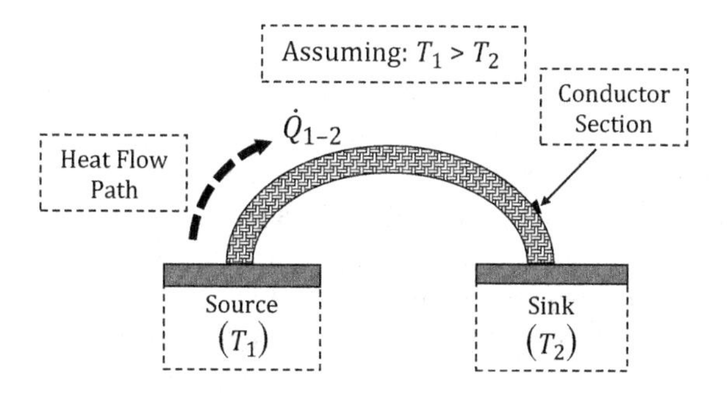

Figure 5.5 Thermal strap connecting heat source and heat sink

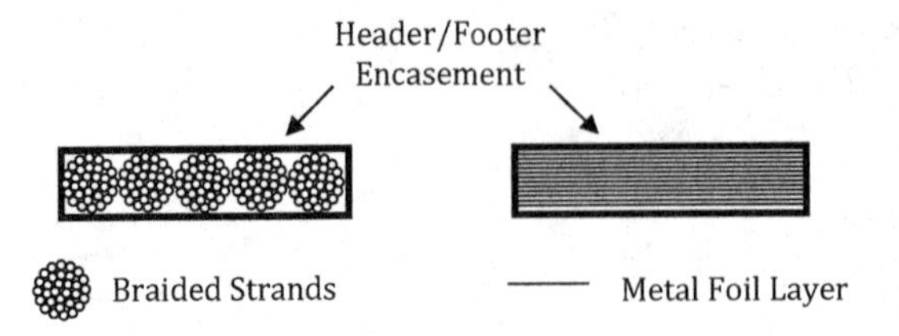

Figure 5.6 View of endpoint cross-sections for braid and foil assembly conductors

Let us examine an example of a thermal strap fabricated from thin metal foils. In aggregate multiple foils stacked together will approximate to the cross-sectional area of a contiguous piece of material (example schematic shown in Figure 5.7). The contiguous material analogue is the ideal case for thermal conduction through a thermal strap of a given geometry. From endpoint 1 to endpoint 2 we know that the effective thermal conductance in the contiguous material case is

$$C_{Contiguous} = \frac{1}{R_{Contiguous}} = \frac{kA}{l} \tag{5.15}$$

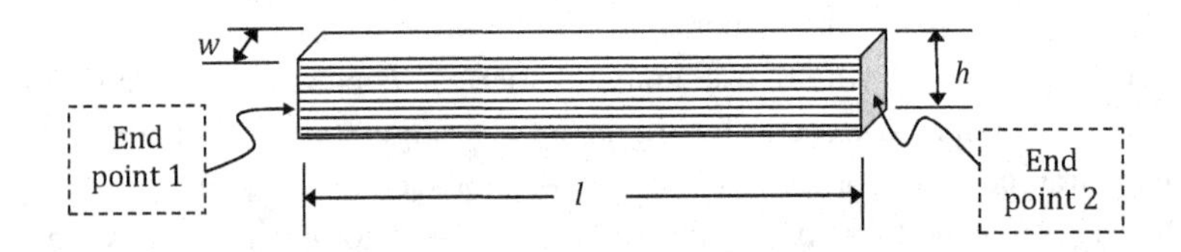

Figure 5.7 Composite foils approximating to a contiguous material conductor

where $A = w \cdot h$. In the case of a thermal strap consisting of several thin foils, Bugby and Marland [8] define the effective thermal conductance as

$$C_{TS} = \frac{1}{R_{TS}} = \left(\frac{kA}{l}\right)\eta_P\eta_S\eta_E \tag{5.16}$$

where η_P, η_S and η_E are scaling factors. These scaling factors are called the packing, shape and endpoint efficiencies respectively. Packing efficiency (η_P) is a measure of how well the foils can be consolidated together in such a way that their comprehensive thermal conductance approximates to the performance of a contiguous material. Shape efficiency (η_S) pertains to the amount by which the foil/braid conductor lengths must extend due to flexing in applications. Last but not least, endpoint efficiency (η_E) is a measure of how well the individual conductor elements attach to the endpoints. Theoretically, each of these efficiencies can be a value of unity. Figure 5.8 shows a schematic of parallel braided thermal straps connecting two mounting flanges.

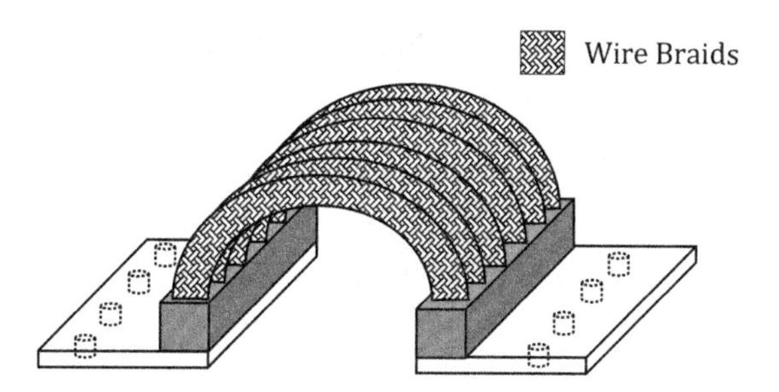

Figure 5.8 Braided thermal strap with header and footer flanges

The product of the cross-sectional area for the contiguous material case and the packing efficiency can be recast in terms of the cross-sectional area for the total number of foils being used. This relation is shown in equation 5.17

$$A\eta_P = N_F \delta_F W_F \tag{5.17}$$

where the variables on the RHS are the number of foils (N_F), the thickness of a single foil (δ_F) and the width of a single foil (W_F). When the foils are packed very tightly, it is not uncommon for η_P to approximate to 1.0. The value for η_S is strongly influenced by the level of tortuosity imposed on the individual conductors in application. For S-bends applied to moderate length foils (i.e., one foot or less), shape efficiency impacts are frequently considered negligible, promoting a value that approximates towards unity. The value for η_E is dominated by the endpoint attachment technique. Standard options for attachment techniques include soldering, swaging and solid-state welding (i.e., diffusion bonding). Solid-state welding is considered the best. Thermal straps with diffusion-bonded end pieces subjected to mild/moderate flexing have demonstrated thermal conductance efficiencies comparable to the contiguous material case when operated at helium temperatures (i.e., 4 K) [9]. However, such high efficiencies depend on the workmanship and on how the thermal strap is implemented. To find the most accurate value for the thermal conductance performance capability of a thermal strap construction, it is customary to test the thermal conductance following fabrication. This practice is especially relevant in cryogenic applications.

Example 3 Thin Foil-Based Thermal Strap Conductance Model Let's say we have 118 foils of 5 mil OFHC ($k_{OFHC} = 380$ W/m · K). Each foil is 150 mm long and 1 inch wide. What is the predicted thermal conductance for this thermal strap?

Solution

We need to find C_{TS}. We will start by incorporating equation 5.16.

$$C_{TS} = \frac{1}{R_{TS}} = \left(\frac{kA}{l}\right)\eta_P \eta_S \eta_E$$

Since the foils are very small, we will assume $\eta_S = 0.7$. Also, we will assume that the endpoint connections are fairly good ($\eta_E = 0.9$). This is a fair assumption if the endpoints are fused together. If we substitute $N_F\delta_F W_F$ for the product of $A\eta_P$ in the conductance equation, we will now have

$$C_{TS} = \frac{1}{R_{TS}} = \frac{k}{l}(N_F\delta_F W_F)\cdot\eta_S\eta_E \tag{5.18}$$

We also know that 1 inch = 2.54 cm and 1 mil = 0.0254 mm. Substituting in these values we will attain

$$C_{TS} = \left(\frac{380\ \mathrm{W/m\cdot K}}{0.150\ \mathrm{m}}\right)(118)(1.27\times10^{-4}\ \mathrm{m})(0.0254\ \mathrm{m})(0.7)(0.9) \tag{5.19}$$

which when calculated out reduces to

$$C_{TS} = 0.6075\ \mathrm{W/K}$$

Next, if we wanted to obtain the total heat transfer occurring through this conductor between two locations with a known ΔT, we would have to invoke the overall heat transfer relation from Chapter 1.

$$\dot{Q}_{TS} = UA\cdot\Delta T$$

In the context of the present problem, we know that this is equal to

$$\dot{Q}_{TS} = C_{TS}\cdot\Delta T \tag{5.20}$$

One could then determine that ΔT = 10°C would result in a heat transfer of 6 W. Alternatively, if a heat removal amount of 5 W was quoted, one could determine that the ΔT using this thermal strap would be 8.23°C.

If wire braids were used, the product of the cross-sectional area for the solid bar ideal case and the packing efficiency would be recast in terms of the cross-sectional braid area required for the comprehensive number of wires being used [8]. That would be

$$A\eta_P = N_{wire}\frac{\pi D_{wire}^2}{4} \tag{5.21}$$

Example 4 Determination of Wire-Based Thermal Strap Dimensions For the C_{TS} value determined in Example 3 (0.6075 W/K), what would be the total number of copper wires required if each wire had a diameter of 0.2 mm and a length of 150 mm? The wire material is OFHC (k_{OFHC} = 380 W/m · K). The shape and endpoint efficiencies are assumed to be η_S = 0.7 and η_E = 0.9.

Solution

From the problem statement and equation 5.16, we know that

$$C_{TS} = 0.6075\ \mathrm{W/K} = \frac{1}{R_{TS}} = \left(\frac{kA}{l}\right)\eta_P\eta_S\eta_E$$

Substituting in the RHS of equation 5.21 for the product of the wire cross-sectional area and the packing efficiency will give a new relation of

$$0.6075\ \mathrm{W/K} = \frac{k}{l} \cdot N_{wire} \frac{\pi D_{wire}^2}{4} \eta_S \eta_E$$

Solving for N_{wire} we would have

$$N_{wire} = \frac{1}{\eta_S \eta_E} \frac{4 \cdot (0.6075\ \mathrm{W/K})}{\pi D_{wire}^2} \cdot \left(\frac{l}{k}\right)$$

or

$$N_{wire} = \frac{1}{(0.7)(0.9)} \frac{4 \cdot (0.6075\ \mathrm{W/K})}{\pi (0.2\ \mathrm{mm})^2} \cdot \left(\frac{150\ \mathrm{mm}}{380\ \mathrm{W/m \cdot K}}\right)$$

This calculation results in a value of N_{wire} = 12,116. If the total number of wires were distributed across eight individual braids then the number of wires per braid would be 1,515.

Thermal straps are often made to order by manufacturers. Vendors will typically quote conductance values associated with different geometries that they manufacture (e.g., 5 W/K, 10 W/K, 15 W/K). Using the conductance/resistance form in Fourier's law of conduction, the user can match the thermal strap with the conductance performance they desire for their thermal configuration. Thermal straps do not enable on/off control. If on/off control is required, an additional heater must be added along the heat flow path to negate thermal energy transfer from the source to the sink. Alternatively, a different thermal control technology may be chosen.

5.4 Boiling Heat Transfer Components in 1-g and Microgravity

The next two sections provide an overview of heat pipes and loop heat pipe technology. Both these types of devices are standard components on contemporary spaceflight platforms. They also are multiphase thermal conductors that operate according to the principles of evaporation, nucleate boiling and condensation. To understand their operating characteristics, we must first establish a good understanding of the phase change process and the nuances inherent to such in a microgravity environment. We start by examining the different heating regimes observed in the pool-boiling process, as well as reviewing online video of boiling processes in 1-g. Video footage of boiling processes and liquid flows in microgravity environments will also be reviewed. Figure 5.9 is an example schematic of the pool-boiling curve for a pure fluid. The boiling curve can be divided into six different sections (each associated with distinct phenomena). As shown in Figure 5.9, numbers are listed at each of the key points along the curve. A description of the heating regimes (I ↔ VI) is provided in Table 5.2.

The heating process for the curve shown in Figure 5.9 follows the path 0-1-2-3-4-5-6. At low heat input (shown in region I), heat transfer is dominated by natural convection occurring within the pooled liquid. Liquid which has undergone heating near the heat input location has an increased temperature (and a corresponding decrease in density). This liquid rises from the heat input location up to the free surface of the pooled liquid, where it cools off. After decreasing in temperature, the now denser liquid falls back to the bottom of the pool near the heat input location. At constant heat input, a recirculation zone is accordingly fostered in the liquid. When the heat input is increased further beyond process point 1 (i.e., into region II), nucleate boiling begins. This is shown in Table 5.2

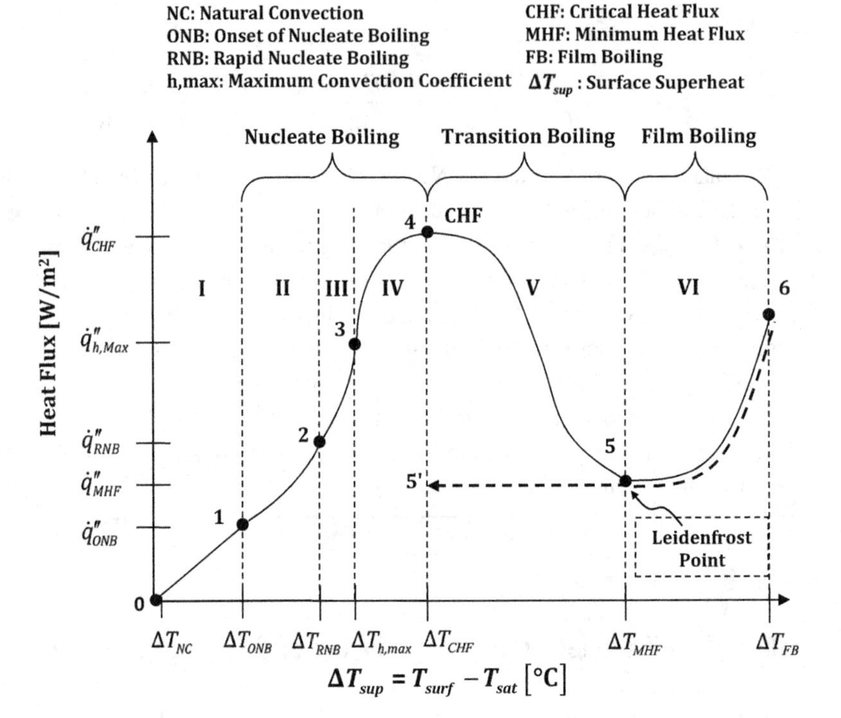

Figure 5.9 Pool-boiling diagram
Source: Adapted with permission of John Wiley & Son's Inc. from *Fundamentals of Heat and Mass Transfer*, Bergman, T., Lavine, A., Incropera, F.P., and Dewitt, D.P., 7th ed., 2019, reference [10]; permission conveyed through Copyright Clearance Center, Inc.

by the production of bubbles originating at and released from the heat input surface. Beyond process point 2 (extending into region III), the rate of bubble nucleation at the heater surface becomes more rapid. This is denoted by the abundance of bubbles released from the heater surface, as shown in Table 5.2. Process point 3, which is an inflection point on the curve, denotes the maximum value for the convection coefficient. In the approach to critical heat flux spanning region IV, rapid nucleation is present once again. However, superheating of the heater surface becomes pronounced and vapor is produced in the form of jets and columns. With continued heating beyond the critical heat flux (CHF) value (i.e., process point 4), the ability to replenish the heater surface with pooled liquid is diminished due to the formation of a sustained vapor layer across

Table 5.2 Pool-boiling curve process points (1-g)

	Pool-boiling curve regions					
	I	II	III	IV	V	VI
Mechanism	Natural convection	Low-heat flux nucleation	Moderate-heat flux nucleation	Near-critical heat flux w/extended superheating	Transition boiling	Film boiling
Schematic	Heater [a]	Heater	Heater	Heater [b]	Heater	vapor Heater [a]
Temperature	$T_{NC} \leftrightarrow T_{ONB}$	$T_{ONB} \leftrightarrow T_{RNB}$	$T_{RNB} \leftrightarrow T_{h,max}$	$T_{h,max} \leftrightarrow T_{CHF}$	$T_{CHF} \leftrightarrow T_{MHF}$	$T_{MHF} \leftrightarrow T_{FB}$
Description	Fluid motion is resultant from a density gradient between the heating location and the free surface of the pooled liquid.	Onset of nucleate boiling occurs. Vapor production is represented by isolated bubbles.	Transition to rapid nucleation. Vapor production represented by jets and columns.	Rapid nucleation with pronounced superheating during the approach to CHF (critical heat flux).	Liquid phase can no longer wet the heater surface continuously and acquire heat. Vapor layer growth begins. Region ends at Leidenfrost point.	Vapor layer blankets the heater surface creating additional thermal resistance between heat input location and pooled liquid.

Sources: [a]Adapted with permission of Taylor & Francis, Inc. from Carey (1992), reference [11], permission conveyed through Copyright Clearance Center, Inc.; [b]Adapted with permission of ASME Journal from Lienhard (1985), reference [12], permission conveyed through Copyright Clearance Center, Inc.

the heater surface. This is known as transition boiling (i.e., region V). When a static vapor layer has been established, MHF (minimum heat flux) is attained (shown as state point 5). Further increases in heat input beyond state point 5 will result in the transfer of heat through this newly formed vapor layer. This phenomenon is known as film boiling (shown as region VI in Figure 5.9). As shown in Table 5.2, phase change now occurs at the liquid/vapor interface as opposed to the solid/liquid interface on the heater surface. If heating were to be stopped at process point 6 and a cooling cycle applied, the path of the cooling cycle would follow 6-5-5′ as the liquid pool is reconditioned. The heat flux observed during this cooling cycle will be noticeably lower than that observed during the heating cycle. The phenomenon associated with the reduction in heat transfer between process points 5 and 5′ is known as boiling hysteresis and is an inherent feature of most boiling systems. On cooling down to a surface superheat value of ΔT_{CHF}, the vapor layer blanketing the surface during transition and film boiling collapses. This will allow liquid to come into direct contact with the heater surface once again and reinstate direct contact boiling. The value associated with the heat flux at process point 5′ will vary depending on the amount of stored thermal energy present in the heater.

Let us now turn to some online video content providing visualization of the pool-boiling phenomenon described above. The University of New Mexico's Institute for Space and Nuclear Power Studies (located at isnps.unm.edu) contains a video library of both flat and enhanced surface pool-boiling studies performed with FC-72. These can be found under the [Publications > Video] header on the website. A review of any of the submerged surface videos shows that as heat input to the heater surface is gradually increased, the generation of bubbles at the heater surface transitions increasingly from isolated bubbles at a low frequency, through rapid bubble generation on the heater surface, to vapor jets and columns until CHF is reached. When the heat input is increased while at CHF, the vapor jets and columns cease immediately and the transition to film boiling occurs. This phenomenon is to be expected in 1-g. However, can one expect the same in microgravity? The separation of the liquid and vapor phases during the bubble growth and departure cycle is promoted by the density gradient experienced within the gravitational reference frame. In 1-g, the body forces acting on the superheated vapor at the heater surface are less than those acting on the saturated liquid, promoting self-separation along the direction of the gravity vector. In the microgravity analogue self-separation of the liquid and vapor phases is substantially reduced.

Figure 5.10 consists of a series of photos illustrating the growth progression of an individual bubble during heating of a liquid pool of perfluoro-n-hexane. The images were taken during the nucleate pool-boiling experiment conducted onboard the ISS (International Space Station) by UCLA's Boiling Heat Transfer Laboratory in 2011. Following incipience of boiling, a small vapor dome emerges (photo a). This is the primary vapor bubble. With additional heat input the primary vapor bubble gradually increases in volume while remaining (approximately) in the same location on the heat input surface (photos b and c). In photo d, a new vapor bubble emerges just to the right of the primary bubble. Further heat input promotes an increase in the size of both bubbles with the smaller vapor bubble moving to the right (photos e and f). Eventually the two vapor bubbles contact one another and the smaller bubble merges with the primary bubble (photo g). The merger of the two bubbles results in the new primary bubble moving just to the right of center on the heater surface. As heat input continues, the new primary bubble moves back towards the center of the heater surface (photo h). Sustained heating (photos i through l) results in continued growth of the primary bubble along with the emergence of smaller bubbles that also merge with the primary vapor bubble over time. As the series of photos show, bubble departure from the heater surface has not occurred. Rather, the vapor bubbles remain attached at the heater surface and tend to coalesce over time [13].

One non-dimensional parameter used to gauge the impact of a gravitational reference frame on liquid/vapor phase separation in pooled liquid boiling processes is the Bond number. By definition, the Bond number is the ratio of gravitational to surface tension forces [14–16]. In equation form it is

$$\mathrm{Bo} = \frac{g(\rho_l - \rho_v)l_c^2}{\sigma_l} \tag{5.22}$$

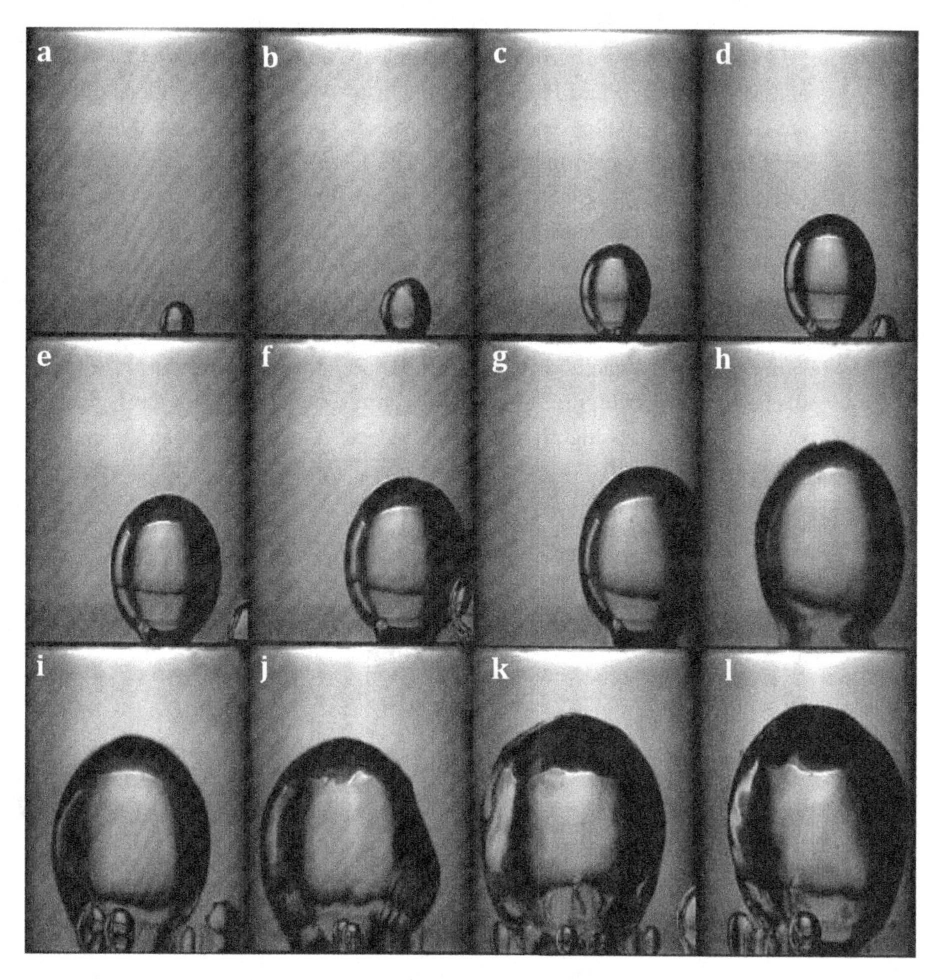

Figure 5.10 Bubble growth in a liquid pool of perfluoro-n-hexane onboard the International Space Station [13]
Source: Photos by Professor Vijay Dhir, University of California Los Angeles

where g is the gravitational constant for the reference frame and l_c is the characteristic length. Gravitational effects are pronounced for Bo $\gg 1$. In the case of microgravity, the numerator approximates to zero (i.e., Bo $\ll 1$) and gravitational effects are considered negligible.

As mentioned previously, the study of pool boiling is a basis for developing a fundamental understanding of the liquid-to-vapor phase change process. However, in spaceflight applications, thermal control hardware designed to cool flight components typically will not consist of a static pool of liquid. Successful operation of the thermal control device will often be dependent upon fluid flow (liquid and vapor phases) processes in the microgravity environment. An understanding of the fundamental aspects of microgravity fluid flow is essential to the design, development and performance assessment of items such as heat pipes, loop heat pipes and liquid-phase storage dewars (cryogenic and non-cryogenic).

Video content pertaining to boiling processes and microgravity fluid flow phenomena may be viewed at Glenn Research Center's Microgravity Two Phase Flow video library [17] and on AIAA's Thermophysics TC YouTube channel, under the Boiling Processes and Microgravity Fluid Phenomena playlists, where the two videos of primary interest are the jet column impacting a flat surface on a post in microgravity and the water spray on a heated surface in microgravity. The jet column video elucidates the fact that in microgravity inertia forces will dominate a flow given the reduction in body forces. Surface tension also has a pronounced effect. This is demonstrated by the adherence of the liquid to the post as opposed to liquid rebound and radial spreading upon contact. The spray cooling videos provide insights into liquid vapor phase change processes that occur in microgravity. As shown in the video, the liquid droplets contacting the heater surface adhere to each other and form liquid domes that are fairly static in position on the heater surface. Figure 5.11 provides comparison photos for the 1-g and microgravity spray cooling cases highlighted in the videos. As shown in the figures, in microgravity

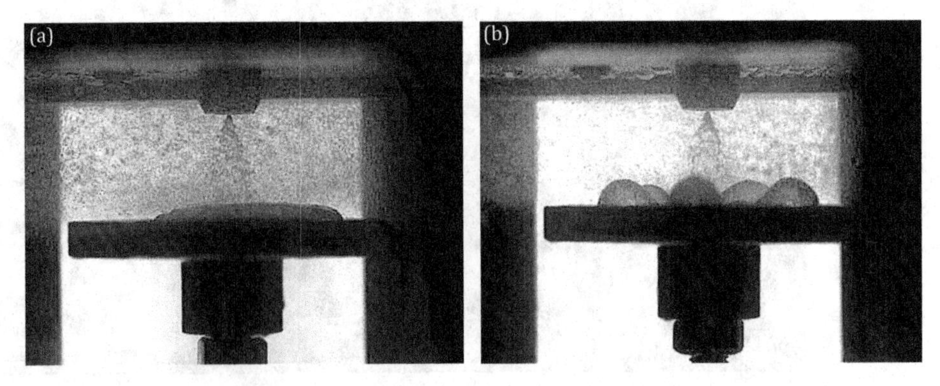

Figure 5.11 Water-based spray cooling as a function of gravity: a) 1-g (Earth gravity), b) μ-g using NASA Glenn 2.2 second drop tower

Source: Golliher et al. (2005), reference [18], photos provided courtesy of ASME w/permission

heightened surface tension effects promote liquid adherence to the heater surface and resident liquid volume either contacting or remaining in close proximity to the heater surface. When it comes to ensuring that cooled liquid is reaching the heater surface, such phenomena can pose challenges that need to be addressed in order for proper performance of liquid-to-vapor phase change devices to be achieved in microgravity.

5.4.1 Heat Pipes

There are many different types of capillary wick heat pipes. The overview in this section focuses on capillary wick heat pipes. The standard capillary wick heat pipe consists of three basic sections that are categorized according to their heat transfer functionality. These are: an evaporator section, an adiabatic section and a condenser section. The interior of a heat pipe contains a working fluid that exists in both the liquid and vapor phases, as well as a wick structure. The wick structure retains liquid and is a key component enabling successful heat pipe operation. The operation of a heat pipe is made possible by the simultaneous occurrence of fluid flow and heat transfer processes internal to the system and at the wall of the encasement structure. The heat exchange sections and the heat and fluid flows occurring during operation are shown in Figure 5.12. During operation, heat is added to the evaporator section. When a sufficient level of superheat is attained, the liquid in the evaporator undergoes phase change (i.e., evaporation and/or boiling). The phase-changed vapor is then released into the core of the heat pipe where it flows past the adiabatic section into the condenser section. In the condenser section, the vapor condenses back into liquid and is absorbed into the wick structure. The wick structure then pumps the liquid back to the evaporator section. Thus, in addition to promoting liquid retention, the wick structure serves as a path for the delivery of liquid from the condenser to the evaporator. The evaporation/condensation cycle that occurs during operation provides continuous cooling to the source of the evaporator's heat input.

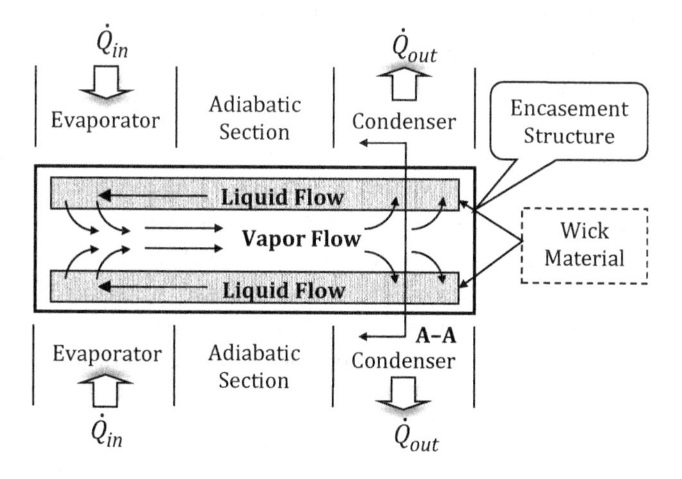

Figure 5.12 Heat pipe sections with liquid and vapor flows

A key measure of how effectively a heat pipe conducts thermal energy from its evaporator to its condenser is the ratio of its on/off thermal conductance. When the heat pipe has turned on and is operating, its thermal conductance is dominated by the evaporation/condensation cycle of the working fluid it is charged with. When it is turned off, thermal conductance is still present. However, it is limited to the value associated with the heat pipe construction materials. Standard values for the thermal conductance on/off ratio can range between 50:1 and 100:1 for room-temperature applications.

One fundamental feature associated with the use of fluid systems in microgravity is fluid management. In a gravitational environment self-separation of the liquid and vapor phases occurs automatically due to the presence of the density gradient between the two phases. However, in microgravity, self-separation of the phases does not occur. Thus, in order to separate phases and ensure that liquid is in the proper location for heat exchange to take place, either a material and/or technique that fosters liquid retention and delivery to the heat exchange location is required. In the case of heat pipes, the wick structure performs the task of liquid management and ensuring that wetting of the heat input surface is occurring. Wick constructions can be made from screens, axial grooves, sintered porous materials, fiber meshes, foams or combinations of any of these. Standard wick materials used in heat pipes include copper, stainless steel, nickel and titanium [19, 20]. Figure 5.13 shows an expanded view of cross-section A–A from Figure 5.12. Three common heat pipe constructions include sintered materials, axial grooves and compound wick structures. Compound wick structures such as axial grooves and screens or axial grooves and sintered pores may also be used. Since the transfer processes associated with the functionality of a heat pipe include heat transfer and fluid flow, the impact upon each of these processes due to the use of different types of wick structures is of interest. One key fluids-based parameter associated with the flow of liquids through porous materials is permeability. Highly permeable structures allow more fluid to flow through them, as they encounter lower flow resistance than that which is observed with the use of low-permeability structures (i.e., structures that foster high flow resistance). Based on the construction cases shown in Figure 5.13, sintered materials tend to foster low permeability. Higher permeability is observed with structures that include axial grooves. With regard to heat transfer, low thermal conductivity sintered materials provide a thermal break between the evaporator and condenser regions. The wick's capillary pumping force, which ultimately drives the transfer of liquid from the condenser to the evaporator, is also a function of the wick structure. The wick structures with the greatest pumping capability are sintered pore structures and compound structures that consist of axial grooves and either sintered pore or screen materials. For a comprehensive listing of wick structure options, please see references [19,21–24]. From a physical perspective, in order to institute flow, the pressure head produced from the capillary wick structure used ($\Delta P_{cap,wick}$) must be greater than or equal to the total pressure drop within the system [21,23,24].

$$\Delta P_{cap,wick} \geq \Delta P_{tot} \tag{5.23}$$

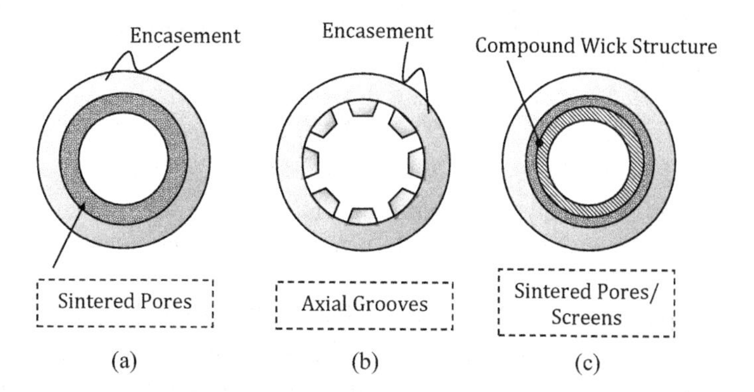

Figure 5.13 Heat pipe wick cross-sections: a) sintered pores; b) axial grooves; c) sintered pores w/screens

The total pressure drop within a heat pipe system (ΔP_{tot}) may be defined as

$$\Delta P_{tot} = \Delta P_{liq} + \Delta P_{vap} + \Delta P_{evap,int} + \Delta P_{con,int} + \Delta P_{g} \tag{5.24}$$

where ΔP_{liq} and ΔP_{vap} are the pressure drop across the wick structure and the vapor in the core respectively. $\Delta P_{evap,int}$ and $\Delta P_{con,int}$ are the pressure drops at the liquid/vapor interfaces due to evaporation and condensation respectively. ΔP_{g} is the pressure drop associated with the body forces acting on the liquid phase in either the radial or axial direction. The dominant contributors to the total pressure drop relation in equation 5.24 are the pressure drops across the wick and the vapor, and that due to body forces when applicable. For a heat pipe with axial grooves and/or a porous wick structure the pressure drop associated with the wick construction used will serve as the limiting resistance to fluid flow in the heat pipe. For a fine pore wick structure with an average pore size on the order of $10^{-5}/10^{-6}$ m, one can easily deduce that the pressure drop across the wick structure will be pronounced relative to the case where a fine pore wick structure is not used (e.g., the axial groove case). Using the Darcy equation for flow in a porous medium [25], the liquid-phase pressure drop across the heat pipe wick structure can be defined as [21,23,24,26]

$$-\frac{\Delta P_{wick}}{\Delta x} = \frac{\mu_l}{K_{wick}} \cdot Vel \tag{5.25}$$

where μ_l is the liquid viscosity [kg/m · s], Vel is the velocity [m/s] of the liquid flow through the media, K_{wick} is the permeability [m^2] for the structure and Δx is the length [m] of the porous medium experiencing the pressure drop [N/m^2]. For a heat pipe wick structure ΔP_{wick} is equivalent to ΔP_{liq} in equation 5.24. This relation is applicable to creeping flows (i.e., Re $\ll$ 1), whereas at high velocities the Darcy–Forchheimer relation is more appropriate. It can be shown analytically that the pressure differential across the wick structure is a function of the mass flow rate by solving the mass flow rate equation

$$\dot{m} = \rho_l \cdot A \cdot Vel \tag{5.26}$$

for velocity and then substituting this new velocity relation into equation 5.25. This would give a new form of

$$-\frac{\Delta P_{wick}}{\Delta x} = \frac{\mu_l \cdot \dot{m}}{\rho_l \cdot A \cdot K_{wick}} \tag{5.27}$$

On inspection it is clear analytically that ΔP_{wick} is a function of $\dot{m}$. The rate of liquid mass flow through the wick structure can be determined from the enthalpy of vaporization conservation relation.

The capillary pumping action which results in liquid flow is due to capillary wicking effects taking place in the individual pores of the structure. These effects are highly dependent upon the surface tension of the working fluid and the surface with which the fluid is in contact at the solid/liquid/vapor interface. In reference to surface tension and solid/liquid/vapor interfaces the terms wetting and non-wetting are often used. A wetting condition for a liquid on a solid surface is defined by the contact angle between the solid/liquid interface and the liquid/vapor interface as measured from the liquid side of the interfaces. Figure 5.14 shows examples of both a wetting and a non-wetting case for a given liquid. In the non-wetting case, the contact angle is obtuse (i.e., $90° < \theta < 180°$). For the wetting case the contact angle is acute (i.e., $\theta < 90°$). For a fine pore wick in a heat pipe with a cylindrical geometry, liquid will advance into the individual pores and wet the walls as shown in Figure 5.15a. Without heat input to the wick structure, the curved surface which defines the liquid/vapor interface (i.e., the meniscus) is static. This interface comprises two individual radii (example shown in Figure 5.15b). When the liquid/vapor interface is in equilibrium, it is defined using the Young–Laplace equation [11,19–22,24].

$$P_{vap} - P_{liq} = \sigma_l \left[\frac{1}{R_{\mathrm{I}}} + \frac{1}{R_{\mathrm{II}}}\right] \tag{5.28}$$

In equation 5.28, σ_l is the liquid surface tension in N/m. The R terms are radii for the spherical sections that combine to create the meniscus. P_{liq} is the liquid pressure and P_{vap} is the vapor pressure. For the case of a heat pipe in operation, the vapor pressure is either at saturation, or slightly superheated relative to the saturation conditions of the working fluid. For small pore radii one can assume $R_{\mathrm{I}} = R_{\mathrm{II}} = R = r_p/\cos\theta$. The standard wick structure will have pore sizes on the order of 10–500 μm. Thus, the small pore assumption applies. When this assumption is applied to the Young–Laplace equation it becomes

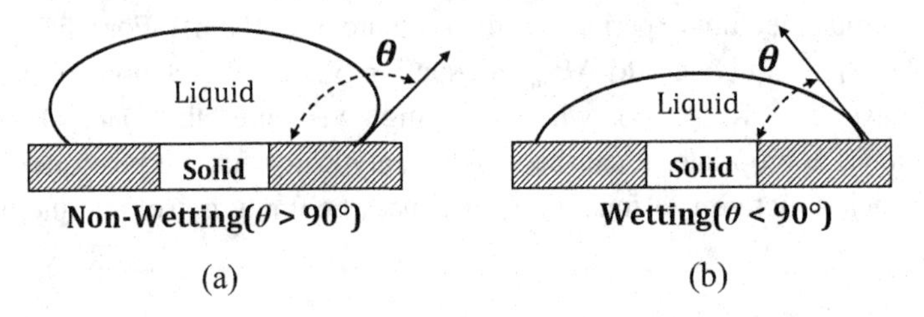

Figure 5.14 Solid/liquid interface surface conditions: a) non-wetting; b) wetting
Source: Katz (2010), reference [2]

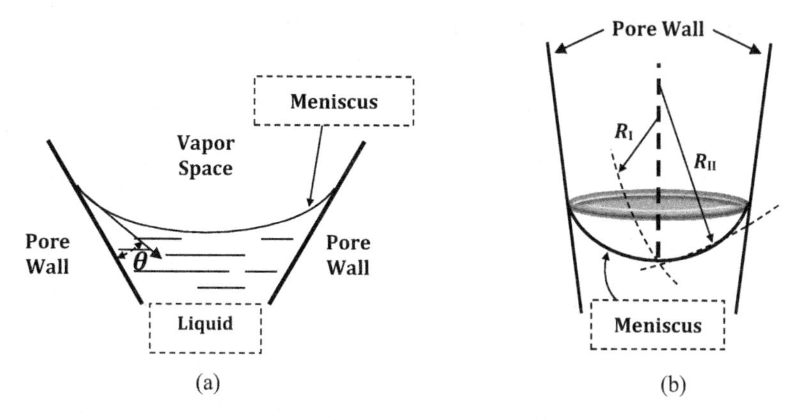

Figure 5.15 Wetting in a porous wick: a) pore contact angle at fluid meniscus (Source: Faghri (1995), reference [24]; permission conveyed through Copyright Clearance Center, Inc.), b) radii for the meniscus at the liquid/vapor interface

$$P_{vap} - P_{liq} = \frac{2\sigma_l}{r_p}\cos\theta \tag{5.29}$$

Equation 5.29 shows that the capillary pressure rise during operation (which is defined as $\Delta P_{cap,wick} = P_{vap} - P_{liq}$) is a function of the surface tension. The maximum capillary pressure rise possible is known as the capillary limit and is reached when full wetting of the pore's solid wall occurs (i.e., the contact angle shown in the figure reduces to zero) (Figure 5.15a). A key question is the levels of pressure rise that can be achieved with various wick pore sizes. Let us explore this with an example calculation.

Example 5 Impact of Pore Diameter Size on Effective Pumping Pressure Taking pure water as the working fluid in a stainless steel wick structure and a water temperature of 20°C, what is the pressure rise ($\Delta P_{cap,wick}$) for pore diameters of 1 cm and 20 μm? Assume that water at 20°C has a surface tension of $\sigma_l = 0.078$ N/m and a contact angle of approximately 70° on stainless steel.

Solution

When the unknowns are inserted into the RHS of the Young–Laplace equation, the pressure rise for a pore of 1 cm is 4.97 N/m^2. For a 20 μm pore the pressure rise is 2,490 N/m^2. Thus, the pressure rise increases by three orders of magnitude as the pore size decreases to the double-digit micron scale. This shows the level of pumping power attainable with capillary wick structures. The capillary pressure rise is functionally dependent on the surface tension. Furthermore, since the heat transfer is a function of the liquid supply to the evaporator (i.e., the mass flow rate), which in turn is a function of the capillary pressure rise in the wick, the heat transfer may be considered a function of the working fluid's surface tension.

Table 5.3 lists some common heat pipe working fluids. Water and acetone are often used for terrestrial applications at room temperature. However, water is not chosen for

Table 5.3 Standard heat pipe working fluids and their boiling points at 1 atm

Fluid	Chemical formula	Boiling temperature [K]
Neon	Ne	27
Ethane	C_2H_6	185
Acetone	CH_3COOH	329
Water	H_2O	373
Anhydrous ammonia	NH_3	240
Methane	CH_4	111.4

space applications due to its high freezing point (0°C) and its volume expansion during conversion to the solid phase. Acetone is not considered viable for space application either because of the temperature range (freezing point to critical point) over which boiling occurs. Ethane and ammonia are most often selected in room-temperature space applications, whereas neon and methane are candidate fluids for cooling to cryogenic temperatures in either terrestrial or space applications.

5.4.1.1 Operational Limits

Heat pipes have several operational limits. From a fundamental thermodynamics perspective, heat pipe operation will occur at saturation conditions. Two temperature limits for the saturation conditions of a working fluid are the critical point and the triple point. The heat pipe will only operate at temperatures offset from these two state points. Figure 5.16 shows the heat pipe limits that can be incurred during operations, as well as the hierarchy of operational limits. Frequently encountered are the capillary limit, the entrainment limit and the boiling limit.

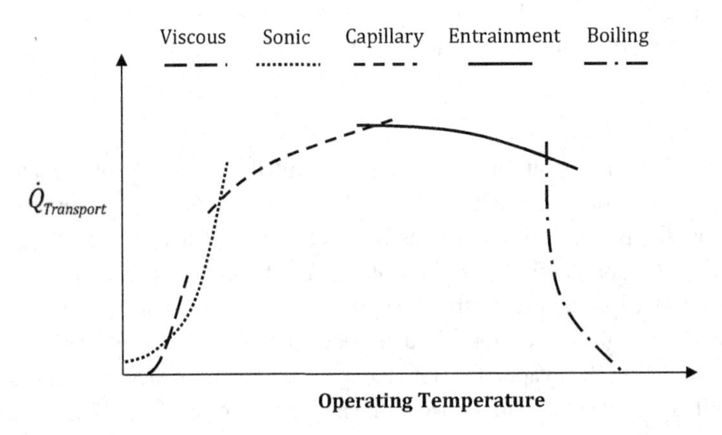

Figure 5.16 Heat pipe operational limits
Source: Peterson (1994), reference [27]; permission conveyed through Copyright Clearance Center, Inc.

The capillary limit is the maximum pumping capability for the wick/fluid combination used in a heat pipe. Fluid flow in a heat pipe will not occur when the total pressure drop loss in the system (explicitly listed in equation 5.24) exceeds the capillary wick's pumping capability. To determine the heat transfer value corresponding to the capillary limit, pressure drop relations for select terms on the RHS of equation 5.24 are determined and substituted back into the relation. Pressure drops due to evaporation and condensation in the heat pipe are typically neglected. Reay et al. [28] also assumed that pressure drop due to vapor flow was negligible. Significant contributory factors in pressure drop include the liquid phase in the wick, and body forces that may be acting upon the liquid phase. When each of the new terms on the RHS of equation 5.24 has been determined, their total can be substituted into equation 5.23 as ΔP_{tot}. The LHS of equation 5.23 can then be recast to reflect the maximum pumping capability (i.e., $2\sigma/r_{eff}$). A mass flow rate can then be determined from the combined relation. The heat transfer is obtained by multiplying the enthalpy of vaporization by the mass flow rate. Given constant liquid properties, an expression for the capillary limit was determined by Reay et al. [28] as

$$\dot{Q}_{cap} = \left[\frac{\rho_l h_{fg} K_{wick}}{\mu_l}\right]\left[\frac{\pi\left(r_o^2 - r_i^2\right)}{l_{eff}}\right]\left[\frac{2\sigma_l}{r_{eff}} - \rho_l g l \cdot \sin\varphi\right] \tag{5.30}$$

where r_{eff} is the effective pore radius of the wick structure, r_i is the inner radius of the wick, r_o is the outer radius of the wick, φ is the inclination angle of the heat pipe condenser side relative to the horizontal and l_{eff} is the effective length of the heat pipe. For a sintered wick material composed of spheres, the permeability (K_{wick}) may be determined as a function of the sphere diameter and the porosity using equation 5.31 [22]. In the relation

$$K_{wick} = \frac{d_s^2 \phi^3}{150(1-\phi)^2} \tag{5.31}$$

d_s is the sphere diameter and ϕ is the porosity of the wick material. Equation 5.31 is applicable for 50×10^{-6} m $< d_s < 3 \times 10^{-4}$ m. For a detailed overview of derivations for the capillary limit, see Brennan and Kroliczek [23], Peterson [21,27], Faghri [24] or Reay et al. [28].

Additional key limits include the entrainment limit and the boiling limit. Entrainment in liquid/vapor multiphase heat transfer systems is a well-studied phenomenon. The heat pipe-specific entrainment limit is reached when vapor in the core of the heat pipe begins to shear a substantial volume of liquid from the wick and liquid replenishment to the evaporator is reduced. This phenomenon is promoted by the presence of high-vapor velocities at the wick's liquid/vapor interface. Cotter [29] was the first to derive the entrainment limit heat transfer in heat pipes by assuming that the We number (the ratio of inertia to surface tension forces) for the vapor flow is equal to unity. In equation form the We number for the present application is defined as

$$\mathrm{We} = \frac{\textit{Inertia Forces}}{\textit{Surface Tension Forces}} = \frac{\rho_v Vel_{avg}^2 d_{h,wick}}{\sigma_l} \tag{5.32}$$

where $d_{h,wick}$ is the hydraulic diameter of the wick structure. It is defined as four times the wick cross-sectional area divided by the wetted perimeter (i.e., $d_{h,wick} = 4A_{wick}/P_{wick}$). From a thermodynamic perspective the product of density and velocity squared is the kinetic energy per unit volume for the vapor flow multiplied by a factor of 2. From a fluids perspective, the numerator is the dynamic pressure of the vapor flow multiplied by a factor of 2 (this is borrowed from the definition of the friction coefficient). The heat transfer due to phase change can be determined from the enthalpy of vaporization conservation relation

$$\dot{Q} = \dot{m} h_{fg} \tag{5.33}$$

In this relation, the mass flow rate is typically associated with liquid. However, since the rate of vapor production (and ultimately mass flow) must equal that of the liquid flow in the wick it can also be used for the vapor produced (i.e., $\dot{m} = \rho_v A_v \cdot Vel_{v,avg}$). Given the entrainment assumption, the average vapor velocity can be determined from the We number equation (equation 5.32) and substituted into the mass flow rate relation. When the new representation of the mass flow rate is incorporated into the enthalpy of vaporization conservation relation, the final form for the entrainment heat transfer is shown in equation 5.34

$$\dot{Q}_{ent} = A_v \cdot h_{fg} \cdot \left[\frac{\sigma_l \rho_v}{d_{h,wick}} \right]^{\frac{1}{2}} \tag{5.34}$$

While the hydraulic diameter was used as the characteristic length in equation 5.32, most representations in heat pipe literature (including the initial derivation by Cotter [29]) use the hydraulic radius. Since the early work of Cotter, numerous representations have been developed. For details on alternate entrainment limit relations, see the work of Marcus [19], Skrabek and Bienert [21], Brennan and Kroliczek [23] or Peterson and Bage [30].

During regular heat pipe operation, liquid-to-vapor phase change resulting from heat input occurs at the flared side of the wick pores (Figure 5.15a). The boiling limit (also known as the heat flux limit) is the point at which heat input to the evaporator is sufficient to promote boiling in the inner portion of the wick's pore structure. Using the pressure balance at the liquid/vapor interface in the wick during bubble formation, growth and collapse, an analytical model can be derived to predict the boiling heat transfer limit. Equation 5.35 is the one developed by Chi [22].

$$\dot{Q}_{boil} = \frac{4\pi l_{eff} k_{eff} \sigma_l T_{vap}}{h_{fg} \rho_v \ln\left(r_{i,wall}/r_{vap}\right)} \left[\frac{1}{r_b} - \frac{1}{r_{eff}} \right] \tag{5.35}$$

where r_b is the average bubble radius at a nucleation site, $r_{i,wall}$ is the inner radius of the pipe wall, r_{vap} is the radius of the vapor space, r_{eff} is the effective radius of the wick pores, T_{vap} is the temperature of the phase-changed vapor in the evaporator and l_{eff} is the sum of the length of the adiabatic section plus half the evaporator and condenser sections.

Evaporator dry-out, a significant reduction in thermal conductance and a loss of temperature control in the evaporator are resultant effects from either the entrainment or boiling limit having been reached. Details pertaining to additional limits (e.g., viscous, sonic and condenser) may be found in *Heat Pipe Design Handbook* [23], *An Introduction to Heat Pipes: Modeling, Testing and Applications* [27], *Heat Pipe Science and Technology* [24], *Handbook of Heat Transfer* [20], "Heat Pipes," in *The Heat Transfer Handbook* [31] or *Heat Pipes: Theory, Design and Applications* [28].

5.4.1.2 Gravity Effects

Capillary wick heat pipes tend to have a moderate to high degree of susceptibility to gravitational fields due to the low pressure heads achievable with wick structures. In μ-g these effects go away. Nonetheless, in the presence of sizeable body forces caused by ground testing or platform accelerations, heat pipes can experience reduced heat transfer performance. Due to the need for ground testing of these systems prior to flight, the impacts associated with orientation testing in a 1-g reference frame must be dealt with. The most favorable orientation for a heat pipe test in 1-g would be horizontal (0° inclination of the evaporator above the condenser). Heat transfer performance in this orientation is the closest analogue to that expected in the μ-g environment. Extreme cases of unfavorable orientation are shown in Figure 5.17. In a gravitational environment, the least favorable orientation is where the liquid has to be pumped against gravity to the evaporator. There are thermal-hydraulic losses associated with pumping against gravity that can marginalize the overall heat transfer from evaporator to condenser. Heat transport capability within a heat pipe will diminish significantly in application before an evaporator orientation of 90° above the condenser is achieved (see Figure 5.17). When the condenser is above the evaporator, the wick structure does not have to pump against gravity. However, pooling in the evaporator can limit the resulting heat transfer to natural convection for an extended degree of superheating. Increased superheating of the evaporator can also occur at start-up [32]. Temperature oscillations during operation may also occur. This is termed "reflux mode" [33] operation. The practitioner should look to avoid unfavorable orientations over long heat transport distances when operating in gravitational environments.

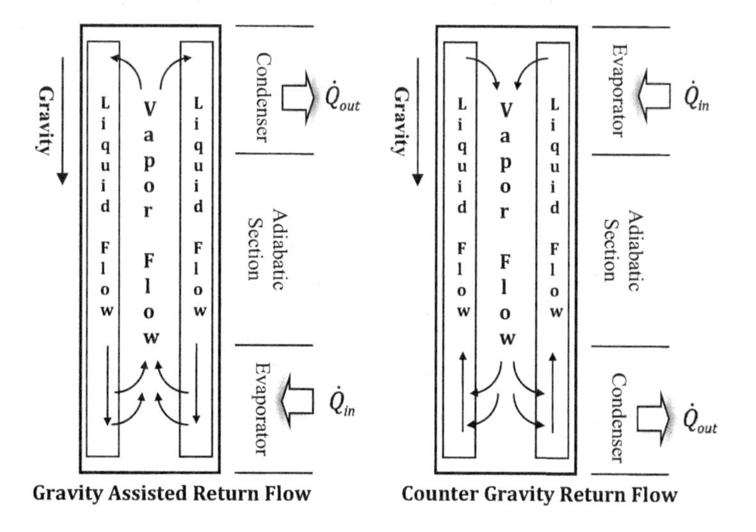

Figure 5.17 Heat pipe extreme orientations

5.4.1.3 Heat Pipe Geometry

The capillary heat pipes we have discussed all consist of a linear flow path. However, heat pipes can operate in an encasement structure with mild/moderate bends within plane and in such case they are known as planar heat pipes (example in Figure 5.18b). However, multiple successive bends resulting in a tortuous flow path and/or placement of the evaporator and condenser out of plane are not advised for 1-g testing due to wick structure liquid pumping limitations.

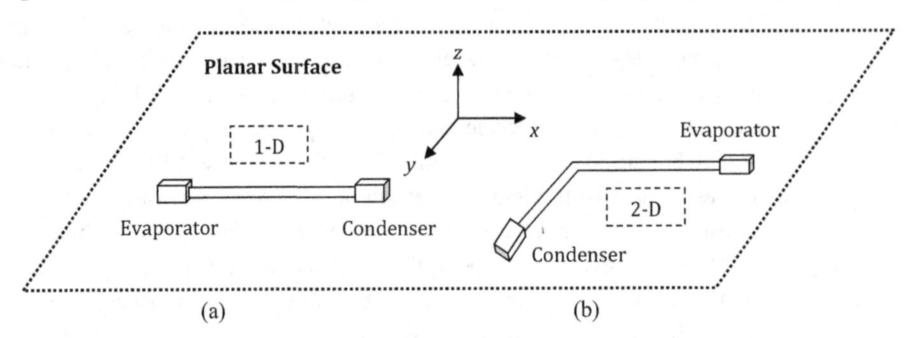

Figure 5.18 Heat pipe configurations: 1D and 2D

5.4.1.4 Variable Conductance Heat Pipes

So far, we have discussed capillary wick heat pipes operating in constant conductance mode. However, one approach to obtaining variable conductance performance in a heat pipe is to use a variable conductance heat pipe (VCHP) (Figure 5.19). The VCHP construction is similar to that of the standard heat pipe with one exception. For the VCHP a reservoir is attached to the condenser end of the encasement. This reservoir is filled with an inert, non-condensable gas (NCG) that remains in vapor phase across the heat pipe's operating range. As heat is applied to the reservoir, the gas expands into the condenser side of the heat pipe. The gas blocks the working fluid from condensing over a portion of the condenser area and effectively reduces the thermal conductance of the heat pipe. Notwithstanding issues such as the charge level and volume of the gas reservoir, the thermal conductance of the heat pipe can be regulated to a desired value as the gas front advances and recedes from the reservoir into the condenser.

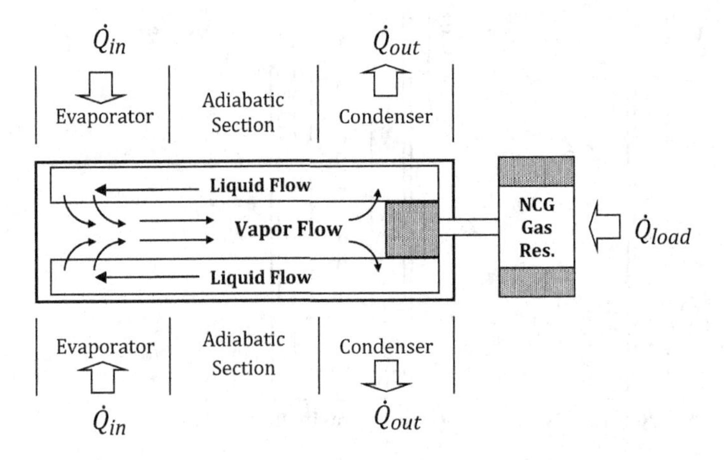

Figure 5.19 Variable conductance heat pipe
Source: Faghri (1995), reference [24]; permission conveyed through Copyright Clearance Center, Inc.

5.4.1.5 Figure of Merit

The enthalpy of vaporization and saturation temperature at 1 atm typically serve as metrics for heat transfer performance when comparing different working fluid options. However, given the fact that the heat transfer process for these devices relies not only upon phase change, but also surface tension, it is not sufficient to compare only enthalpy of vaporization (h_{fg}) values. Two figures of merit have been developed for use as heat transfer performance metrics when comparing different working fluids in heat pipes:

i) Liquid transport factor (zero-g FOM)
ii) Wicking height (1-g FOM).

Assuming that the capillary limit is the performance-limiting factor, in the absence of a gravitational field the capillary pumping heat transfer limit will be proportional to the liquid transport factor [19], which is defined as

$$\frac{\sigma_l \rho_l h_{fg}}{\mu_l} \tag{5.36}$$

The surface tension, density and viscosity used in the calculation all relate to liquid. In addition, all values are taken at saturation conditions. The calculation provides a heat flux in units of W/m^2 and predicts the on-orbit heat flux performance. Note that this calculation neglects the vapor contribution. The wicking height, otherwise known as the 1-g FOM [19,21,23,33], is defined as the liquid surface tension (σ_l) divided by the density (ρ_l). The wicking height is used to compare the sensitivity of a working fluid to

$$\frac{\sigma_l}{\rho_l} \tag{5.37}$$

gravitational effects and to gauge how well a wick will pump a liquid against a gravitational field.

Example 6 Zero-g FOM Determination Determine the zero-g FOM for ethane at a saturation temperature of 250 K.

Solution

Since the solution will require thermophysical properties of ethane under saturated conditions, we will use EES in the solution. Assuming that the saturation temperature is 250 K, the corresponding EES code is

```
{Example problem 5-6: Zero-g FOM calculation}

x_l=0.0

T_sat=250 [K]
P_sat=p_sat(ethane,T=T_sat)

sigma_ethane_l=surfacetension(ethane,T=T_sat)
```

```
rho_ethane_l=density(ethane,T=T_sat,x=x_l)

mu_ethane_l=viscosity(ethane,T=T_sat,x=x_l)

h_fg_ethane=enthalpy_vaporization(ethane,T=T_sat)

FOM=sigma_ethane_l*rho_ethane_l*h_fg_ethane/mu_ethane_l
```

Execution of the code will result in a value of 13.6×10^9 W/m^2.

5.4.1.6 Implementation and Thermal Circuitry

Thus far we have assessed heat pipes from the performance testing perspective (specifically heat transfer and/or heat flux performance). A number of application-based issues are relevant to the application of these systems to actual spaceflight hardware:

i) Orientation effects (these concern gravitational reference frames)
ii) Applicable temperature range
iii) Use of a thermal mass.

Orientation affects were reviewed in Section 5.4.1.2. We now focus on items ii) and iii). For performance testing, a massless heater can easily be used for the thermal energy input (for example, a flexible resistance heater with a mass of a few ounces). However, in real-life applications the heat pipe will most often be attached to an item of non-negligible mass, such as a heat exchanger plate, the exterior of an electronics box, the baseplate of a laser or the housing of a cryocooler compressor. Since the standard geometry consists of a circular cross-section, the heat pipe is often bracketed to a flat surface. Figure 5.20 shows a heat pipe mounted on an electronics box with a saddle bracket. The object serving as the heat source will have a certain amount of mass and specific heat (m and c_p). When heated from a dormant state (0 W at steady-state temperatures), the heat pipe evaporator temperature rises with a slight time lag as the heat has to propagate from the electronics box dissipating components through the structure and into the evaporator. The components through which the heat is dissipated will experience a transient temperature rise. The rate at which the temperature increases with time will be dependent upon the thermal energy storage capacity of the material in question ($m \cdot c_p$). This can be modeled analytically using the sensible heating equation from Chapter 2. This is important because duplicating the actual thermal mass in performance testing provides the most accurate characterization for temperature performance in spaceflight. Operational cases that are of specific interest include start-up, shutdown and steady-state performance.

$$\dot{Q} = \frac{mc_p\Delta T}{\Delta t}$$

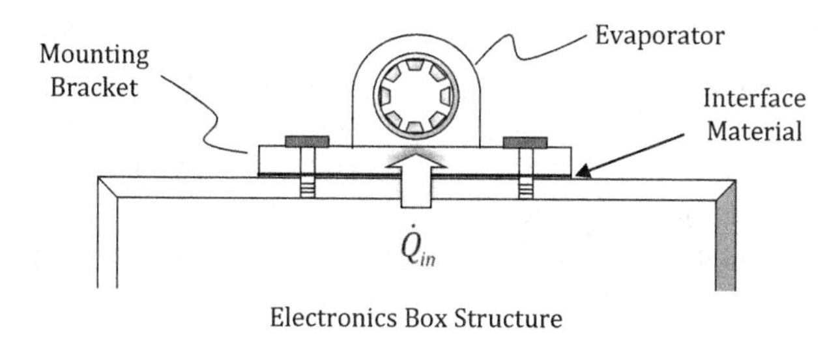

Figure 5.20 Heat pipe mounted on an electronics box by a saddle bracket

Knowledge of the resistance network associated with the heat flow path for a heat pipe is highly pertinent to characterizing heat transfer performance. Thermal resistances associated with a heat pipe include those that are component specific (e.g., thermal resistance through encasement structure) and those that are present at the interface between the heat pipe (condenser and evaporate) and the interface structure at both the source and sink locations. The contribution to thermal resistance resulting from phase change (evaporation and condensation) is negligible. Figure 5.21 shows a heat pipe thermal resistance network. The standard type of capillary wick structure used in most heat pipes is a sintered material. The radial thermal resistance to heat flow through a sintered wick filled with a working fluid (liquid phase) is typically modeled with either serial or parallel paths for heat flow in both the wick structure and the liquid. If the wick structure's porosity is known, effective thermal conductivity can be determined using equation 5.38 for the serial assumption and equation 5.39 for the parallel assumption [21–23].

$$k_{eff,ser} = \frac{k_l k_{wick}}{\phi k_{wick} + k_l(1 - \phi)} \tag{5.38}$$

$$k_{eff,para} = (1 - \phi)k_{wick} + \phi k_l \tag{5.39}$$

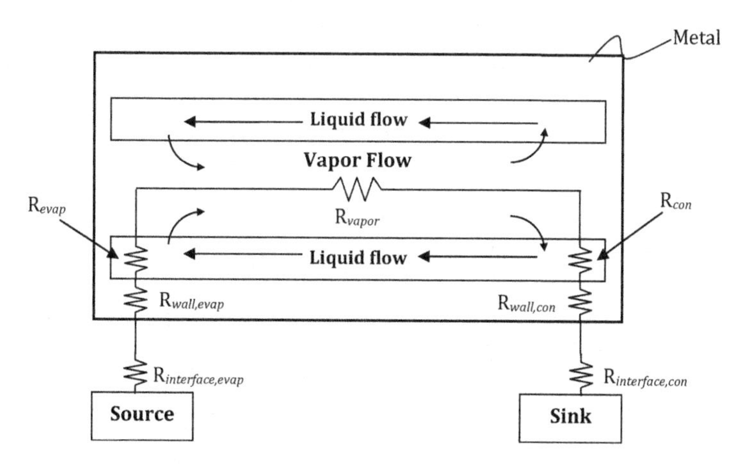

Figure 5.21 Heat pipe resistances
Source: Brennan and Kroliczek (1979), reference [23]

In these relations k_l is the liquid thermal conductivity, k_{wick} is the thermal conductivity of the wick material and ϕ is the porosity of the wick structure. Values determined using the series and parallel assumption are bounding limits for the effective thermal conductivity of the saturated wick structure whereby $k_{eff,ser} \leq k_{eff,actual} \leq k_{eff,para}$. Alternatively, one relation suggested by Dunn and Reay [34] as fairly accurate in predicting effective thermal conductivity of liquid saturated wicks is

$$k_{eff} = k_{wick}\left[\frac{2 + (k_l/k_{wick}) - 2\phi(1 - (k_l/k_{wick}))}{2 + (k_l/k_{wick}) + \phi\,(1 - (k_l/k_{wick}))}\right] \tag{5.40}$$

The equation for thermal resistance in a concentric cylinder can then be applied using k_{eff} as the thermal conductivity [21,22]. This would give the following equation

$$R_{wall,evap} = \frac{\ln(r_o/r_i)}{2\pi L_{evap} k_{eff}} \tag{5.41}$$

where r_o and r_i are the outer radius and inner radius of the wick structure respectively and L_{evap} is the length of the evaporator. Similar constructs can be used to calculate the radial resistance in the evaporator.

Example 7 Effective Thermal Conductivity of a Water-Saturated Titanium Wick

A capillary wick heat pipe with a sintered titanium wick has an outer radius of 11.2 mm and an inner radius of 8.2 mm. Assuming a wick porosity of 0.4 and the wick is filled with water at room temperature (298 K), what is the thermal resistance (radially) through the evaporator section if it has a length of 0.25 m?

Solution

To solve this problem we will use the serial path definition for the effective thermal conductivity of the wick structure (equation 5.38) and the evaporator section's resistance through the wick (equation 5.41). At 298 K, water has a liquid thermal conductivity of 0.6071 W/m · K. Titanium has a thermal conductivity of 21.9 W/m · K at room temperature. Placing all of the known values in the effective thermal conductivity relation would give

$$k_{eff} = \frac{(0.6071\ \mathrm{W/m\cdot K})\cdot(21.9\ \mathrm{W/m\cdot K})}{0.4(21.9\ \mathrm{W/m\cdot K}) + (0.6071\ \mathrm{W/m\cdot K})\cdot(1 - 0.4)}$$

Upon calculation, this gives a value of 1.457 W/m · K. For the thermal resistance, the calculation then becomes

$$R_{wall,evap} = \frac{\ln(11.2\ \mathrm{mm}/8.2\ \mathrm{mm})}{2\pi\cdot(0.25\mathrm{m})\cdot(1.457\ \mathrm{W/m\cdot K})}$$

which results in a value of 0.1362 K/W.

5.4.1.7 Heat Pipe Radiators

Let us revisit the discussion on honeycomb panel structures in Section 2.6 with an emphasis on radiators. One of the primary features of the honeycomb panel radiator was the dispersion of thermal energy across the honeycomb substrate enabled by the proper selection of a high thermal conductivity facesheet material. An alternate approach to the promotion of heat spreading through the radiator structure itself is to use a heat pipe radiator (Figure 5.22) containing heat pipes embedded within a honeycomb panel structure.

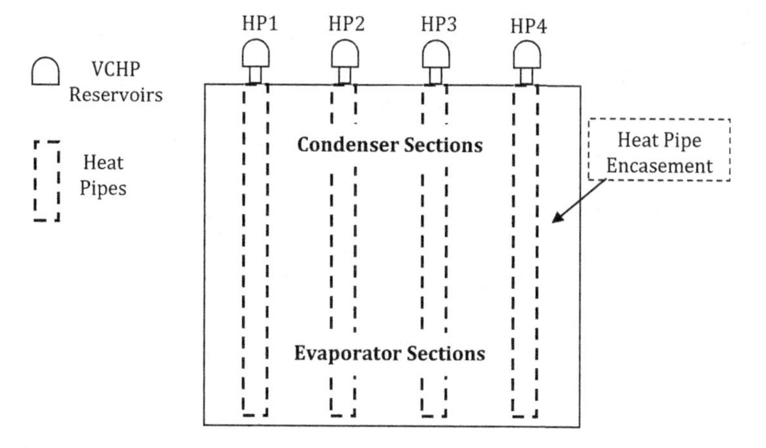

Figure 5.22 Heat pipe radiator
Source: Reproduced w/permission from Aerospace Corporation, *Spacecraft Thermal Control Handbook, Vol. I*, 2002, reference [35]

During heat pipe operation thermal energy is dispersed from the vicinity of the evaporator section towards the condenser section. This promotes a more uniform temperature across the radiator panel facesheet. VCHPs are often used in heat pipe radiators (see Section 5.4.1.4). One of the primary features of VCHPs is their ability to regulate thermal conductance, providing a constant operating temperature for the heat pipe under variable dissipative heat load conditions in the evaporator. This feature, also present in heat pipe radiators, allows the evaporator section to perform as a constant temperature thermal sink for the dissipative item thermally coupled to it on the inboard side of the radiator panel. The disadvantage of heat pipe radiators is that they are considerably greater in mass than the basic honeycomb panel radiator. They are therefore only used when the space platform mass budget is sufficiently robust to accommodate them. They are also considered viable for managing and rejecting large dissipative thermal loads.

5.4.2 Loop Heat Pipes

A loop heat pipe is a closed fluid loop device that acquires heat from a temperature-controlled item and transports thermal energy to a designated heat rejection location. LHP technology was established in Russia in the 1970s and first flown in 1989 on the Russian space platform Granat [36,37]. By the late 1980s the technology had made its way to the United States and was increasingly adopted for use on spacecraft. Today LHPs are commonplace in TCS architectures on both government and commercial satellites worldwide. NASA spaceflight instruments and missions that have successfully flown LHPs include the GLAS (Geoscience Laser Altimeter System) instrument, EOS-AURA TES (Tropospheric Emission Spectrometer) and the SWIFT gamma-ray burst observatory to name a few. While LHPs are a closed fluid loop analogue to heat pipes, they have yet to fully migrate from the space community to terrestrial applications in private industry and the university sector.

5.4.2.1 Fundamentals of Operation

LHP operation is based on the fluid flow and heat transfer processes taking place simultaneously within the closed-loop structure (Figure 5.23). These devices are heat activated using the dissipative thermal energy from the temperature-controlled item in question. The primary heat transfer mechanisms are nucleate boiling and condensation. As shown in Figure 5.23, LHPs consist of three basic closed fluid loop components: an evaporator (or heater), a condenser (or cooler) and a reservoir (compensation chamber). LHPs do not have any moving parts. The key operational states associated with LHPs are start-up, shutdown and steady state, and the characteristics of each state are critical to successful LHP performance.

Since an LHP does not have any mechanical parts to institute flow within the loop, a capillary wick structure is the prime mover of the working fluid in the system. The compensation chamber (CC) supplies liquid to the evaporator for phase change. It also captures (and stores) liquid returning from the condenser/subcooler. The temperature in the CC will be at saturation conditions. The operating temperature of the LHP is established by the CC setpoint temperature [36,37]. Saturation conditions internal to the CC will therefore influence the performance of the entire closed fluid loop. The circular wick cross-section consists of a concentric (yet hollow) structure extending into the evaporator itself (Figure 5.24).

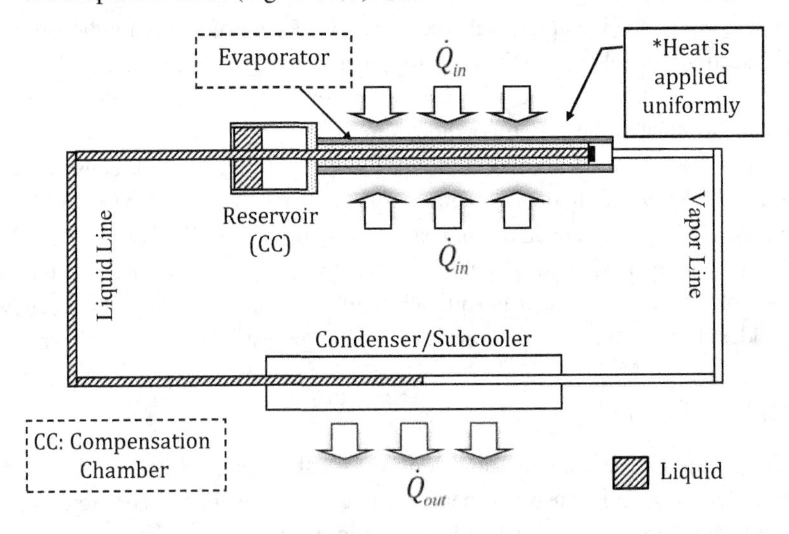

Figure 5.23 Loop heat pipe schematic
Source: Reproduced with permission of SAE International from "Operating Characteristics of Loop Heat Pipes," Ku, J., 1999, reference [36]; permission conveyed through Copyright Clearance Center, Inc.

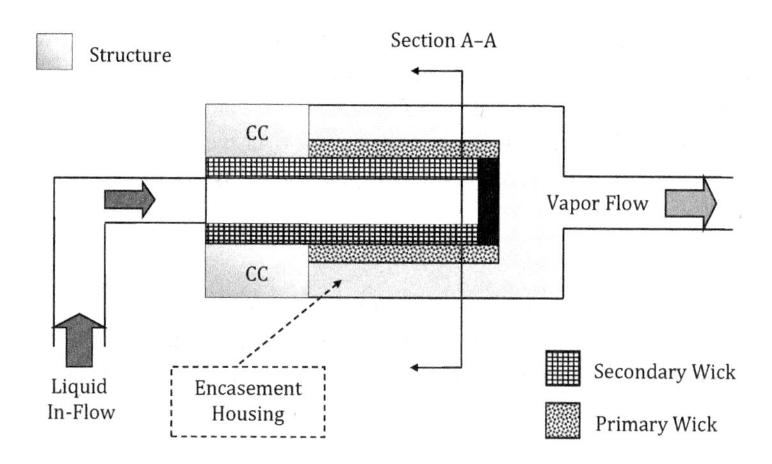

Figure 5.24 Compound wick configuration at the compensation chamber/loop heat pipe intersection

The secondary wick extending into the reservoir's core serves as a liquid feed from the reservoir to the primary wick. Heat input occurs through the evaporator encasement structure, where ribs are in intimate contact with the primary wick (Figure 5.25). Since the primary wick is wetted with saturated liquid, vapor is generated at the wick/structure interface during LHP operation. The interface for the primary and secondary wick structures also has passages leading back to the reservoir in the event that vapor is generated there.

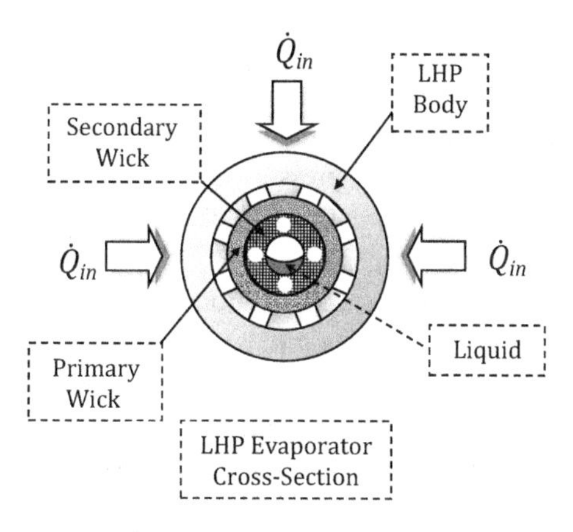

Figure 5.25 Loop heat pipe evaporator section A–A of Figure 5.24
Source: Reproduced with permission of SAE International from "Operating Characteristics of Loop Heat Pipes," Ku, J., 1999, reference [36]; permission conveyed through Copyright Clearance Center, Inc.

Unlike heat pipes, the liquid storage reservoir is separated from the heat input and output locations. Nonetheless, there is a conduction path from the evaporator encasement structure to the reservoir, passing through the wicks. Heat transferred via this path is known as “heat leak.” Since the loop’s setpoint temperature (as well as the h_{fg} for the process) is heavily influenced by the temperature in the reservoir, changes in reservoir temperature will impact the heat transfer. Therefore, standard LHP design seeks to minimize heat leak and provide better user control over the heat transfer process. Low thermal conductivity wick structures (i.e., insulators) are typically chosen for construction of LHPs.

5.4.2.2 Start-Up Considerations, Operation and Shutdown

As heat is applied to the LHP, the evaporator’s encasement structure and ribs (the heat input surface) experiences superheating relative to the saturation temperature of the liquid in the evaporator. After a sufficient level of superheating has occurred, nucleation begins in the evaporator at the primary wick/encasement structure interface. The vapor generated then flows through the gap spaces in the ribs (along the main axis of the evaporator) and towards the vapor line. This occurs simultaneously with the onset of fluid flow through the entire loop. From a fluid hydraulics perspective, since the fluid flow results from nucleation occurring at the wick/structure encasement interface, the capillary pressure rise required to institute flow must overcome the total pressure drop within the LHP (similar to the capillary wick operational requirement for heat pipes). Equation 5.23 also applies here.

$$\Delta P_{cap,wick} \geq \Delta P_{tot}$$

The total pressure drop within the LHP system (ΔP_{tot}) is defined as

$$\Delta P_{tot} = \Delta P_{vap} + \Delta P_{con} + \Delta P_{lrl} + \Delta P_{wick} + \Delta P_{g} \tag{5.42}$$

where ΔP_{vap}, ΔP_{con} and ΔP_{lrl} are the pressure drop in the vapor line, the condenser and the liquid return line respectively. ΔP_{wick} is the pressure drop associated with flow through the wick structure and ΔP_{g} is the pressure drop associated with the body forces acting on the liquid columns internal to the loop.

Standard working fluids used with LHPs in space applications are shown in Table 5.4. These values are all taken at a saturation pressure of 101 kPa. Water is typically not used in space applications with LHPs but is shown here as a reference enabling comparison of its thermophysical properties against other working fluids. Neon is a cryogenic fluid and its successful use in LHPs demonstrates their operational robustness (temperature wise).

Table 5.4 Standard LHP working fluids for space applications

Working fluid	Boiling temperature [K]	h_{fg} [kJ/kg]	σ_l [N/m]
Neon	27	87.8	0.055
Propylene	225	334.8	0.067
R-718 (Water)	373	2,441.5	0.072
R-717 (NH_3)	240	1,165.2	0.021

Start-Up Scenarios

The three most important operational modes of an LHP are start-up, shutdown and regular steady-state operation. Figure 5.26 shows an LHP undergoing start-up and shutdown, with heating to the evaporator initiated at the 1-hour mark. During this time there was a temperature rise in the evaporator and a minor temperature rise in the compensation chamber. Just before the 2-hour mark the loop achieved start-up with a nominal superheat of 20°C. This was denoted by an immediate decrease in the temperature of the evaporator, as well as an increase in the condenser temperature while the working fluid was being pumped through the loop. The vapor line (which with start-up had superheated vapor flowing through it) also increased slightly in temperature and began following the evaporator and CC temperatures. The liquid line, (which then had cooled liquid flowing through it) decreased in temperature and began following the condenser temperature (the condenser being the coldest location throughout the entire loop). The loop was operated for approximately 2 hours and shutdown was initiated at the 4-hour mark by removing heating from the evaporator. Since the saturation temperature of the loop is predominantly controlled by the temperature of the compensation chamber, we can control the saturation temperature of the loop by adding heat to the CC (assuming it is in an inherently cold-biased environment and a dedicated heater is available). A moderate amount of heating can be applied to the CC to execute shutdown, raising the saturation temperature of the loop and eliminating the temperature gradient between the evaporator housing and the liquid. This stops the heat transfer. The shutdown shown in Figure 5.26 was instantaneous and its immediate effect was denoted by the separation of the evaporator, CC and vapor line

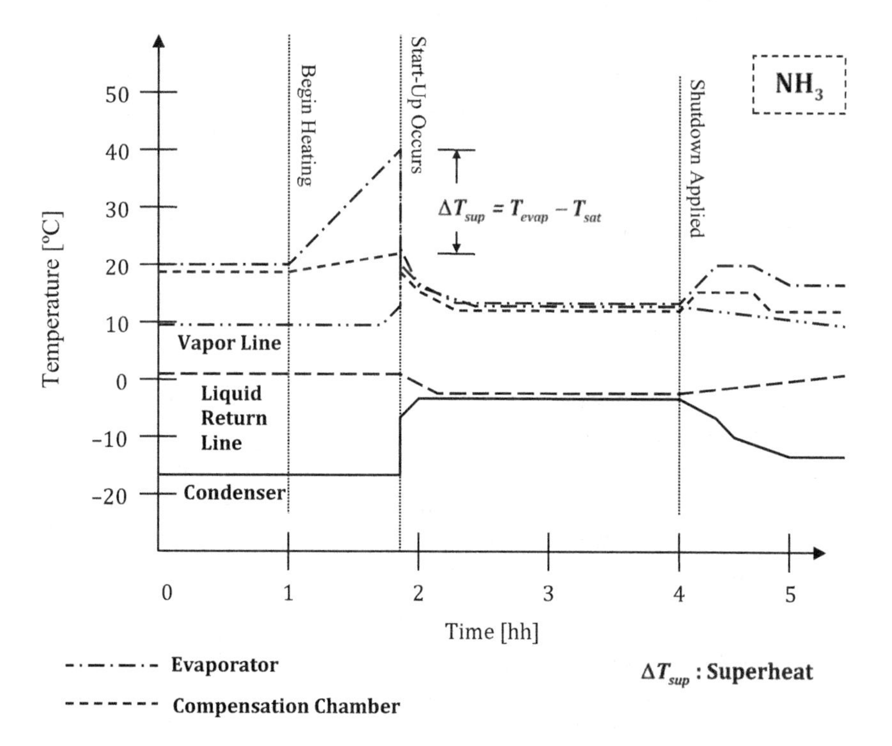

Figure 5.26 Loop heat pipe during standard operation

temperatures, as well as the decrease in the condenser temperature and the separation of the liquid line temperature from it.

How might a failed start-up look? The example in Figure 5.26, which took close to an hour, is an exceptional case since the LHP did actually start eventually. In the example in Figure 5.27, however, heating of the evaporator began at the 1-hour mark. After approximately 4 hours of continuous heating there was no sign of fluid flow in either the evaporator, the CC, the vapor line, the condenser or liquid line (most LHP start-ups occur within 10–15 minutes of the initiation of heating).

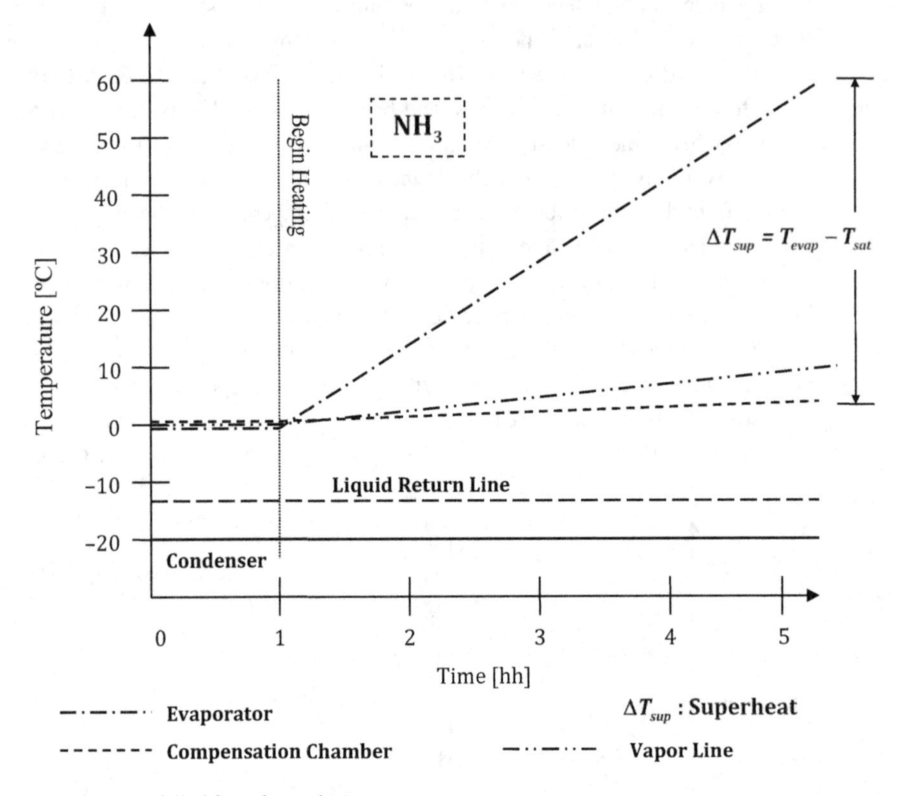

Figure 5.27 A failed loop heat pipe start-up

In addition, the superheat attained over this time was approximately 60°C. This is an extreme amount of superheat compared to the scenario in Figure 5.26, which only took 20°C. For recovery from the failed start-up, heating to the evaporator must be removed and the loop must be allowed to cool down. Once the liquid and evaporator temperature have returned to the desired values the wick must be primed with the working fluid and a second attempt can be undertaken.

Operational Regimes

LHP steady-state operation has two regimes. These are the variable conductance regime and the constant conductance regime. LHP operation in either of these regimes is a function of the competing effects of heating and/or cooling of the CC by the liquid return line fluid and heat leak from the evaporator to the CC. As shown in Figure 5.28, at low heat input just after start-up, the LHP operating temperature self-regulates to lower values as the heat input is increased. From a physical perspective, in this region of the temperature-versus-transport curve, the condenser is not being fully utilized (i.e., vapor front penetration into the condenser does not extend to the condenser exit). This leads to subcooling. Also, the heat leak from the evaporator to the CC is not substantial. These two effects cause a reduction in the evaporator temperature. With additional heat input, this trend will continue until the condenser becomes fully utilized (i.e., the vapor front extends either to or beyond the condenser exit). This, coupled with a pronounced level of heat leak from the evaporator to the CC, will cause the operating temperature to increase as the heat input increases (a constant conductance regime). The transition point (i.e., the inflection point temperature T_{min} in Figure 5.28) between the two operational regimes is a minimum for the evaporator temperature.

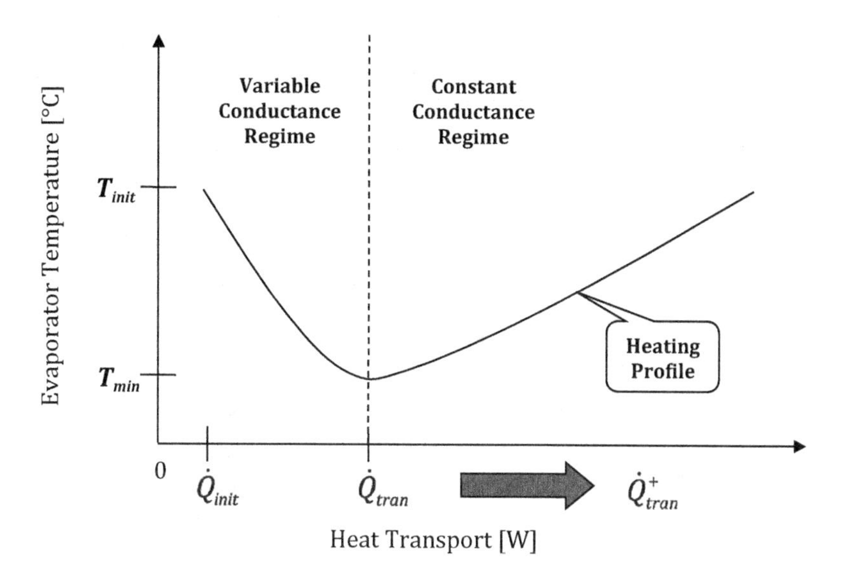

Figure 5.28 Loop heat pipe operational regimes during progressive heating
Source: Chernysheva et al. (2007), pp. 2704–2713, reproduced with permission from Elsevier, reference [38]

Shutdown

Shutdown of an LHP can be achieved by removing the dissipative heat load from the temperature-controlled item and/or applying a nominal amount of heat to the CC to decrease the level of superheat driving the nucleation process in the evaporator. Regardless of the approach used, the practitioner should fully confirm that shutdown has been achieved in order to avoid under-temping of sensitive flight components. A successful shutdown will be denoted by a decrease in the vapor line and condenser temperatures together with an increase in the liquid line temperature. In addition, the evaporator will no longer display temperature control.

Hysteresis

The phenomenon of hysteresis was previously discussed in association with the heating and cooling of a pooled liquid after having transitioned through the Leidenfrost effect. However, this effect occurs not only with pooled liquids after dry-out, but also on some LHP cooling cycles where temperature control is not lost.

5.4.2.3 Orientation Effects in Terrestrial Environments

Due to the capillary wick structure, orientation effects can impact the performance of the LHP in a gravitational environment, as is the case with heat pipes. There are two orientations of interest in the configuration of components in an LHP, tilt and elevation. Tilt is defined as the amount by which the evaporator is rotated above the CC [36,39]. Using a standard flatbed LHP as an example, the CC will be in line with the evaporator and both will be at 0° relative to the horizontal plane (shown as Case 1 in Figure 5.29).

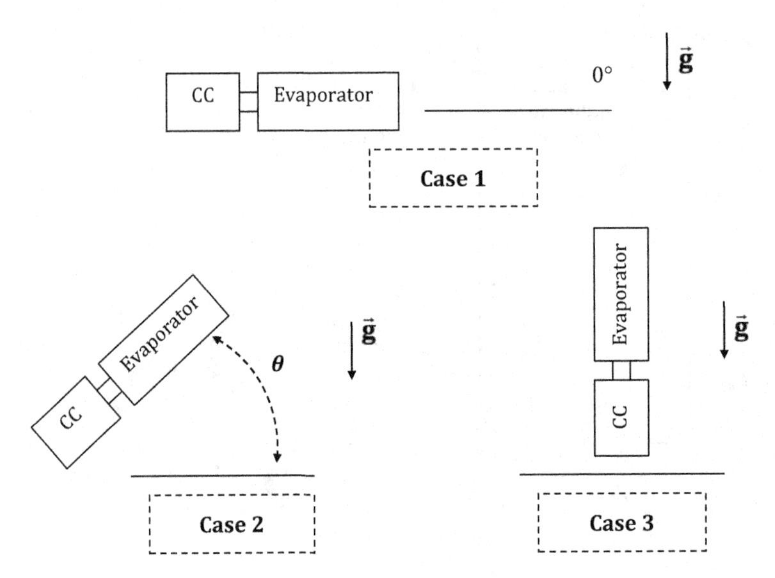

Figure 5.29 Tilt orientations between the CC and the evaporator

As the evaporator is rotated above the CC (Case 2) the function of the secondary wick as a liquid supply mechanism for the primary wick is compromised. The capillary pressure in the evaporator has to rise sufficiently to push the liquid in the secondary wick (and the evaporator core) against gravity. The worst possible tilt orientation would be Case 3 where the CC and evaporator are in line with the gravity vector. Depending on the location of the condenser and its associated liquid, the impacts of these orientations on heat transfer can vary from minor to significant.

Elevation is defined as the amount by which the evaporator is rotated either above or below the condenser relative to the horizontal plane [36,39] (Figure 5.30 Case 1). When not in plane with each other they create an elevation angle (Case 2). When the condenser is lower than the evaporator in a gravitational environment, the liquid column located in the return line and condenser (which has a body force) has to be pumped against gravity in order to sustain flow and ensure liquid wetting of the wick structures, at the expense of the pressure head created from the capillary wick action in the evaporator. This affects mass flow and heat transfer performance. The worst elevation possible is Case 3 where the evaporator is in line with (and above) the condenser. These orientation effects and the respective locations of the evaporator and condenser are significant during 1-g testing of spaceflight hardware in a gravitational environment.

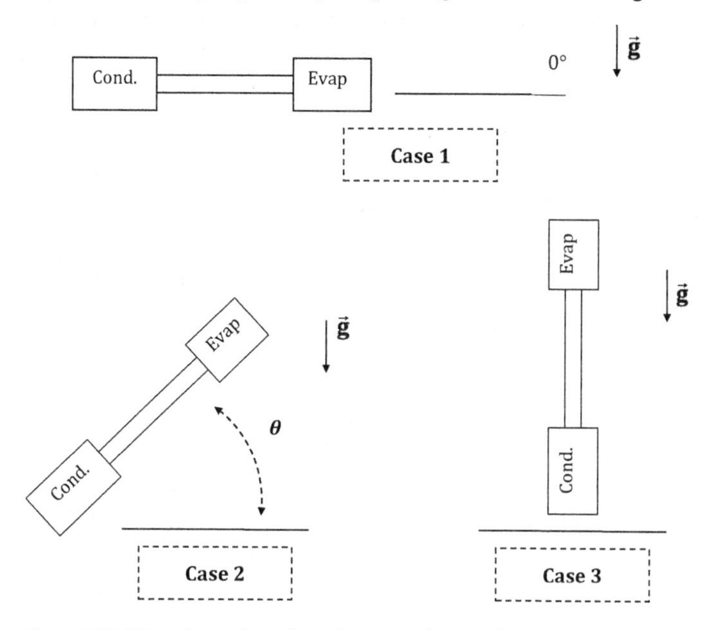

Figure 5.30 Elevation orientations between the condenser and the evaporator

5.4.2.4 Figure of Merit

The capillary wick limit (as discussed in Section 5.4.2.2) applies to the LHP's compound wick structure. As in the case of heat pipes, the LHP operational temperature values are between the triple point and the critical point. The boiling limit (i.e., the point at which sufficient heat enters the evaporator to cause boiling in the inner portion of the wick's pore structure) is also relevant to LHP operation. Practitioners typically rely upon the enthalpy of vaporization and the wicking height as metrics for performance comparison. The enthalpy of vaporization is a thermophysical property of the working fluid. The wicking height, as defined in equation 5.37, is

$$\frac{\sigma_l}{\rho_l}$$

where the surface tension (σ_l) and the density (ρ_l) are for the liquid phases. The wicking height measures the sensitivity of a working fluid to gravity effects. It is a measure of how high a wick can pump liquid against a gravitational field. Thermophysical properties used for this calculation must be taken at saturation conditions.

5.4.2.5 Non-Condensable Gas Effects

As with most multiphase systems, the presence of non-condensable gases (NCGs) can lead either to a reduction in the heat rejection capability of the working fluid in the condenser or to an increase in the saturation temperature at which phase change occurs [36,40] (i.e., the setpoint temperature for the LHP). NCG production typically results from system contamination, working fluid/encasement material incompatibility or poor fluid charging techniques. Overall, the presence of NCGs can alter the nominal heat transfer performance (and/or transport) capability of the loop relative to that for a pure fluid containing negligible amounts of NCG. Working fluid/encasement material compatibility is important such that NCG production does not occur and/or is minimized inside the loop after charging. The closed fluid loop should be free of all possible contaminants and residual materials prior to charging.

5.4.2.6 Heat Transfer Performance Modeling

There are two contemporary approaches to modeling LHP performance. The first is based on heat transfer constructs. From a basic heat transfer perspective, LHPs are thermal conductance devices. Thus, the heat transfer occurring between the evaporator and the condenser can be defined as

$$\dot{Q} = \frac{(T_{evap} - T_{cond})}{R_{LHP}} \tag{5.43}$$

where R_{LHP} is the effective thermal resistance across the LHP. However, heat transfer analysis that incorporates thermal/fluidic effects such as pressure drop due to wall friction and phase change [40–45] can also be performed on each component within the system. Since the boundary conditions for each section of the LHP, as well as the rate of mass flow and the geometric dimensions, are known, the convection coefficient can be determined for flow occurring in the evaporator grooves, the vapor line, the condenser/subcooler and the liquid return line. Newton's law of cooling can then be used to calculate the thermal energy exchange with the environment. The second approach is to perform a thermodynamic analysis [36,43,46–49] similar to the thermodynamic cycle analysis for heat pumps discussed in Section 5.3.1. LHP performance can be modeled by applying energy balances to each of the primary components of the loop. Figure 5.31 shows the state points of interest for an LHP and its components, as well as the potential locations for heat exchange with the environment, while Figure 5.32 highlights the state points in the interior of the evaporator. State point 1 is assigned to the CC and state point 2 is located at the outer encasement of the evaporator housing. In Figure 5.32, state point 3 is the temperature in the vapor grooves immediately adjacent to the primary wick's pores. In Figure 5.31, state point 4 is located at the entrance to the vapor line, whereas state point 5 is at the entrance to the condenser. State point 6 is at the exit of the condenser/subcooler and state point 7 is at the entrance to the CC. Fluid conditions in the CC, the vapor space immediately adjacent to the primary wick and within the condenser prior to the collapse of the vapor front are all at saturation. The energy balance on the evaporator is defined as

$$\dot{Q}_{in} = \dot{Q}_{HL} + \dot{Q}_{2-phase} \tag{5.44}$$

where $\dot{Q}_{in}$ is the total heat applied to the evaporator encasement, $\dot{Q}_{HL}$ is the heat leak from the evaporator through the primary and secondary wick structures into the CC and $\dot{Q}_{2-phase}$ is the energy associated with vaporization of the liquid in the primary wick's

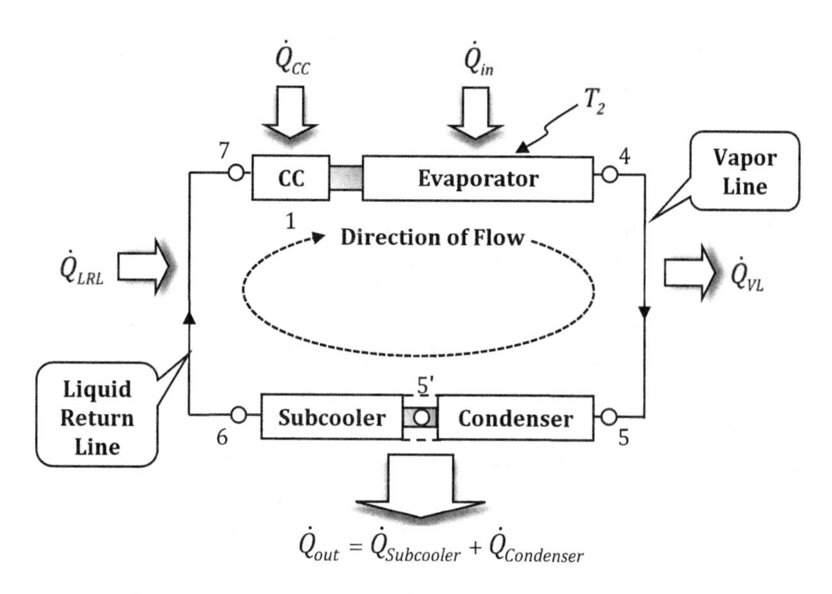

Figure 5.31 Loop heat pipe cycle state points and heat flows

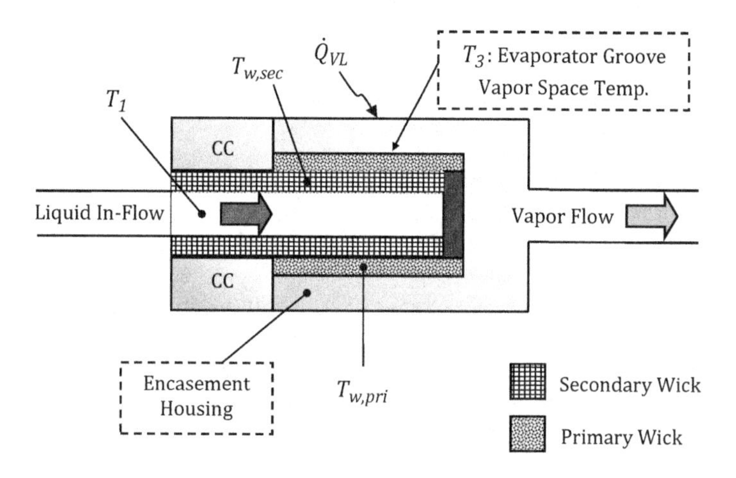

Figure 5.32 Detail of CC and evaporator with corresponding LHP state points

pores. If the vapor line between state points 3 and 4 is experiencing heat loss then the amount of heat loss ($\dot{Q}_{VL}$) occurring may be captured as

$$\dot{Q}_{VL} = \dot{m} \cdot (h_4 - h_5) \tag{5.45}$$

The heat rejection from the condenser/subcooler can be captured in equation form as

$$\dot{Q}_{out} = \dot{m} \cdot (h_5 - h_6) \tag{5.46}$$

If the liquid return line is being heated, the amount of heating may be defined as

$$\dot{Q}_{LRL} = \dot{m} \cdot (h_7 - h_6) \tag{5.47}$$

Assuming that the liquid at state point 7 is subcooled relative to the saturation temperature in the CC, an energy balance for the CC (as shown in Figure 5.31) is

$$\dot{Q}_{CC} = \dot{m}c_p \cdot (T_1 - T_7) - \dot{Q}_{HL} \tag{5.48}$$

where the first term on the RHS is the sensible heating required to heat the subcooled liquid to the saturation temperature in the CC.

Example 8 LHP System Performance Analysis in EES An LHP charged with anhydrous ammonia has a total heat input to its evaporator of 150 W. Heat input to the compensation chamber and liquid return line are 1.5 W and 2.0 W respectively, whereas the heat rejection from the vapor line is 1.0 W. Both the primary and secondary wicks are made of nickel. Each wick has a length of 8 in. and a radial thickness of 10 mm. The outer radius of the primary wick is 0.0254 m. The porosity for the primary and secondary wicks is 0.55 and 0.4 respectively. If the condenser inlet is operating at a temperature of 265 K and the subcooler outlet has a temperature of 260 K, what is the operating temperature of the evaporator and CC?

Solution

We will start by establishing the dimensions of the wick structures.

```
{Example 5-8. Loop heat pipe model}

l_1=2.54*8/100               {primary wick length}
l_2=2.54*8/100               {secondary wick length}
l_eff=l_1/2                  {effective length of evaporator}

r_o_1=0.5*(2.54*2.0/100)     {primary wick outer radius}
r_i_1=r_o_1-(10/1000)        {primary wick inner radius}
A_wick_1=pi*(r_o_1^2-r_i_1^2) {cross-sectional area of primary wick}

r_o_2=r_i_1                  {secondary wick outer radius}
r_i_2=r_o_2-(10/1000)        {secondary wick inner radius}
r_mid_2=(r_o_2+r_i_2)/2      {radial direction midpoint for secondary wick}
A_wick_2=pi*(r_o_2^2-r_i_2^2) {cross-sectional area of secondary wick}

phi_p=0.55                   {primary wick porosity}
phi_s=0.4                    {secondary wick porosity}
```

The thermal conductivity of the primary and secondary wick structures will be based on the temperatures in the evaporator and the CC. While these temperatures arc unknown at this point, we will bookmark the values using function call statements.

```
{Wick structure thermal conductivities}
k_s_p=conductivity(nickel, T=T_3)        {Primary wick}
k_s_s=conductivity(nickel, T=T_1)        {Secondary wick}
```

The effective thermal conductivity of both wick structures can then be defined using equation 5.40.

```
{Wick effective thermal conductivities}
k_eff_p=k_s_p*(((2+k_l_3/k_s_p-2*phi_p*(1-k_l_3/k_s_p)))/(2+k_l_3/k_s_p&
+phi_p*(1-k_l_3/k_s_p)))
k_eff_s=k_s_s*(((2+k_l_1/k_s_s-2*phi_s*(1-k_l_1/k_s_s)))/(2+k_l_1/k_s_s&
+phi_s*(1-k_l_1/k_s_s)))
```

Thermal resistances through the wick structures will span the path length from the evaporator encasement/primary wick interface, through the midpoint of the secondary wick and along the length of the secondary wick. The corresponding EES code for this will be

```
{Wick resistances}
R_wall_p=(2*pi*l_eff*k_eff_p)/(ln(r_o_1/r_i_1)) {Radial through primary wick}
R_1_2=(2*pi*l_eff*k_eff_s)/(ln(r_o_2/r_mid_2))   {Radial through secondary wick}
R_long_2=l_eff/(k_eff_s*A_wick_2)                {Length secondary wick}
```

Next we will include the known values for energy exchange occurring between the environment and the various sections of the LHP. The energy balances for the vapor line and liquid return lines are

```
Energy balance for liquid return line}
Q_dot_LRL=2.0

{Energy balance for vapor line}
Q_dot_VL=1.0
```

The energy balance for the evaporator and the CC are

```
{Energy balance for evaporator}
Q_dot_in=150
Q_dot_in_2ph=Q_dot_in-Q_dot_HL

{Energy balance for CC}
Q_dot_CC=1.5
Q_dot_HL=m_dot*c_p_1*(T_1-T_7)-Q_dot_CC
```

The heat input driving the phase change in the evaporator has been accounted for by subtracting the heat leak to the CC from the total heat input. The heat leak has also been defined as equal to the sensible heating occurring in the liquid returning to the CC minus the heat input to the CC housing. The LHP system-level energy balance will be defined as

```
{LHP system-level energy balance}
Q_dot_out=Q_dot_in+Q_dot_LRL+Q_dot_CC-Q_dot_VL
```

Saturation conditions will be assumed in the CC, the evaporator and the condenser. We will define the liquid and vapor quality as

```
{Saturation conditions}
x_l=0.0
x_v=1.0
```

As for the state points, we will begin by establishing the conditions at the inlet and the exit of the condenser (state points 5 and 5p respectively). We will assume saturated vapor conditions at the inlet and saturated liquid conditions at the exit.

```
{State point 5: condenser entrance}
T_5=265
x_5=x_v
P_5=p_sat(ammonia,T=T_5)
h_5=enthalpy(ammonia,T=T_5,x=x_5)
h_fg_5=enthalpy_vaporization(ammonia,T=T_5)
c_p_5p=cp(ammonia,T=T_5,x=x_5p)

{State point 5p: end of condensation in condenser}
x_5p=x_l
T_5p=T_5
h_5p=enthalpy(ammonia,T=T_5p,x=x_5p)
m_dot=Q_dot_out/(h_fg_5+c_p_5p*(T_5-T_6))
```

As the total heat rejection from the condenser and in the condenser, specific heat of the liquid exiting the condenser and the temperature at the exit of the subcooler are known, the mass flow rate of the fluid can be determined. Heat rejection from the condenser and the subcooler may then be captured as

```
{Energy balance for condenser}
Q_dot_out_2ph=m_dot*h_fg_5

{ Energy balance for subcooler}
Q_dot_out_SC=m_dot*c_p_5p*(T_5p-T_6)
```

The enthalpy and specific heat at the subcooler exit can be defined as

```
{State point 6: subcooler exit}
T_6=260
h_6=h_5p-Q_dot_out_SC/m_dot
c_p_6=cp(NH3,T=T_6)
```

The enthalpy and temperature at the entrance to the CC along the liquid return line may be defined as

```
{State point 7: entrance to CC on liquid return line}
h_7=Q_dot_LRL/m_dot+h_6
T_7=Q_dot_LRL/(m_dot*c_p_6)+T_6
```

The remaining state points of interest are state point 1 and state point 3. Under state point 1 we will define the liquid specific heat and thermal conductivity as

```
{State point 1: CC}
x_1=x_l
k_l_1=conductivity(ammonia,T=T_1,x=x_1)
c_p_1=cp(ammonia,T=T_1,x=x_1)
```

For the purposes of analyzing the present system, we will assume that state point 3 is physically located immediately outside the endpoint of the evaporator's vapor grooves and it is at saturated vapor conditions. We will neglect the superheating defined as having occurred by state point 4 and model the change in the evaporator's and condenser's saturation states as being dominated by heat exchange occurring over the vapor line. This enables determination of the enthalpy, temperature and liquid thermal conductivity at state point 3. Since the temperature at state point 3 and the effective thermal resistance of the compound wick structure are known, the temperature at state point 1 can be determined.

```
{State point 3: evaporator vapor groove}
x_3=x_v
h_3=Q_dot_VL/m_dot+h_5
T_3=temperature(ammonia,h=h_3,x=x_3)
k_l_3=conductivity(ammonia,T=T_3,x=x_3)

R_eff=R_wall_p+R_1_2+R_long_2
T_1=T_3-Q_dot_HL*R_eff
```

Solving the system of equations will result in an evaporator temperature of 272.5 K and a CC temperature of 271.3 K.

The thermodynamic analysis of this example did not require knowledge of the detailed geometry of each of the sections (not including the primary and secondary wick structures). However, the amount of heat exchanged by a particular section with the environment and/or the fluid's state conditions at either the entrance or exit (or both) of the section in question needed to be known. It was also assumed that the vapor front extended to and collapsed right at the condenser exit. In application, LHP systems will consist of a fixed length of plumbing thermally coupled to a cold sink (e.g., a radiator on a spacecraft) that acts as condenser and subcooler combined. Heat transfer analysis techniques that take a multiphase pressure drop in the condenser into account enable variable penetration depths of the vapor front into the condenser to be modeled as a function of the LHP's operating conditions and sink temperature. This is often performed by determining the multiphase convection coefficient for the condensation process and invoking Newton's law of cooling to determine the amount of heat rejection. For more information and examples of this technique, see Launey et al. [43] or Hoang and Kaya [44].

5.5 Thermoelectric Coolers and Generators

Thermoelectric coolers (TECs) and thermoelectric generators (TEGs) are solid-state devices common to spaceflight temperature control and power system architectures. TECs are frequently found on scientific instruments used for localized cooling, whereas TEGs are often used on deep-space missions for thermal-to-electric energy conversion when a large temperature gradient is available. Thermoelectric phenomena for these devices are rooted in simultaneous thermal energy and electrical transfer processes that foster energy exchange with the environment. The processes include the Peltier effect, the Seebeck effect, the Thomson effect, Conduction heat transfer and Joule heating.

5.5.1 Operational Effects

When a current passes through an isothermal junction consisting of dissimilar electrical conductors, a net thermal energy exchange occurs with the environment to which the junction is thermally coupled [50–55] (Figure 5.33). This is known as the Peltier effect. Alternatively, if a temperature gradient is applied across a conductor junction, a voltage potential is produced across the conductor elements (Figure 5.34). This is known as the Seebeck effect. When a temperature gradient exists in a material carrying an electrical current, thermal energy is acquired or rejected from the material. This is known as the Thomson effect. The direction of heat flow in the Peltier and Thomson effects depends upon the direction of current flow [50,53].

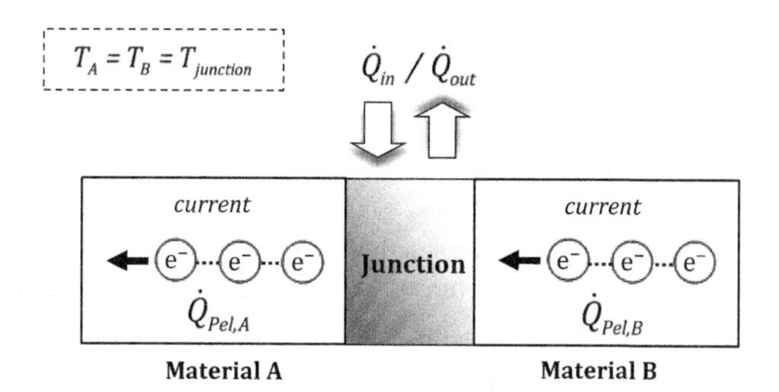

Figure 5.33 Isothermal conductor junction for dissimilar metals A and B

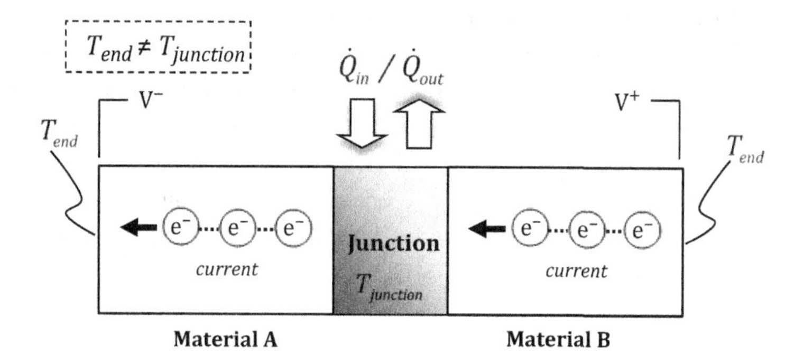

Figure 5.34 Conductor junction exposed to a temperature gradient across materials A and B

Peltier heating of a conductor material is characterized analytically as

$$\dot{Q}_{Pel} = \Pi \cdot I \tag{5.49}$$

where Π is the Peltier coefficient (units of [V]) for the material in question and I is the current (units of [Amps]). An alternate representation of the Peltier heat transfer that includes the Seebeck coefficient is

$$\dot{Q}_{Pel} = \Gamma T J_e A \tag{5.50}$$

where T is the temperature at the isothermal junction, Γ is the Seebeck coefficient in [V/K], J_e is the current density in [A/m^2] and A is the cross-sectional area in [m^2] for the conductor material. Current that passes through a conductor junction from one material to the next is conserved. In equation 5.50, the Seebeck coefficient is a thermophysical property of the conductor material experiencing current flow. When different conductor materials are used to create a junction, according to the Peltier effect, heat transfer along the current's flow path is not conserved. Thus, an energy balance on the junction with the temperature reservoir will include either heat acquisition or heat rejection depending on which material's Seebeck coefficient is greater. Assuming the Seebeck coefficient for material B in Figure 5.33 is greater than that for material A, the energy balance on the junction can be defined as

$$\dot{Q}_{Pel,net} = (\Gamma_{\text{B}} - \Gamma_{\text{A}}) \cdot T J_e A \tag{5.51}$$

The electrostatic potential ($\Delta\phi$ in units of [V]) generated by the Seebeck effect (see Figure 5.34) is characterized using equation (5.52)

$$\Delta\phi = (\Gamma_B - \Gamma_A) \cdot \Delta T \tag{5.52}$$

where ΔT is the temperature difference across the material and Γ is the Seebeck coefficient in [V/K]. Equation 5.53 captures the heating and/or cooling in a conductor material caused by the Thomson effect.

$$\dot{Q}_{Thom} = I\Lambda\Delta T \tag{5.53}$$

In this relation, I is the current in Amps [A], Λ is the Thomson coefficient for the material in units of [V/K] and ΔT is the temperature difference of the material in units of [K]. Details pertaining to the modeling of thermal conduction processes and Fourier's law of conduction were addressed in Chapter 2 and can be referenced there. However, the Ohmic behavior of electrically conductive materials has not yet been addressed. For a conductor material experiencing current flow while operating in the Ohmic (i.e., electrically resistive) state, heating will occur in the conductor material due to its resistance to electrical current. This is known as Joule heating and is captured analytically as

$$\dot{Q}_{Joule} = I^2R_e \tag{5.54}$$

where I is the current in units of Amps [A] and R_e is the material's electrical resistance in units of Ohms [Ω].

5.5.2 Heat Transfer in Thermoelectric Devices

TEC modules consist of paired thermoelectric elements connected together in series. The ends of each element are in thermal contact with the hot and cold sinks. Figure 5.35 shows examples of a single pair of thermoelectric elements (i.e., a couple) in the cooling (Figure 5.35a) and power-generation (Figure 5.35b) configurations. The thermoelectric elements shown comprise individual P-type and N-type semiconductor materials represented by elements ② and ① respectively. In the cooling configuration, a shunt is located on the cold side allowing a path for current flow between the individual elements. At the opposite end, the elements are electrically isolated from each other while in thermal communication with the hot reservoir. The electrical circuit is closed by wiring on the hot side connecting the individual elements and creating a continuous flow path for the current.

Like the cooling configuration, the power-generation configuration (Figure 5.35b) has a shunt connecting the two elements. However, in this case it is located on the hot side of the thermoelectric elements. Once again, the elements are electrically isolated from each other on the side of the thermoelectric elements opposite the shunt. In addition, both are in thermal communication with the cold reservoir. The electrical circuit is closed by wiring containing a resistive element on the cold side.

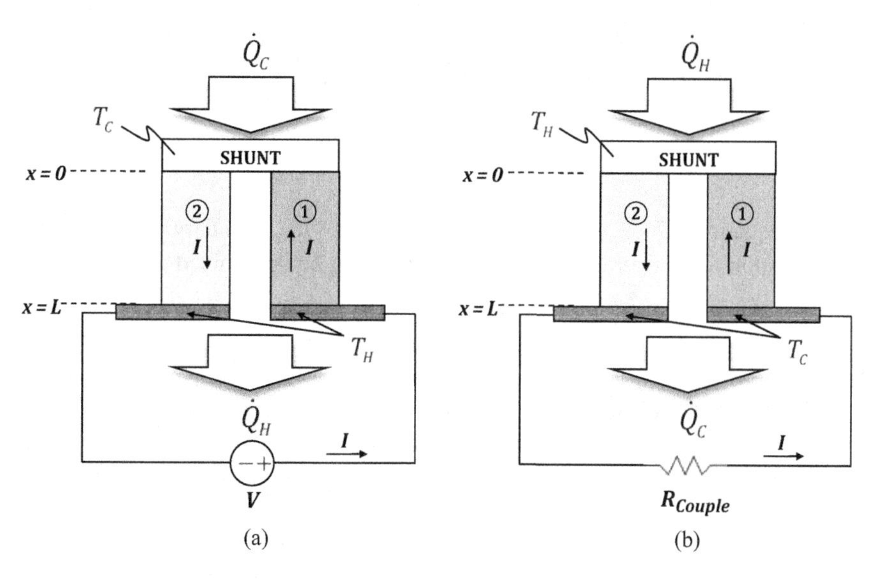

Figure 5.35 Single thermoelectric couple: a) thermoelectric cooler configuration; b) thermoelectric generator configuration

5.5.2.1 Energy Exchange at the Element and Couple Level

The performance capability of a single thermoelectric element is a function of the heat transfer between the element and its boundaries (the reservoirs) and the energy generation within the element itself. Given the reservoirs are at constant temperature, the heat transfer will be highly dependent on the temperature of the elements. Since the heat transfer through the elements is conduction based, 1D conduction analysis techniques (previously covered in Chapter 2) can be applied to determine the temperature distribution across a single element. Assuming x is the primary direction of temperature variation, boundary conditions for the endpoints of an element can be defined as [50,51,53,54,56,57]

i. $T = T_C$ @ $x = 0$
ii. $T = T_H$ @ $x = L$

The governing equation for energy exchange in the element can then be defined as

$$\underbrace{\frac{d}{dx}\left(k\frac{dT}{dx}\right)}_{\substack{Fourier\\Conduction}} + \underbrace{J_e^2\rho_e}_{\substack{Joule\\Heaeing}} - \underbrace{T\frac{d\Gamma}{dT}J_e\left(\frac{dT}{dx}\right)}_{\substack{Thomson\\Effect\\Heating}} = 0 \tag{5.55}$$

where J_e is the current density in [A/m^2] and ρ_e is the semiconductor material's electrical resistivity in [$\Omega \cdot$ m]. Assuming the Seebeck coefficient is constant with respect to temperature, the third term in equation 5.55 drops out leaving the Fourier conduction and Joule heating terms.

$$\underbrace{\frac{d}{dx}\left(k\frac{dT}{dx}\right)}_{\substack{\text{Fourier}\\ \text{Conduction}}} + \underbrace{J_e^2\rho_e}_{\substack{\text{Joule}\\ \text{Heating}}} = 0 \tag{5.56}$$

If equation 5.56 is integrated twice w.r.t. x and the boundary conditions are applied, a relation for temperature as a function of x can be determined as

$$T(x) = -\frac{1}{2}\cdot\frac{I^2\rho_e x^2}{A^2k} + x\cdot\left(\frac{T_H - T_C}{L} + \frac{1}{2}\cdot\frac{I^2\rho_e L}{A^2k}\right) + T_C \tag{5.57}$$

where the current density (J_e) has been recast as the ratio of the current to the conductor cross-sectional area ($J_e = I/A$). The energy balance on the shunt side of the element (which is in thermal equilibrium with the cold reservoir) then becomes

$$\dot{Q}_{C,1} = \Gamma T_C J_e A - \frac{kA}{L}(T_H - T_C) - \frac{1}{2}\cdot\frac{I^2\rho_e L}{A} \tag{5.58}$$

Given that the electrical resistance in a material is defined as a function of the resistivity and the inverse of the conductor's shape factor ($R_e = \rho_e L/A$), equation 5.58 can be recast as

$$\dot{Q}_{C,1} = \Gamma T_C I - \frac{(T_H - T_C)}{R_1} - \frac{1}{2}\cdot I^2 R_{e,1} \tag{5.59}$$

At the pairing level, which includes heat transfer effects from both elements ① and ②, the total cooling at the shunt is

$$\dot{Q}_{C,12} = \Gamma_{12} T_C I - \frac{(T_H - T_C)}{R_{12}} - \frac{1}{2}\cdot I^2 R_{e,12} \tag{5.60}$$

where $\Gamma_{12} = \Gamma_1 - \Gamma_2$ and R_{12} is the total thermal resistance between the hot and cold sinks, considering the elements are in parallel. The total thermal resistance for the two parallel elements can be determined using equation 5.61.

$$\frac{1}{R_{12}} = \frac{1}{R_1} + \frac{1}{R_2} \tag{5.61}$$

The total electrical resistance through the pair of thermoelements ($R_{e,12}$) is defined as

$$R_{e,12} = R_{e,1} + R_{e,2} \tag{5.62}$$

In the case of the TEG, a potential is produced due to the temperature gradient across the thermoelements. When a conductive path that includes a resistance is placed across this potential, current is generated. The current produces Peltier heat transfer at the hot sink whereby heat is acquired. Expressions for the heat transfer can be derived using an approach similar to that used for the TEC. An energy balance at the hot junction of the TEG will have a heat load of

$$\dot{Q}_{H,12} = \Gamma_{12} T_H I + \frac{(T_H - T_C)}{R_{12}} - \frac{1}{2}\cdot I^2 R_{e,12} \tag{5.63}$$

5.5.2.2 Module-Level Energy Exchange

At the module level, equation 5.60 can be recast to reflect thermal contributions from all the paired elements of which the TEC module is comprised. The cooling load at the module level can then be defined as

$$\dot{Q}_{C,mod} = N_{couple} \cdot \Gamma_{12} T_C I - \frac{(T_H - T_C)}{R_{mod}} - \frac{1}{2} \cdot I^2 R_{e,mod} \tag{5.64}$$

where R_{mod} is the total thermal resistance between the hot and cold sinks for all the module's elements in parallel. $R_{e,mod}$ is the total electrical resistance for all the module's elements in series. N_{couple} is the total number of couples that are in the module. The first law of thermodynamics applies to both TEC and TEG modules. Energy balances for both can be analyzed as shown in Figure 5.36. In equation form, the energy balance on the TEC would be

$$\dot{Q}_C + P_{in} = \dot{Q}_H \tag{5.65}$$

where P_{in} is the input power to the TEC module. This is defined as

$$P_{in} = I \cdot (N_{couple} \Gamma_{12} \Delta T + I R_{e,mod}) \tag{5.66}$$

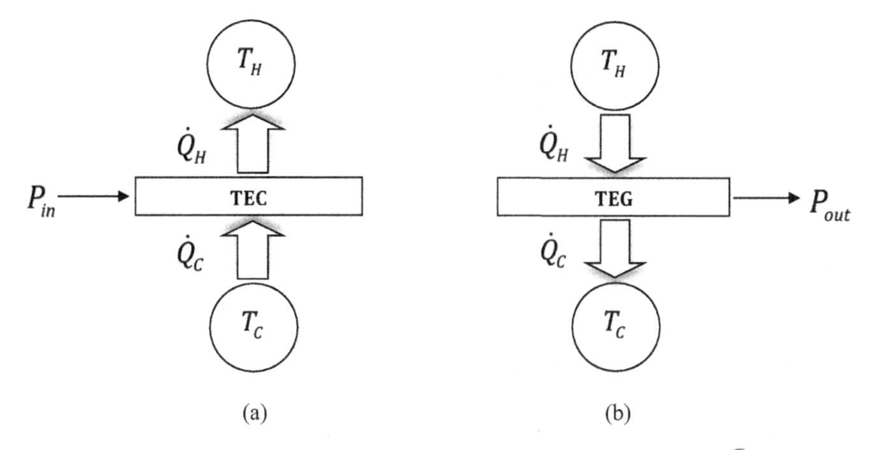

Figure 5.36 Energy balances: a) for thermoelectric cooler; b) for thermoelectric generator

where ΔT is the temperature difference between the hot and cold reservoirs ($T_H - T_C$). Given the definitions for the cooling load (equation 5.64) and the power input (equation 5.66) and using the module-level energy balance, the heat rejection to the hot reservoir can be determined. This TEC relationship is

$$\dot{Q}_{H,mod} = N_{couple} \cdot \Gamma_{12} T_H I - \frac{(T_H - T_C)}{R_{mod}} + \frac{1}{2} \cdot I^2 R_{e,mod} \tag{5.67}$$

Since the TEC is a solid-state heat pump, standard techniques for assessing performance efficiency of a refrigeration cycle apply. The refrigeration COP for a TEC module is defined as [58]

$$COP = \frac{cooling\ capacity}{input\ power} = \frac{\dot{Q}_C}{P_{in}} \tag{5.68}$$

In expanded form, this would be

$$COP = \frac{N_{couple}\Gamma_{12}T_C I - (T_H - T_C)/R_{mod} - 0.5 \cdot I^2 R_{e,mod}}{I \cdot N_{couple}\Gamma_{12}\Delta T + I^2 R_{e,mod}} \tag{5.69}$$

Standard COPs for performance are of the order of 5–8% (sometimes slightly greater), but research into improved material combinations is ongoing. In the case of the TEG, the module-level energy balance is

$$P_{out} = \dot{Q}_H - \dot{Q}_C \tag{5.70}$$

Given that the module-level heat load at the hot junction is

$$\dot{Q}_{H,mod} = N_{couple} \cdot \Gamma_{12}T_H I + \frac{(T_H - T_C)}{R_{mod}} - \frac{1}{2} \cdot I^2 R_{e,mod} \tag{5.71}$$

and the heat rejection at the cold junction is

$$\dot{Q}_{C,mod} = N_{couple} \cdot \Gamma_{12}T_C I + \frac{(T_H - T_C)}{R_{mod}} + \frac{1}{2} \cdot I^2 R_{e,mod} \tag{5.72}$$

the power generated by the module is

$$P_{out} = I \cdot (N_{couple}\Gamma_{12}\Delta T - IR_{e,mod}) \tag{5.73}$$

Since TEGs are power generators, the standard definition for power-cycle efficiency applies. For a TEG, the module-level efficiency is

$$\eta = \frac{useful\ power}{heat\ input} = \frac{I \cdot \left(N_{couple}\Gamma_{12}\Delta T - IR_{e,mod}\right)}{\dot{Q}_H} \tag{5.74}$$

Another metric often used to gauge the efficiency of the thermoelectric conversion process associated with a thermoelectric element is the dimensionless figure of merit (ZT_{avg}). For a given element, the ZT_{avg} value is defined as

$$ZT_{avg} = \frac{T_{avg}\Gamma^2}{\rho_e k} \tag{5.75}$$

where T_{avg} is determined by averaging the values for the hot and cold temperature sinks, k is the thermal conductivity of the material element and ρ_e is the resistivity of the element [54,59,60].

5.5.3 Thermoelectric Cooler Performance Features

This chapter began with a high-level overview of heat transfer technologies analogous to heat pumps followed by a cursory comparison of their performance features with those of the standard mechanical heat pump system. Let us now take a detailed look at TEC performance features. TECs have no moving parts. Like mechanical heat pumps, during operation they provide a temperature lift. This is not observed in HPs, LHPs or thermal straps. TECs require power input in the form of DC current and/or voltage control. TECs also have robust controllability features. Temperatures at the hot and the cold plate are controllable to an accuracy of $\pm 0.1°C$. In addition, bi-polar as well as PID control is capable (e.g., $\pm$voltage) when operating a TEC. Performance parameters for a TEC will be determined by its construction materials, and the quality of its fabrication and assembly.

Figure 5.37 shows a TEC with its construction materials highlighted. The shunt, as well as the P-type and N-type semiconductor materials are all metal. Bismuth telluride is often used as a semiconductor element. The upper and lower plates on the hot- and cold-side junctions are typically ceramic. The semiconductors, the shunt, the circuit contacts and the ceramic plates have to be joined firmly by soldering the shunt and the ceramic parts to the N-type and P-type materials.

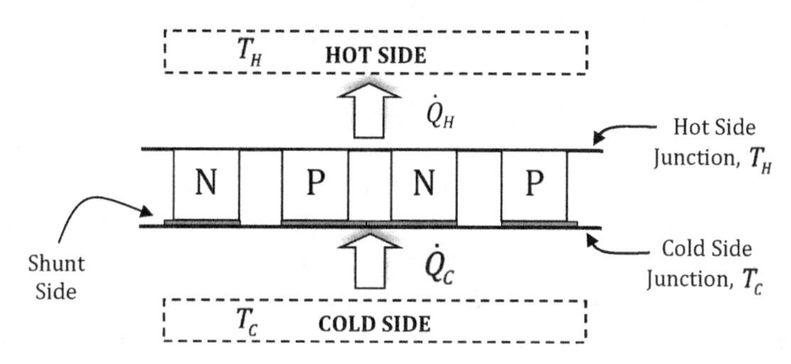

Figure 5.37 Basic TEC module

5.5.3.1 Operational Issues

There are several operational concerns to be addressed when using TECs. Manufacturers typically provide a suggested maximum ΔT between the hot plate and the sink when rejecting to a gas (e.g., air or N_2) or vacuum for space applications. Exceeding the suggested maximum for the ΔT could cause inability to reject heat from the hot side. Failure to reject the required amount of heat from the hot side could lead to excessive ΔT values between the upper and lower plates, creating thermal stresses in the device. A quality bond should be created between the module and the dissipative heat source. One technique used to promote a good interface joint at the heat source is to implement a compression fixture (e.g., bolted pattern or clasp). Figure 5.38 shows a TEC with its hot and cold plates coupled to the respective hot and cold sinks via a compression clamp. Furthermore, since both the upper and lower plates are made of ceramic, which is known to be weak in tension and strong in compression, these devices are subject to marginal cooling performance when exposed to shear or tension loads across the plates. In this configuration an interstitial material (bonding or thermally conductive agent) is placed between the ceramic and the source and sink. However, excessive torqueing can lead to additional stress on the ceramic plates as well as the soldered junctions. Thus, the practitioner should be aware of torque levels for the screws on the clamping device [61]. An alternate, non-clamping approach would be to bond the source/sink to the upper/lower plate. A list of candidate TIMs is given in Table 2.1.

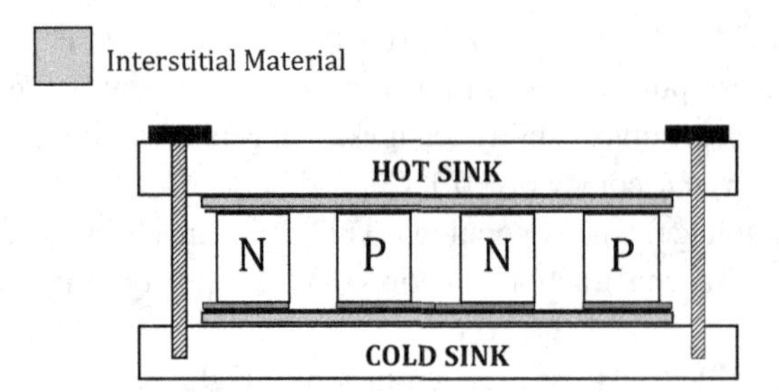

Figure 5.38 TEC with a fastener clamp

5.5.3.2 Applications

The basic application case for a TEC is as an individual module. This is typically referred to as a "single-stage" application, meaning that there is a single cooling stage between the cold and hot sinks. If a single TEC is a single refrigeration cycle, multiple TECs are a compound cycle (otherwise referred to as compound stages). Figure 5.39a shows a TEC compound stage. In this configuration the TECs are stacked one on top of the other. While the term "stacked" is often used, it should be noted that the P–N junctions in the upper stage are physically connected to the junctions in the lower stage in the form of a serial circuit. The total energy rejected from the upper stage will be the summation of contributions from both the upper and lower stages. In real-life terrestrial applications where the practitioner has the benefit of an atmosphere, a finned heat sink that has more surface area for heat exchange with ambient is often applied to the hot sink of the TEC (Figure 5.39b). Without the benefit of an atmosphere (or fluid for heat absorption), either the heat would need to be radiated from the hot sink or there would need to be a thermal coupling between the heat rejection plate and a cold sink. If the heat sink is located remotely from the TEC, the gap will have to be spanned to provide thermal energy exchange between the TEC's heat reject side and the sink. For the example in Figure 5.40, a thermal strap is placed on the upper plate of the TEC and provides a thermal coupling for heat rejection to the sink plate. In order for heat to be

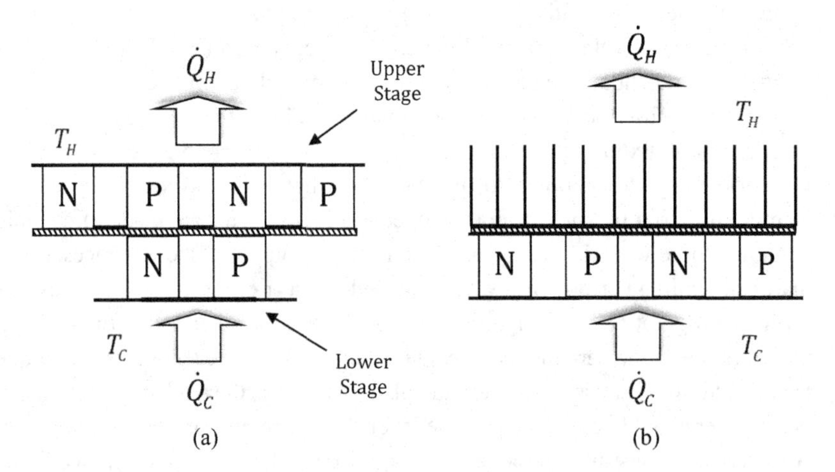

Figure 5.39 Multistage and finned TECs: a) multistage TEC; b) finned TEC for atmospheric applications

rejected from the upper plate in this configuration, temperature gradient requirements between the TEC's upper plate and the cold sink must be satisfied. Specifically, if $T_H > T_{C,2}$ then heat transfer will occur away from the upper plate of the TEC towards the sink plate. However, if $T_H = T_{C,2}$ heat transfer will not occur. If there is no temperature gradient in the direction of the heat sink, the temperature-controlled device for which the TEC/thermal strap combination is providing cooling will not be cooled and consequently may overheat.

Example 9 What is the required thermal strap resistance for a given temperature difference between the upper plate of the TEC and the sink plate for the configuration shown in Figure 5.40?

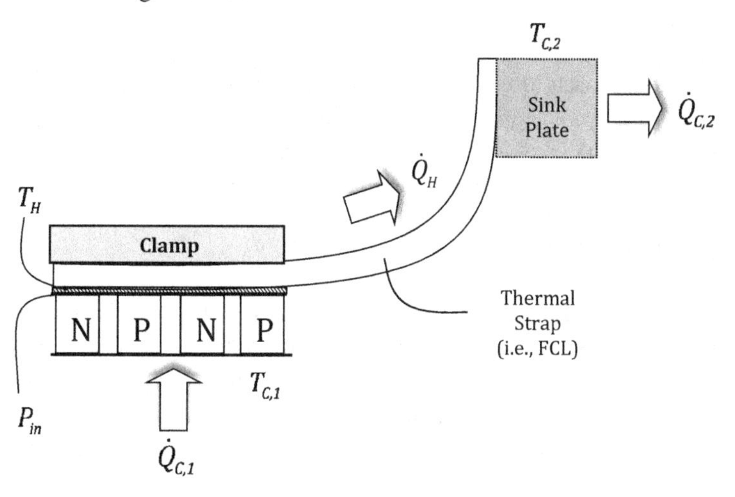

Figure 5.40 TEC located remotely from heat rejection location

Solution

Based on the energy balances for this system the heat rejection from the hot side is

$$\dot{Q}_H = \dot{Q}_{C,1} + P_{in} \tag{5.75}$$

and can be recast as

$$\dot{Q}_H = \dot{Q}_{C,1} + P_{in} = \dot{Q}_{C,2} \tag{5.76}$$

Now let us take a look at the thermal strap and the heat rejection to the sink plate. The heat rejected from the TEC's hot side and carried away to the sink plate may be characterized as

$$\dot{Q}_{C,2} = C_{TS} \cdot \Delta T \tag{5.77}$$

where the thermal conductance (C_{TS}) can be rewritten as $C_{TS} = 1/R_{TS}$. The heat reject to the cold sink can be written as

$$\dot{Q}_{C,2} = \frac{\Delta T}{R_{TS}} = \frac{(T_H - T_{C,2})}{R_{TS}} \tag{5.78}$$

Based on equation 5.76, an equality that includes the cold sink heat reject and the input power can now be established as

$$\dot{Q}_{C,2} = \dot{Q}_{C,1} + P_{in} = \frac{(T_H - T_{C,2})}{R_{TS}} \tag{5.79}$$

Rearranging terms to define the thermal strap resistance will give

$$R_{TS} = \frac{(T_H - T_{C,2})}{(\dot{Q}_{C,1} + P_{in})} \tag{5.80}$$

or, in thermal conductance notation,

$$C_{TS} = \frac{(\dot{Q}_{C,1} + P_{in})}{(T_H - T_{C,2})} \tag{5.81}$$

An alternative approach to using a thermal strap would be to radiate the heat rejected from the TEC's hot side to an effective cold sink. However, in order to implement this approach a good FOV between the TEC's hot side and the cold sink must be present.

5.6 Phase Change Materials

Phase change materials (PCMs) use the latent heat of fusion of a material for heat absorption and release it at a constant temperature. Interest in PCMs for space-based applications began in the 1960s. The potential for passive storage and release of thermal energy onboard a space platform inspired numerous studies of PCM performance capabilities [62–79]. Early spaceflight applications of PCMs included their use on Skylab [66,71,79,80], the Apollo Mission Lunar Rover Vehicles (i.e., 15–17) [66,71], and the Venera Landers (i.e., 11–14) [81–83]. Solar dynamic power generators [84–89], which once were a competing technology for power generation onboard the ISS, also use PCMs. However, contemporary space-based application of PCMs focuses primarily on their use for exploration missions to Venus [90–93] and free-flying space platforms [94–97]. In addition, extended duration experiments such as the phase change material heat exchanger (PCM HX) unit have flown onboard the ISS. Given the previous successful applications of PCMs in the space environment, as well as the space community's sustained interest in them, their use in space-based TCS architectures is only expected to grow in the decades to come.

5.6.1 Fundamentals of Operational Cycles

Figure 5.41 shows the heating profile of a material (from solid to liquid to vapor) along with the primary temperatures and enthalpies at each of the phase interfaces. At the beginning of heat application, the solid material will have an initial temperature that is above the material's triple-point temperature value ($T_{init} > T_{tp}$). Increasing the initial temperature above T_{tp} avoids direct transition into the vapor phase (via sublimation). Heat is added and the temperature of the solid will increase until it reaches melting temperature (T_{melt}). The transition from solid to liquid will occur at this temperature.

The "melt-to-solidification ratio" (x_{MSR}), which gives the amount of melting that has occurred, is defined as the ratio of the mass of liquid (m_l) to the total mass of the material (i.e., the sum of the solid and liquid present, $m_s + m_l$). In equation form this is

$$x_{MSR} = \frac{m_l}{m_s + m_l} \tag{5.82}$$

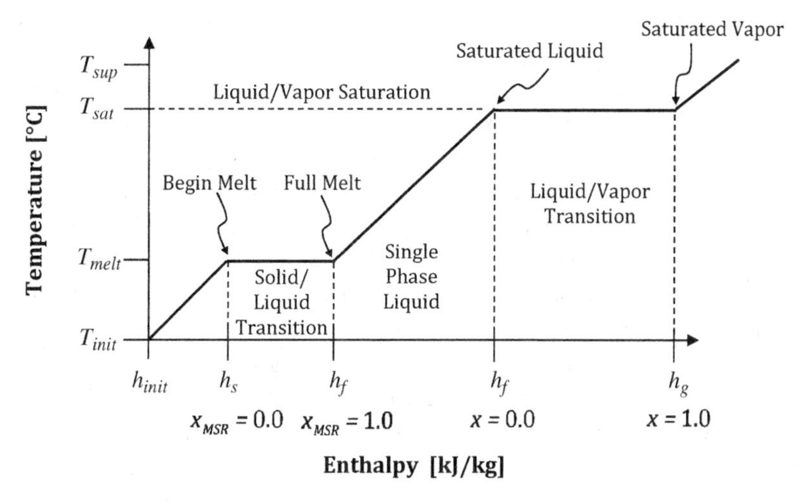

Figure 5.41 Solid/liquid/vapor heating cycle for a material
Source: Reproduced with permission of Saunders Publishing, from *Physics for Scientists and Engineers,* Serway, R., 1986, reference [97]; permission conveyed through Copyright Clearance Center, Inc.

The enthalpy of the material at the beginning of melt (at $x_{MSR} = 0$) is h_s. When sufficient heat has been absorbed by the material such that all the solid is converted to liquid, full melt ($x_{MSR} = 1.0$) has been achieved. Given sustained heat input, the single-phase liquid will undergo temperature increase until it reaches the saturation temperature (T_{sat}) that corresponds with the system pressure. The material is now in the saturated liquid state and has an enthalpy of h_f and a quality of $x = 0$. At this point, liquid-to-vapor phase transition occurs at a constant temperature. When all the liquid has been converted to vapor, the fluid is in the saturated vapor state which is denoted by an enthalpy value of h_g and a quality of $x = 1$. Continued heating of the single-phase vapor will result in further temperature increases. The two-phase regions of the schematic (solid/liquid and liquid/vapor) provide an opportunity to store thermal energy at constant temperature. The energy difference between the beginning and end of transition is the amount of thermal energy a PCM is capable of storing in these different phases. The melt/freeze segment (see Figure 5.41) is the basis for the operation and application of PCMs and applies to almost all fluids.

5.6.2 Features and Modules

PCMs are typically selected for use with temperature-controlled items when the radiator area provided does not reject enough thermal energy for the item to operate within its temperature limits. They are also used when temperature oscillations are present and damping is required. PCM applications make use of the material's thermal energy storage capacity in the solid phase when melting. The PCM is housed in a container (Figure 5.42) that typically is thermally coupled to a radiator. On the opposite end of the PCM's radiator interface is the thermal interface with the energy source (e.g., E-box, battery, laser, etc.). During operation, waste heat from the temperature-controlled item is directed through the encasement structure and into the PCM itself. Heat rejection from the radiator surface ($\dot{Q}_{out}$) takes place whether the source is generating heat or is dormant. Depending on the FOV of the radiator, radiant heat from environmental sources and/or onboard components may be incident upon the radiator surface. These are captured as ($\dot{Q}_{in}$). Ideally, for temperature control of the energy source, the net heat will be transferred away from the radiator surface rather than towards it.

The primary criteria for selecting a PCM for a specific application are the enthalpy of fusion (Δh_{sf}) and the melt temperature (T_{melt}). The enthalpy of fusion determines how much energy per unit mass can be stored. The melt temperature is a good estimate of the operational temperature of the source. However, T_{melt} should be low enough to accommodate temperature rise that may occur at the interface between the liquid PCM and the energy source during heating. The thermal conductivity of the PCM in the solid and liquid phases should also be considered. Under ideal performance conditions the PCM would melt and solidify at constant temperature during the heating and cooling cycle. However, as there are two distinct phases during both the heating and cooling cycles, and heat is transferred through both phases, there will be a temperature gradient. Higher-conductivity PCMs help to minimize the gradient and better approximate to the ideal performance case. In addition, in the ideal performance scenario, the PCM will have reversible heat transfer and temperature during melting and solidification. Last but not least is the choice of container material. Compatibility between the PCM and the container is important in order to avoid chemical reactions leaving by-products in the PCM or causing degradation of the container's structural integrity on sustained contact with the PCM.

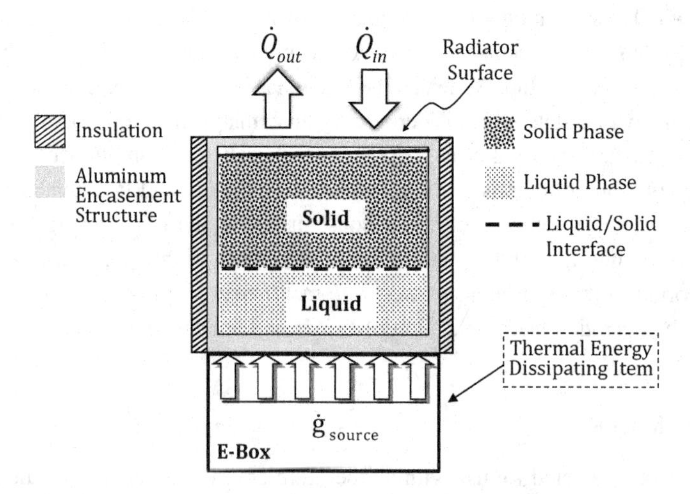

Figure 5.42 PCM (at partial melt) and encasement container

Thermophysical properties for PCMs such as glycerol may be obtained using the properties tables within EES. The most frequently used PCMs in space applications to date have been paraffin waxes (each having a chemical formula of C_nH_{2n+2}). Table 5.5 lists the primary thermophysical properties for some paraffin waxes, as well as a few

Table 5.5 Common phase change materials

Material name	T_{melt} [K]	Δh_{sf} (kJ/kg)	Density (kg/m^3) solid	liquid	Thermal conductivity (W/m · K) solid	liquid	Specific heat (kJ/kg · K) solid	liquid
n-Undecane	248	141	–	772[3]	0.1496[1]	0.1496[1]	–	1.53[1]
($C_{11}H_{24}$)								
n-Dodecane	261	211	800[1]	750[1]	0.1488[1]	0.1488[1]	1.45[1]	1.53[1]
($C_{12}H_{26}$)								
n-Tridecane	267	155	847[3]	756[1]	0.1496[1]	–	1.45[1]	1.53[1]
($C_{13}H_{28}$)	279	228	825[2]	771[2]	0.150[2]	0.150[2]	2.07[2]	2.07[2]
n-Tetradecane	290	237	835[2]	776[2]	0.1505[1]	0.1505[1]	1.45[1]	1.70[1]
($C_{14}H_{30}$)								
n-Hexadecane								
($C_{16}H_{34}$)								
n-Heptadecane	295	213	–	–	–	0.146[5]	–	–
($C_{17}H_{36}$)								
n-Octadecane	301	244	810	770	0.1505[1]	0.1505[1]	1.45[1]	1.70[1]
($C_{18}H_{38}$)								
n-Nonadecane	305	187	–	–	–	0.151[5]	1.45[1]	1.69[1]
($C_{19}H_{40}$)								
n-Eicosane	310	246	856[2]	778[2]	0.150[2]	–	1.44[1]	1.94[1]
($C_{20}H_{42}$)	291.1	198.7	1264	1263	0.284	0.284	2.381	2.386
Glycerol ($C_3H_8O_3$)	303.3	296	1573	1431	0.82[6]	0.594[4]	1.8	2.8
Lithium Nitrate								
Trihydrate								

[1] Values republished w/permission of NASA from *A Design Handbook for Phase Change Thermal Control and Energy Storage Devices,* reference [72].
[2] Values republished from *Phase Change Materials Handbook*, w/permission from Lockheed Martin Corporation, Bethesda, MD, reference [66].
[3] Values reprinted w/permission from Hust, J., and Schramm, R., "Density and Crystallinity Measurements of Liquid and Solid n-Undecane, n-Tridecane, and o-Xylene from 200 to 350K," *Journal of Chemical & Engineering Data,* Vol. 21, No. 1, pp. 7–11. Copyright year 1976 American Chemical Society, reference [99].
[4] Values @294 K, reprinted w/permission from Shamberger, P.J., and Reid, T., "Thermophysical Properties of Lithium Nitrate Trihydrate from (253 to 353) K," *Journal of Chemical & Engineering Data*, Vol. 57, pp. 1404–1411. Copyright year 2012, American Chemical Society, reference [101].
[5] Values republished from Velez, C., Zarate, J., and Khayet, M., "Thermal Properties of n-Pentadecane, n-Heptadecane and n-Nonadecane in the Solid/Liquid Phase Change Region," *International Journal of Thermal Sciences*, Vol. 94, pp. 139–146, reference [100]. Copyright © 2015, Elsevier Masson SAS, All rights reserved.
[6] Values @308 K, reprinted w/permission from Shamberger, P.J., and Reid, T., "Thermophysical Properties of Lithium Nitrate Trihydrate from (253 to 353) K," *Journal of Chemical & Engineering Data*, Vol. 57, pp. 1404–1411. Copyright year 2012, American Chemical Society, reference [101].

other common space-based PCMs. Table values denoted by a series of dashes reflect that the thermophysical property listed is not readily available at the PCM melt temperature, according to the literature. For an extensive list of PCMs, see *Spacecraft Thermal Control Handbook* [98], "Thermal Control by Freezing and Melting" [64], *Phase Change Materials Handbook* [66] or *A Design Handbook for Phase Change Thermal Control and Energy Storage Devices* [72].

5.6.3 Heat Transfer Analysis

For a given mass of PCM, heating during the melt process can be characterized by the conservation relation for the enthalpy of fusion, shown in equation 5.83

$$\dot{Q}_{melt} = \dot{m} h_{sf} \tag{5.83}$$

where

$\dot{m}$: The rate of solid melting [kg/s]
h_{sf}: Enthalpy of fusion for the PCM material [J/kg]

This is captured in Figure 5.41 in energy per unit mass form as the difference between $x_{MSR} = 0$ and $x_{MSR} = 1.0$ while at a constant temperature of T_{melt}. Rearranging variables (i.e., dividing both sides of the equal sign by the enthalpy of fusion) in equation 5.83, the melt rate of the PCM solid ($\dot{m}$) for a known amount of heat input can be determined. The total mass melt (or change) for a given period of time can be calculated by separating the time parameter from the melt rate (in the denominator) and then multiplying both sides of the equal sign by the time change. The total mass change then becomes

$$m = \frac{\Delta t \cdot \dot{Q}_{melt}}{h_{sf}} \tag{5.84}$$

If heating of the PCM includes sensible heating of the solid up to the melt temperature, the total heat absorbed by the PCM at full melt ($x_{MSR} = 1.0$, $h = h_f$) is characterized as

$$\dot{Q}_{s\text{-}l} = \dot{m}\left[c_{p,solid}\left(T_{melt} - T_{Solid,init}\right) + h_{sf}\right] \tag{5.85}$$

Let us revisit the E-box/PCM/radiator configuration shown in Figure 5.42. Assume the radiator is facing deep space and has no environmental loading on its surface. For a component flying onboard a space platform in Earth orbit, the time taken for heating will typically be less than the time for a single orbit. Heat transfer from the E-box to the PCM will occur in pulses. Figure 5.43 shows the length of time for a heat pulse sequence. A key parameter commonly associated with pulsed heating is the duty cycle, which is the ratio of the time heating is applied to the time from start to completion of a single cycle. For space platforms in Earth orbit, the cycle would typically be the time taken for a single Earth orbit.

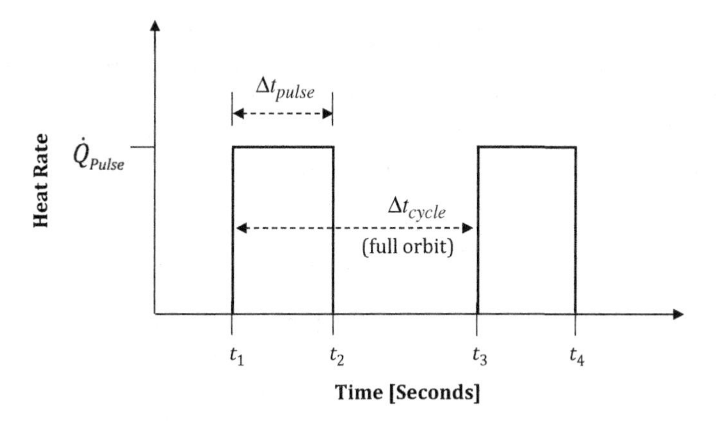

Figure 5.43 Pulse heating cycle
Source: Reproduced w/permission from Aerospace Corporation, *Spacecraft Thermal Control Handbook*, Vol. I, 2002, reference [98]

The duty cycle for the pulsed heating shown in Figure 5.43 can be defined in equation form as

$$\xi_{source} = \frac{\Delta t_{pulse}}{\Delta t_{cycle}} \tag{5.86}$$

Assuming that heat dissipation from the E-box, radiation to space and environmental loading on the radiator are taking place, the radiator area required for use with the PCM can be determined by taking an energy balance on the radiator for a single cycle (shown in equation 5.87)

$$\dot{Q}_{out} \cdot \Delta t_{cycle} = \dot{g} \cdot \Delta t_{pulse} + \dot{Q}_{in} \cdot \Delta t_{load} \tag{5.87}$$

where Δt_{load} is the duration for environmental loading. As for Δt_{pulse}, it is expected that the environmental loading time will be shorter than the full cycle. Expanding the LHS of equation 5.87 gives

$$\sigma \varepsilon A_{surf} T_{melt}^4 \Delta t_{cycle} = \dot{g} \cdot \Delta t_{pulse} + \dot{Q}_{in} \cdot \Delta t_{load} \tag{5.88}$$

where A_{surf} is the radiator surface area. The radiator temperature has been set to T_{melt} in this relation to reflect the full melt condition. Assuming that sensible heating of the liquid-phase PCM does not take place after full melt, T_{melt} will be the maximum temperature attained on the radiator surface. Solving equation 5.88 for A_{surf} will result in

$$A_{surf} = \frac{\dot{g} \cdot \Delta t_{pulse} + \dot{Q}_{in} \cdot \Delta t_{load}}{\sigma \varepsilon T_{melt}^4 \Delta t_{cycle}} \tag{5.89}$$

This can be recast as

$$A_{surf} = \frac{1}{\sigma \varepsilon T_{melt}^4} \cdot \left[\dot{g} \cdot \xi_{source} + \dot{Q}_{in} \cdot \xi_{load}\right] \tag{5.90}$$

where ξ_{load} is the duty cycle for environmental thermal loading. If a PCM were to undergo repeated full melt/solidification cycles, the energy change as a function of time associated with the PCM would follow the form of the sawtooth curve shown in Figure 5.44. The melt-to-solidification ratio (defined previously as equation 5.82) at any given point in time during the cycle can be recast as

$$x_{MSR} = \frac{\dot{Q}_{net} \cdot \left(t_{final} - t_{init}\right)}{E_{Max}} \tag{5.91}$$

where $\dot{Q}_{net}$ is determined from the net energy balance on the PCM.

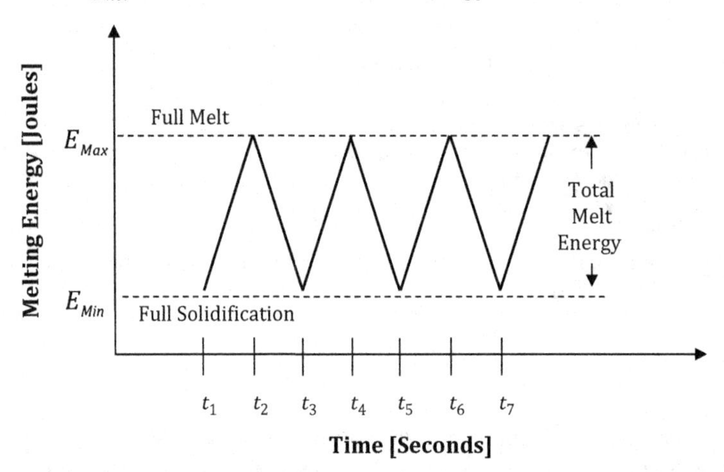

Figure 5.44 Pulse heating cycle melt energy
Source: Adapted w/permission from Aerospace Corporation, *Spacecraft Thermal Control Handbook,* Vol. I, 2002, reference [98]

Example 10 Consider a space platform that contains an E-box dissipating 700 W during operation. The E-box is mounted to a fixed-size radiator that can only reject 200 W. Assume the radiator surface is not receiving any environmental thermal loading. If glycerol (Δh_{sf} = 199 kJ/kg) were used as a PCM (positioned between the E-box and the radiator) to absorb the heat that cannot be rejected by the radiator while the E-box is in operation, how much mass of glycerol would be required to achieve full melt over an E-box operation period of 45 minutes?

Solution

From Table 5.5 we know that glycerol has a melting temperature of 18°C and the enthalpy of fusion for this PCM is given in the problem statement as 199 kJ/kg. 45 minutes is equal to 2,700 seconds. Using the mass melt relationship from equation 5.84, we can calculate the mass requirement as follows:

$$m = \frac{\Delta t \cdot \dot{Q}_{melt}}{h_{sf}}$$

$$m = \frac{(2,700 \text{ s}) \cdot (500 \text{ W})}{(199{,}000 \text{ J/kg})} = 6.78 \text{ kg}$$

Example 11 A mass of 5 kg of n-Octadecane is thermally coupled to a laser baseplate. The laser is pulsing 500 W (continuous thermal energy) every 40 minutes at durations of 20, 30 and 25 minutes. Assuming that the radiator coupled to the PCM/laser combination is continuously rejecting 200 W in both the "on" and "off" states, what is the melt-to-solidification ratio throughout the three pulse heating cycles?

Solution

To determine x_{MSR}, the maximum energy capable of storage must be determined. From Table 5.5 we know that n-octadecane has a melting point of 28°C and an enthalpy of fusion of 244 kJ/kg. Using the mass provided in the problem statement and the thermophysical properties, E_{max} can be calculated as

$$E_{max} = m_{PCM} \cdot \Delta h_{sf} \tag{5.92}$$

$$E_{max} = (5\ \text{kg})(24{,}400\ \text{J/kg}) = 1{,}220{,}000\ \text{Joules}$$

Cycle 1 will include 20 minutes of heat dissipation from the laser immediately followed by a 20-minute period with the laser in the "off" state. Assuming that the PCM is in the solid phase at the start of heating, x_{MSR} at the end of the 20-minute heating portion of the cycle can be determined as

$$x_{MSR} = \frac{(500\ \text{W} - 200\ \text{W})(1{,}200\ \text{s})}{1{,}220{,}000\ \text{Joules}} = 0.29$$

For the portion of the cycle with the laser in the non-operational state, x_{MSR} can be calculated as

$$x_{MSR} = \frac{(500\ \text{W} - 200\ \text{W})(1{,}200\ \text{s}) - (200\ \text{W})(1{,}200\ \text{s})}{1{,}220{,}000\ \text{Joules}} = 0.098$$

The energy from the heating of the PCM has been included in the "off"-state determination in order to capture the time history of the value for x_{MSR}. These calculations can be repeated for cycles 2 and 3 using EES. Thermophysical properties for the PCM, as well as the mass, will be defined at the beginning of the EES code as

```
{Example 5-11: Melt-to-solidification ratio for series of heating cycles}

{PCM thermophysical properties and features}
m=5       [ kg]                {Mass of PCM}
DELTAh_sf=244000               {Enthalpy of fusion, n-octadecane}
E_max=m* DELTAh_sf             {Energy capable of storage at Full melt}

{Initial values}
Q_dot_in[ 1] =0.0              {Initial heat input}
Time[ 1] =0                    {Time at start of first cycle}
lapse[ 1] =0                   {Time elapsed since start of first cycle}
x_MSR[ 1] =0.0 {Initial melt-to-solidification ratio}
```

Initial values for the arrays that will be used in the code are specified as well. Next we will establish the heat inputs and heat rejections from the PCM in both the "on" and "off" states for the laser.

```
{Laser heat dissipations}
Q_dot_on=500 [W]
Q_dot_off=0 [W]

Q_dot_in[2]=Q_dot_on; Q_dot_in[3]=Q_dot_off         {Cycle 1}

Q_dot_in[4]=Q_dot_on; Q_dot_in[5]=Q_dot_off         {Cycle 2}

Q_dot_in[6]=Q_dot_on; Q_dot_in[7]=Q_dot_off         {Cycle 3}
```

The radiator heat rejection and the duration (in units of minutes) for the "on" and "off" mode is

```
{Radiator heat rejection}
Q_dot_out=200 [W]

{Time durations in minutes}
Time[2]=20  [s]
Time[3]=20  [s]
Time[4]=30  [s]
Time[5]=10  [s]
Time[6]=25  [s]
Time[7]=15  [s]
```

Last but not least, the iterative loop that calculates the melt-to-solidification ratio on both the heating and cooling portion of each cycle is

```
{Iterative Loop}
ITER=7

Duplicate K=2,ITER
 lapse[K]=lapse[K-1]+Time[K]
 x_MSR[K]=x_MSR[K-1]+(60*(Q_dot_in[K]*Time[K]-Q_dot_out*Time[K]))/E_max
End
```

Execution of the program will generate a table of values that include the variables x_{MSR}, time and lapse. Figure 5.45 provides a plot of x_{MSR} for the heating and cooling profile spanning the three cycles described in the problem statement.

The analysis techniques discussed so far have focused on radiator sizing and calculation of the net heat flow through the PCM during cyclic thermal loading. However, since heat transfer in and out of the PCM volume is the main driver for melting and solidification it is also worthy of overview. The standard heat transfer analysis

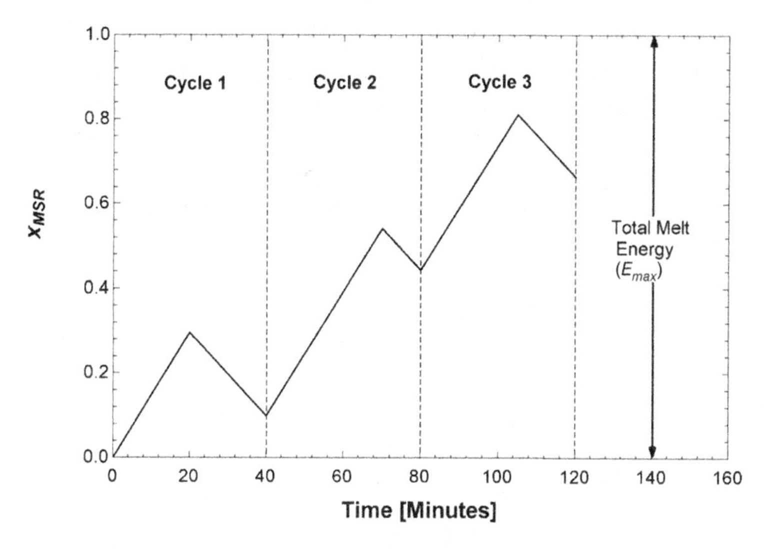

Figure 5.45 Melt-to-solidification ratio for pulse heating cycle

technique applied to thermal energy storage systems incorporates a semi-infinite assumption along the primary direction for heat flow in and out of the storage volume. In the example in Figure 5.46, the temperature at the boundary where heat inflow and outflow is occurring is calculated, whereas the temperature at the opposite end (i.e., at $z = \infty$) is assumed to be constant (i.e., $T_\infty = T_{melt}$). The PCM problem with these boundary conditions is known as the Stefan problem and is well studied throughout the heat transfer community.

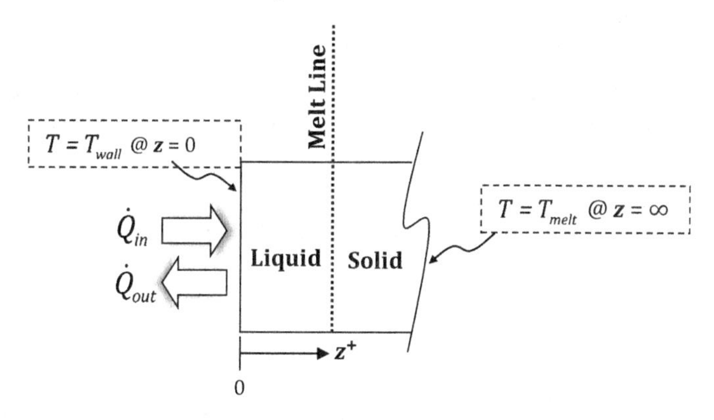

Figure 5.46 Stefan problem heat flow

However, the space-based PCM application deviates from the Stefan problem in the boundary conditions used. The finite length of the container along the primary direction of heat flow, as well as the thermal masses attached at each boundary, suggest an alternate approach to modeling of the PCM is required. Figure 5.47 shows the side view (assuming a uniform depth into the page) of a PCM volume during transition between full solidification and full melt. The liquid and vapor phases are separated by the melt line. Assume that the PCM is all solid at the initiation of heating. In this start condition

x_{MSR} = 0 and the melt line is at position z = 0. As heat is applied to the container boundary at z = 0, the melt line will translate from z = 0 to z = L (full melt condition) given sustained heating. The volume of each phase, along with the key thermophysical properties, will be updated with the translation of the melt line. The full melt condition corresponds to x_{MSR} = 1.0. When heating is removed from the wall at z = 0, solidification of the fully melted PCM will begin and the melt line will translate in reverse from z = L to z = 0 (full solidification).

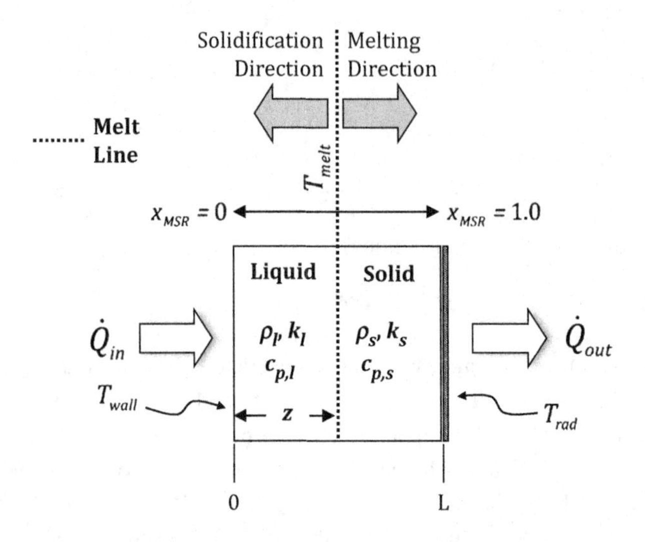

Figure 5.47 Profile of PCM liquid and solid phases during transition

Figure 5.47 assumes that heat rejection from the radiator surface at $z = L$ will be occurring throughout heating and cooling. Fourier's law of conduction applies to both the liquid and solid phases (if present) during both melting and solidification. When $x_{MSR} < 1.0$ the temperature inequality $T_{wall} > T_{melt} > T_{rad}$ holds and the heat rejected away by the radiator is equal to the conduction heat transfer through the solid phase. The transient energy balance on the radiator then becomes

$$m_{rad}c_{p,rad} \cdot \frac{(T_{rad,i+1} - T_{rad,i})}{\Delta t} = k_s A_{surf} \cdot \frac{(T_{melt} - T_{rad,i+1})}{(L - z)} - \dot{Q}_{out} \tag{5.93}$$

where the i + 1 and i subscripts on the radiator temperature (T_{rad}) denote timesteps. Similarly, the transient energy balance at the interface between the PCM container wall and the E-box (at z = 0) will be

$$m_{int}c_{p,int} \cdot \frac{(T_{int,i+1} - T_{int,i})}{\Delta t} = \dot{Q}_{in} - k_l A_{surf} \cdot \frac{(T_{wall} - T_{melt})}{z} \tag{5.94}$$

When full melt is achieved, $T_{rad} = T_{melt}$. When full solidification has been attained, $T_{int} = T_{melt}$.The next example uses the energy balance relations established for each phase of the PCM during two heating and cooling cycles.

Example 12 A cubic shaped E-box (side length of 17 cm with a wall thickness of 5 mm) is periodically operated onboard a space platform. In the "on" state, the E-box has a dissipation of 66.8 W. One full side of the E-box is to be coupled to a high-efficiency radiator with a coating emissivity of ($\varepsilon = 0.98$). However, due to size constraints for a potential radiator, the project leadership wishes the use of a PCM with a reduced radiator size to be explored.

i) What would be the required radiator size for heat rejection to space (w/no environmental loading) when not using a PCM if the radiator surface were to be operated at 270 K?
ii) Assuming a radiator surface area of 19 cm × 19 cm (5 mm thickness of Al-6061) with the specified emissivity coating and a surface temperature of 267 K, what would be the duty cycle for a PCM experiencing a pulse heat load of 66.8 W?
iii) If n-Tridecane is to be used as the PCM for the E-box (made from Al-6061), determine (and show) the heating profile for both the E-box/PCM interface and the radiator surface (given in part ii) over two full cycles if the duration of the heat pulses is approximately 405 seconds.

Solution

The surface area required for heat rejection of 66.8 W from the E-box can be determined using the flat plate energy balance methodologies reviewed in Chapter 4. Equation 5.90 can be rearranged to solve for ξ_{source}. Since no environmental heating is being received by the radiator surface, $\xi_{load} = 0$. However, the solution for part (iii) requires more extensive analysis and the use of a software tool. To simplify the process of obtaining answers to all three parts of the problem, a single EES code will be developed. We will begin with the syntax for constants and features of the radiator surface.

```
{Example 5-12: PCM Melt/solidification cycle using n-Tridecane}

{Constants}
sigma=5.67e-8
h_sf_Tridecane=155*1000    Enthalpy of fusion in J/kg, n-Tridecane}
T_melt=267 [ K]            {Melt temperature for n-Tridecane}

{Pulse heat load}
Q_dot_pulse=66.8

{Radiator features}
eps_rad=0.98               {High emissivity radiator surface}
A_rad=(19/100)*(19/100)    {m^2}
```

The calculation of the surface area will also be included in the code.

```
{Radiator size determination for Part a)}
A_baseline=Q_dot_pulse/(eps_rad*sigma*(270^4))          {m^2}
l_baseline=A_baseline^0.5  {Radiator side length for a square surface}
```

The area of the radiator used for the PCM/radiator configuration should be noticeably less than that of the baseline case determined in part (i). The duty cycle can be determined using the following code.

```
{Pulse heat load}
xi_source=A_rad*eps_rad*sigma*(T_melt^4)/Q_dot_pulse {Duty cycle
                                                        calculation}

{Time durations}
Pulse=405                                  {Pulse time in seconds}
Pulse_min=Pulse/60                         {Pulse time in minutes}

Cycle=Pulse/xi_source                      {Cycle time in seconds}
Cycle_min=Cycle/60                         {Cycle time in minutes}
```

As shown in the syntax, the cycle time is calculated using the pulse time and the duty cycle. Next we will focus on the PCM itself. The dimensions of the E-box were given in the problem statement. For the PCM to be used, we will assume that the cross-sectional area of the PCM will be the same as that of the E-box. We will capture this, as well as the thermophysical properties for n-Tridecane (Table 5.5) in the following.

```
{PCM features}
A_PCM=(17/100)*(17/100)

rho_l=756        [kg/m^3] {density of liquid PCM}
rho_s=847        [kg/m^3] {density of solid PCM}

k_l=0.150        [W/m-K]  {thermal conductivity of liquid PCM}
k_s=0.150        [W/m-K]  {thermal conductivity of solid PCM}

c_p_l=1.53*1000           {specific heat in J/kg-K}
c_p_s=1.45*1000           {specific heat in J/kg-K}
```

The total mass required for use in the PCM/radiator configuration over a full melt/solidification cycle must also be determined. The PCM, which is thermally coupled to the radiator, will reject thermal energy via the radiator throughout the cycle. However, melting will take place when the E-box is in its "on" state and dissipating heat to the PCM. The thermal energy associated with the melting process will be the difference between the PCM's incoming and exiting thermal energy over the duration of a single pulse. To determine the required mass, the rejected thermal energy is taken at the full melt case. The product of the net thermal energy and the pulse time divided by the enthalpy of fusion for n-Tridecane will give the mass required. As shown in the following syntax, the mass can then be used to determine the maximum thermal energy storage capability.

```
{Mass and maximum energy determination}
Q_dot_rej=eps_rad*A_rad*sigma*T_melt^4  {heat rejection from
                                         radiator at full melt}

mass=((Q_dot_pulse-Q_dot_rej)*Pulse)/h_sf_Tridecane  {PCMmass requirement}
E_max=mass*h_sf_Tridecane    {Max energy storage capable during heating}

L=mass/(rho_l*A_PCM)              {Length required for PCM container}
```

In order to perform the transient analysis required for temporal determination of the radiator surface and E-box/PCM interface temperatures, multiple variables must be recorded. These variables include the E-box energy dissipation, heat rejection from the radiator, elapsed time, melt-to-solidification ratio, stored energy in the PCM, melt line location, E-box/PCM interface temperature, radiator temperature and masses for both liquid and solid phases. Each of these variables will be recorded by establishing arrays for each variable in an iterative loop. However, each variable must be initialized for time $t = 0$ seconds. Syntax for initialization of these variables is shown in the following lines.

```
{Initial array settings}
g_dot_in[1]=Q_dot_pulse                 {Thermal dissipation from E-box}

Q_dot_out[1]=eps_rad*A_rad*sigma*T_rad[1]^4 {Heat rejection from radiator}


m_s[1]=L*A_PCM*rho_s                    {Initial mass of solid PCM}

E_sum[1]=DELTAt*Q_dot_out[1]            {Energy accumulated in PCM}

z[1]=0.0                                {Initial location of melt line}

T_rad[1]=T_melt-((L-z[1])/(k_s*A_PCM))*eps_rad*sigma*A_rad*T_rad[1]^4
{Initial temperature}

T_int[1]=T_melt                         {Initial wall temperature}

lapse[1]=0.0                            {Initial elapsed time in minutes}

x_MSR[1]=0.0 {Initial melt-to-solidification ratio} {Initial mass of liquid}
```

Transient modeling of the E-box/PCM interface temperature and the radiator temperature require knowledge of the thermal storage capacity of both. This is shown in the energy balance relations in equations 5.93 and 5.94. The mass of the radiator and the interface plate for the E-box and PCM have been determined and substituted into the code.

```
{Al-6061 Thermophysical properties}
c_p_plate=cp(aluminum_6061, T=T_int[1])        {Al-6061 specific heat}
rho_plate=density(aluminum_6061, T=T_int[1])     {Al-6061 density}

{Radiator plate mass determination}
th_plate=(5/1000)                {thickness of radiator plate in mm}
Vol_rad=th_plate*A_rad           {volume of radiator plate}
m_rad=rho_plate*Vol_rad          {mass of radiator plate}
m_plate=m_rad                    {mass of E-box/PCM interface plate}
```

The iterative loop used to calculate the temperatures for the E-box equivalent thermal mass at the interface and the radiator at each timestep is

```
{Iterative loop}
DELTAt=5.0                          {Seconds}

ITER=1080                           {Maximum iteration count}

Duplicate K=1,ITER

  T_rad[K+1]=T_rad[K]+(DELTAt/(m_rad*c_p_plate))*((k_s*A_PCM*&
    (T_melt-T_rad[K+1]))/(L-z[K+1])-eps_rad*sigma*A_rad*T_rad[K+1]^4)

  Q_dot_out[K+1]=eps_rad*A_rad*sigma*T_rad[K+1]^4

  g_dot_in[K+1]=heat(K,Q_dot_pulse) {E-box dissipation}

  m_s[K+1]=m_s[K]-((g_dot_in[K+1]-Q_dot_out[K+1])*DELTAt)/h_sf_Tridecane

  T_int[K+1]=T_int[K]+(DELTAt/(m_Ebox*c_p_plate))*((g_dot_in[K&
    +1]-(k_l*A_PCM)*(T_int[K+1]-T_melt)/(z[K+1])))

  lapse[K+1]=K*DELTAt/60              {Total elapsed time in minutes}


  z[K+1]=L-m_s[K+1]/(A_PCM*rho_s)      {Update for melt line}

  E_sum[K+1]=E_sum[K]+(g_dot_in[K+1]-Q_dot_out[K+1])*DELTAt

  x_MSR[K+1]=E_sum[K+1]/E_max          {melt-to-solidification ratio}

End
```

The timestep has been set to five seconds. The variable ITER provides the maximum number of 5-second increments to be included in the loop. The thermal resistance between the edge and center of the radiator and the E-box interface have not been included explicitly since the PCM material will serve as the dominant thermal resistance to heat flow in both energy balance relations. The function entitled "heat" determines whether or not heat dissipation from the E-box is occurring as a function of the loop count K (and/or time). The syntax for the function "heat" shown below should be placed at the top of the code in the equations window.

```
Function heat(A,B)
 If (A<81) Then                         {Cycle 1 pulse on}
  Y:=B
 Endif

 If (A=81) or (A>81) Then               {Cycle 1 pulse off}
  Y:=0
 Endif

 If (A>540) and (A<614) Then            {Cycle 2 pulse on}
  Y:=B
 Endif

 If (A=614) or (A>614) Then             {Cycle 2 pulse off}
  Y:=0
 Endif

 heat:=Y                    {Assignment of pulse heat load value}
End
```

Execution of the code will result in a solution to part i) of 0.2365 m^2. If the radiator area were to be square then a single side would have a length of 48.6 cm. This length is much greater than the side length of the square radiator designed for in the PCM/radiator case (i.e., 17 cm). For the pulse heat load and radiator specifications given in part ii, the duty cycle will be 15.3%. This will result in a total cycle time of 45 minutes. The total time allocated to the pulse heat load was 6.87 minutes. However, in the EES code this was treated as a maximum. The actual times at which the heat load was placed in the "off" state (shown in the "heat" function) were adjusted to ensure that the melt-to-solidification ratio did not exceed a value of unity. The temperature-versus-time plot for the radiator and the E-box/PCM interface are shown in Figure 5.48. Both the interface and radiator temperatures gradually increase with the E-box in the "on" state. The interface temperature gradually increases from 267 K to a peak of 309 K. Heat dissipation is reduced to zero immediately after the peak temperature is reached. The interface temperature then gradually decreases back down to 267 K. The radiator temperature increases from 254 K to a peak of 264 K. Following its peak there is an extrapolated cooling period during which the stored thermal energy is radiated away. The interface temperature is consistently higher than the radiator temperature. Relative to the melt temperature of the PCM, the temperature oscillation for the interface has an amplitude of approximately 40 K. The temperature oscillation on the radiator is significantly less (approximately 10 K). The pronounced amplitude on the interface's temperature oscillations is an artifact of the thermal mass used in the interface's transient energy balance. Given the use of a larger thermal mass (e.g., the entire E-box), the amplitude of the oscillations will reduce. When the 6.87-minute mark of the cycle is reached, the E-box is placed in the "off" state and thermal dissipation to the PCM ceases. Shortly afterwards the temperatures begin to decrease as solidification of the melted PCM takes place. By the end of the first cycle, both the radiator and interface temperatures have nearly reconditioned to their initial values observed prior to the start of the heating cycle. After 45 minutes, the pulse heating cycle starts again and the full cycle is repeated. Using the temperature-versus-time data collected, the temperature of the composite-phase PCM can be determined as a function of time with the aid of the melt-to-solidification ratio, as shown in equation 5.95.

$$T_{PCM,avg} = x_{MSR} \cdot T_{liq,avg} + (1 - x_{MSR}) \cdot T_{sol,avg} \tag{5.95}$$

where $T_{liq,avg}$ is the arithmetic average of T_{int} and T_{melt}. The value for $T_{sol,avg}$ is the arithmetic average of T_{rad} and T_{melt}.

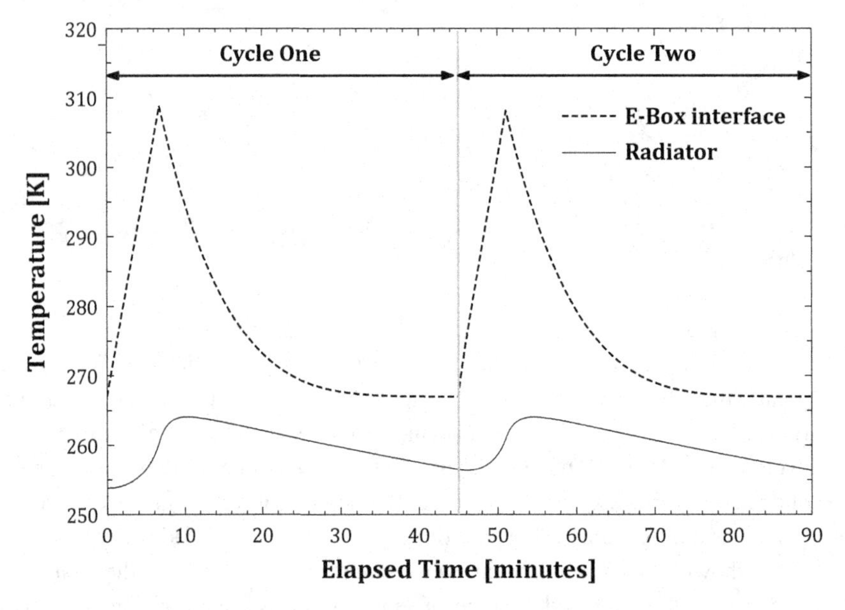

Figure 5.48 Heating/cooling profiles for radiator and E-box/PCM interface

The use of a PCM increases the effective thermal storage capacity of the item being temperature controlled, reducing its temperature excursion during heating and cooling. Nonetheless, the heat transfer to and from the melt line (where phase change takes place) is limited by the thermal conductivity of the liquid and solid phases. For most PCMs, the thermal conductivity of the liquid and solid phases is fairly low (see values in

Table 5.5). While the temperature of the interface (and ultimately the E-box) was controlled using the PCM, a higher effective thermal conductivity for the phases of the PCM can achieve better approximation of the interface and radiator temperatures with the melt temperature. A common technique used to increase the effective thermal conductivity of PCMs is filler materials, for example, metal foams and metal honeycomb structures. For analysis techniques for PCMs with metal fillers, see *Spacecraft Thermal Control Handbook* [98] or *Phase Change Materials Handbook* [66].

References

[1] Johnson, D., Johnson, J., and Hilburn, J., 1989, *Electric Circuit Analysis*, Prentice-Hall, Inc., Upper Saddle River, NJ

[2] Katz, J., 2010, *Introductory Fluid Mechanics*, Cambridge University Press, New York, NY

[3] Donabedian, M., 2002, "Cryocooler Overview," *Spacecraft Thermal Control Handbook* Vol. II: Cryogenics, edited by Donabedian, M., The Aerospace Press and The American Institute of Aeronautics and Astronautics, Inc., Chap. 7, pp. 121–133

[4] Moran, M.J., Shapiro, H.N., Boettner, D.D., and Bailey, M.B., 2014, *Fundamentals of Engineering Thermodynamics*, John Wiley & Sons, Hoboken, NJ

[5] Hendricks, R.C., Peller, I., and Baron, A., 1972, "Joule Thomson Inversion Curves and Related Coefficients for Several Simple Fluids," NASA TN D-6807, Lewis Research Center

[6] Bejan, A., 1988, *Advanced Engineering Thermodynamics*, John Wiley & Sons, Hoboken, NJ

[7] Domanski, P., Didion, D., and Doyle, J., 1992, "Evaluation of Suction Line-Liquid Line Heat Exchange in the Refrigeration Cycle," International Refrigeration and Air Conditioning Conference. Paper no: 149

[8] Bugby, D., and Marland, B., 2003, "Flexible Conductive Links," *Spacecraft Thermal Control Handbook* Vol. II: Cryogenics, edited by Donabedian, M., The Aerospace Press and The American Institute of Aeronautics and Astronautics, Inc., Chap. 14, pp. 327–346

[9] Kimball, M.O., and Shirron, P.J., 2015, "Ultra-flexible thermal bus for use in the Astro-H Adiabatic Demagnetization Refrigerator," 3rd International Planetary Probe Workshop, June 28–July 2, Tucson, AZ

[10] Bergman, T., Lavine, A., Incropera, F.P., and Dewitt, D.P., 2017, *Fundamentals of Heat and Mass Transfer*, John Wiley & Sons, Hoboken, NJ

[11] Carey, V., 1992, *Liquid Vapor Phase Change Phenomena*, Taylor & Francis, New York, NY

[12] Lienhard, J.H., 1985, "On the two Regimes of Nucleate Boiling," *Journal of Heat Transfer*, Vol. 107, pp. 262–264

[13] Dhir, V., Warrier, G., Aktinol, E., Chao, D., Eggers, J., Sheredy, W., and Booth, W., 2012, "Nucleate Pool Boiling Experiments (NPBX) on the International Space Station," *Microgravity Science Technology*, Vol. 24, pp. 307–325

[14] Hewett, G.F., 1988, "*Boiling,*" *Handbook of Heat Transfer*, 3rd edn., edited by Rohsenow, W.M., Hartnett, J.P., and Cho, Y., McGraw-Hill, New York, NY, Chap. 15, pp. 15.1–15.168

[15] Tong, L.S., and Tang, Y.S., 1997, 2nd edn, *Boiling Heat Transfer and Two-Phase Flow*, Taylor & Francis, New York, NY

[16] Monti, R., 2001, "Fluid Science Relevance in Microgravity Research," *Physics of Fluids in Microgravity*, edited by Monti, R., Taylor & Francis, New York, New York, Chap. 1, pp. 1–20
[17] McQuillen, J., 2003, "Two Phase Flow in a Microgravity Environment," NASA Glenn Research Center, www1.grc.nasa.gov/space/iss-research/
[18] Golliher, E., Zivich, C., and Yao, S.C.., 2005, "Exploration of Unsteady Spray Cooling for High Power Electronics at Microgravity Using NASA Glenn's Drop Tower," ASME Summer Heat Transfer Conference, July 17–22, San Francisco, CA, Paper no: HT2005–72123
[19] Marcus, B.D., 1971, "Theory and Design of Variable Conductance Heat Pipes: Control Techniques," 2nd report, Contract #: NAS2–5503, TRW, Inc
[20] Peterson, G.P., 1998, "Heat Pipes," *Handbook of Heat Transfer*, 3rd edn, edited by Rohsenow, W.M., Hartnett, J.P., and Cho, Y., McGraw-Hill, New York, Chap. 12, pp. 12.1–12.20
[21] Skrabek, E.A., and Bienert, W.B., 1972, "Heat Pipe Design Handbook," Contract #: NAS9–11927, Dynatherm Corporation
[22] Chi, S.W., 1976, *Heat Pipe Theory and Practice*, McGraw-Hill, New York, NY
[23] Brennan, P.J., and Kroliczek, E.J., 1979, "Heat Pipe Design Handbook," Contract #: NAS5–23406, B&K Engineering, Inc
[24] Faghri, A., 1995, *Heat Pipe Science and Technology*, Taylor & Francis, New York, NY
[25] Kaviany, M., 1995, *Principles of Heat Transfer in Porous Media*, Springer, New York, NY
[26] Tien, C.L., 1975, "Fluid Mechanics of Heat Pipes," *Annual Review of Fluid Mechanics*, Vol. 7, pp. 167–185
[27] Peterson, G.P., 1994, *An Introduction to Heat Pipes: Modeling, Testing and Applications*, John Wiley & Sons, Inc., New York, NY
[28] Reay, D., Kew, P., and McGlen, R., 2014, *Heat Pipes: Theory, Design and Applications*, 6th edn, Elsevier, Inc., Amsterdam, Netherlands
[29] Cotter, T.P., 1967, "Heat Pipe Startup Dynamics," Proceedings of the SAE Thermionic Conversion Specialist Conference, Palo Alto, CA
[30] Peterson, G.P., and Bage, B.K., 1991, "Entrainment Limitations in Thermosyphons and Heat Pipes," *Journal of Energy Resources Technology*, Vol. 113, No. 3
[31] Ochterbeck, J., 2003, "Heat Pipes," *Heat Transfer Handbook* Vol. I, edited by Bejan, A. & Kraus, A., Wiley, Interscience, Inc., New York, NY, Chap. 16, pp. 1181–1230
[32] Drolen, B., 2001, "Reflux Performance of Spacecraft Heat Pipes during Ground Test," 39th AIAA Aerospace Sciences Meeting & Exhibit, January 8–11, Reno, NV
[33] Prager, R.C., Nikitkin, M., and Cullimore, B., 2002, "Heat Pipes," *Spacecraft Thermal Control Handbook* Vol. I: Fundamental Technologies, edited by Gilmore, D.G., The Aerospace Press and The American Institute of Aeronautics and Astronautics, Inc., Chap. 14, pp. 489–522
[34] Dunn, P.D., and Reay, D.A., 1982, *Heat Pipes*, 3rd edn, Pergamon Press, Oxford, England
[35] Gilmore, D.G., 2002, "Radiators," *Spacecraft Thermal Control Handbook* Vol. I: Fundamental Technologies, edited by Gilmore, D.G., The Aerospace Press and The American Institute of Aeronautics and Astronautics, Inc., Chap. 6, pp. 207–222
[36] Ku, J., 1999, "Operating Characteristics of Loop Heat Pipes," The 29th International Conference on Environmental Systems, Society of Automotive Engineers, July 12–15, Denver, CO

[37] Maydanik, Y.F., 2005, "Loop Heat Pipes," *Applied Thermal Engineering*, Vol. 25, pp. 635–657
[38] Chernysheva, M.A., Vershinin, S.V., and Maydanik, Y.F., 2007, "Operating Temperature and Distribution of a Working Fluid in LHP," *International Journal of Heat and Mass Transfer*, Vol. 50, pp. 2704–2713
[39] Kaya, T., Ku, J., Hoang, T., and Cheung, K., 1999, "Investigation of Low Power Start-up Characteristics of a Loop Heat Pipe," Space Technology and Applications International Forum, pp. 799–804
[40] Nikitkin, M., Bienert, W., and Goncharov, K., 1998, "Non-Condensable Gases and Loop Heat Pipe Operation," International Conference on Environmental Systems, Danvers, MA, SAE Paper no: 981584
[41] Kaya, T., Hoang, T.T., Ku, J., and Cheung, M.K., 1999, "Mathematical Modeling of Loop Heat Pipes," 37th AIAA Aerospace Sciences Meeting and Exhibit, January 11–14, Reno, NV, Paper#: 99-0477
[42] Kaya, T., and Hoang, T.T., 1999, "Mathematical Modeling of Loop Heat Pipes and Experimental Validation," *Journal of Thermophysics and Heat Transfer*, Vol. 13, No. 3, pp. 314–320
[43] Launay, S., Sartre, V., and Bonjour, J., 2008, "Analytical Model for Characterization of Loop Heat Pipes," *Journal of Thermophysics and Heat Transfer*, Vol. 22, No. 4, pp. 623–631
[44] Hoang, T.T., and Kaya, T., 1999, "Mathematical Modeling of Loop Heat Pipes with Two-phase Pressure Drop," 33rd Thermophysics Conference, June 28–July 1, Norfolk, VA, Paper#: 99-3488
[45] Singh, R., Akbarzadeh, A., Dixon, C., and Mochizuki, M., 2007, "Novel Design of a Miniature Loop Heat Pipe Evaporator for Electronic Cooling," *Journal of Heat Transfer*, Vol. 129, pp. 1445–1452
[46] Chernysheva, M.A., and Maydanik, Y.F., 2009, "Heat and Mass Transfer in Evaporator of Loop Heat Pipe," *Journal of Thermophysics and Heat Transfer*, Vol. 23, No. 4, pp. 725–731
[47] Bienert, W., and Wolf, D., 1995, "Temperature Control with Loop Heat Pipes: Analytical Model and Test Results," Proceedings of the 9th International Heat Pipe Conference, May 1–5, Albuquerque, NM
[48] Hamdan, M., Gerner, F.M., and Henderson, H.T., 2003, "Steady State Model of a Loop Heat Pipe (LHP) with Coherent Porous Silicon (CPS) Wick in the Evaporator," 19th Annual IEEE Semiconductor Thermal Measurement and Management Symposium, Institute of Electrical and Electronics Engineers, March 11–13, Piscataway, NJ, pp. 88–96
[49] Launay, S., Sartre, V., and Bonjour, J., 2007, "Parametric Analysis of Loop Heat Pipe Operation: A Literature Review," *International Journal of Thermal Sciences*, Vol. 46, pp. 621–636
[50] Kraus, A., and Bar-Cohen, A., 1983, *Thermal Design Analysis and Control of Electronic Equipment*, Hemisphere Publishing Company, Washington DC
[51] Chen, G., 2010, *Nanoscale Energy Transport and Conversion*, Oxford University Press, Inc., Oxford, England
[52] Rowe, D.M., 2006, "General Principles and Basic Considerations," *Thermoelectrics Handbook Macro to Nano*, edited by Rowe, D.M., CRC Press, Taylor & Francis, Inc., New York, NY, Chap. 1, pp. 1–14
[53] Lee, H., 2010, *Thermal Design: Heat Sinks, Thermoelectrics, Heat Pipes, Compact Heat Exchangers and Solar Cells*, John Wiley & Sons, Inc., New York, NY

[54] Tian, Z., Lee, S., and Chen, G., 2014, "A Comprehensive Review of Heat Transfer in Thermoelectric Materials and Devices," *Annual Review Of Heat Transfer*, Vol. 17, pp. 1–64
[55] Kraftmakher, Y., 2005, "Simple Experiments with a Thermoelectric Module," *European Journal of Physics*, Vol. 26, pp. 959–967
[56] Lee, H., 2013, "The Thomson Effect and the Ideal Equation on Thermoelectric Coolers," *Energy*, Vol. 56, pp. 61–69
[57] Fraisse, G., Ramousse, J., Sgorlon, D., and Goupil, C., 2013, "Comparison of Different Modeling Approaches for Thermoelectric Elements," *Energy Conversion and Management*, Vol. 65, pp. 351–356
[58] Simons, R.E., Ellsworth, M.J., and Chu, R.C., 2005, "An Assessment of Module Cooling Enhancement with Thermoelectric Coolers," *Transactions of the ASME*, Vol. 127, pp. 76–84
[59] Xuan, X.C., 2002, "Optimum Design of a Thermoelectric Device," *Semiconductor Science and Technology*, Vol. 17, pp. 114–119
[60] Zhang, H.Y., 2010, "A General Approach in Evaluating and Optimizing Thermoelectric Coolers," *International Journal of Refrigeration*, Vol. 33, pp. 1187–1196
[61] Laird Technologies, *Thermoelectric Handbook,* London, United Kingdom
[62] Fixler, S.Z., 1965, "Analytical and Experimental Investigation of Satellite Passive Thermal Control Using Phase-Change Materials (PCM)," Unmanned Spacecraft Meeting, March 1–4, Los Angeles, CA
[63] Fixler, S.Z., 1966, "Satellite Thermal Control Using Phase-Change Materials," *Journal of Spacecraft and Rockets*, Vol. 3, No. 9, pp. 1362–1368
[64] Grodzka, P., and Fan, C., 1968, "Thermal Control by Freezing and Melting," NAS8–21123
[65] Grodzka, P., 1970, "Thermal Design and Development of a Planetary Probe: Pioneer Venus Large Probe," AIAA 8th Aerospace Sciences Meeting, January 19–21, New York, NY, Paper #: 70-12
[66] Hale, D., Hoover, M., and O'Neill, M., 1971, *Phase Change Materials Handbook*, Contract #: NASA CR-61363, Lockheed Missiles and Space Company, Huntsville, AL
[67] Schelden, B.G., and Golden, J.O., 1972, "Development of a Phase Change Thermal Control Device," AIAA 7th Thermophysics Conference, April 10–12, San Antonio, TX, Paper #: 72-287
[68] Abhat, A., and Groll, M., 1974, "Investigation of Phase Change Material (PCM) Devices for Thermal Control Purposes in Satellites," AIAA/ASME 1974 Thermophysics and Heat Transfer Conference, July 15–17, Boston, MA, Paper #: 74-728
[69] Keville, J.F., 1976, "Development of Phase Change Systems and Flight Experience on an Operational Satellite," AIAA 11th Thermophysics Conference, July 14–16, San Diego, CA, Paper #: 76-436
[70] Abhat, A., 1976, "Experimental Investigation and Analysis of a Honeycomb-Packed Phase Change Material Device," AIAA 11th Thermophysics Conference, July 14–16, San Diego, CA, Paper #: 76-437
[71] Humphries, W., 1974, *Performance of Finned Thermal Capacitors*, NASA Technical Note D-7690, NASA, Washington DC
[72] Humphries, W., and Griggs, E., 1977, *A Design Handbook for Phase Change Thermal Control and Energy Storage Devices*, NASA Technical Paper 1074, NASA, Washington DC
[73] Grodzka, P.G., Picklesimer, E., and Conner, L.E., 1977, "Cryogenic Temperature Control by Means of Energy Storage Materials," AIAA 12th Thermophysics Conference, June 27–29, Albuquerque, NM, Paper #: 77-763

[74] Hennis, L., and Varon, M., 1978, "Thermal Design and Development of a Planetary Probe: Pioneer Venus Large Probe," 2nd AIAA/ASME Thermophysics and Heat Transfer Conference, May 24–26, Palo Alto, CA

[75] Peck, S., and Jacobsen, D., 1979, "Performance Analysis of Phase Change Thermal Energy Storage/Heat Pipe Systems," AIAA 14th Thermophysics Conference, June 4–6, Orlando, FL, Paper #: 79-1096

[76] Richter, R., and Mahefkey, E.T., 1980, "Development of Thermal Energy Storage Units for Spacecraft Cryogenic Coolers," AIAA 18th Aerospace Sciences Meeting, 14–16, Pasadena, CA, Paper #: 80-0145

[77] Busby, M.S., and Mertesdorf, S.J., 1987, "The Benefit of Phase Change Thermal Storage for Spacecraft Thermal Management," AIAA 22nd Thermophysics Conference, June 8–10, Honolulu, HI, Paper #: 87-1482

[78] Stafford, J., and Grote, M.G., 1971, "Thermal Capacitor, Liquid Coolant-to-Phase Change Material Heat Exchanger, for the NASA Skylab I Airlock Module," AIAA 6th Thermophysics Conference, April 26–28, Tullahoma, TN, Paper #: 71-429

[79] Stafford, J.L., and Grote, M.G., 1972, "NASA Skylab I Airlock Module Thermal Capacitor," *Journal of Spacecraft and Rockets*, Vol. 9, pp. 452–453

[80] Tillotson, B., 1991, "Regolith Thermal Energy Storage for Lunar Nighttime Power," AIAA/NASA/OAI Conference on Advanced SEI Technologies, September 4–6, Cleveland, OH, Paper #: 91-3420

[81] Lorenz, R., Bienstock, B., Couzin, P., and Cluzet, G., 2005, "Thermal Design and Performance of Probes in Thick Atmospheres: Experience of Pioneer Venus, Venera, Galileo and Huygens," 3rd International Planetary Probe Workshop, June 27–July 1, Anavyssos, Attiki, Greece

[82] Del Castillo, L.Y., Kolawa, E., Mojarradi, M., Johnson, J., and Hatake, T., 2005, "High Temperature Electronics for Atmospheric Probes and Venus Landed Missions," 3rd International Planetary Probe Workshop, June 27–July 1, Anavyssos, Attiki, Greece

[83] Ocampo, A., Saunders, S., Cutts, J., and Balint, T., 2006, "Future Exploration of Venus: Some NASA Perspectives," Chapman Conference: Exploring Venus as a Terrestrial Planet, February 13–16, Key Largo, FL

[84] Roschke, E., 1986, "Solar Dynamic Systems for Spacecraft Power Applications," 24th Aerospace Sciences Meeting, January 6–9, Reno, NV, Paper #: 86-0382

[85] Archer, J., and Diamant, E., 1986, "Solar Dynamic Power for the Space Station," AIAA/ASME 4th Joint Thermophysics and Heat Transfer Conference, June 2–4, Boston, MA, Paper #: 86-1299

[86] Wu, Y., Roschke, E., and Birur, G., 1988, "Solar Dynamic Heat Receiver Thermal Characteristics in Low Earth Orbit," 26th Aerospace Sciences Meeting, January 11–14, Reno, NV, Paper #: 88-0472

[87] Secunde, R.R., Labus, L.L., and Lovely, R.G., 1989, *Solar Dynamic Power Module Design, NASA Technical Memorandum 102055*, NASA Lewis Research Center, Cleveland, OH

[88] Hall, C.A., Glakpe, E.K., Cannon, J.N., and Kerslake, T.W., 1998, "Modeling Cyclic Phase Change and Energy Storage in Solar Heat Receivers," *Journal of Thermophysics and Heat Transfer*, Vol. 12, pp. 406–413

[89] Hall, C.A., Glakpe, E.K., Cannon, J.N., and Kerslake, T.W., 2000, "Thermal State-of-Charge of Space Solar Dynamic Heat Receivers," *Journal of Propulsion and Power*, Vol. 16, No. 4, pp. 666–675

[90] Pauken, M., Kolawa, E., Manvi, R., Sokolowski, W., and Lewis, J., 2006, "Pressure Vessel Technology Developments," 4th International Planetary Probe Workshop, June 27–30, Pasadena, CA

[91] Pauken, M., Emis, N., and Watkins, B., 2007, "Thermal Energy Storage Technology Developments," Space Technology and Applications International Forum, February 11–15, Albuquerque, NM

[92] Pauken, M., Emis, N., Van Luvender, M., Polk, J., and Del Castillo, L., 2010, "Thermal Control Technology Developments for a Venus Lander," Space, Propulsion and Energy Sciences International Forum, February 23–25, Laurel, MD

[93] Pauken, M., Li, L., Almasco, D., Del Castillo, L., Van Luvender, M., Beatty, J., Knopp, M., and Polk, J., 2011, "Insulation Materials Development for Potential Venus Surface Missions," 49th AIAA Aerospace Sciences Meeting, January 4–7, Orlando, FL, Paper #: 2011-256

[94] Choi, M., 2012, "Using Pre-melted Phase Change Material to Keep Payload Warm without Power for Hours in Space," 10th International Energy Conversion Engineering Conference, July 30–August 1, Atlanta, GA, Paper #: 2012-3894

[95] Choi, M., 2013, "Using Paraffin with −10°C to 10°C Melting Point for Payload Thermal Energy Storage in SpaceX Dragon Trunk", 49th AIAA/ASME/SAE/ASEE Joint Propulsion Conference, January 14–17, San Jose, CA

[96] Choi, M., 2014, "Phase Change Material for Temperature Control of Imager or Sounder on GOES Type Satellites in GEO," 44th International Conference on Environmental System, July 13–17, Tucson, AZ

[97] Serway, R.A., 1986, *Physics for Scientists & Engineers*, Saunders College Publishing, Philadelphia, PA

[98] Bledjian, L., Hale, D.V., Hoover, M.J., and O'Neill, M.J., 2002, "Phase-Change Materials," *Spacecraft Thermal Control Handbook* Vol. I: Fundamental Technologies, edited by Gilmore, D.G., The Aerospace Press and The American Institute of Aeronautics and Astronautics, Inc., Chap. 11, pp. 373–404

[99] Hust, J., and Schramm, R., 1976, "Density and Crystallinity Measurements of Liquid and Solid *n*-Undecane, *n*-Tridecane, and *o*-Xylene from 200 to 350K," *Journal of Chemical & Engineering Data*, Vol. 21., No. 1, pp. 7–11

[100] Velez, C., Zarate, J., and Khayet, M., 2015, "Thermal Properties of n-Pentadecane, n-Heptadecane and n-Nonadecane in the Solid/Liquid Phase Change Region," *International Journal of Thermal Sciences*, Vol. 94, pp. 139–146

[101] Shamberger, P.J., and Reid, T., 2012, "Thermophysical Properties of Lithium Nitrate Trihydrate from (253 to 353) K," *Journal of Chemical & Engineering Data*, Vol. 57, pp. 1404–1411

[102] Xiang, H.W., Laesecke, A., and Huber, M.L., 2006, "A New Reference Correlation for the Viscosity of Methanol," *Journal of Physical and Chemical Reference Data*, Vol. 35, No. 4

Problems

5.1 An R410a (R32/125 with a 50/50 composition) heat pump cycle with a mass flow rate of 0.045 kg/s has an evaporator temperature of 0°C. The refrigerant exits the condenser as saturated liquid and is under saturated vapor conditions at the exit of the evaporator. The system is operating at a pressure ratio (i.e., P_{high}/P_{low}) of 1.8. Assuming isentropic compression across the compressor, determine

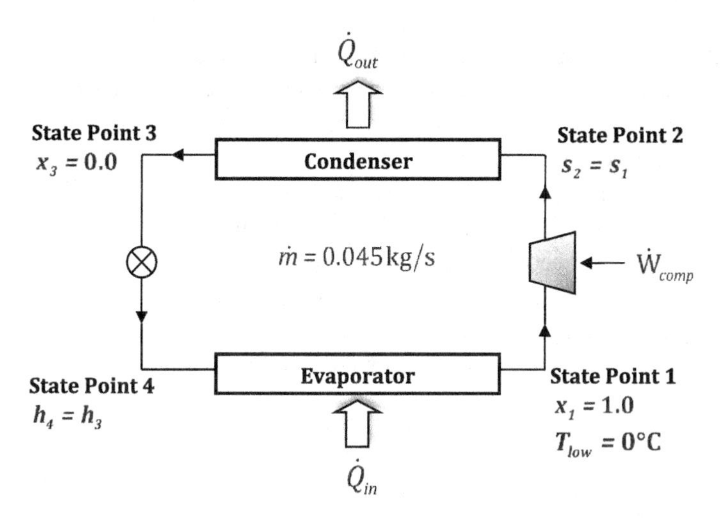

Figure P5.1 R410a vapor compression cycle

i) The compressor work input if the compressor efficiency is 0.9
ii) The heat rejection from the condenser
iii) The actual coefficient of performance for heat rejection for the cycle.

5.2 An R32 heat pump cycle with a mass flow rate of 0.05 kg/s is operating at an evaporator temperature of 0°C. The condenser exit conditions are saturated liquid, whereas the fluid is saturated vapor at the exit of the evaporator. The compressor work input is 0.8 kW.

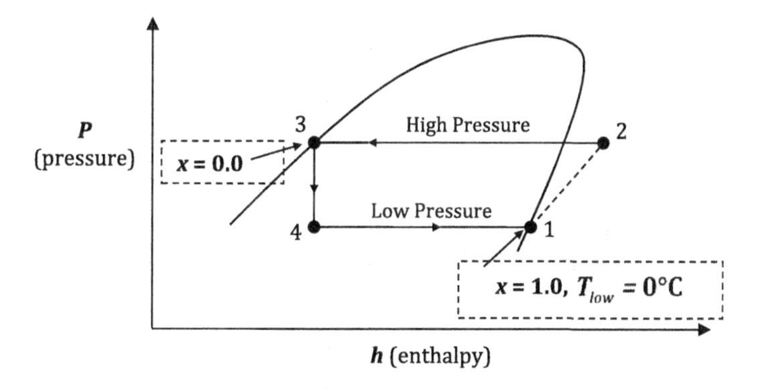

Figure P5.2 Pressure–enthalpy diagram for an R32 heat pump cycle

i) Assuming a pressure ratio (P_{high}/P_{low}) of 3.0 for the cycle, what is the heat rejection load from the condenser and the actual coefficient of performance for refrigeration in the cycle?
ii) Assuming a variation in the pressure ratio between 0.1 and 7.0, what is the actual COP for refrigeration? Please provide a plot.
iii) For a variation of the pressure ratio from 0.1 to 7.0, what is the condenser temperature? Please provide a plot.

5.3 An R744 (CO_2) heat pump cycle with a mass flow rate of 0.002 kg/s is providing temperature control to a flight component on a space platform that is thermally coupled to the evaporator. The evaporator is operating at a constant temperature of −5.0° C. The refrigerant exits the evaporator as fully saturated vapor. On exit from the condenser, the refrigerant is fully saturated liquid. The condenser is thermally coupled to the space platform's radiator which has a surface area of 1.0 m^2 that is viewing deep space. The interface resistance between the condenser and the inner surface of the radiator is 1.0 K/W. The active surface area of the radiator is coated with carbon black paint NS-7. Assuming a pressure ratio of 2.25 and 0.125 kW power input to the compressor, what is the actual refrigeration COP and the temperature of the radiator?

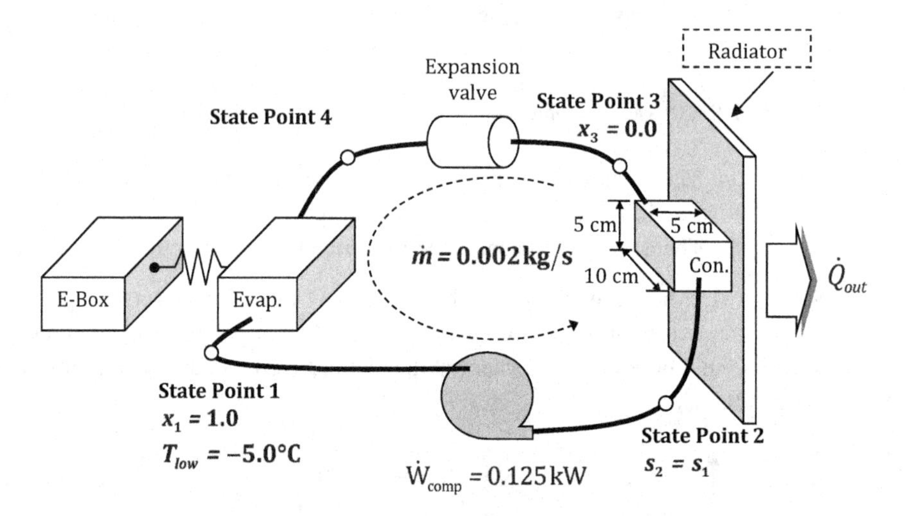

Figure P5.3 R744 heat pump cycle with an E-box/radiator thermal coupling

For the radiator you can assume a single plate of Al-6061, 5 mm thick. Using the geometry in the figure, the condenser can be modeled as a solid made of Al-6061 as well.

5.4 An R744 (CO_2) refrigeration cycle with a LLSL-HX has a mass flow rate of 0.012 kg/s and an evaporator temperature of −10°C. The refrigerant exits the condenser as saturated liquid and is under saturated vapor conditions at the exit of the evaporator. Given a pressure ratio of 1.65, a compressor work input of 0.190 kW and a LLSL-HX efficiency of 98%, determine the heat rejection from the condenser and the actual COP for refrigeration.

5.5 An E-box dissipating 32 W is thermally coupled to a temperature sink (−10°C) via a thermal strap (example in Figure P5.5). The thermal strap endpoints have a mating surface area of 1.75 × 1.75 $inches^2$ contacting the E-box and temperature sink structure. The interface conductance at these locations is 10,000 $W/m^2 \cdot K$. The thermal strap foil specifications are shown in Figure P5.5.

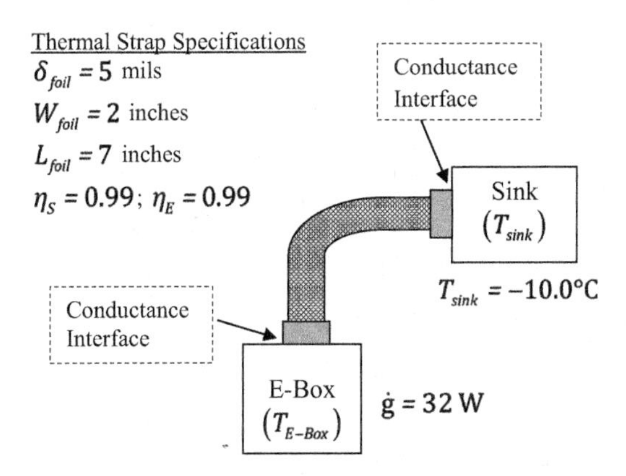

Figure P5.5 Thermal strap thermal coupling between an E-box and a temperature sink

i) Determine the effective conductance of a thermal strap that has 48 OFHC (k_{OFHC} = 400 W/m · K) foils.

ii) Determine the operational temperature of the E-box using the OFHC thermal strap conductance calculated in part (i).

iii) If the thermal strap were made of Al-1100 ($k_{1100,\ Al}$ = 220 W/m · K) foils, what would be the operational temperature of the E-box (assuming the same endpoint efficiency and interface conductance)?

iv) How many Al-1100 foils would be required to match the thermal conductance determined in part (i) when using OFHC foils?

5.6 A thermal strap is made using 3 mil foils of OFHC/RRR = 150. Each foil has a width of 1 inch and a length of 200 mm. Assuming that the shape efficiency is 0.9 and the endpoint efficiency is 0.99,

i) How many foils are required to transfer 10 W across a temperature gradient of 30° C?

(Note: The foil thermal conductivity can be based on a temperature of 296 K.)

ii) Imagine that the thermal strap fabricated in part (i) based on your inputs is now being assessed for use on a cryogenic application. Assuming you want to transfer 250 mW, what is the required temperature difference?

(Note: Take the thermal conductivity for the thermal strap modeled as 20 K.)

5.7 A 12-inch-long thermal braid (Figure P5.7) made of OFHC wires attaches two endpoints operating at temperatures of T_1 = 30°C and T_2 = 5°C. For the braid specifications listed below:

k_{OFHC} = ***400 W/m · K*** $\quad \eta_S = 0.99$
$N_{wire} = 100$ $\quad \eta_E = 0.99$
d_{wire} = 2 mm

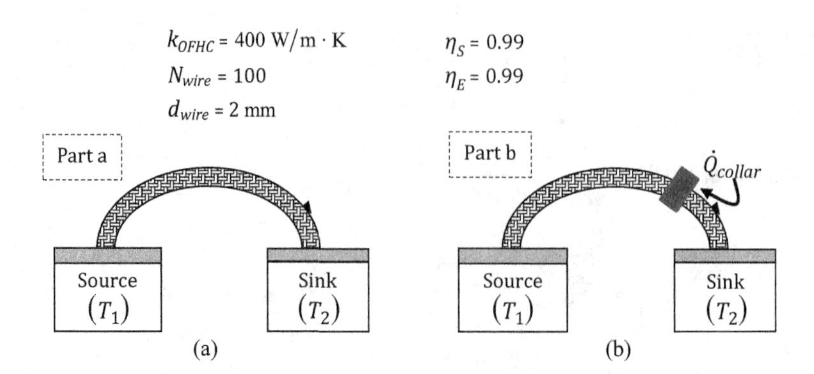

Figure P5.7 Thermal braid: a) basic; b) with heat input collar

i) What is the heat transfer?

ii) If a heated collar is to be placed around the braid at a distance of 3.5 inches from the sink surface, how much heat must be supplied to nullify heat flow coming from the source surface while maintaining a temperature of T_1 = 30°C?

5.8 Two plates are thermally coupled to one another via six parallel braided thermal straps individually oriented in a "C" configuration. Both the upper and lower plates (Al-6061) are square with a thickness of 1.9 cm. The braids are made from OFHC and consist of 91 wires, each having a length of 18 cm. The shape and endpoint efficiency of each braid is 0.99 and 0.95 respectively. The endpoint mating surfaces are 2.7 cm × 2.7 cm. Assume the top plate has a temperature of 308 K and is transferring 165 W to the bottom plate. If the contact pressure between the endpoints and the plates is 1 MPa in vacuum (see Table 2.2),

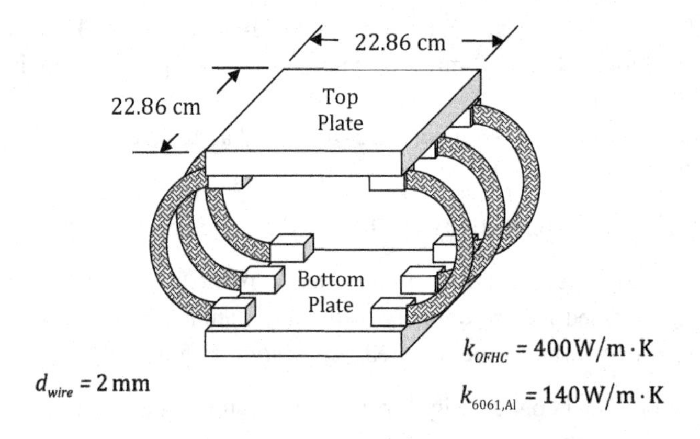

Figure P5.8 Multi-conductor thermal coupling between two plates

i) Determine the temperature of the bottom plate

ii) Determine how many braids are required to raise the bottom plate temperature to 270 K.

5.9 Two E-boxes are thermally coupled to a temperature sink operating at 2°C via a network of rigid and flexible conductors (Figure P5.9). Each E-box is cubic in shape with a side length of 20 cm and a wall thickness of 5 mm. E-box 2 ($\dot{g} = 10$ W) is immediately attached to a cross-beam the same width as the E-box (into the page), a thickness of 5 mm and a length of 40 cm. E-box 1 ($\dot{g} = 12$ W) is coupled to the top of the cross-beam. Immediately adjacent to the coupling location is a pair of thermal straps (identical to one another) that couple the cross-beam to the temperature sink. The pressure at the interface between the thermal strap endpoints and their opposing surface may be assumed as 5 MPa for OFHC in vacuum (see Table 2.2 for values). The E-boxes and cross-beam are made of Al-6061 ($k_{6061-\mathrm{Al}} = 155.3$ W/m · K), whereas the thermal straps are made of OFHC ($k_{OFHC} = 390$ W/m · K). Assuming 1-D heat transfer between the thermal straps and the cross-beam, as well as the thermal straps and the temperature sink, with the thermal strap endpoints having a contact surface area of 3.3 cm × 3.3 cm:

TS 1 Specifications	**TS 2 & 3 Specifications**
$\delta_{foil} = 5$ mils	$\delta_{foil} = 5$ mils
$W_{foil} = 2.54$ cm	$W_{foil} = 2.54$ cm
$L_{foil} = 12.75$ cm	$L_{foil} = 11.0$ cm
$\eta_S = 0.99$; $\eta_E = 0.99$	$\eta_S = 0.9$; $\eta_E = 0.99$
$N_{foils} = 50$	$N_{foils} = 50$

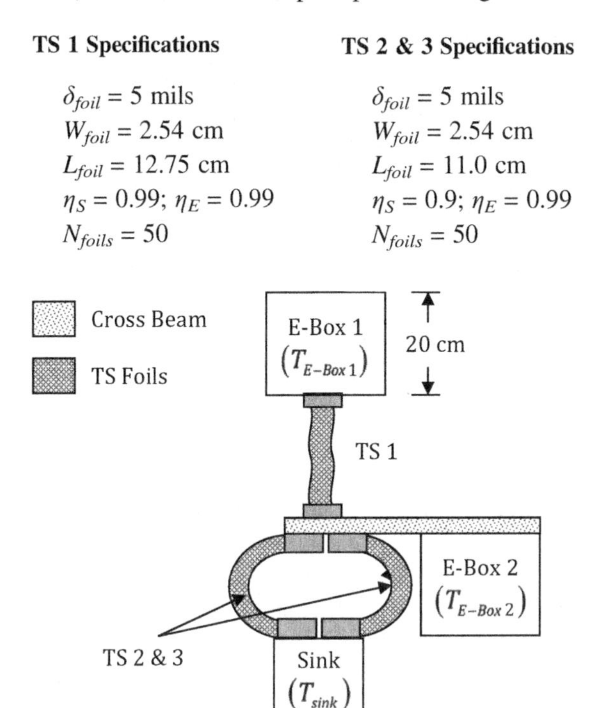

Figure P5.9 Compound TS configuration

i) What is the temperature of the interior wall of E-box 1 and E-box 2 using the hollow square shape factor from Table 2.5?

ii) How many foils would be required for use in thermal strap 1 to operate the interior wall of E-box 1 at 40°C?

5.10 Three electronics boxes, each with dissipations, are interconnected via two OFHC thermal straps (Figure P5.10). The E-box network is thermally coupled to a temperature sink operating at −11°C via 8 OFHC thermal braids (in parallel). Thermal straps 1 and 2 are made of 50 and 60 foils respectively with an interface conductance at the contact point areas (5.2 cm × 5.2 cm each) of 8,000 W/m^2 · K. The interface conductance at the contact point areas (6.5 cm × 2.0 cm each) for the parallel braids is 10,000 W/m^2 · K. For an OFHC thermal conductivity of 390 W/m^2 · K, what is the temperature of the E-boxes?

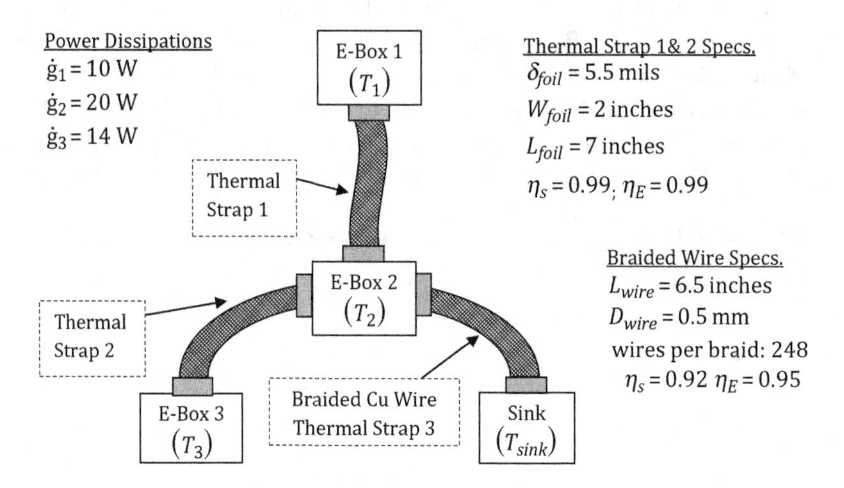

Figure P5.10 Wishbone configuration thermal strap coupling

5.11 A design team you are working with has an electronics box that requires additional heat removal. The team has decided to use a heat pipe to increase the conductance of the thermal energy away from the electronics box to the radiator. The manufacturer of the heat pipe has provided the option of using either ammonia, ethane or methanol as the working fluid. You have been asked to determine which working fluid will provide the best heat transfer for a temperature range of −5°C to 50°C (where appropriate). Using tabulated data, determine which working fluid you would expect to perform better over this temperature range. Please provide a plot of the heat pipe Zero-g FOM as a function of saturation temperature as the basis of your determination. Include the performance curve for pure water in the plot.

Data for each of the fluids may be found in EES or at webbook.nist.gov/chemistry/fluid/. For additional information on the thermophysical properties of methanol, see reference [102]

(Note: The NIST webbook provides a download capability for tabulated data in comma-delimited format for ease of manipulation and calculations.)

5.12 Please determine the Bo number for helium, oxygen, ammonia and water at a saturation pressure of 1 atm while in a 1-g reference frame. You can assume the characteristic length for the phase change surface is 3.0 cm. How does the Bo number vary between each of these fluids?

5.13 Assume you have a heat pipe (Figure P5.13) charged with anhydrous ammonia (NH_3) that is physically coupled (i.e., bolted) to a sink operating at a temperature of −20°C. Both the evaporator and condenser segments have a length of 33 cm and an interface resistance of 0.15 K/W with the source and sink. The heat pipe has an encasement that is a circular cross-section (1.0 inch O.D. tube with a wall thickness of 0.049 inches) made of Al-6061 and a sintered metal wick (nickel) with a wall thickness of 2.5 mm and a porosity of 0.33. If the heat pipe is carrying 120 W (evaporator to condenser) what is the temperature at the source?

(Note: Use $R_{vapor} = 9.0 \times 10^{-4}$ K/W and assume the thermophysical properties for NH_3 are at room temperature.)

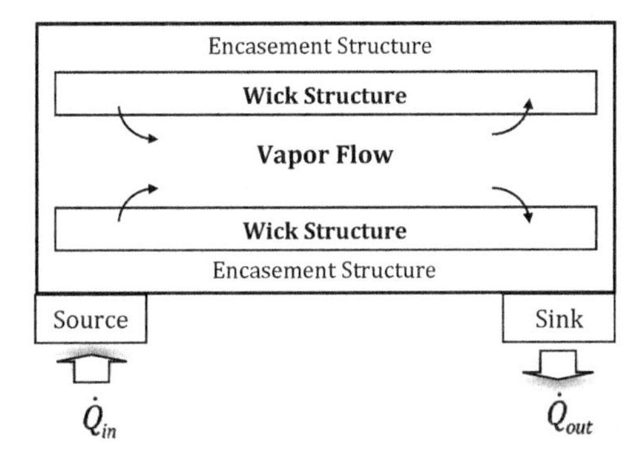

Figure P5.13 Heat pipe schematic

5.14 An ethane heat pipe with a total length of 0.6 m is operating at a saturation temperature of 205 K. The outer encasement has a diameter of 2.54 cm with a wall thickness of 1.5 mm. The wick structure is made of 304 stainless steel with an effective pore diameter of 300 μm and has a radial thickness of 5.1 mm.

i) Determine the entrainment limit.

ii) Assuming parallel effective thermal conductivity in the wick, a porosity of 0.58, a bubble radius of 3.0×10^{-7} m at the nucleation sites, an evaporator length of 0.22 m and a condenser length of 0.22 m, what is the boiling limit?

iii) Assuming the heat pipe is in the horizontal orientation with respect to gravity, what is the capillary limit? (Hint: the diameter for the sintered spheres may be approximated as $d_s = r_{eff}/0.41$ [21].)

iv) If the evaporator is elevated above the condenser by an angle of 2° relative to the horizontal, what is the new capillary limit?

5.15 Figure P5.15 shows a plot of temperature vs. time for an LHP. A series of actions was taken regarding the operation of the device. You have been asked to determine what took place with the LHP in the time frame specified on the plots.

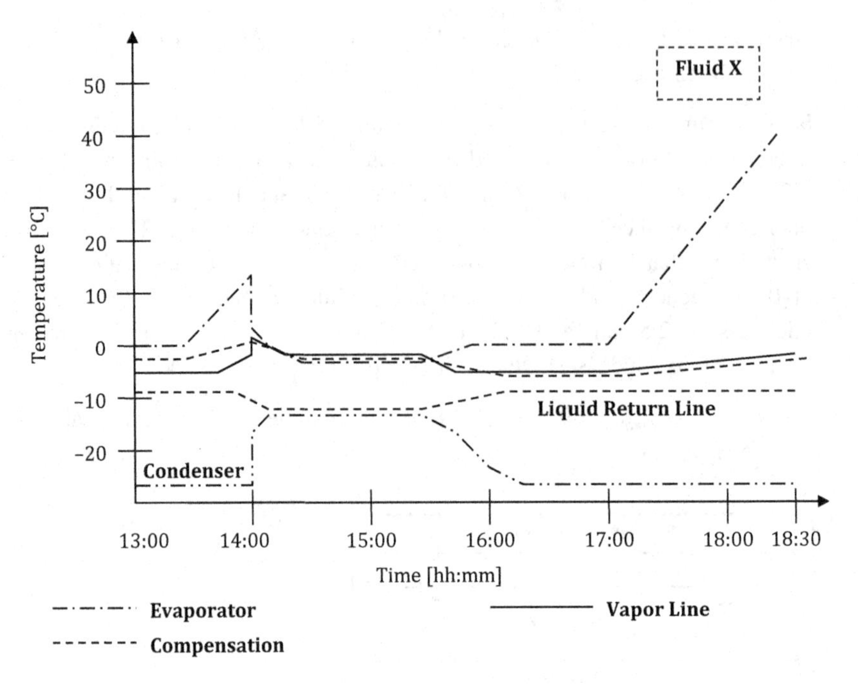

Figure P5.15 LHP temperature vs. time plot

Please state and/or identify:

i) The time/s for initiation of heating.

ii) LHP start-up if it has occurred. Please state time of start-up.

iii) Failed LHP start-up if it has occurred. Please list times associated with failed start-up.

iv) LHP shutdown if it has occurred. Please list time of shutdown.

v) Describe how you would perform a shutdown of the LHP at 15:00 hours.

5.16 A hardware configuration for an electronics box onboard a satellite instrument is shown in Figure P5.16a. In this configuration the electronics box is dissipating 125 W during operation. Only 4% of the thermal energy dissipated by the E-box travels through insulation separating it from a mounting plate. The rest flows into an LHP evaporator which is thermally coupled to a honeycomb radiator. A temperature-versus-time plot for the component-level operation of the LHP (with equivalent thermal mass attached) is provided.

i) Based on the heat transfer performance plot provided, what is the thermal resistance of the LHP during operation (evaporator to condenser)?

ii) Assuming the E-box is an aluminum cube (Al-6061) with a side length of 30 cm and the condenser is in thermal equilibrium with the radiator, what is the temperature of the E-box during operation of the LHP?

(**Note**: Resistance through the evaporator housing to the interface with the E-box can be neglected in the calculations.)

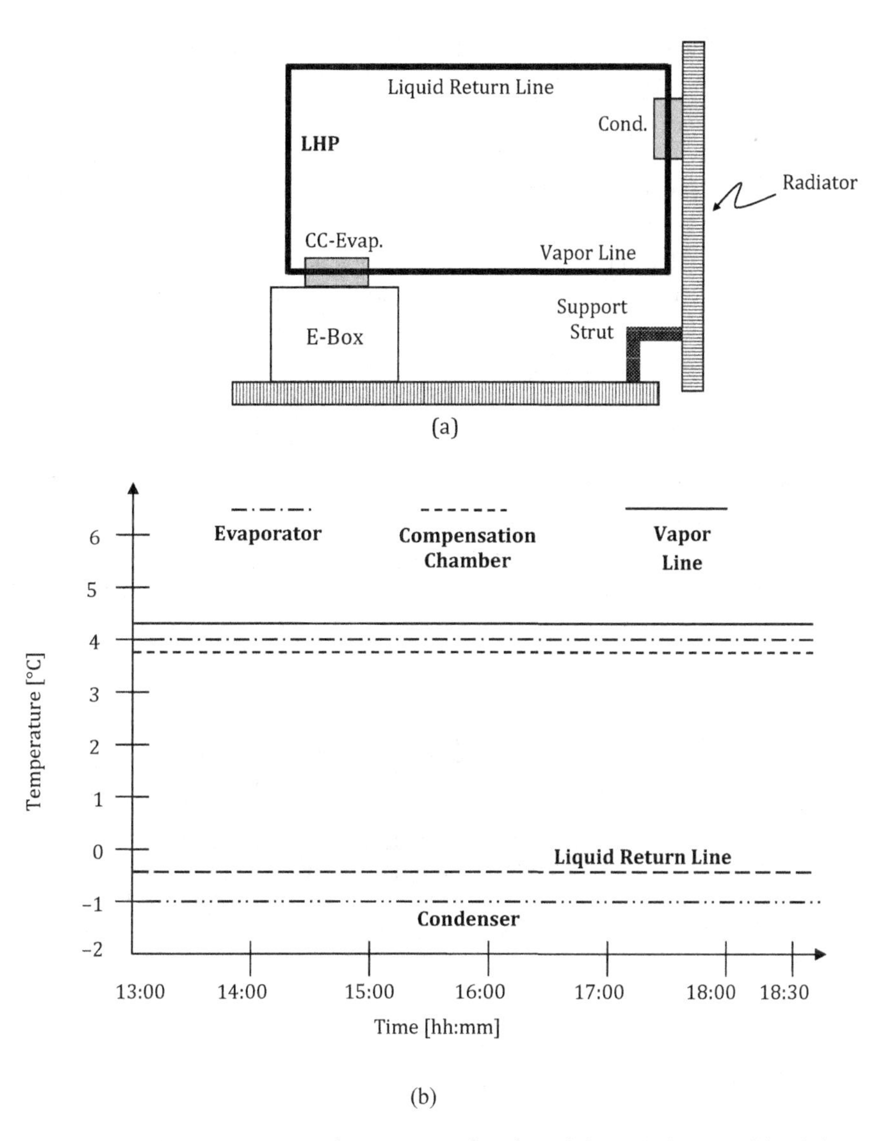

Figure P5.16 LHP temperature performance as a function of time: a) diagram of local thermal architecture; b) LHP temperature vs. time plot at 120 W heat input

5.17 An LHP charged with propylene has an evaporator operating temperature of 295 K and is rejecting a total of 203 W from the condenser and subcooler combined. Heat input to the compensation chamber is measured as 3.0 W, whereas the liquid return line is experiencing a uniform heat flux load of 3.342×10^{-2} W/cm^2. The liquid return line is made of 3/8-inch tubing and has a length of 20 cm. Heat rejection from the vapor line is 0.75 W. Both the primary and secondary wicks are made of nickel and have a length of 9.5 inches. Both wicks have a radial thickness of 10 mm. The outer radius of the primary wick is 0.0254 m. The permeability of the primary and secondary wicks is 2.515×10^{-12} m^2 and 1.128×10^{-8} m^2 respectively. The primary and secondary wick pore diameters are 50 μm and 800 μm respectively. Assuming the temperature of the liquid exiting the subcooler is 270 K, what is the CC temperature and the heat load being applied at the evaporator?

Hints:

- The diameter for the sintered spheres in each wick may be approximated as $d_s = r_{eff}/0.41$ [21].
- Variable limits must be set for the temperatures and the wick porosity values (e.g., 0.1 ↔ 0.6) in order for the EES solver to converge.

5.18 A TEC module consisting of 36 couples (Bi_2Te_3 material elements) is rejecting heat to its hot side at a temperature of 50°C. During operation at a current load of 6.8 amps, the cold side is cooled to a temperature of 25°C. The electrical conductivity for the P- and N-type materials is $\sigma_{e,p} = 700 \times 10^2\ \Omega^{-1} \cdot m^{-1}$ and $\sigma_{e,n} = 900 \times 10^2\ \Omega^{-1} \cdot m^{-1}$ respectively. The Seebeck coefficients for the elements are $\Gamma_p = 200 \times 10^{-6}$ V/K and $\Gamma_n = -215 \times 10^{-6}$ V/K. If the thermal conductivity for the two element types are $k_p = 0.75$ W/m · K and $k_n = 1.9$ W/m · K, what is the cooling load and the COP of the TEC module for the thermoelement dimensions shown in Figure P5.18?

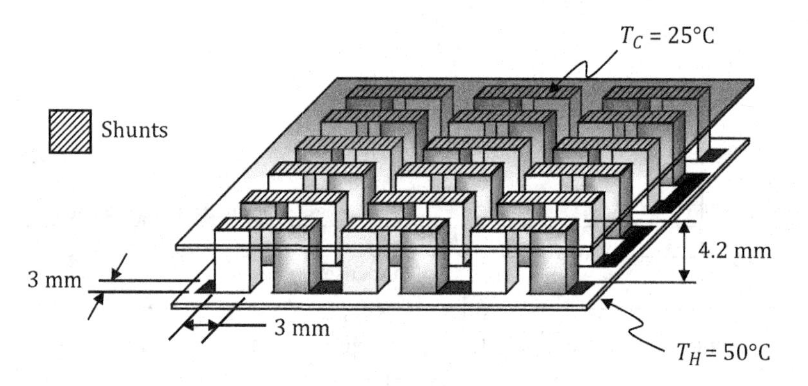

Figure P5.18 TEC with thermoelement dimensions

5.19 A Bi_2Te_3 TEC module carrying 7.578 amps is operating with a cold-side temperature of 20°C. Each element in the TEC has base dimensions of 3.5 mm × 3.5 mm and a height of 5.0 mm. If the elements have electrical conductivities of $\sigma_{e,p} = \sigma_{e,n} = 650 \times 10^2\ \Omega^{-1} \cdot m^{-1}$, Seebeck coefficients of $\Gamma_p = -\Gamma_n = 212 \times 10^{-6}$ V/K and thermal conductivities of $k_p = k_n = 1.0$ W/m · K, determine:

i) How many couples (i.e., N_{couple}) are required to provide 21 W of cooling to the cold side if the rejection temperature (on the hot side) is 62°C?

ii) Using equation 5.60 and Max/Min theory, determine a relation for the maximum current and calculate an i_{max} value for a single couple based on the thermophysical properties listed in the problem statement.

iii) Using equation 5.60, and the i_{max} relation determined in part (ii) as the current, derive a relation for the maximum temperature difference in the module and calculate this value. (Hint: ΔT_{max} can be calculated using the relation for $\dot{Q}_C$ when it is set to a value of 0 and the current equals i_{max}.)

5.20 A Bi_2Te_3 TEC module comprises 49 couples containing thermoelectric elements with electrical conductivities of $\sigma_{e,p} = \sigma_{e,n} = 990 \times 10^2\ \Omega^{-1} \cdot m^{-1}$, Seebeck coefficients of $\Gamma_p = -\Gamma_n = 200 \times 10^{-6}$ V/K and thermal conductivities of $k_p = k_n = 1.5$ W/m · K. For this TEC module:

i) Provide a plot for the cooling load as a function of current applied for thermoelement A/l ratios of 0.5 mm, 1.0 mm, 1.5 mm and 2.0 mm. The variation in current should be between 1.0 amp and 6.0 amps. Each A/l case is with the TEC operating at a cold-side temperature of 300 K and a $\Delta T = 30$ K. How is the 0.5 mm case different (w.r.t. current) from each of the other cases?

ii) Provide a plot for the cooling load as a function of ΔT between the TEC plates for thermoelement A/l ratios of 0.5 mm, 1.0 mm, 1.5 mm and 2.0 mm. The variation in ΔT should be between 5 K and 55 K. Each A/l case is with the TEC operating at a cold-side temperature of 300 K and a current of 5 amps. Describe the difference in the cooling load performance for each of the cases.

iii) Provide a plot for the COP as a function of cold-side temperature for thermoelement A/l ratios of 0.5 mm, 1.0 mm, 1.5 mm and 2.0 mm. The variation in the cold-side temperature should be between 283 K and 313 K. Each A/l case is with the TEC operating at a ΔT of 30 K and a current of 5 amps. How does the COP compare for each of the cases?

5.21 A Bi_2Te_3 TEC module is mounted on top of an E-box (see the configuration in Figure P5.21). Heat rejection on the hot side of the TEC is performed by way of an OFHC thermal strap that is coupled to a remote temperature sink operating at 3°C. The thermal resistance between the hot side of the TEC and the thermal strap is 0.25 K/W, whereas the thermal resistance between the thermal strap and the temperature sink is 0.1 K/W. The dimensions of the TEC's elements are 3.5 mm × 3.5 mm at the base with a height of 4.6 mm. The TEC's electrical conductivities are $\sigma_{e,p} = \sigma_{e,n} = 620 \times 10^2\ \Omega^{-1} \cdot m^{-1}$. The elements' Seebeck coefficients are $\Gamma_p = -\Gamma_n = 210 \times 10^{-6}$ V/K and the thermal conductivities are $k_p = k_n = 0.8$ W/m · K.

i) For the thermal strap specifications listed below, what is the input current required to provide 18 W of cooling at a temperature of 22°C on the TEC cold side?

TS Foil Thermophysical Properties, Dimensions and Specifications

Thermal conductivity: 398 W/m · K	Foil width: 1.8 inches
Number of foils: 47	Foil length: 6.5 inches
Shape efficiency: 0.99	Foil thickness: 8.5 mils
End efficiency: 0.99	

Note: 1 inch = 2.54 cm and 1 mil = 0.0254 mm

ii) Upon review of the TEC/TS configuration proposed, it is determined that 47 layers of foils will produce a thermal strap that is fairly bulky and hard to mount directly on the hot side of the TEC. An alternative approach is proposed that includes placing a heat spreader on the hot side of the TEC and a "dog house" enclosure around the TEC/heat spreader (Figure P5.21) that is temperature controlled to 3°C. The interior wall of the dog house is coated with Black

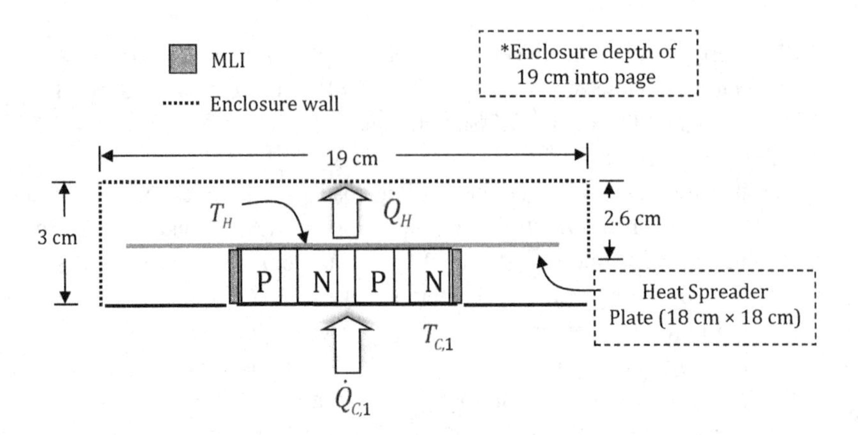

Figure P5.21 TEC internal to a "dog house" enclosure

Anodize (Sample 1). If the TEC's hot-side heat-spreader plate has an area of 18 cm × 18 cm coated with ZOT (w/potassium silicate) and the current is reduced to 4.0 amps, what is the cooling load on the TEC cold side if the temperature is still maintained at 22°C?

(Note: You can assume the MLI surrounding the TEC elements has ideal performance.)

5.22 An array of PbTe TEG modules is sandwiched between a hot and a cold plate. Each element has base dimensions of 14.0 mm × 14.0 mm and a height of 8.6 mm. The electrical conductivities for the TEG elements are $\sigma_{e,p} = \sigma_{e,n} = 222.22 \times 10^2\ \Omega^{-1} \cdot \text{m}^{-1}$, whereas the Seebeck coefficients are $\Gamma_p = -\Gamma_n = 214 \times 10^{-6}$ V/K and the thermal conductivities are $k_p = k_n = 1.48$ W/m · K. The hot and cold panels are stable at temperatures of 420 K and 263 K respectively. If each TEG consists of 36 couples and the sandwich panel is populated with a total of 16 modules, what is the power output and the panel's resistive load for a current of 5.4 amps? You can assume that each module on the panel is wired in series.

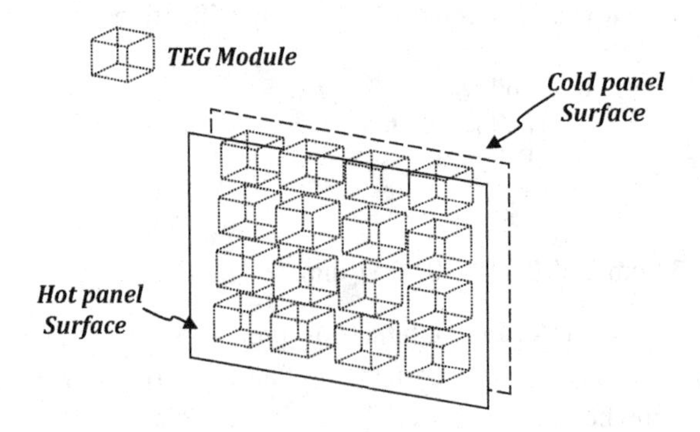

Figure P5.22 4 × 4 TEG module configuration

5.23 A mission designer approaches you with the idea of using the TEG sandwich panel defined in Problem 5.22 onboard a space platform flying in the plane of the Earth's elliptical orbit at a distance of 1 AU from the Sun. It is proposed that heating and cooling of the hot and cold plates should be provided by environmental thermal loads and sinks as opposed to artificial heating and cooling. However, the platform is at a sufficient distance from the Earth for IR loading to be neglected. The sandwich panel is mounted on the space platform via a low thermal conductivity truss. In the flight configuration, the hot-side plate is fully illuminated (continuously) by the sun, whereas the cold-side plate is continuously viewing deep space. Both plates have a surface area of 1.0 m^2. If the hot panel's Sun-viewing side is coated with Maxorb and the cold panel's anti-Sun side is coated with Martin Black Velvet paint:

i) What is the power output for the 16 TEG modules (each having 36 couples) using the element specifications from Problem 5.22 and a current of 1.2 amps?

ii) How might the power output be increased beyond the value determined in (i)?

(Note: Edgewise thermal losses to space from the TEG module's element sidewalls can be neglected in the calculation.)

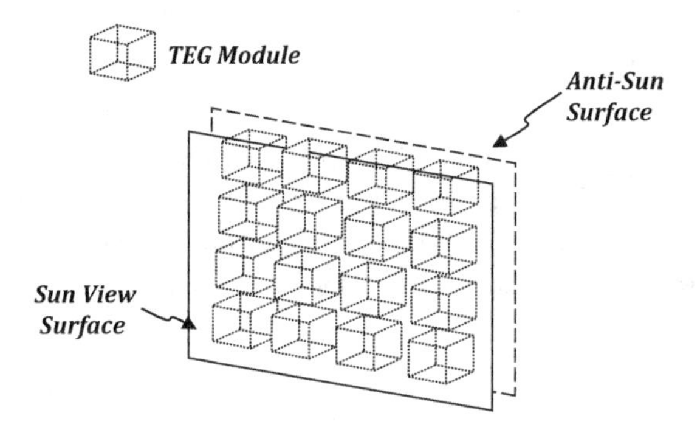

Figure P5.23 4 × 4 TEG module receiving solar loading

5.24 A radiator surface is receiving a heat pulse of 300 W applied every 90 minutes for a duration of 45 minutes. The radiator is pointed towards deep space and has a surface coating of Z306 Black Polyurethane paint (3 mils).

i) Assuming no environmental loading on the radiator surface, what is the radiator area required to maintain a melt temperature (T_{melt}) of 263 K?

ii) What paraffin PCM most closely matches the melt temperature requirement?

iii) If the radiator is no longer viewing deep space, but has an FOV whereby it receives an environmental load of 50 W, what would be the radiator area required?

5.25 A small satellite is flying a laser that contains electronics sinked to a radiator via 1.5 kg of a PCM (n-Eicosane). The radiator is coated with Z93 white paint and is nadir facing. The orbit for the smallsat has a period of 90 minutes and consists of an inclination of 66.6° with a beta angle of 0°.

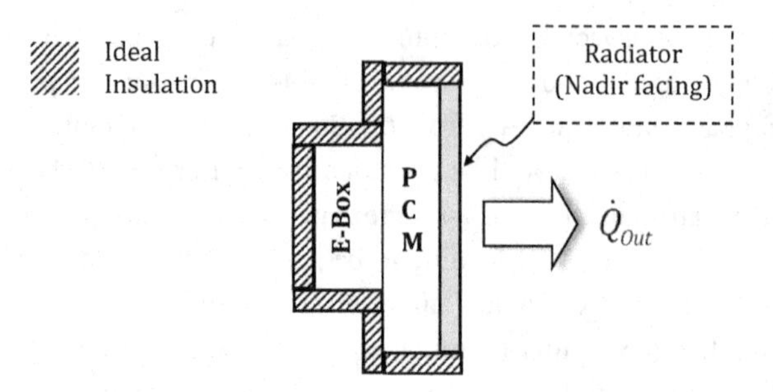

Figure P5.25 PCM w/nadir-facing radiator

i) What is the maximum thermal energy able to be stored during a complete melt event for the 1.5 kg?

ii) The project scientists have decided that they would like to fire the laser every 10 minutes (500 W dissipation) while in eclipse for sequential durations of 3, 4, 6 and 5 minutes. Assuming an Al-6061 radiator with an area of 0.475 m^2 (thickness of 5 mm), determine and plot x_{MSR} for the PCM and the radiator temperature over the four cycles in question. State whether full melt has been achieved at any point in time on the plot. In your calculations, assume $\Delta t = 2$ seconds.

5.26 An E-box that is pulsing heat loads of 250.2 W is coupled on one side to a mass of n-Tetradecane. On the opposite side of the PCM container is a radiator (5 mm thick) coated with Chemglaze A276 white paint. The duration of the heat pulse and the full cycle are 696 and 3,600 seconds respectively. Both the radiator and PCM cross-sectional areas are 0.4 cm × 0.4 cm.

i) Determine and plot two full cycles for the temperatures on the radiator and the E-box/PCM interface if the interface is modeled as having the same dimensions as the radiator and an initial temperature of T_{melt}. Assume both the radiator and the interface are made of Al-6061.

ii) Determine and plot two full cycles for the temperatures on the radiator and the E-box/PCM interface if the interface is modeled as having an equivalent thermal mass equal to that of the E-box with an initial temperature of T_{melt}. Assume the E-box is cubic in shape with a side length of 0.4 cm and a wall thickness of 5 mm. Both the radiator and the E-box are made of Al-6061.

6 Sensors, Instrumentation and Test Support Hardware

6.1 Introduction

In most engineering applications, physical measurements are relied upon to gauge the performance and state of health (SOH) of a system during operation, as well as to collect data. The level of detail for any measurement depends on how critical it is; it follows that determination of physical phenomena will only be as good as the quality of the measurements taken during the design, build, test and flight of space systems. Quality measurements providing insight into the SOH of the system, as well as phenomena occurring in the surrounding environment, can be used to identify performance issues should they arise. Sensors, instrumentation and test support equipment (ground based) are the tools used to validate system performance in a space-relevant environment prior to launch, as well as to perform measurements during actual space flight. This chapter provides an overview of some measurement technologies, guidelines and associated hardware used in the test and operation of spaceflight systems.

6.2 What Is a Sensor?

Key devices used to monitor the SOH of spacecraft and collect data on phenomena in space are temperature sensors and detectors. A sensor is a device used to detect a discrete value of a measurable quantity for correlation into a quantity of interest. The process of taking a base measurement for correlation with a desired quantity can be found in nature. The human eye detects radiation in the visible spectrum of the EM (electromagnetic) wave scale, and processes these waveforms into images by the optic nerve and the brain. The human ear detects sound waves through the vibration of the inner ear's bone structure, and the brain processes the vibrations for correlation to audible sounds. In the case of spaceflight systems quantities of interest may include mass flow, temperature, pressure and photon-based thermal energy absorption.

6.2.1 Temperature-Based Sensors

During spaceflight and ground testing, temperature sensors are used to determine the temperature of specific items that have prescribed operational temperature ranges in both contexts. Temperature predictions from analytical and/or numerical heat transfer modeling are validated prior to launch by experimentation and testing. Temperature sensors measure either the voltage or the electrical resistance across the sensor (specifically the leads). A common type of temperature sensor that relies on voltage measurements is the thermocouple (TC). Thermocouples consist of two dissimilar metals joined together at a junction (Figure 6.1). Exposure of the junction to a temperature within a prescribed temperature range (depending on the metals used) produces a voltage potential across the junction that can be read by a voltage potentiometer device (recall the Peltier effect discussed in Chapter 5). The TC voltage potential, measured in mV, is resolved to a temperature value by comparing it to a reference voltage potential at a known temperature. Resistance-based temperature sensors measure the sensor element's electrical resistance at temperature. As the total electrical resistance across the sensor is known, the temperature can be determined by reference to a pre-defined temperature vs. electrical resistance curve for the sensor. Examples of sensor elements that resolve temperature as a function of electrical resistance include thermistors (TMs), platinum resistance thermometers (PRTs) and Cernox™ thermometers. These types of sensors consist of two wire leads joined together at a measurement junction (Figure 6.2). Materials used for the leads will have a well-defined resistance-to-temperature relationship. The leads are typically 1.0 to 2.0 inches long and are usually attached to extension wire for remote sampling. The mathematical relationship most often used to establish the temperature function is the Steinhart–Hart equation (equation 6.1) where B_1, B_2 and B_3 are constants unique to the sensor element. Each of these constants can be determined by solving a system of equations that include the sensor's electrical resistance and true temperature value at three equilibrium points.

$$\frac{1}{T} = B_1 + B_2 \cdot \ln(R_e) + B_3 \cdot \left[\ln(R_e)^3\right] \tag{6.1}$$

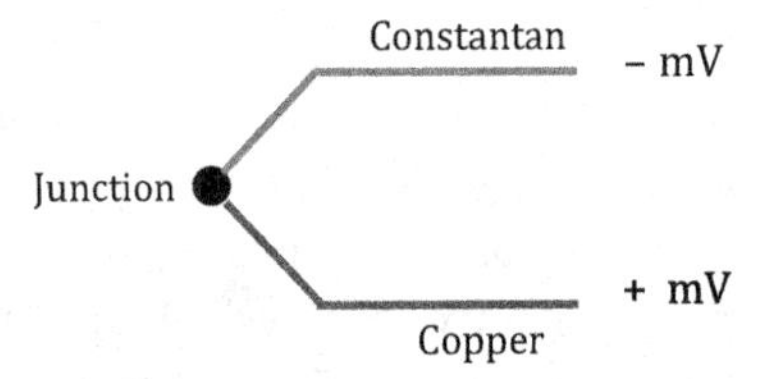

Figure 6.1 T-type thermocouple junction

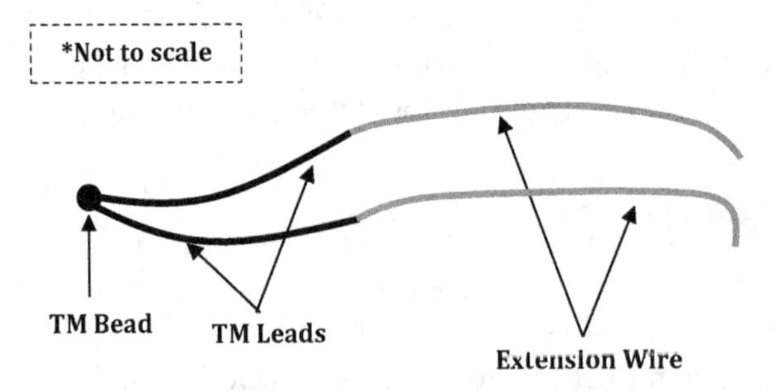

Figure 6.2 Thermistor bead with extension wire

Once the constants are determined, a curve for the temperature as a function of resistance can be established. Figure 6.3 shows such a curve for the Omega™ Corporation's 44033 series TMs.

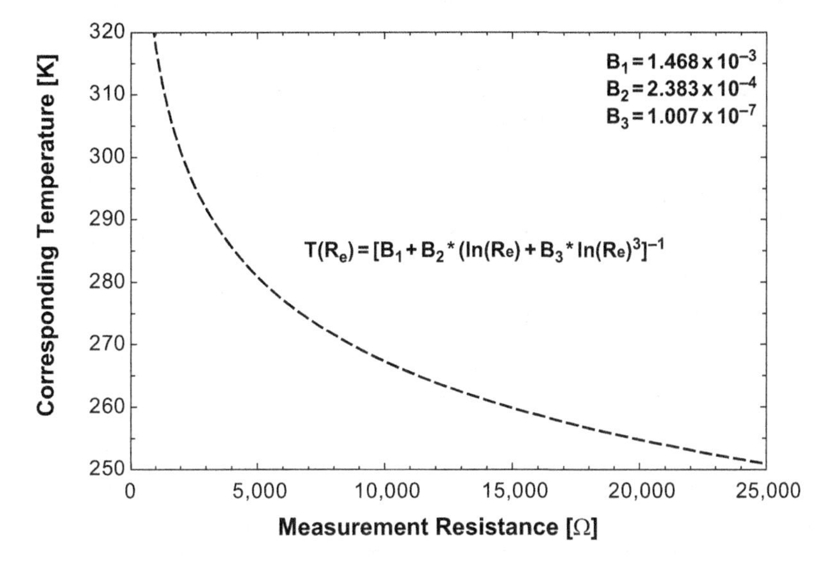

Figure 6.3 Resistance-based temperature curve for Omega™ 44033 series TM

Table 6.1 lists standard temperature sensor types with their temperature ranges and relative accuracy of measurement. Several types of TC are available on the market, spanning sensing ranges between −270°C and 1,372°C with measurement accuracies of ±1°C.

TCs are routinely used for ground tests but are not used on spaceflight systems for on-orbit temperature telemetry due to their performance features and complexity in integration. In general, TCs produce a noisy voltage signal and are less accurate than other types of measurement devices. They also require a reference voltage and signal-conditioning electronics and are hence much less appealing than temperature sensors that rely on resistance measurements. For further information on thermocouple and thermistor types not referenced in the table along with sheathing and extension wire options, see Omega Corporation, reference [1].

TMs measure a smaller temperature range than TCs but are more accurate, and are frequently used on space systems. PRTs, Silicon Diode and Cernox™ thermometers are frequently chosen for use with cryogenic temperature measurements. PRTs can measure temperatures down to 14 K, whereas Silicon Diode and Cernox™ thermometers can perform measurements as low as 1.4 K and 100 mK respectively. Measurements taken by these sensors will be least accurate towards the higher end of their detectable temperature range. The most accurate results can be achieved using the manufacturers' temperature-vs.-resistance calibration curves. Magnetic fields can be an issue at low temperatures. Cernox™ thermometers are negligibly susceptible to magnetic fields down to temperatures as low as 1 K. PRTs and Silicon Diode thermometers are more susceptible and should not be used for temperature measurements in the presence of a magnetic field below 30 K and 60 K respectively [3]. For details of more low-temperature sensors, extension wire, sensor packaging options and best practices, see Lakeshore Cryotronics, Inc., references [3] and [4].

Table 6.1 Standard temperature sensors

Sensor name		Type of measurement	Maximum applicable temperature range	Measurement accuracy
[1]**TC** [1]	Type J	milliVolts	−210°C–1,200°C	±1°C
	Type K	milliVolts	−270°C–1,372°C	±1°C
	Type T	milliVolts	−270°C–400°C	±1°C
	Type E	milliVolts	−270°C–1,000°C	±1°C
	Type N	milliVolts	−270°C–1,300°C	±1°C
[2]**TM** [2]	Omega 44000 Series	Resistance 0.9 Ω–142 kΩ	−80°C–150°C	±0.1°C/ ±0.2°C
[2]**PRT** [3]		Resistance	14 K–873 K	±10 mK[§]
[2]**Silicon Diode** [3]		Voltage	1.4 K–500 K	±12 mK[‡]
[2]**Cernox™** [3]		Resistance	100 mK–420 K	±5 mK[‡]

[§] Measurement accuracy quoted for calibrated sensors at 20 K.
[‡] Measurement accuracy quoted for calibrated sensors at 1.4 K.
[1] Reproduced with permission of Omega Engineering, Inc., Norwalk, CT, 06854, www.omega.com.
[2] Reproduced with permission of Lake Shore Cryotronics, Westerville, OH, 43082, www.lakeshore.com.

Example 1 Temperature Sensor Selection The in-flight operational temperature of a space-flight detector that is being designed must be 5 K ± 50 mK while measurements are being taken. Suggest a resistance-based temperature sensor to measure the detector housing.

Solution

For the resistance-based measurements in question, Cernox™ sensors will be able to reach a temperature of 5 K. In addition, Cernox™ sensors have a measurement uncertainty of ±5 mK. This level of uncertainty is far more accurate than the requirement quoted (±50 mK) as the upper limit. Thus, Cernox™ temperature sensors are a viable option for application.

6.2.2 Detectors

Detectors are advanced sensors that measure either energy or energy flux. Detectors such as the microscale heater array (Figure 6.4a) are used as thermal energy sources for heat transfer performance testing of boiling and/or flow boiling processes. The microscale heater array has a heat transfer surface area of 7.0 mm × 7.0 mm [5], and comprises 96 individual heaters (Figure 6.4b) which enable the heat input to be spatially resolved at specific locations on the heater surface. Each heater segment contains a platinum resistor element. Dissipative power delivered to each heater element is regulated by a dedicated bridge circuit [5]. Since $\rho_e = \rho_e(T)$, measuring the resistance in each heater element enables the temperature of the element to be determined. As with the temperature sensors discussed in Section 6.2.1, this is done with the aid of a calibration curve (i.e., temperature as a function of resistance) which is determined prior to actual use during experimentation and data collection.

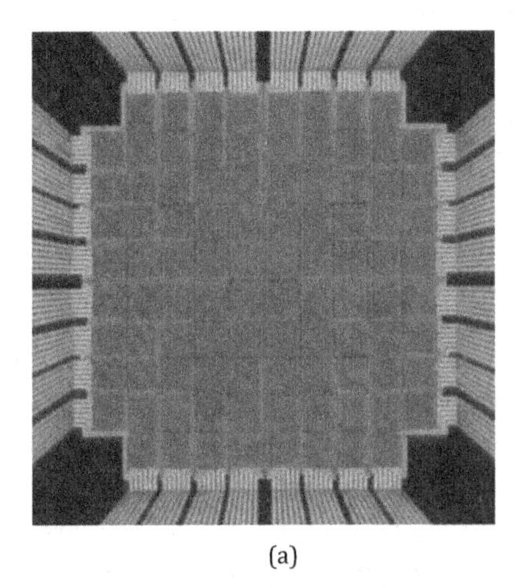

(a)

	96	95	94	93	92	91	90	89	
65	37	64	63	62	61	60	59	58	88
66	38	17	36	35	34	33	32	57	87
67	39	18	5	16	15	14	31	56	86
68	40	19	6	1	4	13	30	55	85
69	41	20	7	2	3	12	29	54	84
70	42	21	8	9	10	11	28	53	83
71	43	22	23	24	25	26	27	52	82
72	44	45	46	47	48	49	50	51	81
	73	74	75	76	77	78	79	80	

(b)

Figure 6.4 Phase change heat transfer laboratory microscale heater array [5]: a) heater array; b) individualized heater elements on the microscale heater array
Source: Reprinted with permission of Elsevier from Horacek, B., Kiger, K., and Kim, J., "Single Nozzle Spray Cooling Heat Transfer Mechanisms," *International Journal of Heat and Mass Transfer,* Vol. 48 (2005), pp. 1425–1438.

Two advanced detector technologies used in the space-based community to measure radiant thermal energy are bolometers [6–8] and calorimeters [8–10]. The first bolometer was designed by Professor S.P. Langley in the late 1800s to measure the radiant energy from the Sun [11]. Langley's bolometer design consisted of thin metal strips wired to a Wheatstone bridge and a galvanometer to measure the balance between the two sides. One strip absorbed solar radiation and the other was protected from the Sun's rays. The resistance of the irradiated material changed, generating an imbalance in the bridge circuit which deflected the galvanometer. The amount of change in the resistance of the metal strip was proportional to its change in temperature [11]. Contemporary bolometer configurations consist of a photon absorber material, a constant temperature sink, a thermal coupling between the absorber and the temperature sink and temperature sensors for both (Figure 6.5). The thermal resistance for the coupling between the absorber and the sink can be determined given knowledge of the material used and the geometry of the coupling. The absorber material is tailored to the wavelength of interest. Photons impinging the absorber material provide an increase in the absorber's thermal energy, along with a corresponding increase in temperature. If the mass and specific heat of the absorber material are known, the thermal storage capacity (i.e., mc_p) can be determined. The initial energy increase associated with the photons received by the absorber material can be determined as

$$\Delta E_{abs} = (mc_p) \cdot \Delta T_{abs} \tag{6.2}$$

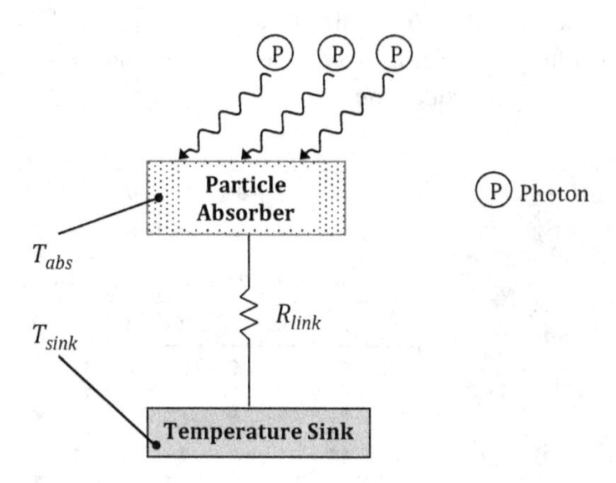

Figure 6.5 Bolometer configuration at the detector level

Since the absorber is thermally coupled to the constant temperature sink, the absorber temperature will decay back to its original value once photon capture in the absorber ceases.

Calorimeters have a similar configuration to that shown for bolometers in Figure 6.5. However, calorimeter measurements typically involve the capture and measurement of single photons in the absorber material. Advanced calorimeters such as Micro-Kinetic Inductance Detectors (MKIDs) use absorber material operating in the superconducting state, which is much more sensitive to changes in energy levels resulting from photon capture. An overview of superconductivity is given in Chapter 7. For further information on MKIDs, see Enss and McCammon [10].

6.3 Heaters

Spaceflight thermal control systems often use resistance heaters to prevent components from falling below their assigned survival and lower operational temperature limits. Resistance-based heaters dissipate thermal energy when a voltage potential is applied to the resistive element. Standard types of resistors used in the aerospace industry include flexible heaters (also known as patch heaters), cartridge heaters and wirewound power resistors.

6.3.1 Types of Heaters

Flexible heaters (Figure 6.6) include a wirewound resistive element placed in an encapsulant material such as polyimide with two power lead extensions protruding from the encapsulant. Flexible heaters come in a variety of shapes, resistances and sizes (footprints). The total power output will vary as a function of each parameter and the voltage potential applied. They come with the option of pre-applied PSA (pressure-sensitive adhesive) on the heater's attachment side. When using PSA, the heater should not be operated at heat fluxes in excess of the manufacturer's suggested limit (on the order of 3.1 W/cm^2). An alternate adhesive TIM such as Stycast can be used to increase the practical watt density operational limit if required. Discontinuities between the heater and the mounting surface at the contact interface can lead to localized burnout of the resistance element during heater operation. Bubbles must not be present when flexible heaters are being applied to their mounting surface.

Figure 6.6 Flexible heater in encapsulant material

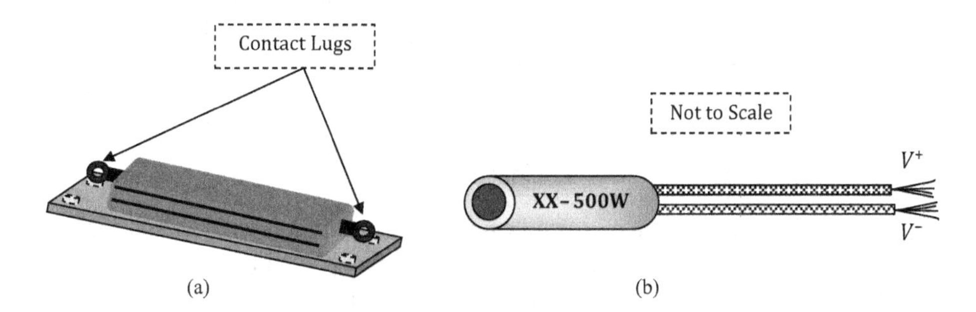

Figure 6.7 a) Wirewound power resistor; b) cartridge heater

Figure 6.7a and b show a wirewound power resistor and a cartridge heater. Wirewound power resistors contain a wrapped resistor element inside a metal housing; typically the housing is made of aluminum, with a designated mounting surface and through holes for fastening the resistor housing to the mounting surface. Electrical lugs protrude from the interior on both sides of the housing. Power resistors come in variable power ratings and heat-exchange surface areas. Cartridge heaters contain a resistance element encapsulated in an insulator with power leads extending out of one side. The cylindrical resistor/insulator structure is housed in a metal sheath which allows for operation at elevated temperatures. Cartridge heaters come in variable sheath sizes and power dissipation ratings. In application, the cartridge heater is embedded in a structural enclosure whose inner surface area matches the sheath's outer surface area, creating a snug fit when the cartridge heater is seated. Regardless of the heater type used, for a given resistance value, the dissipative power can be characterized as

$$P = \frac{(V^+ - V^-)^2}{R_e} \tag{6.3}$$

Given that the standard spacecraft bus voltage is approximately 28 VDC, the resistance required on the heater element can be calculated if the desired dissipative heat load is known.

Example 2 Power Dissipation for a Resistive Heater A flexible heater with a resistance rating of 4.5 Ω has 9 VDC applied across its leads. What is the dissipative power of the heater during operation?

Solution

As the resistance and voltage are known, the dissipative heater power can be determined using equation 6.3. Placing values in the relation obtains

$$P = \frac{(9\ \text{V})^2}{4.5\ \Omega} = \frac{81\ \text{V}^2}{4.5\ \Omega} = 18\ \text{W}$$

Thus, 18 W of thermal energy is dissipated from the heater with an application of 9 VDC. To achieve a larger thermal energy dissipation at this same potential, a heater element with a smaller resistance value must be used.

For more information on heater product options, see the websites of Minco Products Inc. (www.minco.com), Watlow (www.watlow.com) and Omega™.

6.3.2 Temperature Control Techniques

Active thermal control of an item is performed using a temperature sensor, a controller and a device to apply either heating or cooling to the monitored item according to predetermined setpoint temperatures. The controller will determine whether or not the item being sensed requires heating or cooling based on the value read by the sensor. As the location of the temperature sensor is critical for obtaining an accurate reading, it is best placed either on or in close proximity to the item being monitored. When the controller determines that additional heating/cooling is required, a switch is closed (either electrically or mechanically) to supply power to the heating/cooling device until the item being sensed returns to its desired operational temperature range. The difference between the setpoint temperatures at which the switch opens and closes is referred to as the deadband (Figure 6.8). Controllers can be either electronic, thermomechanical or software based and can operate using either an on/off or proportional integral derivative (PID) method. Standard laboratory temperature controllers (e.g., Omega™ CNi series) and LabView VI programs are terrestrial-based examples of electronic and software controllers respectively. Thermomechanical controllers are usually traditional thermostats containing a thermomechanical coil calibrated to open and close a mechanical switch at pre-set temperature values.

Since space is colder than the desired operational temperature ranges for most onboard components, spaceflight control schemes most often include heating. The heater is activated by the controller based on a lower temperature bound; once the item being monitored is back within its desired temperature range, the switch is opened and

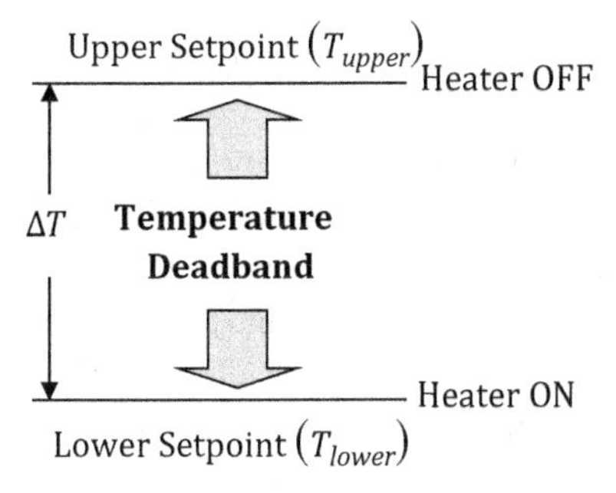

Figure 6.8 Deadband temperature variation between setpoints

heating is no longer applied. The heater control configuration in Figure 6.9 consists of a power distribution box, a controller, an enclosure being temperature controlled, a thermistor for sensing the temperature of the enclosure and a heater. The controller performs continuous sensing of the enclosure temperature. Leads from the power distribution box provide power to the enclosure's heater by way of the controller once the switch is closed. The switch reopens when the enclosure reaches the upper level setpoint temperature. The duty cycle of an on/off switched heater is defined as

$$Duty\ Cycle\ \% = \frac{\Delta t_{ON}}{\Delta t_{ON} + \Delta t_{OFF}} \cdot 100 \tag{6.4}$$

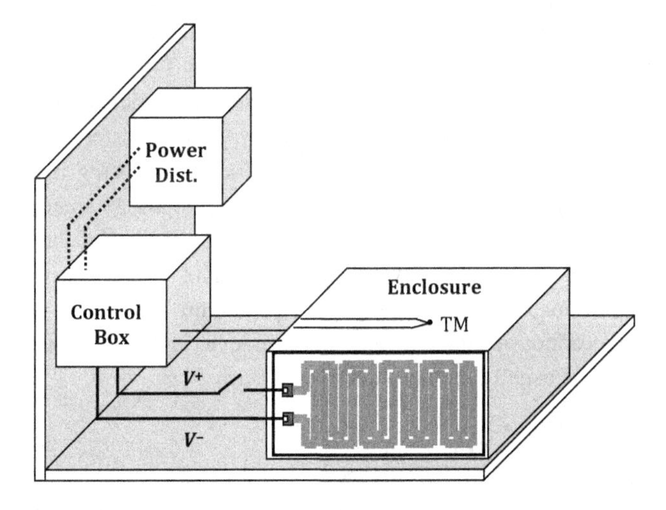

Figure 6.9 Standard heater/controller configuration

A heater repeatedly placed in the on state for 10 minutes followed by 20 minutes in the off state is described as having a duty cycle of 33.3%. The average power for the heater dissipation would then be

$$P_{avg} = \frac{\Delta t_{ON}}{\Delta t_{ON} + \Delta t_{OFF}} \cdot P_{ON} = Duty\ Cycle \cdot P_{ON} \tag{6.5}$$

The operational duty cycle of a heater should be appreciably less than 100%.

Controls can also be applied to TECs for cooling. If the item the TEC is thermally coupled to for cooling has a temperature above its desired operational range, the controller switch closes, turning on power to the TEC and instituting cooling. Similarly, a controller which is in a closed-circuit state for a heater on an LHP's CC could be opened, allowing the LHP to turn on and provide cooling to the temperature-controlled item once the CC temperature reduces.

6.4 Test Support Equipment

In order to establish adequate thermal performance of spaceflight systems in a relevant environment, ground-based thermal testing is performed inside a pressure vessel under vacuum conditions. Spaceflight thermal testing is undertaken using mechanical vacuum pumps to achieve pressure levels near those that apply in space. The next section looks at pressure vessels and the characteristics of mechanical vacuum pump systems.

6.4.1 Vacuum Systems

Pressure vessels can be constructed to operate at pressures either below (sub-atmospheric) or above (pressurized) atmospheric conditions (Figure 6.10). Since the pressure in space is on the order of 10^{-11} Pa, space is considered as being at hard vacuum conditions. Pressure vessels explicitly used to test the thermal performance of spaceflight systems in a vacuum environment are known as T-Vac (thermal vacuum) chambers. Mechanical vacuum pump systems are used in conjunction with T-Vac chambers to attain vacuum levels whereby convective heat transfer effects can be negated from the thermal performance of the system (or object) being tested ($\leq 10^{-6}$ Torr). Vacuum pump systems use a staged pumping methodology to reach desired vacuum levels. The first stage consists of a roughing pump that carries the pressure in the vessel from ambient down to crossover pressure ($\approx$1 Pa), when control passes to a turbopump. With the aid of the turbopump, an equilibrium pressure is reached and lower vacuum levels cannot be attained. The pressure level in T-Vac testing of spaceflight thermal systems is usually on the order 10^{-7} Torr.

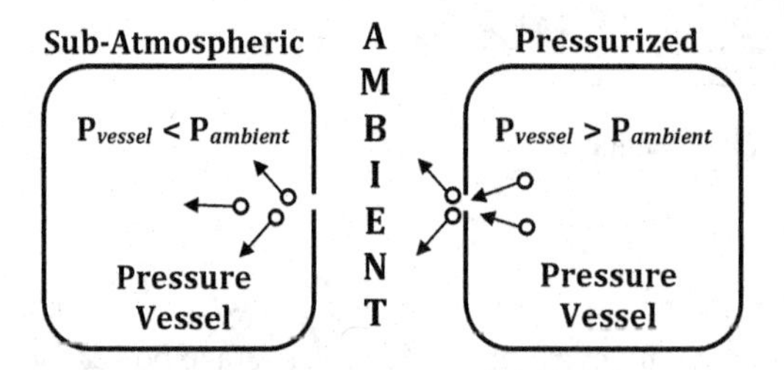

Figure 6.10 Sub-atmospheric vs. pressurized vessel conditions

When a vacuum pump system is connected to a sealed vessel, the pump's sump line serves as a pressure sink for the volume it is pulling vacuum against. Ultimately, gas volume present in the vessel will be evacuated through the sump line and its associated volume flow rate will be $\dot{Vol}_{pump}$. The ability of a mechanical vacuum pump to pull vacuum on a sealed volume is measured in terms of throughput ($\dot{Z}_{pump}$), which is defined as the product of the pressure and the volume flow rate with units of [Pa · m^3/s]. In equation form it is

$$\dot{Z}_{pump} = P(t)\cdot \dot{Vol}_{pump} \tag{6.6}$$

In this equation the pressure is shown as a function of time in order to reflect that it is applicable at all times during the pumpdown process. From a thermodynamics perspective, it is the flow work of the vacuum pump system per unit time. The time required to transition a chamber of a known volume (*Vol*) between two specific pressure levels using a vacuum pump operated at constant speed is referred to as the pumpdown time. The pumpdown time is defined mathematically as [12–15]

$$\Delta t_{pumpdown} = \frac{Vol}{\dot{Vol}_{pump}} \cdot \ln\left[\frac{P_{init}}{P_{final}}\right] \tag{6.7}$$

Equation 6.7 can be applied to both the continuum and molecular flow regimes. When a chamber is being pumped down from atmospheric pressure, the pump will evacuate not only the gas within the chamber but also gas desorbing from interior surfaces, permeating through the chamber walls and diffusing from materials within the chamber. Gasses resulting from leaks (either through or on the interior of the chamber's structural wall) may also be present. When the pressure of gasses in the chamber can no longer be overcome by the suction capability of the vacuum pump, equilibrium with the system throughput has been attained.

While diffusion, desorption and permeation may be present, the capability of achieving desired pressure levels in vacuum chambers is often limited by leaks. The leak rate can be defined in equation form as

$$X_{leak} = \frac{\left(P_{final} - P_{init}\right)}{\Delta t}\cdot Vol \tag{6.8}$$

where X_{leak} is in units of [Pa · m^3/s]. The standard set-up for a vacuum pump system attached to a T-vac chamber is shown in Figure 6.11. The vacuum pump system is physically coupled to a sump port on the T-vac chamber through a plumbing line of known dimensions (i.e., cross-sectional area and length). The speed for the vacuum pump system $\left(\dot{Vol}_{pump}\right)$ is typically assigned to the pump inlet location. Due to flow losses in the transit of the evacuated gasses from the sump port to the pump inlet, the speed of the volumetric flow at the sump port will be different than that at the inlet. The relationship between the speed of the volumetric flow at the sump port and the vacuum pump inlet is

$$\frac{1}{\dot{Vol}_{sump}} = \frac{1}{\dot{Vol}_{pump}} + \frac{1}{C_{vac,path}} \tag{6.9}$$

where $\dot{Vol}_{sump}$ is the effective speed of the flow at the sump port and $C_{vac,path}$ is the conductance of the gas flow between the chamber sump port and the pump inlet [16]. For a segment of interest along the flow path, the conductance is defined as being the throughput divided by the pressure drop.

$$C_{vac,path} = \frac{\dot{Z}}{\Delta P} \tag{6.10}$$

The vacuum path conductance ($C_{vac,\ path}$) accounts for all resistances impeding the flow of the evacuated gasses between the sump port and the pump inlet [12–14,16]. As shown in Figure 6.11, this conductance will include both the aperture of the sump port along the T-Vac chamber inner wall and the vacuum line itself.

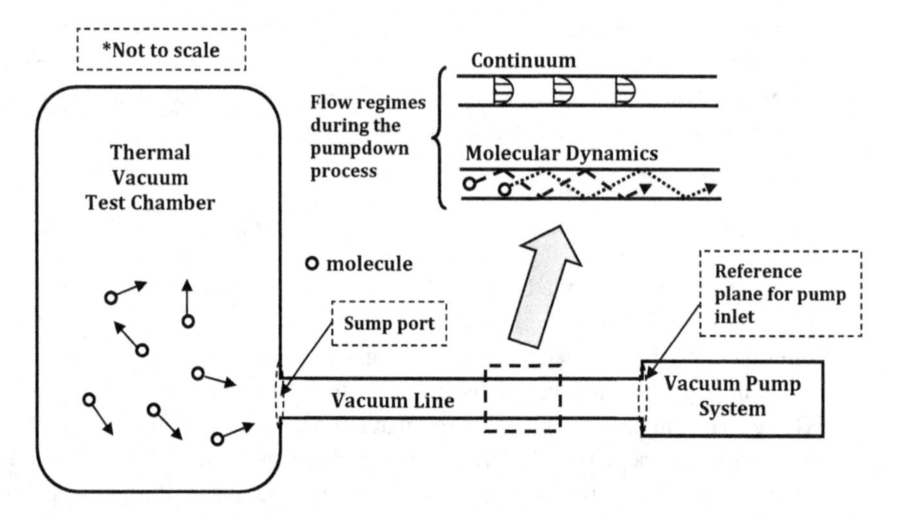

Figure 6.11 T-vac/vacuum pump system configuration

At the start of pumpdown from ambient pressure, flow through the vacuum line will be in the continuum regime. However, as the pressure in the T-Vac chamber and vacuum line decreases, the mean free path of the molecules will eventually become much greater than the characteristic length of the vacuum line (i.e., Kn > 1). At this point the flow through the vacuum line will be based on molecular dynamics. Standard fluids-based relations for Poiseuille flow can be applied to the vacuum line to determine the effective conductance while the system is operating in the continuum regime. Under the molecular flow regime, the aperture conductance is defined as

$$C_{aperture} = Vel_{par} \cdot \frac{A}{4} \tag{6.11}$$

where A is the cross-sectional area of the aperture and Vel_{par} is the average particle velocity using Maxwell–Boltzmann theory [12,14–16]. In equation form this is expressed as [17]

$$Vel_{par} = \sqrt{\frac{8k_B T}{\pi m_{par}}} \tag{6.12}$$

where k_B is Boltzmann's constant, T is the temperature in Kelvin and m_{par} is the particle mass (atomic weight/Avogadro's number) for the gas being pumped out. The vacuum line conductance is defined as

$$C_{vac,path} = TP \cdot Vel_{par} \cdot \frac{A}{4} \tag{6.13}$$

where TP is the probability of particle transit (also known as the transmission probability) through the vacuum line [12–14,16]. The factor TP can vary depending on cross-sectional geometry and tortuosity of the vacuum line. For round vacuum lines with large aspect ratios (length/diameter > 5,000) [16] the probability factor can be approximated as $TP = 4D/3l$.

Example 3 Vacuum Line Conductance as a Function of Throughput A vacuum line has a throughput of 0.075 Pa · m^3/s. Determine the flow conductance if the pressure drop spanning the vacuum line is 620 Pa.

Solution

With throughput of $\dot{Z} = 0.075 \text{ Pa} \cdot \text{m}^3/\text{s}$ and the pressure drop spanning the line being 620 Pa, equation 6.10 is used to determine the conductance as

$$C_{vac,path} = \frac{0.075 \text{ Pa} \cdot \text{m}^3/\text{s}}{620 \text{ Pa}} = 1.2 \times 10^{-4} \text{m}^3/\text{s}$$

Example 4 Pump Speed Requirement A pressure vessel with a volume of 100 liters is at 101 kPa. A vacuum pump is attached to the chamber and pumpdown is initiated. Assuming the vacuum pump operates at a constant speed of 0.5 liters/s, how long does it take to pull the chamber down to 10.1 kPa?

Solution

The time for pumpdown in either the continuum or molecular flow regime is expressed as

$$\Delta t_{pumpdown} = \frac{Vol}{\dot{Vol}_{pump}} \cdot \ln\left[\frac{P_{init}}{P_{final}}\right]$$

Substituting in values from the problem statement for each of the variables, we then have

$$\Delta t_{pumpdown} = \frac{100 \text{ liters}}{0.5 \text{ liters/s}} \cdot \ln\left[\frac{101 \text{ kPa}}{10.1 \text{ kPa}}\right]$$

When calculated, this results in a pumpdown time of 460.5 seconds or 7.68 minutes.

For additional probability factors for various alternate geometries, see O'Hanlon [16], Pfeiffer [12,13] or Oerlikon [14]. A list of analytical relations used to predict transmission probability as a function of flow regime is provided by Barron [15].

6.4.2 Leak Checking

Some leaks will always occur in test chambers operating at vacuum pressures. In order to eliminate convective heat transfer effects, the total gas inputs through leaks must be sufficiently small for vacuum levels on the order of $\leq 10^{-6}$ Torr to be maintained throughout testing. Leaks that compromise the ability to achieve these vacuum levels must be identified and mitigated during test set-up in a process known as leak checking. For practitioners who routinely uses vacuum systems leak checking is an art form that applies to both sub-atmospheric and pressurized chamber vessel conditions. Areas and/or features on a chamber vessel that are highly susceptible to leakage include feed-throughs, bolted interfaces, seals and gaskets. Leaks are categorized as either gross or fine according to the volume rate of gas flowing through them. As the leak rate may be sensitive to the direction of the pressure differential across the leak (sub-atmospheric vs. super-atmospheric), leak checking should be performed with the pressure differential in the same direction as it will be during actual testing. Gross leaks in pressurized systems with relatively small volumes can be identified by applying a liberal amount of soapy water to the suspected area and observing the formation of bubbles. However, a leak detector system is usually chosen for fine leaks in pressurized systems. In vacuum systems, a detector system normally is used to check the test volume for leaks in a sub-atmospheric state. Standard leak detectors use He-4, although most contemporary leak detectors can also detect He-3 and H_2. Leak detection systems can be operated in either "sniffer" mode or "housing unit" mode. In sniffer mode the interior volume of the vacuum vessel is injected with the detection gas. A "sniffer extension arm" with a plumbing connection back to the detector housing unit is then probed around the exterior volume of the vessel from bottom to top. Since the standard detection gasses have relatively light molecules, they will rise when released into ambient. In housing-unit mode, the suction port line for the leak detector is connected to a sump port on the vacuum vessel. The detection gas is released onto the exterior surface of the chamber vessel from top to bottom. If the detection gas can work its way into the vessel it will be pulled into the sump line and registered as a detection on the unit itself.

6.5 Error Analysis

Standard error analysis can be applied to measurable quantities such as voltage, resistance, temperature, heat and heat flux. For a quantity that is a dependent variable, the error in the measurement will be a function of the independent variables the measurement was based on. If the measurement variable X comprises independent variables a, b, and c then $X = X(a, b, c)$. The independent variables will often have a certain prescribed amount of error. If X is calculated from measured values of a, b and c then the error inherent to the calculation is

$$\delta X = \pm\sqrt{\left(\frac{\partial X}{\partial a}\delta_a\right)^2 + \left(\frac{\partial X}{\partial b}\delta_b\right)^2 + \left(\frac{\partial X}{\partial c}\delta_c\right)^2 + \ldots} \tag{6.14}$$

This root sum square error (RSS) analysis technique is otherwise known as the Kline and McClintock [18] method and is widely used throughout the engineering community.

Example 5 Error Calculation for Longitudinal Heat Transfer Measurement Heat transfer measurements are being taken along the longitudinal axis of a solid copper rod (k_{OFHC} = 398 W/m · K). The rod has a diameter of 2.54 cm and a length of 15 cm. The heat transfer along the rod is determined using Fourier's Law for conduction with the aid of discrete temperature measurements spaced 10 cm apart. The temperature measurements are performed by placing the temperature sensors in a small circular intrusion that extends from the rod's perimeter to its centerline axis. If the accuracy of the sensor location is ±1 mm and K-type thermistors are used as the sensor elements, determine the error associated with the heat transfer measurements for a ΔT of 40°C.

Note: You can assume the cross-sectional area and the thermal conductivity of the copper have no error.

Solution

Fourier's Law for conduction (as applied in this case) is

$$\dot{q} = \frac{k_{OFHC} A \Delta T}{\Delta z}$$

The RSS error relation will be

$$\delta q = \pm\sqrt{\left(\frac{\partial q}{\partial z}\delta_z\right)^2 + \left(\frac{\partial q}{\partial(\Delta T)}\delta_{\Delta T}\right)^2}$$

Using the Fourier's Law relation, the partial derivative terms can be recast as

$$\frac{\partial \dot{q}}{\partial z} = \left(-\frac{k_{OFHC} A \Delta T}{(\Delta z)^2}\right) \text{ and } \frac{\partial \dot{q}}{\partial(\Delta T)} = \left(\frac{k_{OFHC} A}{\Delta z}\right)$$

Placing these terms into the RSS error relation, we have

$$\delta q = \pm\sqrt{\left(-\frac{k_{OFHC} A \Delta T}{\Delta z^2} \cdot \delta_z\right)^2 + \left(\frac{k_{OFHC} A}{\Delta z} \cdot \delta_{\Delta T}\right)^2}$$

Before values for the variables can be substituted, the cross-sectional area for heat flow within the rod must be determined. The cross-sectional area of the rod is

$$A = \pi r^2 = \frac{\pi D^2}{4} = \frac{\pi}{4}(0.0254\text{m})^2 = 5.0 \times 10^{-4}\ \text{m}^2$$

Based on the error associated with K-type TMs, the maximum possible error for a temperature difference between two discrete measurement locations will be $\delta_{\Delta T} = 0.2$ K. For the error in the distance between the temperature sensor measurement locations, the maximum error will be $\delta_{\Delta z} = 2.0$ mm. Placing all known values in the RSS relation obtains

$$\delta q = \pm\sqrt{\left(-\frac{(398\,\text{W/m}\cdot\text{K})(5.0\times10^{-4}\text{m}^2)(40^\circ\text{C})}{(10\,\text{cm})}\cdot(2.0\,\text{mm})\right)^2 + \left(\frac{(398\,\text{W/m}\cdot\text{K})(5.0\times10^{-4}\text{m}^2)}{(10\,\text{cm})}\cdot 0.2\text{K}\right)^2}$$

Calculating this out will result in an error value of $\delta q = 0.43$ W.

References

[1] "ANSI and IEC Color Codes for Thermocouples, Wire and Connectors," *Thermocouples*, Omega Engineering, Stamford, CT

[2] "Thermistor Elements," *Thermistors*, Omega Engineering, Stamford, CT

[3] "Sensor Selection Guide," *Sensors*, Lake Shore Cryotronics, Inc., Westerville, OH

[4] "Temperature Sensor Selection Guide," *Sensors*, Lake Shore Cryotronics, Inc., Westerville, OH

[5] Horacek, B., Kiger, K., and Kim, J., 2005, "Single Nozzle Spray Cooling Heat Transfer Mechanisms," *International Journal of Heat and Mass Transfer*, Vol. 48, pp. 1425–1438

[6] Moseley, S.H., Mather, J.C., and McCammon, D., 1984, "Thermal Detectors As X-ray Spectrometers," NASA-TM-86092

[7] Alsop, D.C., Inman, C., Lange, A.E., and Wilbanks, T., 1992, "Design and Construction of High-sensitivity Infrared Bolometers for Operation at 300 mK," *Applied Optics*, Vol. 31, No. 31, pp. 6610–6615

[8] Richards, P.L., 1994, "Bolometers for Infrared andMmillimeter waves," *Journal of Applied Physics*, Vol. 76, No. 4, pp. 1–24

[9] Kraus, H., 1996, "Superconductive Bolometers and Calorimeters", *Superconducting Science and Technology*, Vol. 9, pp. 827–842

[10] Enss, C., and McCammon, D., 2008, "Physical Principles of Low Temperature Detectors: Ultimate Performance Limits and Current Detector Capabilities," *Journal of Low Temperature Physics*, Vol. 151, pp. 5–24

[11] Langley, S.P., 1880, "The Bolometer and Radiant Energy," *Proceedings of the American Academy of Arts and Sciences*, Vol. 16, pp. 342–358. doi:10.2307/25138616

[12] "Introduction to Vacuum Technology," *Vacuum Technology Book Vol. II*, Pfeiffer Vaccum GmBH, Asslar, Hesse, Germany

[13] "Basic Calculations," *Vacuum Technology Book Vol. II*, Pfeiffer Vaccum GmBH, Asslar, Hesse, Germany

[14] Umrath, W., Adam, H., Bolz, A., Boy, H., Dohmen, H., Gogol, K., Jorisch, W., Mönning, W., Mundinger, H.-J., Otten, H.-D., Scheer, W., Seiger, H., Schwarz, W., Stepputat, K., Urban, D., Wirtzfeld, H.-J., and Zenker, H.-J., 2016, "Fundamentals of Vacuum Technology," Oerlikon, Cologne, Germany

[15] Barron, R.F., 1985, *Cryogenic Systems*, 2nd edn, Oxford University Press, New York, NY

[16] O'Hanlon, J.F., 2003, *A User's Guide to Vacuum Technology*, 3rd edn, John Wiley & Sons, Inc., New York, NY

[17] Serway, R.A., 1986, *Physics for Scientists & Engineers*, Saunders College Publishing, Philadelphia, PA
[18] Kline, S.J., and McClintock, F.A., 1953, "Describing Uncertainties in Single Sample Experiments," *Journal of the Society of Mechanical Engineers*, Vol. 56, No. 1, pp. 3–8
[19] "Polyimide Thermofoil™ Heaters," Minco Products Inc., http://catalog.minco.com/catalog3/d/minco/?c=fsearch&cid=3_1-polyimide-thermofoil-heaters

Problems

6.1 A temperature sensor is required to measure the temperature of a pair of magnetic torque bars located on a spacecraft while inflight. The torque bars (which generate a moderate magnetic field for platform orientation purposes) are expected to be operating between 100 K and 300 K. A resistance-based reading at a minimum accuracy of 0.1 K is also required. What options are available based on Table 6.1?

6.2 Following successful assembly of an instrument, T-vac testing is to be performed. Recent thermal model predictions are showing what appears to be a large temperature gradient (approximately 40°C) in an area of the instrument that has a temperature gradient requirement of 15°C. There are no telemetry sensors in the area with the potential gradient and the sensor scheme cannot be changed. Consider how the temperature gradient issues can be addressed during testing.

6.3 A honeycomb deck is separated into six different planar zones (A through F). Two of the zones on the plane of the deck contain electronics boxes (Figure P6.3). It is assumed that each E-box is operating at the same temperature as the zone of the deck on which it is mounted. The deck itself has a planar temperature gradient requirement that must be satisfied during operation of the E-boxes. Four temperature sensors are available to be placed on the honeycomb deck. To which zones should they be attached in order to determine the E-box temperatures, as well as the planar gradient across the deck?

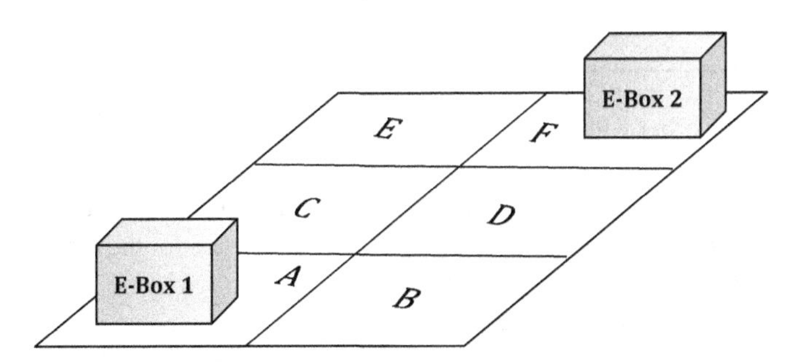

Figure P6.3 Honeycomb deck with temperature zones and E-boxes

6.4 An LHP is being used to cool a laser E-box which is thermally coupled to the LHP evaporator. The LHP condenser is thermally coupled to a radiator that serves as a dedicated thermal sink for the LHP (Figure P6.4). The E-box that will be used for telemetry read-out of temperature sensors on the LHP only has three channels available. Using the numbering convention for locations on the LHP (Figure P6.4), determine the three best locations for application of the temperature sensors.

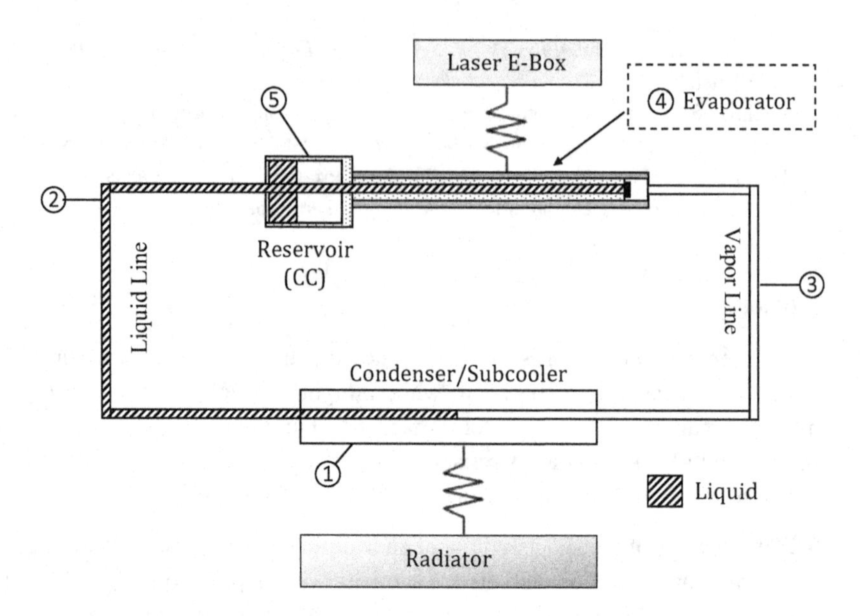

Figure P6.4 Example LHP with select temperature sensor locations

6.5 A sensor element made of an unknown material has temperature and corresponding resistance measurements as shown in Table P6.5. For a temperature span of 300 K to 220 K, determine a resistance-based temperature function using the Steinhart–Hart relation.

Table P6.5 Measured resistance as a function of temperature for sensor X

T [K]	R_e [Ω]
300	556.4
260	660
230	1,156

6.6 A particle absorber with a thermal storage capacity of 1.5×10^{-11} J/K is acquiring 100 photons every millisecond. The photons have an approximate wavelength at the center of the X-ray spectrum. The absorber is thermally coupled to a temperature sink which is held at a constant temperature of 2.0 K. The resistance for the thermal coupling is 1.33×10^{9} K/W. Determine the total thermal energy acquired by the absorber and the temperature rise in the absorber after 25 milliseconds of photon collection. Assume that the absorber is in equilibrium with the temperature sink at the start of collection.

Note: See Chapter 3 regarding energy of a photon as a function of wavelength.

6.7 A 40 W heater is being cycled using on/off control. The heater is activated every 30 minutes for a duration of 12 minutes.

i) Determine the duty cycle for heater operation.

ii) Determine the value for average power during a single cycle of heater activation.

6.8 A temperature-controlled object resides inside a "dog-house" enclosure with a view to space. The object has a weak conductive coupling to its mounting structure and a strong radiative coupling to the inner wall of the dog house. Cyclical heating is applied to the dog house to maintain the temperature of the object above −10°C. The resultant effect is a temperature vs. time profile for the object, as shown in Figure P6.8. For the plot shown, determine the duty cycle of the heating applied to the dog house.

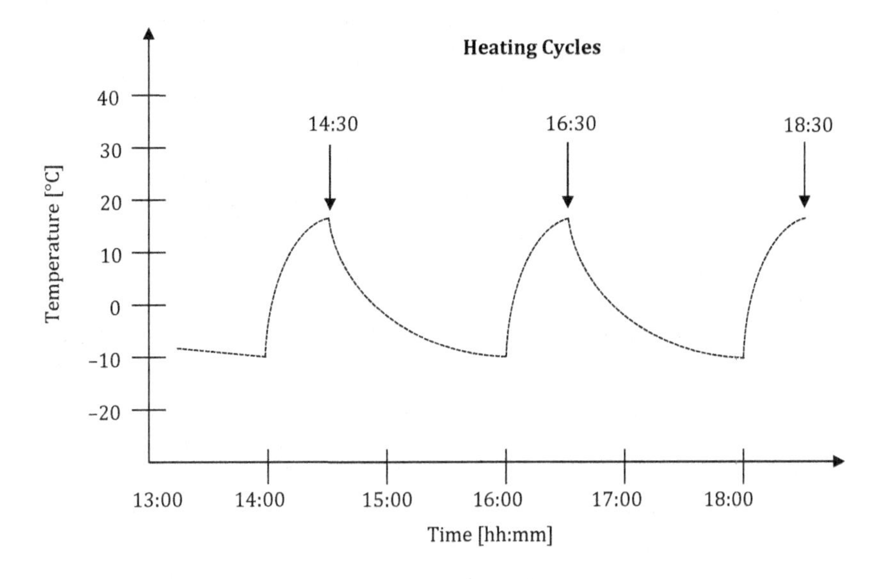

Figure P6.8 Temperature vs. time plot for item inside a "dog house"

6.9 Preliminary temperature predictions show that a Star Tracker assembly needs 20 W of heater power during spaceflight. The manufacturer that supplies the project's heaters is running low on stock and only has flexible heaters with resistance ratings of 3.5 Ω, 6.0 Ω and 12 Ω.

i) For a voltage potential of 12 VDC, determine the duty cycle required to supply 20 W average heater power during a 90-minute orbit for each heater available.

ii) What is the best heater option?

iii) What is the least practical heater option and why?

6.10 An E-box is flying on a future earth observing satellite. On-orbit temperature predictions show that an additional 40 W (peak) heat input needs to be supplied to the E-box while in eclipse for it to stay above the low-temperature limit. The standard heater available to the project is a flexible resistance heater encapsulated in polyimide (see Table P6.10). The only space available for population of heaters on the E-box is a 4-inch × 6-inch flat surface area.

i) Given that standard spacecraft voltage is 28 VDC, which HK series heater (or heaters) are suitable? If more than one is required, please state how many heaters.

ii) If using more than one heater should they be wired in series or parallel?

Table P6.10 Candidate polyimide heaters

Minco product number	Dimensions (inch × inch)	Resistance (Ω)
HK6901	0.5 × 2.0	100.11
HK6909	1.0 × 4.0	23.59
HK6912	2.0 × 3.0	14.73
HK6914	2.0 × 5.0	157.08
HK6917	3.0 × 5.0	100.04

Source: "Polyimide Heater Online Catalog," Minco Products, Inc., reference [19].

6.11 A cylindrical vacuum chamber with a diameter of 0.75 m and a length of 1 m is attached to a vacuum pump system via a short round vacuum line. The vacuum line contains an inline valve for sealing the sump port of the chamber from the vacuum pump. The vacuum pump is operated with the inline valve open until the pressure inside the chamber reaches a value of 10^{-3} Pa. When this pressure level is reached, the inline valve is closed shut and the vacuum pump is turned off to check if there is a slow leak on the system. After a period of 8 hours, the pressure level in the chamber has risen to a value of 0.1 Pa.

i) Assuming the seal on the inline valve is not failing, determine the leak rate for the chamber.

ii) A vacuum system operated in the presence of a leaking vessel will ultimately achieve an equilibrium pressure. Assuming the leak in part i) is equal to the product of the volumetric flow rate at the chamber's sump port and the chamber's equilibrium pressure, determine the volumetric flow rate required to achieve an equilibrium pressure of 1.33×10^{-3} Pa.

6.12 A mechanical vacuum pump is operating in the molecular flow regime. The pump is connected to a chamber via a 0.5-inch diameter vacuum line that has a transmission probability of 0.405. Assuming the gasses in the vacuum line are at 22°C and the pump is operating at a speed of 300 liters/s, determine the effective pump speed at the sump port aperture on the chamber.

6.13 A mechanical vacuum pump is attached to a cylindrical T-Vac chamber (uniform temperature of 21°C) that has a diameter of 1.5 m and a length of 2.5 m. The aperture at the chamber's sump line location has a diameter of 1.5 inches. Assuming the pumpdown system has crossed over to the molecular flow regime and is operating at a speed of 45 liters/s:

i) Determine the conductance at the aperture and the corresponding time for pumpdown from a pressure level of 0.1 Pa to 10^{-4} Pa assuming the reference plane for the pump is at the aperture.

ii) Determine the conductance for a vacuum line with a diameter of 1.5 inches and a length of 76.2 cm, as well as the corresponding time for pumpdown from a pressure level of 0.1 Pa to 10^{-4} Pa. Assume a transmission probability of 0.05949.

6.14 A high vacuum pump system is connected to a cylindrical T-Vac chamber via a vacuum line 30 m long. The T-Vac chamber has a diameter of 2.0 m and a length of 3.0 m. Three candidate vacuum lines with diameters of 0.5 inches, 0.75 inches and 1.0 inch are being considered for use. Provide a comparison plot of the time required to pull the chamber pressure down from 0.1 Pa to 10^{-5} Pa (i.e., the molecular flow regime) as a function of vacuum line diameter for average pump speeds of 50, 75 and 100 liters/s. The probability for transmission of the air molecules through the vacuum line can be approximated using the relation for long round plumbing. Assume that the molecules internal to the vacuum line are at a temperature of 20°C.

6.15 A project is using Omega™ Brand 10,000 Ω thermistor wire sensors for temperature measurements (product number TH-44031-40-T). The Steinhart–Hart equation constants are $B_1 = 1.032 \times 10^{-3}$, $B_2 = 2.387 \times 10^{-4}$ and $B_3 = 1.580 \times 10^{-7}$. The sensor leads themselves are approximately 2 inches long. The distance between the measurement location and the signal-processing electronics for the temperature measurements is approximately 1 meter. The leads are soldered to extension wire to span the distance. The extension wire used is multi-strand 22-gauge tinned copper wire (diameter of 0.8 mm).The task is to validate that the use of the extension wire does not increase the uncertainty of the temperature measurement beyond 0.3°C.

i) Assuming an electrical resistivity of $1.7 \times 10^{-8}\ \Omega \cdot \text{m}$ for the tinned copper wire, determine the resistance at room temperature (i.e., 298 K) for the wire length specified.

ii) Using the RSS method, calculate the uncertainty in the resistance for the thermistor measurements, assuming that the resistance measurement for the 22-gauge extension wire has an accuracy of approximately +/− 20% between 243 K and 313 K.

iii) If the thermistor leads have a measurement uncertainty of ±0.1°C, determine the total uncertainty for temperature measurements given the resistance uncertainty calculated in part ii).

7 Fundamentals of Cryogenics

7.1 Background

Cryogenics is an expansive technical topic that can easily fill a standalone text under the same title. This chapter is not intended to replace the extensive technical content that may be found in contemporary cryogenics engineering texts [1–6]. Its goal is, rather, to provide the reader with some technical insights into basic cryothermal topics that arise in the design of cryogenic systems for space applications.

The Meriam-Webster dictionary defines cryogenics as the branch of science that deals with the production of low temperatures and the corresponding effects upon matter [7]. However, the relative term "low temperatures" must be defined. The boundary temperature for the beginning of the cryogenic temperature regime (when going cold) was established by the NBS (National Bureau of Standards) as −150°C (123 K) [2]. This temperature will therefore be used as the starting point for what is referenced as a cryogenic temperature.

7.1.1 The Challenge of "Going Cold"

The designer of a cryogenic system has two goals. The system must cool the temperature-controlled item down to desired operational temperatures and maintain them for extended periods of time. Cryogenic systems in an Earth-based terrestrial environment must be colder than ambient. The nominal temperature in space is approximately −270°C. Items operating within the cryogenic regime at temperatures well above 3 K will be warm biased compared to space. Nonetheless, the heat-rejection capability of low-temperature surfaces is significantly less than that of surfaces operating in the 270–300 K temperature range. Onboard temperature sinks passively rejecting heat to space have limited application below 80 K. There are also thermophysical challenges associated with cooling objects to low temperatures ($\leq$123 K). A system becomes more ordered at the molecular level as it cools down and approaches absolute zero [8,9]. Since entropy is a measure of the amount of order (and/or disorder) in a system or object, the second law of thermodynamics applies. In equation form, that law is

$$\Delta S = \frac{\Delta Q}{T} \tag{7.1}$$

where the numerator is the thermal energy transferred in and/or out of the system and T is the temperature at which the energy is transferred across the system boundary [10]. For any given value of heat being transferred, a continual decrease in temperature will clearly only increase the change in entropy required to cool to lower temperature states. Thus, cooling of an object becomes increasingly difficult at successively lower temperatures.

7.1.2 Cryogenics Engineering from a Multi-disciplinary Perspective

Prior to the advent of mechanical cryocoolers in the mid-1970s, the cooling of objects to cryogenic temperatures was mostly performed using cryogenic fluids in either the liquid or solid phase. The fluid was used to acquire the sensible heat of the object (and/or system) in order to cool it down to the desired operational temperatures. For a given cooling application, the cryogenic fluid selected had either a saturation temperature at 101 kPa (1 atm) or a triple point that closely approximated to the desired operational temperature for the item of interest.

All laboratory test apparatus and spaceflight structures require mechanical assembly of individual parts. At the parts level, the mechanical properties of materials can change with variation in temperature. Structural integrity and strength of materials under mechanical loads at temperature must be considered to avoid plastic deformation or fracture of parts. At the assembly level, interfaces (and associated fastening techniques) need to accommodates material CTE (coefficient of thermal expansion) effects as well as potential loss of mechanical load when cooling down to temperature. Although conduction and radiation will be present, nominal thermal budgets are orders of magnitude less (i.e., mW scale) in cryogenic cooling systems than in room temperature cases. At cryogenic temperatures, interface conductances have a pronounced effect upon heat flows within a thermal circuit and are always modeled. In addition, when working with electrical wiring, improper selection of wire material can lead to undesired heat flows, and/or joule heating in the system. Finally, low-temperature physics has shown that phenomena in matter vary as an object gets closer to absolute zero. Multiple phenomena present at low temperature (e.g., superfluid He, superconductivity, magnetocaloric effect) that are not present at room temperature can be manipulated and integrated into temperature control systems.

It is clear from the above that the design and development of a cryogenic cooling system requires knowledge of multiple engineering disciplines, including: thermophysical properties of cryogenic fluids; mechanical behavior of cryogenic materials; thermal and electrical design at cryogenic temperatures; and low-temperature physical phenomena. Cryogenics engineering is therefore multi-disciplinary in nature and a useful skills developmental activity for aspiring systems engineers.

7.2 Materials at Cryogenic Temperatures

7.2.1 Cryogenic Fluids

Three primary techniques are used to achieve cryogenic temperatures in space-based systems:

i) Cryogenic materials (and their associated thermophysical properties)
ii) Specialized low temperature heat pump systems (i.e., Cryocoolers)
iii) Passive radiative heat rejection.

In recent decades cryocoolers have emerged as a standard technique for cooling of items on space-based platforms to cryogenic temperatures. Historically, cooling was performed using a combination of cryogenic materials and highly efficient radiator designs [11–14]. Specific to cryogenic materials, exposure of an item to a liquid-phase cryogen with low boiling temperature (at standard pressure) will provide cooling to the object. If an object initially at room temperature is placed in a bath of a liquid-phase cryogen (Figure 7.1) then the cryogen boils off, effectively cooling the object until it is in thermal equilibrium with the bath. Cryogenic fluids are mainly made from one or more of eight elements. As shown in Figure 7.2, the first four elements (He, Ne, Ar and Kr) of the noble gas family (family VIII) are all cryogens. They are also inert fluids. Hydrogen, nitrogen, oxygen and fluorine all form compounds that are in the gaseous state above 123 K at 101 kPa. These gas-phase compounds do not liquefy until cooled to their saturation temperature. With the exception of oxygen (O_2) and liquid fluorine (F_2), each of these cryogens is clear in the liquid phase. Both H_2 and Ne have been used for space-based cooling applications, whereas LH_2 and LO_2 have been used in propellant systems. For health and safety reasons, practitioners should always reference a fluid's safety data sheet (SDS) to learn hazards associated with handling it, and the appropriate safety precautions. Both liquid and gaseous oxygen are highly reactive with hydrocarbons [2]. Great care should therefore be taken when working with this element. Some materials, such as carbon steels, have been known to experience embrittlement when exposed to H_2 for extended periods [1,2]. Human exposure to pure fluorine (gas or liquid phase) should be avoided. N_2 is the most frequently used cryogen for ground testing. It is non-toxic, inexpensive and abundantly available (note: air is 78% nitrogen by composition). The most commonly used cryogen for space-based cryogenic cooling systems is He. Last but not least, to avoid

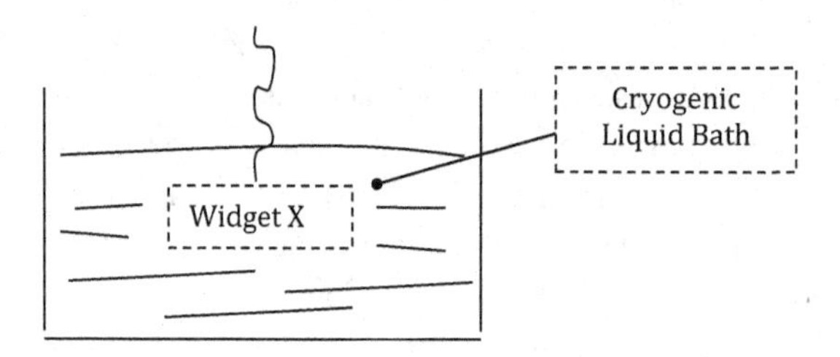

Figure 7.1 Cool-down of a warm object placed in a temperature bath

Periodic Table of Elements

I	II											III	IV	V	VI	VII	VIII
H																	He
Li	Be											B	C	N	O	F	Ne
Na	Mg											Al	Si	P	S	Cl	Ar
K	Ca	Sc	Ti	V	Cr	Mn	Fe	Co	Ni	Cu	Zn	Ga	Ge	As	Se	Br	Kr
Rb	Sr	Y	Zr	Nb	Mo	Tc	Ru	Rh	Pd	Ag	Cd	In	Sn	Sb	Te	I	Xe
Cs	Ba	Lu	Hf	Ta	W	Re	Os	Ir	Pt	Au	Hg	Tl	Pb	Bi	Po	At	Rn
Fr	Ra	Lr	Unq	Unp	Unh	Uns	Uno	Une									

Figure 7.2 Periodic table highlighting elements common to cryogenic fluids

oxygen deficiency, there should be proper ventilation when working with any cryogenic fluid in confined spaces.

When designing cryogenic systems to be cooled using liquid-phase cryogens, the saturation temperature at 101 kPa serves as the metric for selection of the cryogen to use. For solid cryogens, the triple-point temperature is the metric that serves as the basis for selection of the cryogen. Table 7.1 lists the saturation temperature (at 101 kPa) and triple-point temperature for the cryogens highlighted in the periodic table shown in Figure 7.1. Methane, a commonly used cryogen, has been added to the table. As shown in the table, the coldest saturation temperatures are observed in the helium isotopes, whereas CH_4 and Kr have saturation temperatures just under 123 K. The triple-point temperatures for most of the cryogens shown are slightly lower than the saturation temperatures. However, oxygen and fluorine have triple points that are significantly lower than their saturation temperature at 101 kPa. Triple points for helium are listed in the table as "n/a". These will be discussed in Section 7.3.1.

Table 7.1 Standard cryogenic fluids and their associated saturation temperatures

Cryogenic fluid	Saturation temperature at 101 kPa [K]	Triple-point temperature [K]
He^3	3.2	n/a
He^4	4.2	n/a
Hydrogen (H_2)	20.39	13.95
Neon	27.2	24.56
Nitrogen (N_2)	77.34	63.14
Fluorine (F_2)	85.2	53.48
Argon	87.4	83.8
Oxygen (O_2)	90.1	54.3
Methane (CH_4)	111.0	90.67
Krypton (Kr)	119.78	115.91

7.2.2 Thermal Expansion Effects

Objects exposed to initial and final temperature conditions spanning a large gradient will experience thermal expansion and/or contraction effects. In the case of cryogenic systems, mechanical assembly of structures is performed at room temperature. Since the beginning of the cryogenics temperature regime is 177 K lower than room temperature, thermal contraction of assembled parts can be expected to take place when cooling down to operational temperatures. The amount of linear thermal contraction occurring during cool-down can be predicted. For a solid, the coefficient of linear expansion is defined in equation form as

$$CTE = \frac{1}{\Delta T} \cdot \left(\frac{\Delta L}{L_{init}} \right) \tag{7.2}$$

where L_{init} is the original length of the item in question and ΔL is the change in length due to the change in temperature [1,2,5,9]. The amount of linear expansion occurring over a large temperature gradient is defined as

$$\frac{\Delta L}{L_{init}} = \int_{T_{ref}}^{T} CTE\, dT \tag{7.3}$$

where T_{ref} denotes a reference temperature [1,6]. Determination of the amount of contraction taking place within assembled parts and interfaces, and/or bolted joints, at the time of cool-down can be used to predict thermally induced stresses in both cases. The linear expansion (i.e., $\Delta L/L_{init}$) at 77 K for a few commonly used cryogenic materials is shown in Table 7.2. Of the materials listed in the table, Invar undergoes the least amount of temperature-based contraction, whereas PTFE experiences the greatest. The CTE observed in Al-6061 is comparable to that of stainless steel 304. Copper, Ti 6Al-4V and Nickel Steel Fe 2.25 Ni all have comparable CTE values when cooled to 77 K from room temperature. For a detailed overview of analytical relations used to determine thermal expansion and contraction within materials, see reference [16].

Table 7.2 Linear expansion for common cryogenic materials at 77 K†

Material	Linear expansion ($\Delta L/L_{init}$)
Al 6061	-1.8×10^{-5}
Ti 6Al-4V	-7.515×10^{-6}
Stainless Steel 304	-1.296×10^{-5}
Invar	-1.889×10^{-6}
Copper	-7.945×10^{-6}
Nickel Steel Fe 2.25 Ni	-8.813×10^{-6}
PTFE	-8.998×10^{-5}

† Values tabulated from NIST's material properties website correlations [15].
Reference temperature for values is 293 K.

7.2.3 Materials Selection and Mechanical Strength

Materials commonly used in cryogenic systems design may be either metals or non-metals. Metals can be either pure or alloys. Non-metals consist of ceramics, polymers and composites. Thermal conductivity as a function of temperature for some common pure metals and alloys is shown in Figure 7.3a, and that for commonly used polymers and composites in Figure 7.3b. As shown in the figures, OFHC and aluminum have the highest thermal conductivities relative to other metals at any given temperature. Ti 15-3-3-3 and Ti 6AL-4V have the lowest thermal conductivities. In the case of polymers and composites, the thermal conductivity for Vespel is fairly constant down to approximately 15 K. G10, nylon, PTFE, S-glass and polystyrene undergo a slight decrease in thermal conductivity with decreasing temperature. T300 has a larger reduction in thermal conductivity with decreasing temperature than that displayed by the other materials.

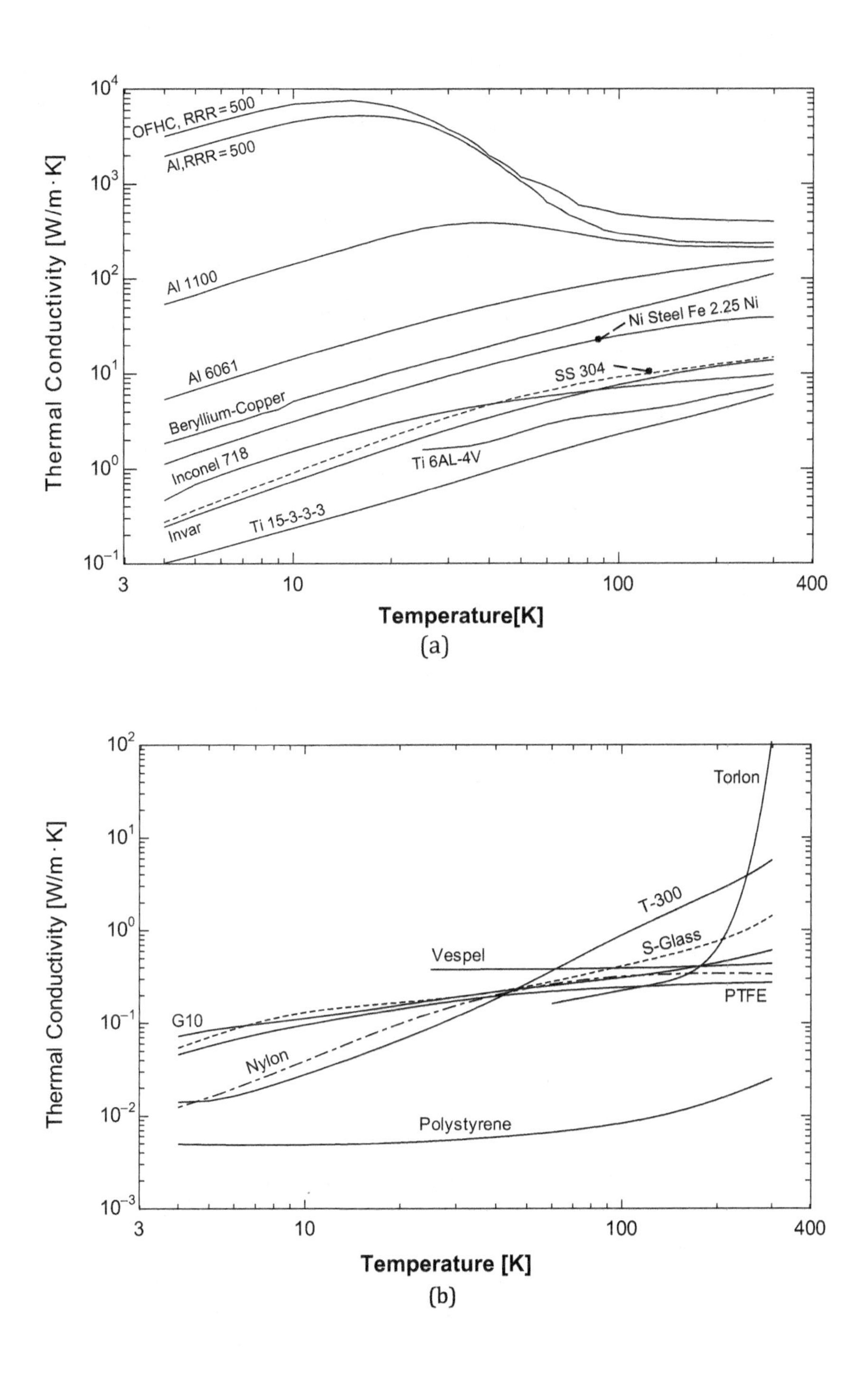

Figure 7.3 Thermal conductivity of materials used in cryogenic systems: a) metals and alloys; b) polymers and composites

Mechanical properties of a material will also change with temperature. When a mechanical assembly is at cryogenic temperatures its components must be able to withstand the mechanical loads distributed throughout the assembly while avoiding plastic deformation and fracture. This applies to both static and vibrational loading conditions experienced during ground-based assembly and test, launch and actual spaceflight. The yield strength (normal) is the basis for designating a material's strength, and in cryogenic applications it generally increases as the temperature of the material decreases. The amount by which the yield strength increases (relative to room temperature) is an inherent feature of the material itself. Ductile materials are favored and materials that become brittle when exposed to cryogenic temperatures avoided near or at operational temperature. The selection of materials must also take into account the intended thermal performance of any given part within the overall assembly. The sometimes competing effects associated with structural and thermal performance of a part are often considered in the design of cryo-radiator and cryogen storage tank support structures.

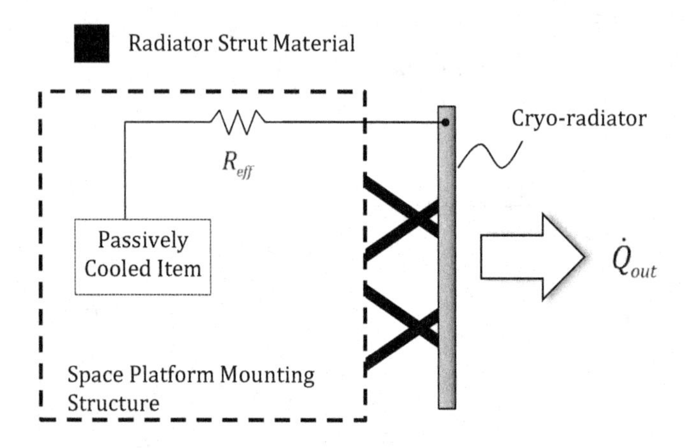

Figure 7.4 Cryo-radiator/truss structure attached to a space platform

Figure 7.4 shows a cryogenic radiator mounted to a space platform via a support structure (a truss). The dashed line represents the space platform structure. The radiator is thermally coupled to an inboard component that is being passively cooled. The radiator is conductively decoupled from the body of the spacecraft via low thermal-conductivity supports. If parasitic thermal loading of the radiator occurred, the operational temperature of the radiator would increase and the ability to obtain the inboard component's target temperature would be compromised. To avoid this, the FOV of the radiator's active surface is chosen to avoid environmental sources and thermal backloading. In addition, the struts must be able to support the weight of the radiator while also minimizing the conductive heat flow (heat leak) from the spacecraft. Ideally it will be made from a high-strength material that has low thermal conductivity. The strength-to-thermal-conductivity (STC) ratio measures the mechanical and thermal performance of candidate strut materials. In equation form this is defined as [1,2,5]

$$STC = \frac{Yield\ Strength}{Thermal\ Conductivity} = \frac{S_y}{k} \tag{7.4}$$

where S_y is the yield strength [MPa] and k is the thermal conductivity. These are based on the primary axis for the mechanical load and the heat flow being the same. Both values are taken at the same temperature. The larger this number, the greater the material strength per unit thermal conductivity. The yield strength can be calculated from the CTE and the material's E_{mod} (modulus of elasticity) for a known change in temperature using equation 7.5 [5]

$$S_y = E_{mod} \cdot CTE \cdot \Delta T \tag{7.5}$$

Table 7.3 lists various materials used as radiator and cryogen storage tank struts. The values listed are taken at 77 K (i.e., temperature of LN_2 at 101 kPa).

Table 7.3 Candidate strut materials for cryogenic applications at 77 K

Material	Yield strength (S_y) [MPa]	Thermal conductivity [W/m · K]	Type of material
Ti 6Al-4V	[1]1482	[2]3.459	Metal
Stainless Steel 304	[3]428.6‡	[2]7.9	Metal
Stainless Steel 316	[3]418.2‡	[2]7.321	Metal
T300 CFRP	[4]278‡	[5]0.5639	Fiber-reinforced plastic
G10	[6]67.4‡	[2]0.2798	Fiberglass epoxy laminate
Torlon	[7]374	[5]0.19	Polyamide-imide
S-Glass	[8]6,620†	[5]0.3354	Glass composite fibers

‡ Yield strength values have been determined as 0.2% of the modulus of elasticity.
† Yield strength values approximated using tensile strength-to-yield strength ratio of 1.25.
[1] Republished with permission from Nagai, K., Yuri, T., Ogata, T., Umezawa, O., Ishikawa, K., Nishimura, T., Mizoguchi, T., and Ito, Y., "Cryogenic Mechanical Properties of Ti 6Al-4V Alloys with Three Levels of Oxygen Content," *ISIJ International*, Vol. 31, No. 8 (1991), pp. 882–889, reference [17].
[2] Thermal conductivity value sourced from NIST's cryogenic material properties website [15].
[3] Modulus of Elasticity value sourced from NIST's cryogenic material properties website [15].
[4] Republished with permission of Springer Science from Weiss, W., "Low temperature properties of carbon fiber reinforced epoxide resins," *Nonmetallic Materials and Composites at Low Temperatures 2*, Edited by Hartwig, G. and Evans, D., 1982, pp. 293–309, reference [18]; permission conveyed through Copyright Clearance Center, Inc.
[5] Reprinted with permission of Elsevier from Tuttle, J., Canavan, E., and Jahromi, A., "Cryogenic Thermal Conductivity Measurements on Candidate Materials for Space Missions," *Cryogenics*, Vol. 88 (2017), pp. 36–43, reference [19].
[6] Reprinted with permission from Springer Nature from Kasen, M.B., MacDonald, G.R., Beekman, D.H., and Schramm, R.E., "Mechanical, Electrical, and Thermal Characterization of G-10CR and G-11CR Glass-Cloth/Epoxy Laminates Between Room Temperature and 4K," *Advances in Cryogenics*, Vol. 26 (1980), pp. 235–244, reference [20].
[7] Reprinted with permission of Elsevier from McKeen, L.W., *The Effect of Temperature and Other Factors on Plastics and Elastomers*, 3rd edn, 2014, reference [21].
[8] Republished with permission of AGY Corporation from Hartman, D., Greenwood, M.E., and Miller, D.M., *High Strength Glass Fibers*, 1996, Aiken, South Carolina, reference [22].

7.2.4 Emissivity of Metals

The optical properties of a material surface (emissivity, absorptivity and reflectivity) depend on the structural features of the surface and phenomena occurring immediately beneath the surface at the molecular scale. From an optical physics perspective, the emissivity of metals has been shown to be strongly coupled to the transport capability of electrons residing within the skin depth. The majority of analytical models predicting the emissivity of metal surfaces are based on Drude single electron (DSE) theory [23,24]. DSE theory assumes that free electrons near the surface of the material can migrate through the lattice in the presence of an electric field [25]. Their motion (and ultimately their transport) through the lattice is regulated by collision events with impurities and the lattice due to phonons. Several analytical models have been developed to predict emissivity based on DSE theory. One model commonly accepted as producing accurate values for emissivity of metals near room temperature is the Parker and Abbott relation [26,27]:

$$\varepsilon(T) = 0.7666(\sigma_{dc} \cdot T)^{0.5} - (\sigma_{dc} \cdot T)[0.309 - 0.889 \cdot \ln(\sigma_{dc} \cdot T)] - 0.015(\sigma_{dc} \cdot T)^{1.5} \quad (7.6)$$

The emissivity has functional dependence on the material's DC electrical conductivity and temperature. With regard to the cryogenic temperature regime, most models based on DSE theory (including the Parker and Abbott model) have demonstrated marginal success in accurately predicting emissivity values compared to experimental data. In an effort to improve accuracy of low-temperature predictive models based on DSE theory, Reuter and Sondheimer [28] developed an alternate model based on ASE anomalous skin effect (ASE) theory. In addition to collision events occurring within the material lattice, ASE theory takes into account spatial variation in the electric field present in the material [25]. Using ASE theory, Domoto et al. [29] define the total hemispherical emissivity as

$$\varepsilon(T) = 2.47 \cdot Vel_{Fermi}\left(\frac{3}{4} + \frac{\sqrt{3}}{\sigma_{dc}}\right) \cdot \left(\frac{\kappa}{\kappa + 1}\right) \quad (7.7)$$

where

$$\kappa = \frac{3}{2}\left[\frac{2}{3 \cdot Vel_{Fermi}}\left(\frac{3\pi m_{par}}{\mu_o e^2 N'_e}\right)^{1/2}\left(\frac{k_B T}{h}\right)\right]^{2/3} \text{ and } Vel_{Fermi} = \frac{Vel_{par}}{c}$$

In the equation, N'_e is the electron number density, k_B is Boltzmann's constant, h is Planck's constant, μ_o is the magnetic permeability of the material and m_{par} is the mass of the particle (electron). The variable e denotes the charge of a single electron. The fermi velocity is defined as the ratio of the average particle velocity (i.e, the velocity of the electrons in question) to the speed of light. The average particle velocity can be determined from equation 6.12. For additional information on ASE theory and its application to emissivity at low temperature, see Toscano and Cravalho [25] or Domoto et al. [29].

7.2.5 The Residual Resistivity Ratio

The correspondence between a material's electrical resistivity and its thermal conductivity (Wiedemann–Franz law) was covered in Chapter 1 (equation 1.3). Recall that thermal conductivity has contributions from both the material's lattice vibrations (phonons) and the migration of electrons. In metals at low temperatures, the magnitude of the phonon contribution significantly reduces and the thermal conductivity is dominated by that of electron migration. However, electron migration within a metal can be impeded by impurities present in the material. The purer a material is, the greater the potential for free electron migration and ultimately a higher thermal conductivity. For any given metal, different grades of the material will have different levels of impurity. The residual resistance ratio (RRR) gauges the purity of different grades of materials. The RRR value is the ratio of a material's electrical resistance at room temperature to its electrical resistance at helium temperatures (4.2 K) [1,5]. In equation form this ratio is written as

$$RRR = \frac{R_e|_{290\mathrm{K}}}{R_e|_{4.2\mathrm{K}}} \tag{7.8}$$

The larger the RRR value of a particular grade of material, the better its thermal conductivity at low temperatures.

7.3 Transfer Processes at Low Temperature

The rules pertaining to transfer processes occurring at or near room temperature were covered in detail in Section 5.2. However, these rules can be altered at low temperatures due to changes in material-based phenomena at the molecular level. The following subsections overview three phenomena specific to low temperatures that are relevant to fluids, electrical conductors and magnetic phenomena at cryogenic temperatures.

7.3.1 Superfluid Helium

Helium comes in two different isotopes (He^3 and He^4). While He^3 is rare, He^4 is easily obtained and commonly used. References to helium hereafter will be assumed to refer to the He^4 isotope unless otherwise specified. Identifying some features of standard heat transfer fluids is necessary for understanding the uniqueness of helium as a fluid. Fluids are commonly categorized as either Newtonian or non-Newtownian (e.g., plastic, pseudo-plastic, dilatant). The majority of fluids used for heat transfer applications are Newtonian. From a fluid mechanics standpoint, when a Newtonian fluid is exposed to shear, the constant of proportionality between the shear and the strain rate is the viscosity [30]. From a thermodynamics standpoint, standard Newtonian fluids can freeze when cooled sufficiently. As shown in the P–T (pressure–temperature) diagram in Figure 7.5, they can exist in three different phases (solid, liquid and vapor) that converge at the triple point. However, the phase diagram for He^4 (Figure 7.6) is quite different. Helium does not freeze until it is compressed to 2,525 kPa (i.e., 25 atm).

Thus, there is no triple point for helium. At each pressure level below 2,525 kPa there is a corresponding temperature at which the transition into what is known as a superfluid state occurs. This transition point, referred to as the lambda point, occurs at a temperature of 2.172 K at 101 kPa. While the diagram shown is representative of the He^4 isotope, a similar phenomenon occurs in the He^3 isotope. In addition, superfluid helium (otherwise known as He-II) has unique thermofluidic transport properties.

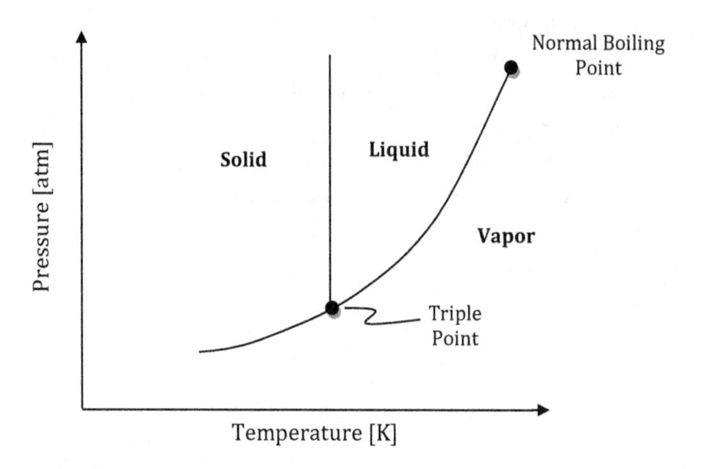

Figure 7.5 Phase diagram for a standard pure fluid

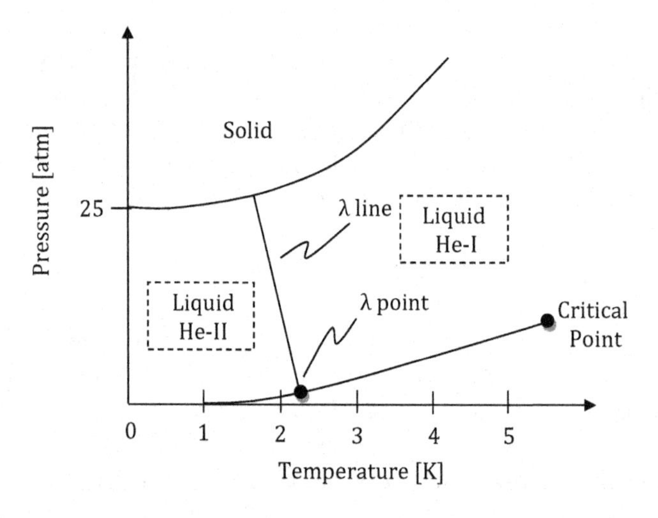

Figure 7.6 He^4 phase diagram
Source: Barron, R., *Cryogenic Systems*, 2nd edn, 1985, reference [1].

Superfluid helium has an approximate viscosity of zero, so fluid flow systems that use it are considered to be ideal flows. Its thermal conductivity is on the order of 90,000 W/m · K [1,4], whereas for He-I (i.e., non-superfluid helium) it is approximately 0.024 W/m · K. The inability of helium to freeze at 101 kPa and its unique thermofluidic properties in the superfluid state beg the question of why this phenomenon occurs with helium and not other basic Newtonian fluids. The answer can be found in the physics of the fluid.

Borrowing from constructs based on the kinetic theory of gases, the classical physics interpretation of matter and its motion is that the temperature of a system of molecules corresponds to the average KE of the system. The temperature is often considered a measure of the KE. This interpretation is applicable to both individual molecules and molecular systems (e.g., fluids). The total kinetic energy (or energy) of a molecule can be captured in equation form as

$$KE_{tot} = KE_{Translation} + KE_{Rotation} + KE_{Vibration} + KE_{E\text{-}M,Force} \tag{7.9}$$

where the total is the sum due to translation, rotation, vibration and electromagnetic forces that may be present. Above absolute zero, matter has inherent thermal energy that manifests itself as random motion of atoms and dynamic forces between particles associated with them [9]. Thus, if a molecule cools down to absolute zero it is motionless and in a state of zero entropy, and should lose all of its KE. However, as evidenced by the behavior of helium when approaching 0 K, this does not happen at 101 kPa. An alternate framework to explain the behavior of helium when close to absolute zero is provided by quantum theory.

According to the fundamental principles of quantum physics, particles are never at rest; they contain a minimum energy level that they do not fall below, regardless of temperature. The minimum energy level at 0 K is known as zero-point energy (ZPE). Helium does not freeze at 101 kPa because its minimum energy level (i.e., its ZPE) is large enough to prevent it from condensing unless highly compressed [3]. Due to this quantum-based behavior, helium is known as a "quantum" fluid. Other quantum fluids include H_2 and Ne. However, the minimum energy level for these fluids is not substantially large, such that freezing is prevented at 101 kPa.

7.3.2 Superconductivity

An alternative to electrical conductors operating in the Ohmic state on wiring applications is superconducting elements. The transfer process inherent to a material element operating in the superconducting state is superconductivity. Superconductivity is a phenomenon occurring in materials below a transitional temperature in which several of the electrons within the material pair up to form what are known as Cooper pairs. These pairs exist at low energy states (much lower than those of unpaired electrons) [1,4,31]. Unlike the unpaired electrons, Cooper-paired electrons are not carriers of thermal energy. When in the superconducting state the only heat carriers within the material are the remaining unpaired electrons.

In the superconducting state the material in question does not demonstrate electrical resistance according to Ohm's Law. The electrical resistance of the material reduces to values on the order of $10^{-6}\,\Omega$ [31]. This significantly reduces any joule heating that may be present in the element during current flow. The use of superconducting materials is highly favored when operating at single- and low double-digit temperatures (Kelvin scale). Three primary metrics (and/or characteristics) are associated with superconductivity phenomena in materials: the critical temperature (T_C), the critical magnetic field (H_C) and the critical current (I_C). The critical temperature is the temperature value below which superconductivity occurs. The critical magnetic field is the value for application of a magnetic field in which the material loses its superconducting capability. The critical current is the maximum current feasible while still maintaining the material in the superconducting state. Upon losing the superconducting state, the material returns to the Ohmic state (i.e., normal electrical resistance). This occurrence is known as "quenching."

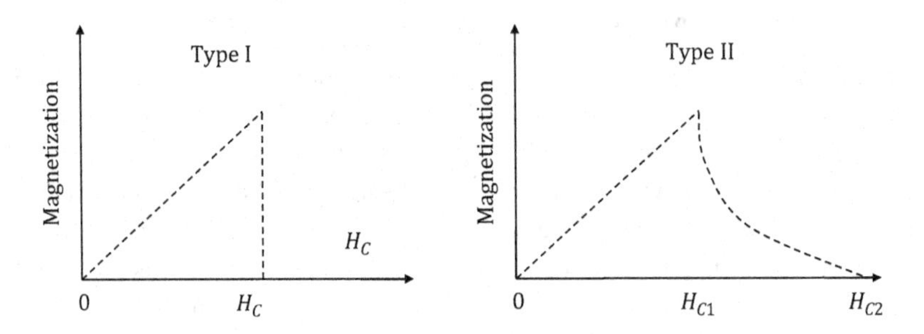

Figure 7.7 Type I and type II superconductor critical field phenomena
Source: Flynn, T.M., *Cryogenic Engineering*, 2nd edn, 2005, reference [2]; permission conveyed through Copyright Clearance Center.

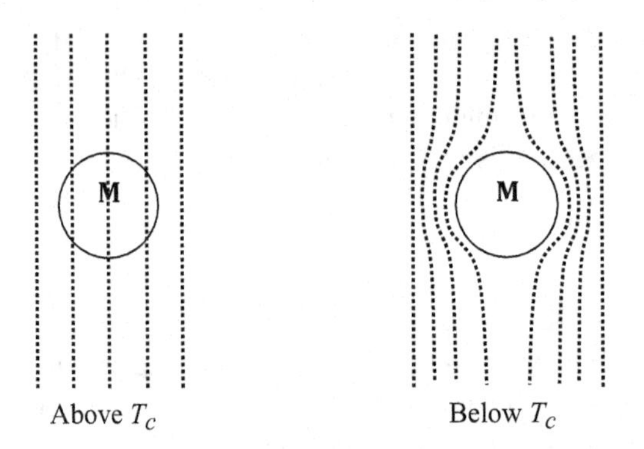

Figure 7.8 Magnetic field line cases for superconducting and non-superconducting sphere in magnetic field
Source: Barron, R., *Cryogenic Systems*, 2nd edn, 1985, reference [1].

There are two types of superconductors. Type I superconductors experience a single H_C whereby an instantaneous transition back to the normal (Ohmic) resistance state occurs. Type II superconductors undergo a gradual transition to the non-superconducting state between two H_C values (i.e., an upper and a lower bound).

When in the superconducting state magnetic fields that have a magnitude less than the H_C value for the material are repelled (Figure 7.8). This phenomenon is known as the Meissner Effect [32]. The Meissner Effect is often illustrated by table-top magnetism demonstrations involving magnetic levitation. Figure 7.9 is a photo of a small magnetic

BSCCO (Bismuth strontium calcium copper oxide) cube suspended above a superconducting YBCO (yttrium barium copper oxide) magnetic disk, both of which are in a petri dish filled with LN_2.

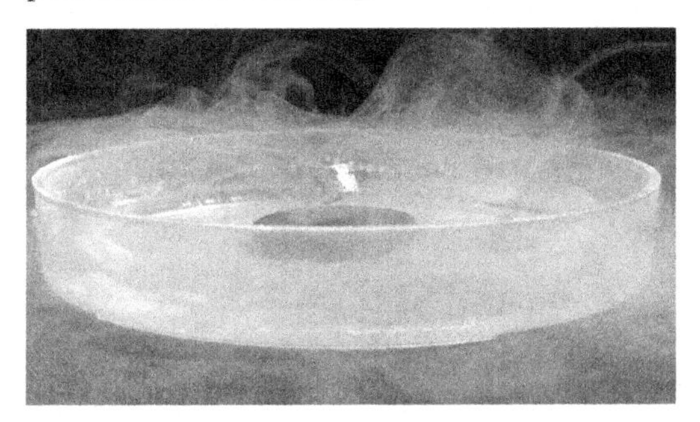

Figure 7.9 Superconductivity petri-dish demonstration using a YBCO cube repelled above a disk of BSCCO in the superconducting stage
Source: Photo courtesy of Dr. Xiaoyi Li, NASA Goddard Space Flight Center.

Since the T_C for both of these materials is greater than 77 K, the Meissner Effect is present in both. The repelling of the materials' magnetic fields results in the BSSCO magnet hovering above the YBCO disk. Because the population of thermal energy carrying electrons is significantly reduced in superconductors, the thermal conductivity due to electron migration reduces and approximates to that of an insulator. Superconducting transition temperatures for some common materials are shown in Table 7.4.

Table 7.4 T_C values for common materials

Material	Critical temperature [K]
Aluminum	1.19
Cadmium	0.55
Mercury	4.15
Niobium	9.2
Zinc	0.85
Lead	7.19

Source: Roberts, B.W., *Superconductive Materials and Some of Their Properties*, 1966, reference [33].

7.3.3 The Magnetocaloric Effect and ADRs

Magnetism in a material is based on phenomena occurring at the molecular scale. Figure 7.10 illustrates an atom with its nucleus and electrons. Each of the electrons has a magnetic moment associated with it. Unpaired electrons in an atom have magnetic moments which can foster magnetism at the macro scale. There are three types of magnetic materials: ferromagnetic, paramagnetic and diamagnetic. In ferromagnetic materials atomic magnetic dipoles align in the presence of a weak magnetic field and remain aligned after the field is removed. In a paramagnetic material, the alignment of the atomic magnetic dipoles is susceptible to magnetic fields. However, the magnetic dipoles become mis-aligned when the field is not present [9]. Diamagnetic materials have no permanent magnetic dipole moment either in or outside the presence of a field.

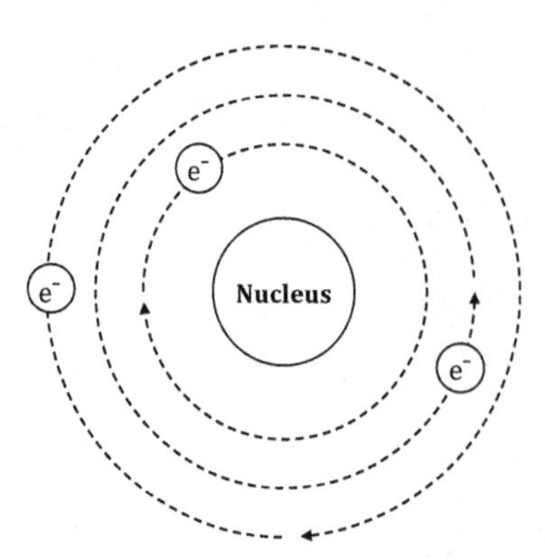

Figure 7.10 Electron orbits around a nucleus

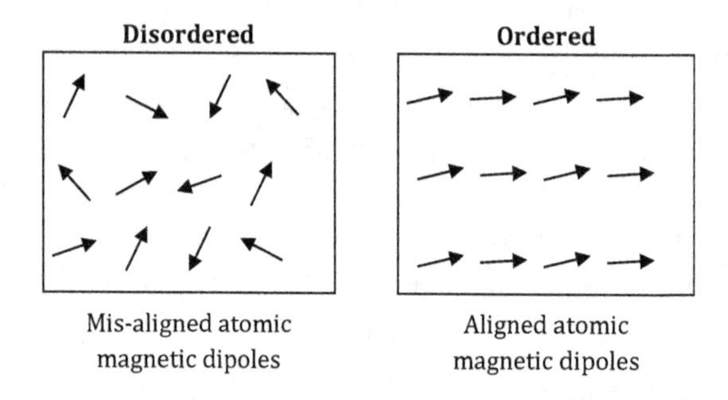

Figure 7.11 Aligned vs. misaligned moments
Source: Reproduced with permission of Saunders Publishing from Serway, R., *Physics for Scientists and Engineers*, 1986, reference [9]; permission conveyed through Copyright Clearance Center, Inc.

As shown in Figure 7.11, when a magnetic field is applied to a magnetic material (ferromagnetic and/or paramagnetic) the material's magnetic dipoles become ordered. Since entropy is a measure of the disorder in a system, application of the magnetic field while providing cooling will reduce the entropy. The coupling between the order and the application of the magnetic field is termed the magnetic disorder entropy.

Figure 7.12 highlights the magnetic disorder entropy effect at low temperature with and without a magnetic field and a thermal coupling to an external temperature sink applied. The system (which includes the paramagnetic material) is precooled via the ambient cold sink to a starting temperature T_1. The magnetic field is at intensity level H_0 (this is typically zero field intensity). Afterwards, a heightened magnetic field (H_1) is applied to perform isothermal magnetization. This is shown as steps $1 \rightarrow 2$ in Figure 7.12. The system becomes ordered while the heat generated in the magnet is absorbed by the cold sink to which the system is coupled.

In steps $2 \rightarrow 3$ the paramagnetic material is isolated from the cold sink and the field intensity is set back to its initial value of H_0. Because the system was isolated from the ambient cold sink, it became adiabatic. The ordering created by the magnetization

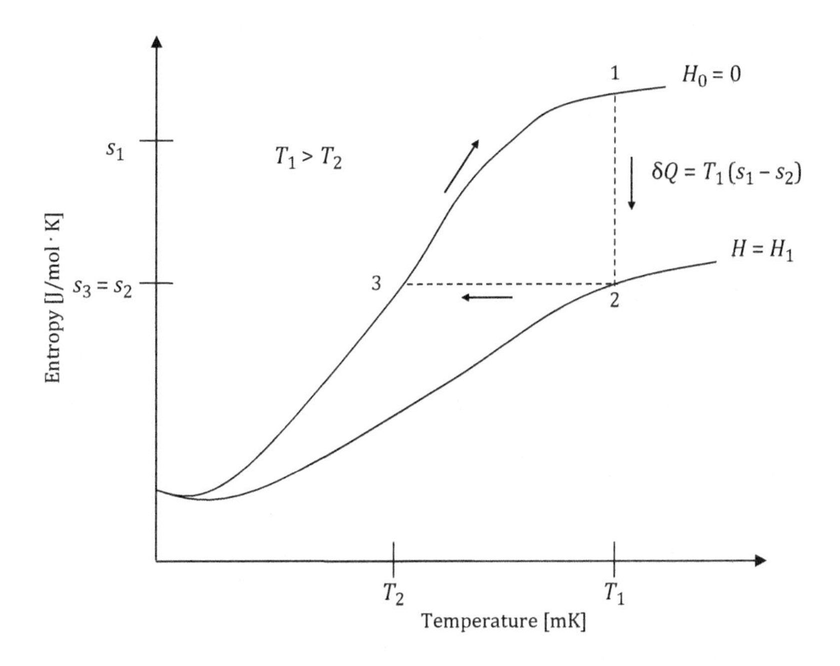

Figure 7.12 Magnetization and temperature decrease for a paramagnetic material
Source: Bejan, A., *Advanced Engineering Thermodynamics*, 1988, reference [34]; permission conveyed through Copyright Clearance Center, Inc.

remains constant (i.e., isentropic) during steps 2 → 3. As a result, the temperature must decrease. The final step (3 → 1) increases the material back to its initial disordered state via heat acquisition at a magnetic field intensity H_0. At this point, the thermodynamic cycle has been completed. Thermal energy acquired at state point 3 has effectively been lifted to temperature T_1 at state point 1. The cooling effect fostered by the cycle shown in Figure 7.12 is known as the magnetocaloric effect. This effect occurs at liquid helium temperatures and provides further cooling to sub-Kelvin levels. It was first observed by Langevin [35] and later by Weiss and Piccard [36]. The application of the magnetocaloric effect to cooling objects was first proposed by Debye [37] and later demonstrated by Giaque [38] in a laboratory setting.

Were this phenomenon to be captured in the context of a cycle, the magnetocaloric process could be repeated multiple times to create a low-temperature magnetic heat pump effect. Adiabatic demagnetization refrigeration (ADR) systems use the magnetocaloric effect in the context of a cycle to cool items down to sub-Kelvin temperatures. Because the magnetocaloric effect occurs at helium temperatures, ADR technology is primarily used for cooling of low temperature detectors (LTDs). Magnetocaloric materials that allow for sizeable magnetic disorder entropy effects at higher temperatures than those quoted here have yet to be discovered and researched. ADR technology is increasingly used within the astrophysics community. The most recent spaceflight mission to use an ADR system was Astro-H [39]. It launched in spring of 2016 and demonstrated continuous cooling down to 30 mK in space. For additional information on ADR technology, see Tishin and Spichkin [40].

7.4 Design Features of Cryogenic Systems

This section reviews common design techniques to avoid thermal loading of cryogenically cooled items and practices that avoid the creation of large limiting resistances along heat flow paths used to remove thermal energy from cryogenically cooled items.

7.4.1 Temperature Staging

Temperature staging is commonly used to get to low temperatures. Segments of an assembly located along the heat leak path leading to the cryogenic volume of interest are operated at gradually descending temperatures to control the amount of thermal energy transferred to the lowest stage (and ultimately the cryogenically cooled item). Figure 7.13 shows a composite structure with a large temperature gradient from end to end. The composite conductor is made of four different materials (stainless steel, titanium, G10 and OFHC). The conductive path has a cross-sectional area of 0.2 m × 0.2 m and a fixed temperature of 290 K on the warm end. Each of the material sections in the composite member has a length of 0.25 m. In this example we will establish three different temperature stages spanning a temperature gradient of 210 K to 65 K. Given that the warm end of the composite member is at a constant temperature of 290 K, 159.3 W of thermal energy must be removed from stage 1 in order to achieve a temperature of 210 K. By removing an additional 64.8 W at the stage 2 location, a temperature of 130 K can be achieved. Finally, by removing 3.17 W at the stage 3 location, a temperature of 65 K can be attained. The remaining thermal energy to be managed in the OFHC segment would be a function of the operational temperature at T_{Low}. Without temperature staging between the 290 K endpoint and the 65 K stage, the total heat removal required to get down to 65 K would take place at the lowest temperature value (65 K). Techniques for removing hundreds of watts of thermal energy at 65 K are limited to cryogenic fluids which are not always practical for use in every situation. With 1-D conduction-based temperature staging, the effective heat load transferred into the OFHC was significantly reduced. Cryogenic systems (space-based and terrestrial) that use temperature staging often include both conduction- and radiation-based heat-removal techniques enabled by active cooling systems such as cryocoolers.

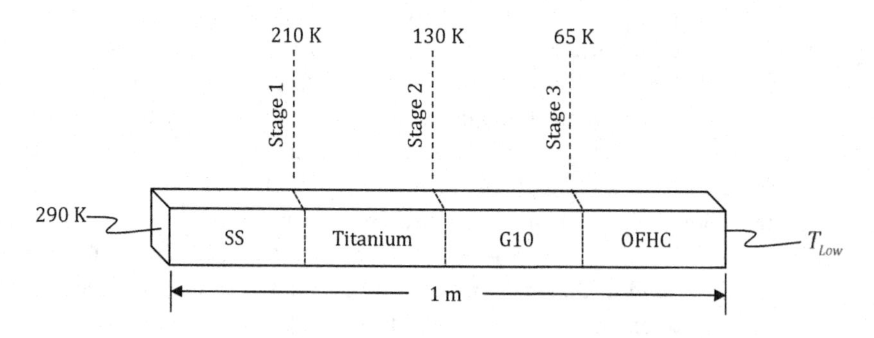

Figure 7.13 Staged temperature control of a conductive heat flow path

7.4.2 Wiring

Electrical wiring is used in spaceflight systems to supply power to components such as mechanisms and heaters, as well as for read-out of onboard sensors (e.g., temperature, pressure and flow). The system's cryothermal energy budget must take account of thermal energy conduction through wiring, as well as thermal energy generation. Wire harnesses (i.e., bundles of wire pairs) often span temperature gradients on the order of a couple of hundred degrees. Figure 7.14 illustrates a cryostat configuration. Assuming the test volume is at uniform temperature, the wall of the innermost portion of the cryostat has a temperature of 75 K. Heat from the temperature gradient at the endpoints flows through wiring that physically connects the 300 K segment to the 75 K segment. Joule heating may also be present in the wires. In Section 5.5, joule heating was defined in equation 5.54 as a thermal energy input resulting from current flow present in a conductor material when operating in the Ohmic state. An alternate form of this equation is

$$\dot{Q}_{Joule} = I^2 R_e'' \cdot \left(\frac{L_{wire}}{A_{wire}}\right) \tag{7.10}$$

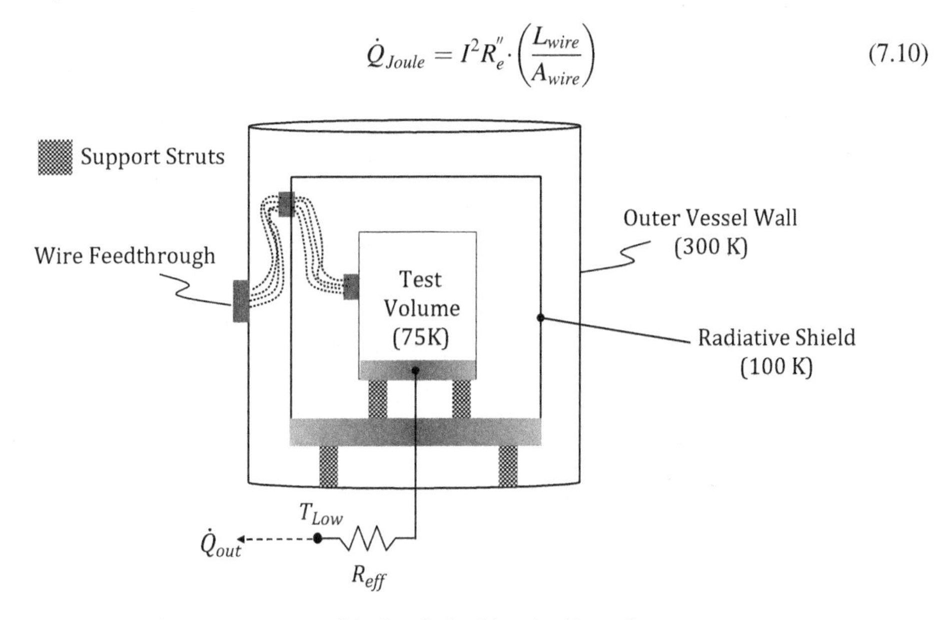

Figure 7.14 Laboratory cryostat with electrical wiring feedthroughs

where the thermal energy dissipation is dependent upon the magnitude of the current flow (I), the resistivity of the conductor material (R_e''), the cross-sectional area of the wire (A_{wire}) and the length of the wire (L_{wire}). Since there is a variation in temperature spanning the length of the wire and the thermal conductivity is a function of temperature, the conduction-based heat leak along the wire is calculated using the Fourier–Biot equation [41,42]:

$$\dot{Q}_{wire} = \frac{A_{wire}}{L_{wire}} \int_{T_C}^{T_H} k(T)dT \tag{7.11}$$

The variables T_H and T_C are the values for the temperature at the hot and cold end of the wire respectively. When calculating conductivity integrals at cryogenic temperatures it is standard practice to use 4 K as the lower limit for integration, therefore the integrand in equation 7.10 may be recast as

$$\dot{Q}_{wire} = \frac{A_{wire}}{L_{wire}} \left[\int_{4\text{K}}^{T_H} k(T)dT - \int_{4\text{K}}^{T_C} k(T)dT \right] \tag{7.12}$$

Electrical resistivity, size and thermal conductivity must be considered in the choice of wiring. Common materials used for cryogenic wiring applications include brass, copper, nichrome, phosphor bronze, platinum, tungsten and stainless steel. The amount of thermal energy actually carried to the lower temperature segments through electrical wiring can be mitigated by heat sinking (thermal anchoring), in which length segments of the wiring are placed in intimate contact with temperature stages of the cryogenic architecture before reaching the endpoint connected to the low-temperature zone of interest. For a detailed overview of thermal anchoring techniques, see Ekin [5] or Barron [1].

7.4.3 Interfaces and Bolted Joints

Heat flows in cryogenic systems can often be on the order of milliwatts (or microwatts at extremely low temperatures). Good thermal conductance across an interface joint enables efficient removal of thermal energy along designated heat flow paths within a cryogenic system. However, the potential negative impacts on thermal performance of poor conductances across an interface joint are more severe in cryogenic systems due to the low magnitude of the heat flows. The slightest deterrent to heat flow could mean the desired operational temperature cannot be reached. Effective pressure at the interface, as well as low-temperature effects on TIMs (when used) are key to the design of interface joints for operation at cryogenic temperatures. TIMs, gaskets and/or o-ring materials should be ductile at the operational temperature of the interface. The only TIMs from the list in Chapter 2 that are viable for use at cryogenic temperatures are indium, Stycast, Apiezon N and GraFoil. In addition, at helium temperatures and below, gold-plated bare joints rather than TIMs are commonly used. Thermal performance of an interface joint is highly dependent upon the quality of the joint created during fabrication and assembly. The most accurate values for interface joint thermal performance are obtained via post assembly conductance measurements that capture workmanship features. For an overview of thermal conductance data as a function of low temperature and applied forces, see Salerno and Kittel [43] or Ekin [5].

7.4.4 Thermal Design Margins

Two key metrics often used to assess whether a TCS can meet performance requirements are the operational temperatures for select components and the overall thermal budget to be managed. In the design phase of a TCS, if the operational temperatures are at or near 300 K, the performance margin is typically accounted for using the high/low temperature limits (i.e., yellow and red limits). The temperature difference between these limits and the nominal temperature for the item in question is the margin.

During testing the actual temperature of the component must be shown to fall within the margin limits in a flight-like thermal environment. When the margin is calculated, additional heat load is added to the best estimate of the thermal budget to account for errant heat loads that may not have been identified during the design phase. The comprehensive cooling system must be able to manage the sum of the thermal load (including the thermal margin) while cooling to and operating at desired temperature levels. In the early stages of the design of a cryogenic cooling system, 100% margin is applied to the best estimate of the cryothermal budget. Since milliwatt scale heat flows (at temperature) are common in cryogenic systems, early identification of errant heat leak paths prior to actual assembly and testing at low temperatures can be a challenge. The 100% margin rule accommodates these heat flows and insures they can be thermally managed by the cooling system/technique being implemented. Throughout the engineering process, as the design becomes more refined the magnitude of the margin will reduce. A minimum margin level of 30% should be reached prior to launch of the spaceflight system.

7.5 Standard Methods for Cool-Down

Cool-down processes typically use either a cryogenic material or a mechanical cryocooler to reach the desired low-temperature levels. This section reviews design considerations associated with the use of stored cryogens, as well as analysis techniques and performance features for some common spaceflight cryocooler cycles.

7.5.1 Stored Cryogen Systems

When temperature is controlled by stored cryogens, the fluid used may be in either the liquid or solid phase. This applies to both terrestrial and spaceflight systems dedicated to cooling optics, electronics and sensors. The decision whether or not to use a cryogen in liquid or solid phase will be driven by the desired operational temperature for the system being cooled. For cooling applications with liquid-phase stored cryogens, a liquid saturation temperature of 101 kPa is used as the standard reference for the operational temperature of the item to be cooled. For cooling applications with solid-phase stored cryogens the triple point of the fluid is the upper limit for the operational temperature of the item.

Stored cryogens used for cooling purposes are finite in volume. Heat leak into the stored cryogen will promote its phase change and eventually lead to full transition of the stored cryogen into the vapor phase. The duration over which the stored cryogen can maintain its storage phase (either liquid or solid) used for cooling purposes is known as its hold time. In space applications the hold time determines the period over which scientific data may be collected. Thus, the practitioner will want to maintain the cryogen in its storage phase for as long as possible. Heat leak into the storage volume may be the result of structural conductance, wiring conductance, joule heating, magnetic fields or radiative thermal loads. The total cryothermal energy budget must allow for the system's heat leak contributions. Figure 7.15 shows a standing dewar, used for ground-based laboratory testing, which has multiple potential heat leaks. There is a thermal conduction path from the container's outer vessel wall to the dewar inner wall. Assuming the outer wall surfaces are in equilibrium with room temperature, low-conductivity materials along the heat flow path leading to the inner vessel are desirable. The largest surface area adjacent to the inner vessel wall is the insulation space. Insulations typically opted for include foam or a simple vacuum. The wiring shown also may serve as a heat input. Finally, joule heating may be present but can be avoided by using low-resistivity conductors, such as phosphor bronze, for the wiring. Optimized wire sizes as a function of the wire material can be found in Barron [1] and Ekin [5].

Suppose that LNe is to be stored in the ground-based dewar. Before the dewar can be filled with liquid-phase cryogen the system must be cooled to a temperature that will not cause rapid boil-off of the liquid in either the transfer line or the dewar. The saturation temperature of Neon is 27.2 K at 101 kPa. Prior to filling, both the dewar and transfer line itself are initially at equilibrium with ambient (i.e., 298 K). Thus, the transfer line and the dewar are superheated relative to the LNe temperature. On initial transfer of LNe to the dewar, boil-off of LNe will occur in the transfer line. In order to prevent pressurization of the dewar a vent path must be provided to release the superheated gas from the inner vessel of the dewar to ambient. Once the sensible heat of the transfer line is removed, the LNe will undergo boil-off inside the inner vessel. After the inner vessel has cooled down sufficiently to prevent rapid boil-off of the saturated LNe at 1 atm, pooling of the liquid will begin inside the inner vessel. It is expected that once filled, the dewar will remain charged with liquid-phase neon for a couple of days, during which time testing will take place. As the temperature gradient is approximately 272.8 K (outside to inside), even the best insulated systems will have some degree of heat leak. Thus, when the system is at temperature (i.e., $\leq$123 K) the temperature gradient relative to ambient is negative. How can hold time for a specific fluid charge volume be determined, given knowledge of the heat leak?

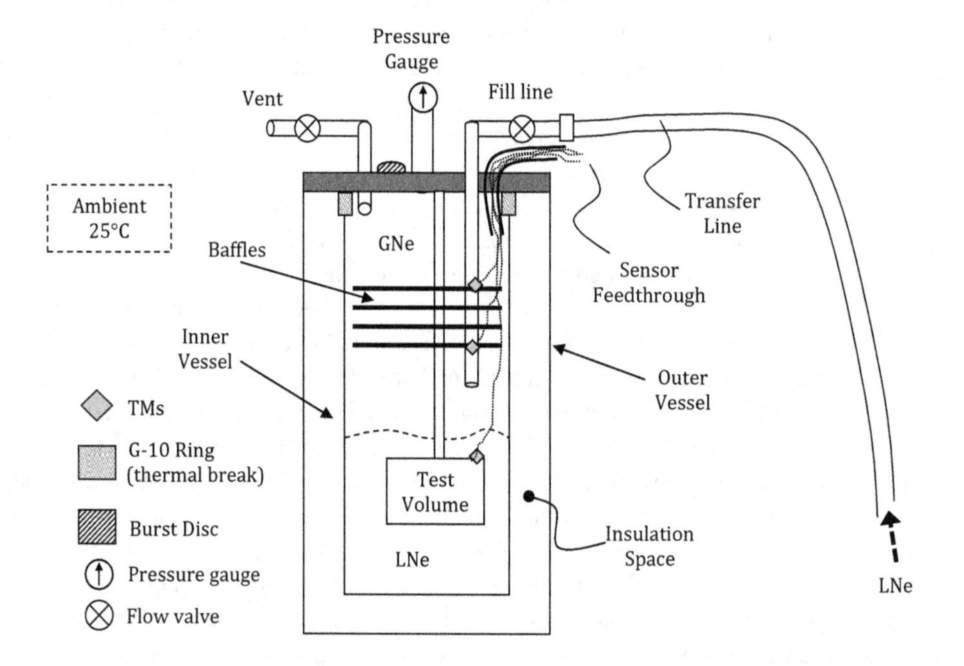

Figure 7.15 Ground-based standing dewar with vacuum gap space

Example 1 Cryogen Hold Time as a Function of Charge Volume and Heat Leak A test dewar (Figure 7.15) that is filled with 20 liters of LNe at 1 atm. The desired hold time for the test volume at temperature is 4 months. If the effective heat leak through the test dewar is 900 mW, can the desired 4-month hold time be achieved?

Solution

For the fluid in the state conditions defined in the problem, the density and enthalpy of vaporization are ρ_{LNe} = 1,207 kg/m^3 and h_{fg} = 85.75 kJ/kg respectively. Also, we know that 1 liter = 0.001 m^3. Thus, the volume of the liquid charge is Vol = 20 liters = 0.02 m^3. Since one day equals 86,400 seconds, 4 months equates to 10,368,000 seconds. Using the equation for latent heat of vaporization, the hold time can be solved for given the charge volume, saturation conditions and magnitude of the heat leak quoted.

$$\dot{Q}_{leak} = \frac{m \cdot h_{fg}}{\Delta t}$$

Solving for Δt we get

$$\Delta t = \frac{m \cdot h_{fg}}{\dot{Q}_{leak}}$$

$$\Delta t = \frac{\rho_{LNe} \cdot Vol \cdot h_{fg}}{\dot{Q}_{leak}} \tag{7.13}$$

Substituting in values for each of the variables in the RHS of equation 7.13

$$\Delta t = \frac{(1,207\ \mathrm{kg/m^3}) \cdot (0.02\ \mathrm{m^3}) \cdot (85.75\ \mathrm{kJ/kg})}{(900\ \mathrm{mW})}$$

Solving for Δt gives

Δt = 2,300,005.55 seconds
= 638.9 hours
= 26.6 days
= 0.89 months based on a 30 day/month estimate

What would be the hold time for the heat leak if it could be decreased to 150 mW? Repeating the calculation for hold time will give a duration of

Δt = 13,800,033.33 seconds
= 3,833.34 hours
= 159.7 days
= 5.32 months based on a 30 day/month estimate

Standard space-based science missions seek to perform at least 6 months of measurements. This suggests that for a spaceflight system flying a stored cryogen the dewar should have a heat leak allowing a minimum hold time of 6 months. A heat leak of 150 mW is very low for the type of ground-based dewar construction shown in

Figure 7.15. A more realistic estimate would be on the order of a couple of watts. Nonetheless, for the combination of charge level and heat leak used in the example, the hold time was lengthened significantly with the reduction in heat leak relative to the initial case with a heat leak of 900 mW. Were the heat leak into the system actually 900 mW, the desired 4-month hold time could be achieved by overcompensating on the charge volume to accommodate for the boil-off expected. Overcompensating on the charge level by adding more cryogen is fairly straightforward. To find the amount by which the charge volume would need to be increased, the relation developed for hold time would need to be solved for the volume required.

$$Vol = \frac{\Delta t \cdot \dot{Q}}{\rho_{LNe} \cdot h_{fg}} \tag{7.14}$$

To date numerous spaceflight systems have flown stored cryogens for purposes of optics, electronics and detector cooling. Table 7.5 lists notable stored cryogen missions

Table 7.5 Stored cryogen spaceflight missions

Mission	Cryogen/s	Volume (liters)	Launch year	Lifetime (months)
SESP-72 (Space Experiments Support Program) test A [44,45]	Solid CO_2	–	1972	7
SESP-72 (Space Experiments Support Program) test B [44,45]	Solid CO_2	–	1972	8
NIMBUS 6/LRIR (Limb Radiance Inversion Radiometer) [46]	Liquid CH_4/Solid NH_3	–	1975	7
NIMBUS 7/LRIR (Limb Radiance Inversion Radiometer) [46]	Liquid CH_4/Solid NH_3	–	1978	12
HEAO-B (High Energy Astrophysics Observatory B) [44]	Liquid CH_4/Solid NH_3	–	1978	11
HEAO-C (High Energy Astrophysics Observatory C) [44]	Liquid CH_4/Solid NH_3	–	1979	8
IRAS (Infrared Astronomy Satellite) [47,48]	He-II	450	1983	10
COBE (Cosmic Background Explorer) [49]	He-II	650	1989	10.5
SHOOT (Superfluid Helium On Orbit Transfer)† [50]	He-II	207	1991	0.5
ISO (Infrared Space Observatory) [51,52]	He-II	2,300	1995	30
SPIRIT III (Spatial Infrared Imaging Telescope III) [53]	Solid H_2	944	1996	10
NICMOS (Near Infrared Camera and Multi-object Spectrometer) [54]	Solid N_2	120	1997	24
Spitzer Space Telescope [55]	He-II	360	2003	66
GP-B (Gravity Probe B) [56]	He-II	2,441	2004	17
WISE (Wide-Field Infrared Survey Explorer) [57–59]	Liquid H_2/Solid H_2	32/120	2009	8
HERSHEL Space Telescope [60,61]	He-II	2,300	2009	47

† Shuttle payload bay experiment that terminated prior to full boil-off of the stored cryogen.

that have flown since the 1970s with the cryogen used, the charge volume at the time of launch, the year the mission launched and the lifetime of the stored cryogen during spaceflight. He-II, hydrogen (liquid and solid), solid N_2, liquid CH_4, solid CO_2 and solid NH_3 have been the fluids of choice for almost all stored cryogen spaceflight missions to date. Since each mission had a slightly different dewar design, heat leak into the stored cryogen tank varied. Nonetheless, as the phase change conditions, initial charge volume and lifetime in flight are known, the heat leak into the stored cryogen over the lifetime of the mission can be determined. As the table shows, SPITZER had the longest hold time in flight followed by HERSHEL, ISO and HST/NICMOS. The shortest hold time in flight is attributed to SHOOT. However, it should be noted that this mission was a space-shuttle payload bay experiment that ended prior to full boil-off of the stored cryogen flown.

An alternative to overcompensating on the charge level would be to reduce the heat leak into the system. Techniques for reducing heat leak fall into two different categories: reduced boil-off (RBO) and zero boil-off (ZBO) [62–69]. Both techniques focus on intercepting thermal energy transferred along the heat leak path before it reaches the stored cryogen (as illustrated in Figure 7.16). RBO systems reduce the heat leak, but do not eliminate it completely. Material selection, insulation scheme and active cooling are important for reducing heat leak. Active cooling may consist of a thermodynamic vent system (TVS), sprays or vapor-cooled shields. TVSs release phase-changed vapor present in the inner vessel to ambient. Since temperature increases with pressure, this helps to maintain a lower temperature for the stored cryogen, but at the cost of loss of cryogen.

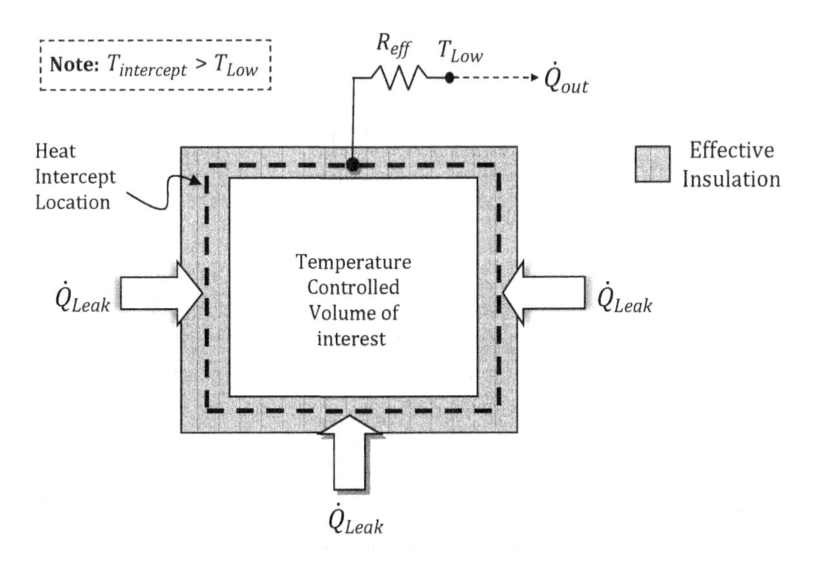

Figure 7.16 Temperature-controlled volume with effective insulation

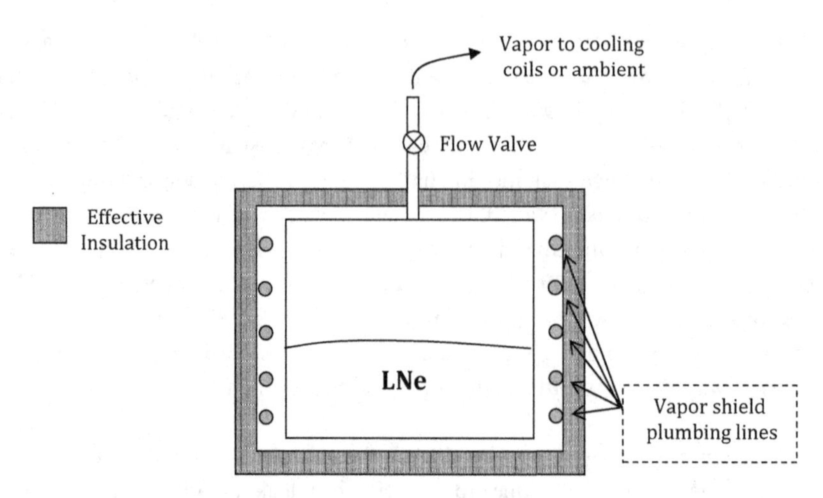

Figure 7.17 Vapor-cooled shield configuration

For vapor-cooled shields, the vapor phase of the fluid passes through cooling coils attached to an intermediate shield positioned between the inner and outer vessel (Figure 7.17). Ultimately, placing a temperature-controlled shield between the inner and outer vessel walls helps to reduce boil-off of the stored cryogen due to radiative and conductive heat leak on the inner vessel wall.

Figure 7.18a and b illustrate the internal and external configuration of a typical spaceflight dewar. The structure is mounted to the space platform via low thermal conductivity struts (Figure 7.18a). The shape factor for the struts is designed to carry the mechanical load of the dewar and its internal components, while also providing a thermal break for conductive heat flow from the platform to the dewar. The struts attach to the dewar via a coupling ring. The internal construction (Figure 7.18b) consists of the inner vessel tank, which is the storage location for the cryogenic material, and a single shield to mitigate radiative heat leak into the inner vessel. The inner vessel attaches to the inner wall of the vapor-cooled shield (VCS) via low thermal-conductivity structure mounts. In Figure 7.18b, the plumbing line on the VCS circumscribes the shield nine times before the phase-changed vapor is released to ambient. The space between the outer vessel and the shield, as well as the shield and the inner vessel, is at vacuum conditions. Plumbing for the vapor line from the inner vessel to the VCS is not shown, nor are electrical wiring, the fill line within the inner vessel, temperature sensors, pressure sensors, pressure relief valves and burst discs. However, they are standard for dewar construction. Depending on the desired temperature of the cryogenic material during its hold period, as well as environmental thermal loading, multiple radiation shields and/or MLI may be used in a dewar construction to mitigate heat leak.

Figure 7.19 shows heat flows present in the dewar/VCS thermal circuit illustrated in Figure 7.18b. The heat leak entering the dewar at the outer vessel will result in both conductive and radiative heat transfer to the VCS.

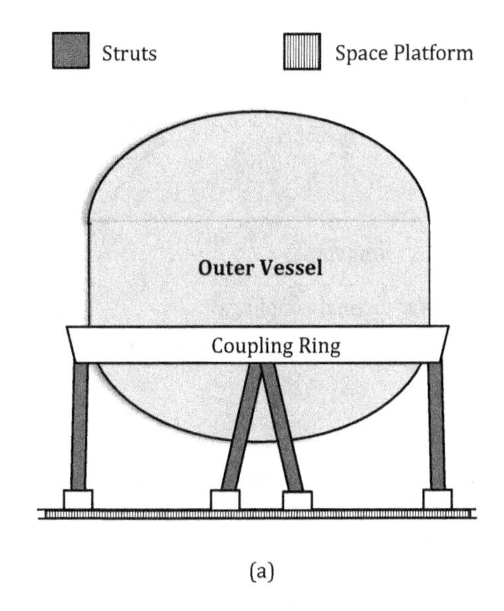

(a)

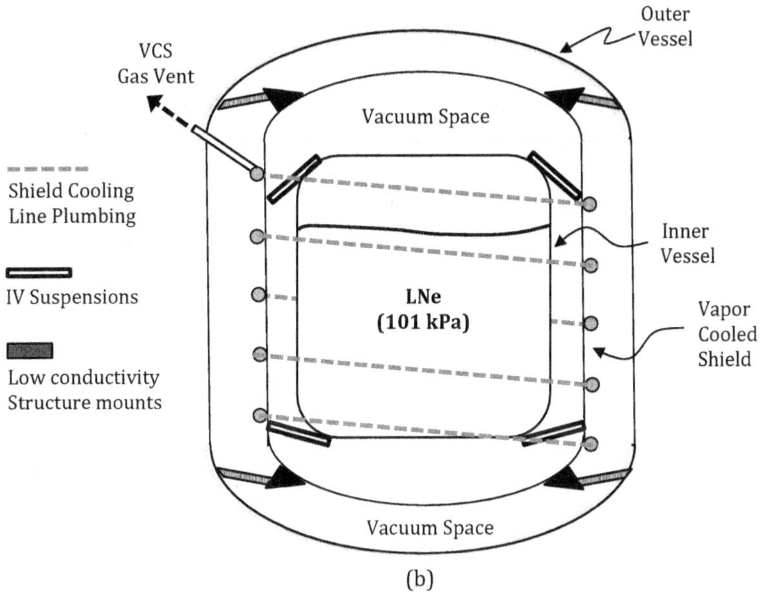

(b)

Figure 7.18 Spaceflight dewar for stored cryogens: a) outer vessel with low thermal-conductivity struts; b) internal configuration

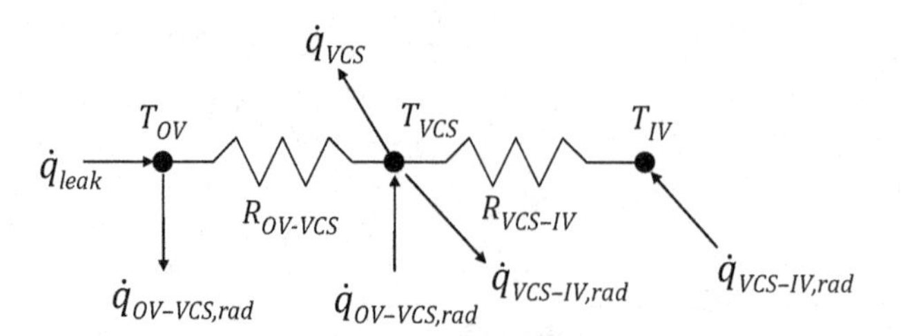

Figure 7.19 VCS dewar thermal couplings

Example 2 Cryogen Hold Time Using Vapor-cooled Shield Assume that in the LNe dewar system shown in Figure 7.18, the dimensions for the inner vessel, the shield and the outer vessel wall are

Inner vessel (IV)	Vapor-cooled shield (VCS)	Outer vessel (OV)
h = 30 cm	h = 32 cm	h = 34 cm
r = 15 cm	r = 18 cm	r = 21 cm

The interior wall of the outer vessel, the outer wall of the inner vessel and both sides of the VCS have an emissivity of $\varepsilon = 0.1$. Eight support struts made of Ti 6AL-4V span the gap between the outer vessel and the shield, as well as the shield and the inner vessel. The vapor line tubing for the shield has a diameter of 1.5 cm and makes nine passes around the shield wall before being vented to ambient. If the outer vessel is passively cooled to 175 K and the inner vessel wall temperature is 35 K, determine the heat leak into the inner vessel and the hold time associated with it for 20 liters of LNe maintained at a pressure of 101 kPa.

Solution

Assuming the vapor line on the VCS makes 9 circuits around the shield, the total length of the tube plumbing will be approximately 8.5 m. The energy balance on the outer vessel may be captured as

$$\dot{Q}_{leak} = \dot{Q}_{OV-VCS,cond} + \dot{Q}_{OV-VCS,rad} \tag{7.15}$$

The energy balance on the VCS is

$$\dot{Q}_{leak} - \dot{Q}_{VCS} = \dot{Q}_{VCS\text{-}IV,Total} \tag{7.16}$$

where $\dot{Q}_{VCS}$ is the convective cooling applied to the VCS from the phase-changed vapor vented from the inner vessel and $\dot{Q}_{VCS\text{-}IV,Total}$ is the total thermal energy reaching the inner vessel. The vapor entering the VCS plumbing from the inner dewar will be at saturated vapor conditions. Assuming an internal flow constant heat-flux boundary condition on the heat exchange surface, the mean temperature of the vapor at the exit of the plumbing line can be modeled as

$$T_{exit} = T_{inlet} + \frac{\dot{Q}''_{VCS} \cdot A_{tube}}{\dot{m} c_{p,vapor}} \tag{7.17}$$

where T_{inlet} is the mean temperature of the vapor at the plumbing inlet location. The value for $\dot{Q}_{VCS}$ can then be determined by rearranging variables in equation 7.17 and multiplying the energy flux term by the total surface area of the plumbing line experiencing cooling from the vapor. The new relation then becomes

$$\dot{Q}_{VCS} = \dot{Q}''_{VCS} \cdot A_{surf} = \dot{m} c_{p,vapor}(T_{exit} - T_{inlet}) \tag{7.18}$$

The thermal energy that reaches the inner vessel will drive the phase-change process. The energy balance for the pooled liquid in the inner vessel will be

$$\dot{Q}_{VCS\text{-}IV,Total}\Big|_{T_{sat}} = \dot{m} h_{fg} \tag{7.19}$$

Assuming that the vapor exiting the VCS plumbing is in equilibrium with the shield, Equations 7.18 and 7.19 can then be used to recast equation 7.16 as

$$\dot{Q}_{leak} = \dot{m} c_{p,vapor}(T_{VCS} - T_{inlet}) + \dot{m} h_{fg} \tag{7.20}$$

We will assume the interior surface of the outer vessel has full view to the outer surface of the shield. The radiative thermal energy exchange between the outer vessel and the shield is

$$\dot{Q}_{OV\text{-}VCS,rad} = \frac{\sigma(T_{OV}^4 - T_{VCS}^4)}{\left[\frac{1 - \varepsilon_{OV}}{\varepsilon_{OV} A_{OV}} + \frac{1}{A_{OV} F_{OV\text{-}VCS}} + \frac{1 - \varepsilon_{VCS}}{\varepsilon_{VCS} A_{VCS}}\right]} \tag{7.21}$$

The conductive heat load from the outer vessel to the VCS can then be captured as

$$\dot{Q}_{OV\text{-}VCS,cond} = \frac{(T_{OV} - T_{VCS})}{R_{OV-VCS,struts}} \tag{7.22}$$

Similar relations can be developed for the conductive and radiative heat loads from the VCS to the inner vessel.

Using saturation-based thermophysical properties for LNe at 101 kPa, for an outer vessel wall temperature of 175 K and an inner vessel wall temperature of 35 K, the heat leak at the inner vessel wall is 3.186 W. This would result in a hold time of 7.5 days. An improved design that would approximate closer to the 6-month metric would include additional shields and/or MLI. A ZBO cooling system approach could also be implemented with the aid of a cryocooler to intercept heat at the radiation shield or via temperature conditioning of the inner vessel wall.

7.5.2 Mechanical Cryocoolers

Historically, spaceflight missions that required components to be cooled to cryogenic temperature levels used passive cooling techniques as well as stored cryogens.

Neglecting the charge volume used for the mission and the required cooling power, a stored cryogen will have limited lifetime in space determined by how long it can maintain its initial phase. Parasitic heat loads coming in through the dewar/tank construction materials are always present. Thus, even if the loads are on the milliwatt scale, use of a stored cryogen is term limited because once the fluid has changed phase, re-conditioning of the fluid in space is not practical. Cryocoolers are important because they provide continuous on-demand cooling to cryogenic temperatures and are an alternative to the traditional approach of cooling with stored cryogens. By 2019, more than 150 missions had successfully flown cryocoolers and their use on spaceflight cryogenic systems is steadily increasing.

Cryocoolers are mechanical heat pump systems that cool down to cryogenic temperatures. Several types of cryocooler cycles are used to achieve cryogenic temperatures in ground-based and space-based applications. Regardless of the type of cryocooler, they all use some combination of a compressor, heat exchanger/s, expansion device and refrigerant fluid. He^4 is frequently used as the refrigerant. Cryocooler cycles are generally categorized as either oscillatory or continuous, according to the plumbing configuration and the fluid flow path during operation.

Oscillatory flow systems include reciprocating fluid flow within segments of the closed fluid loop, whereas in continuous flow systems the refrigerant in the closed fluid loop has a continuous flow path. Oscillatory flow systems incorporate a regenerative heat exchanger within the refrigeration cycle, whereas continuous flow systems use a recuperative HX to improve cycle efficiency. A regenerative heat exchanger is a densely packed bed that stores and releases thermal energy through its pore wall structure. Particles used in regenerators are typically made of materials with good specific heat properties. Variations on oscillatory flow cryocooler cycles include Stirling and Pulse Tube cycles. Most contemporary ground-based and spaceflight Stirling and Pulse Tube cryocoolers use a linear piston/cylinder assembly for compression of the refrigerant. Figure 7.20 shows a Stirling cryocooler cycle. The cycle begins with the piston compressing the refrigerant. Heat from the compression is released via a thermal coupling to a heat exchanger positioned just outside the entrance to the regenerator. As the piston head achieves full stroke, the expander head retracts and travels along its cylinder in the opposite direction. The net flow of the working fluid is through the regenerator material and into the expander section. With the expansion of the gas comes

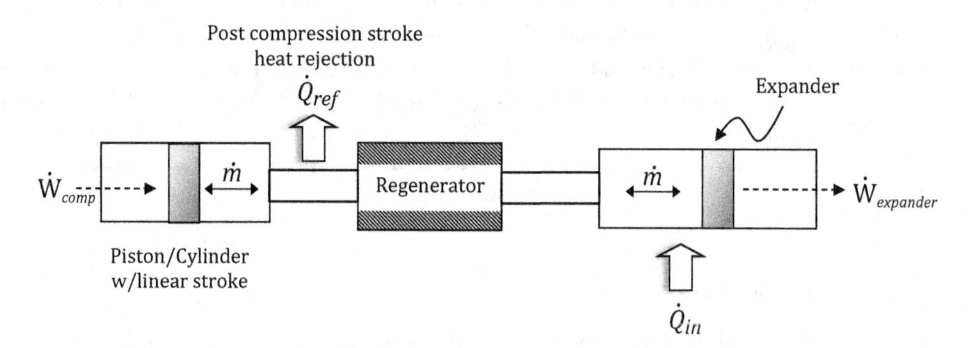

Figure 7.20 Stirling cryocooler cycle

a reduction in temperature. Heat is then acquired locally (via thermal coupling to the cooled object), and the expander head compresses the newly conditioned refrigerant while the compression piston retracts. This creates a net flow of the working fluid in the direction of the piston cylinder. This cycle is then repeated continuously for the net transfer of heat [1,2,70,71].

A Pulse Tube cryocooler is shown in Figure 7.21. In this oscillatory cycle the reservoir stores a portion of the refrigerant charge at a nominal pressure throughout the cycle [70,72]. The piston adiabatically compresses the refrigerant, prompting mass flow in the direction of the reservoir. As the compressed gas flows towards the Pulse Tube, heat is released just before the entrance to the regenerator. As in the Stirling cycle, the regenerator provides thermal energy storage for some of the gas's sensible heat while in transit to the Pulse Tube. While the Pulse Tube is receiving high-pressure gas from the compressor's forward stroke, gas inside the Pulse Tube flows into the reservoir until the two pressures are in equilibrium. Heat rejection at the warm end of the Pulse Tube takes place during the mass transfer into the reservoir. As the piston head in the compressor reverses stroke, the refrigerant in the Pulse Tube undergoes adiabatic expansion. With the expansion of the gas comes a reduction in temperature. The net flow in the system is now from the reservoir towards the compressor. As the net mass flow in the direction of the compressor is occurring, heat is acquired from the temperature-controlled item at the cold side of the Pulse Tube. When the piston head reaches full stroke and reverses, the cycle repeats itself. The thermo-mechanical configuration highlighted here is just one of many variations used for Pulse Tube cycles. For details on alternate variations of Pulse Tube cycles, see Radebaugh [72].

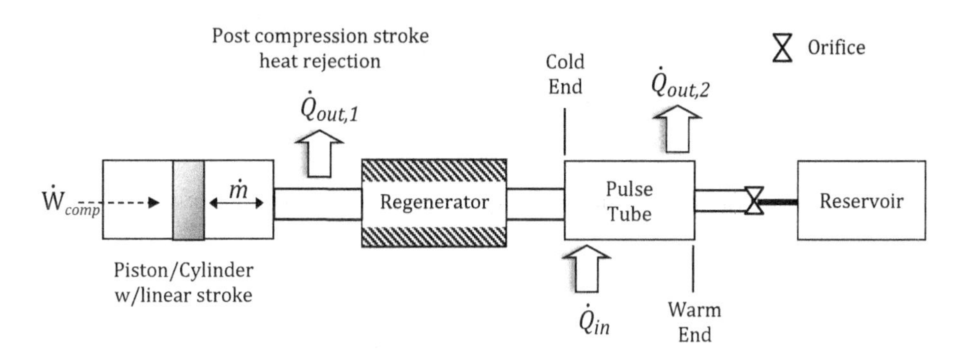

Figure 7.21 Pulse Tube cryocooler cycle

In a recuperative heat exchanger system, thermal energy exchange occurs between two opposing fluid streams at different temperatures within the recuperator. Following the compression stroke, cooling of the gas occurs through the aftercooler heat exchanger. The high-pressure gas from the compressor pre-heats the low-pressure side return line gas. The refrigeration cooling effect occurs at the heat exchanger following the J-T (Joule–Thomson) expansion [1,2,71,73]. The cooling capacity here is denoted as $\dot{Q}_{in}$ in Figure 7.22.

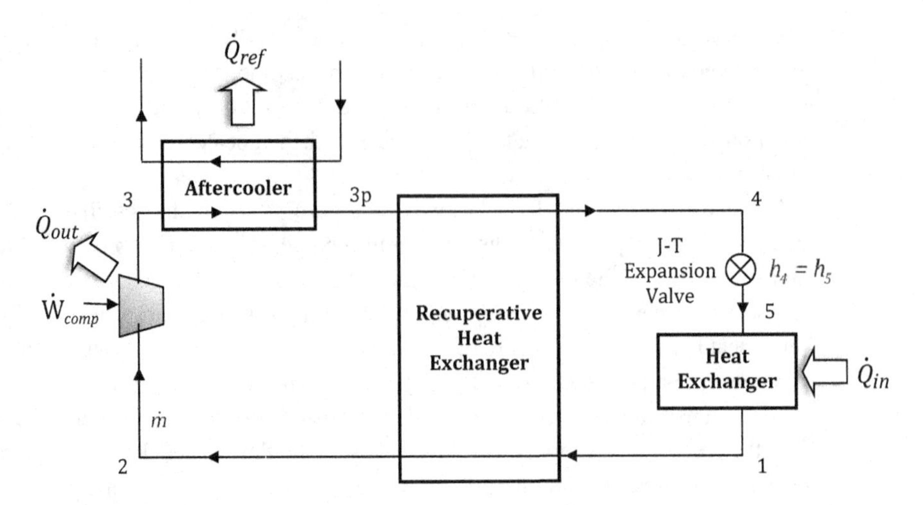

Figure 7.22 Recuperative heat exchanger system (J–T cryocooler cycle)

The configuration of the thermo-machinery components in the J-T cryocooler cycle is similar to those included in the vapor compression cycle with the LLSL heat exchanger (see Section 5.3.1). However, there are two exceptions. In the J-T cryocooler cycle the compression of the refrigerant gasses is considered an isothermal, internally reversible process. The energy balance for the compressor would then be

$$\dot{W}_{comp} = \dot{m} \cdot \left[T_{avg}(s_2 - s_3) + (h_3 - h_2)\right] \tag{7.23}$$

In practical applications, the temperature at state point 3 may not be equal to the temperature at state point 2 [1,2,73]. The average temperature for these two state points is then used as an approximation of the isothermal compression assumption. The expansion process across the J-T valve is isenthalpic, resulting in

$$h_4 = h_5 \tag{7.24}$$

The energy balance on the low-pressure side heat exchanger is

$$\dot{Q}_{in} = \dot{m} \cdot (h_1 - h_5) \tag{7.25}$$

Because the cycle is designed to cool to cryogenic temperature levels on the low side, the effectiveness of the recuperative heat exchanger is a key aspect of the cycle's performance. Unlike the case for the standard vapor compression cycle that makes use of an LLSL-HX, the effectiveness of the recuperative heat exchanger is defined using the ratio of the actual enthalpy difference for the return gas to the enthalpy difference for the return gas assuming full temperature lift to the value for the gas at the recuperator inlet on the high-pressure side. For the cycle shown in Figure 7.22, the recuperator effectiveness is defined as

$$Eff_{recup} = \frac{h_2|_{T=T_2} - h_1}{h_2|_{T=T_{3p}} - h_1} \tag{7.26}$$

Finally, the energy balance applied to the recuperative heat exchanger will be

$$h_{3p} - h_4 = h_2 - h_1 \tag{7.27}$$

Figure 7.23 illustrates the Reverse-Brayton cryocooler cycle. In the Reverse-Brayton system the flow path of the refrigerant during operation is once again continuous. Post-compression heat rejection occurs via the aftercooler before entry to the recuperative heat exchanger. The refrigerant expands via a turbine expander. Heat acquisition on the low-pressure side of the cycle takes place after the turbine expansion step [1,2,71,74].

Since heat rejection from the refrigerant is taking place via the aftercooler, the energy balance on the compressor simplifies to

$$\dot{W}_{comp} = \dot{m}(h_3 - h_2) \tag{7.28}$$

The energy balance across the aftercooler heat exchanger then becomes

$$\dot{Q}_{rej} = \dot{m}(h_3 - h_4) \tag{7.29}$$

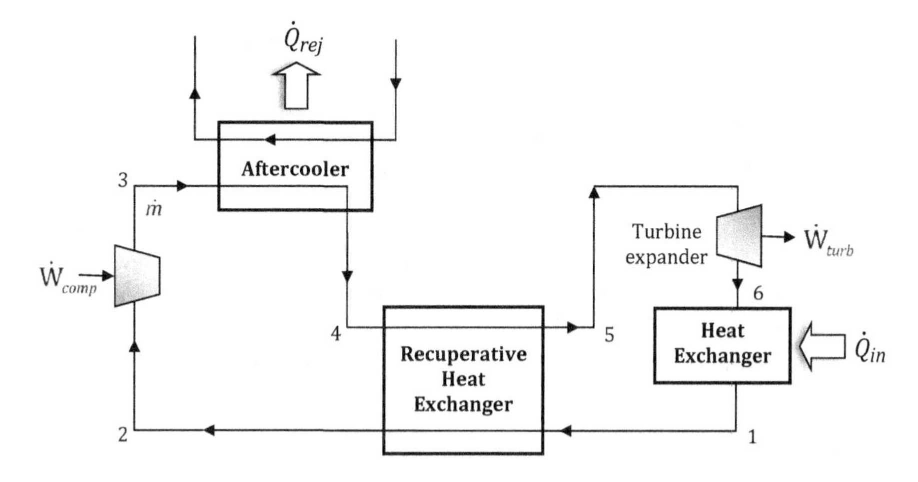

Figure 7.23 Reverse-Brayton cryocooler cycle

The energy balance for the turbine is then

$$\dot{W}_{turb} = \dot{m}(h_5 - h_6) \tag{7.30}$$

The refrigeration load on the low-side heat exchanger (Figure 7.23) is

$$\dot{Q}_{in} = \dot{m} \cdot (h_1 - h_6) \tag{7.31}$$

Once again, the effectiveness of the recuperative heat exchanger is a key aspect of overall cycle performance. The relation for the recuperative heat exchanger effectiveness in the Joule–Thomson (J-T) cycle also applies to the present cycle. The recuperative heat exchanger effectiveness and energy balance are

$$Eff_{recup} = \frac{h_2|_{T=T_2} - h_1}{h_2|_{T=T_4} - h_1} \tag{7.32}$$

$$h_4 - h_5 = h_2 - h_1 \tag{7.33}$$

Since cryocoolers are actually heat pump systems, similar to room temperature vapor compression cycles, they will have generalized performance metrics (e.g., efficiency and COP) that can be used to gauge their performance relative to other cryocooler systems. Two performance metrics that apply to cryocoolers are the thermodynamic COP and the fraction of the Carnot efficiency that the cycle is operating at [75]. The thermodynamic COP is defined as

$$COP_{Thermo} = \frac{refrigeration\ load}{electrical\ input\ power - I^2 R_e} \tag{7.34}$$

where the I^2R_e values represent the drive motor losses on the compressor mechanism [75]. This COP value is a measure of the cooler's ability to convert work done on the gas into net cooling power. However, cooler efficiencies are typically expressed as a percentage of the Carnot efficiency which is defined as

$$\%COP_{Carnot} = \frac{COP_{Thermo}}{COP_{Carnot}} \tag{7.35}$$

Recall that the Carnot efficiency based on the reservoir boundary temperatures was defined in Chapter 5 (equation 5.10) as

$$COP_{Carnot} = \frac{T_L}{T_H - T_L}$$

Cryocoolers typically perform at somewhere between 3% and 20% [75] of their Carnot efficiency depending on the cycle used, the cooling load and the cold-side temperature. An additional metric used for cryocoolers is specific power (S.P.), which is defined as

$$S.P. = \frac{Total\ Input\ Power}{Cooling\ Power} \tag{7.36}$$

and is a measure of the electrical power required per watt of cooling load that the cycle can produce.

Example 3 Cryocooler Power Input A cryocooler is acquiring a refrigeration load of 500 mW at 40 K. The cryocooler's compressor is rejecting heat at a temperature of 300 K. If the cooler operates at 6% of the Carnot efficiency for the temperature boundaries specified, how much input power is required?

Solution

The Carnot efficiency is defined as

$$COP_{Carnot} = \frac{T_L}{T_H - T_L}$$

Since the cryocooler is operating at 6% of its Carnot efficiency, the actual COP would be

$$COP_{Thermo} = 0.06 \cdot \left(\frac{T_L}{T_H - T_L}\right)$$

Based on the definition of specific power, we know that

$$S.P. = \frac{Total\ Input\ Power}{Cooling\ Power} = \frac{1}{COP_{Thermo}}$$

The specific power can be recast as

$$S.P. = \frac{1}{\left[0.06 \cdot \left(\frac{T_L}{T_H - T_L}\right)\right]}$$

This relation can also be rewritten as

$$\frac{1}{\left[0.06 \cdot \left(\frac{T_L}{T_H - T_L}\right)\right]} = \frac{Input\ Power}{500\ \text{mW}}$$

Solving for the input power results in

$$Input\ Power = \frac{0.5\ \text{W}}{\left[0.06 \cdot \left(\frac{40\ \text{K}}{300\ \text{K} - 40\ \text{K}}\right)\right]}$$

$$Input\ Power = 54.15\ \text{W}$$

Thus, using the ratio as defined by the specific power, it can be determined that 108.3 W of input power will be required for every watt of cooling power achieved.

Example 4 Argon-Based Joule–Thomson Cryocooler Cycle A Joule–Thomson cryocooler cycle (Figure 7.22) is using argon (R740) with a mass flow rate of 3.0×10^{-4} kg/s as the refrigerant. The argon gas has a compressor inlet condition of 300 K at 101 kPa and an outlet temperature of 335 K. The recuperator for the cycle has an effectiveness of 0.98. Assuming that the argon is at saturated conditions, leaving the heat exchanger on the low side of the cycle, determine the work input, the temperature of the argon entering the low-side heat exchanger, the cooling capacity on the low side and the specific power if the system has a compression ratio of 165.

Solution

To determine the values for the power input, the temperature at which cooling is applied and the specific power for the system, the thermophysical properties of argon must be known. An EES code will be developed and used to model the cycle. The code will begin by defining constants which were listed in the problem statement.

```
{Example Argon Joule-Thomson cryocooler cycle}

m_dot=0.0003          [kg/s]

P_low=101             [kPa]
P_ratio=165
P_high=P_ratio*P_low
```

State point 1, which is located at the exit of the low-side heat exchanger, is captured in EES as

```
{State point 1}
x_1=1.0                    {Quality at exit of heat exchanger}
P_1=P_low
h_1=enthalpy(Argon, x=x_1, P=P_1)
T_1=temperature(Argon,x=x_1,P=P_1)
```

State points 2 and 3 are the compressor inlet and outlet respectively. These are captured in EES as

```
{Statepoint 2}
P_2=P_low
T_2=300 [K]
h_2=enthalpy(Argon,T=T_2,P=P_2)
s_2=entropy(Argon,T=T_2,P=P_2)

{State point 3}
P_3=P_high
T_3=335 [K]
h_3=enthalpy(Argon, T=T_3, P=P_3)
s_3=entropy(Argon, T=T_3,P=P_3)
```

Since the state points at the inlet and exit of the compressor have been defined, the compressor work can now be calculated.

```
{Compressor}
T_avg=0.5*(T_2+T_3)
W_dot_comp=m_dot*T_avg*(s_2-s_3)-m_dot*(h_2-h_3)
```

T_{avg} has been defined as the average temperature of the argon between state points 2 and 3. Next, state points 3p, 4 and 5 must be established. State point 3p is the recuperator's high pressure-side entrance, whereas state point 4 is the recuperator's high pressure-side outlet. In EES these state conditions are captured as

```
{State point 3p; Recuperator high-pressure side inlet}
P_3p=P_high
```

```
T_3p=temperature(Argon,h=h_3p,P=P_3)

{State point 4}
P_4=P_high
h_4=enthalpy(Argon,T=T_4,P=P_4)
```

Between state points 4 and 5 the fluid undergoes expansion at constant enthalpy. Thus, state point 5 can be captured as

```
{State point 5}
P_5=P_low
h_5=h_4
T_5=temperature(Argon,h=h_5,P=P_5)
```

The effectiveness of the heat exchange across the recuperator, as well as the energy balance for this component of the cycle, is

```
{Recuperator}
Eff_recup=0.98
h_2_max=enthalpy(Argon, T=T_3p, P=P_2)
h_2-h_1=Eff_recup*(h_2_max-h_1)

h_3p-h_4=h_2-h_1
```

The energy balance on the aftercooler and the low-side heat exchanger are

```
{Aftercooler heat exchanger}
Q_dot_after=m_dot*(h_3-h_3p)

{Low-side heat exchanger cooling capacity}
Q_dot_in=m_dot*(h_1-h_5)
```

Finally, the performance metrics for the cycle (i.e., SP, percent COP of Carnot) are captured as

```
{Cycle performance metrics}

T_high=T_avg
T_low=T_5

SP=W_dot_comp/Q_dot_in                  {Specific power}

COP_Thermo=1/SP                         {Thermodynamic COP}
COP_Carnot=T_low/(T_high-T_low)         {Carnot COP}
Percent=COP_thermo/COP_Carnot           {% of Carnot COP}
```

Before running the code we will set the lower limit for each temperature variable to 0 K, as well as set the initial guess values to 200 K in the variables block. On execution of the code, it is determined that the work input is 100.2 W. The cooling capacity of the cycle is 7.54 W at 87.3K. The cycle is performing at 19.8% of Carnot.

Example 5 Reverse-Brayton Cryocooler Cycle A Reverse-Brayton cryocooler cycle charged with neon (R720) has a mass flow rate of 1.8×10^{-3} kg/s. The compressor has an efficiency of 90% and an inlet condition of 282 K at 101 kPa. The temperature at the entrance of the low-side heat exchanger is 57 K. For a pressure ratio of 1.55 and a recuperator effectiveness of 0.99, determine the work input to the cycle, the heat rejection via the aftercooler, the specific power and the percent COP if the cooling capacity at the low-side heat exchanger is 4.8 W.

Solution

As with the solution for example 4, EES will be used to model the Reverse-Brayton cryocooler cycle. Constants listed in the problem statement are captured as

```
{Example Reverse-Brayton cryocooler cycle using R720 (neon)}
m_dot=0.0018                    [kg/s]

{Pressures} P_low=101      [kPa]
P_ratio=1.55
P_high=P_ratio*P_low
```

Since the temperature and pressure at state points 2 and 6 are known, we will define these state points in EES first.

```
{State point 2}
T_2=282        [K]
P_2=P_low
h_2=enthalpy(Neon, T=T_2, P=P_2)
s_2=entropy(Neon, T=T_2,P=P_2)

{ State point 6}
P_6=P_low
T_6=57        [K]
h_6=enthalpy(Neon,T=T_6,P=P_6)
```

Assuming isentropic compression between state points 2 and 3, the condition at state point 3 can be defined as

```
{State point 3}
P_3=P_high
s_3=s_2
h_3=enthalpy(Neon,s=s_3,P=P_3)
T_3=temperature(Neon,s=s_3,h=h_3)
```

Given knowledge of the compressor efficiency, state points 2 and 3 can be used to determine the isentropic work done by the compressor. This can then be used to determine the actual compressor work when divided by the compressor efficiency.

```
{Compressor}
eta=0.9
W_dot_comp_isen=m_dot*(h_3-h_2)
W_dot_comp=W_dot_comp_isen/eta
```

The work input calculated can then be used to determine the actual conditions at the state point located at the exit of the compressor (i.e., state point 3p).

```
{State point 3p}
P_3p=P_high
h_3p=W_dot_comp/m_dot+h_2
T_3p=temperature(Neon,h=h_3p,P=P_3p)
```

The cooling capacity at the heat exchanger on the low side of the cycle is

```
{ Cooling capacity}
Q_dot_in=4.8 [ W]
```

Given knowledge of the cooling capacity on the low side of the cycle, as well as the enthalpy at state point 6, the conditions at state point 1 can be captured as

```
{State point 1}
P_1=P_low
h_1=Q_dot_in/m_dot+h_6
T_1=temperature(Neon,P=P_1,h=h_1)
```

The effectiveness of the recuperative heat exchanger can then be used to determine the temperature at state point 4.

```
{Recuperator}
Eff_recup=0.99
h_2_max=enthalpy(Neon,T=T_4,P=P_2)
h_2-h_1=Eff_recup*(h_2_max-h_1)
h_5+h_2=h_1+h_4
```

The enthalpy at state point 4 can be referenced in the thermophysical properties tables of EES.

```
{State point 4}
P_4=P_high
h_4=enthalpy(Neon,T=T_4,P=P_4)
```

The energy balance on the recuperator aided in the determination of the enthalpy at state point 5. The temperature and pressure at state point 5 are

```
{State point 5}
P_5=P_high
T_5=temperature(Neon,h=h_5,P=P_5)
```

The energy balances on the aftercooler and the turbine are

```
{Aftercooler}
Q_dot_rej=m_dot*(h_3p-h_4)

{Turbine}
W_dot_turb=m_dot*(h_5-h_6)
```

Finally, the performance metrics are

```
{Performance metrics}

T_high=T_3
T_low=T_6

SP=W_dot_comp/Q_dot_in                   {Specific power}

COP_Thermo=1/SP                          {Thermodynamic COP}
COP_Carnot=T_low/(T_high-T_low)          {Carnot COP}
Percent=COP_Thermo/COP_Carnot            {Percent of Carnot COP}
```

The lower limit for each temperature variable will again be set to 0 K, and the initial guess values to 200 K in the variables block. Execution of the code will result in a work input of 111.4 W. The heat rejection from the aftercooler is 107.2 W. The specific power and the percent Carnot for the cycle are 23.2% and 17% respectively.

Standard cryocooler specification data sheets usually list the temperature cooled to on the low side of the cycle under a zero cooling load condition. The temperatures that can be cooled to in the presence of a heat load on the cycle's low side will actually be slightly higher than the zero-load case with the same power input. For actual cryocooler performance under a given cooling load, reference should be made to its performance curves, which can be obtained from the cryocooler manufacturer. Figure 7.24 illustrates a plot containing input power, cooling capacity of the cycle and the temperature at which heat is being acquired on the low side.

If the cooling capability for cryocooler Z at 60 K with 100 W of power input applied to the compressor is needed, the intersection of the horizontal line crossing the 100 W mark on the vertical axis and the 60 K cooling load line should be located. From this point a vertical line is drawn to the intersection with the horizontal axis to determine the cooling load. In the example in Figure 7.24, the cooling load is 3 W.

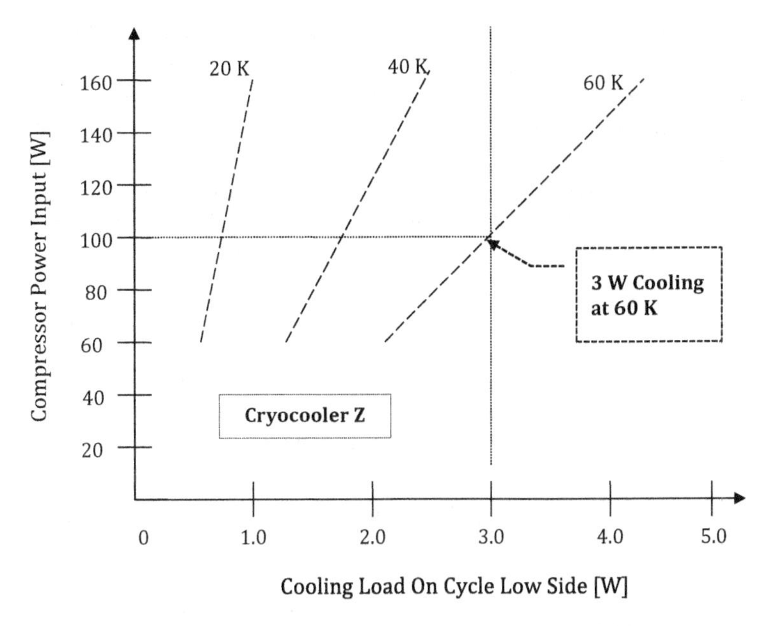

Figure 7.24 Cryocooler performance plot at variable heat-acquisition temperatures

Cryocoolers have been shown to have reliable performance in the space environment. However, their use on space platforms requires integrated design efforts across subsystems. When a cryocooler is used on a space platform, the power resources necessary to run the cryocooler must be allowed for in the power system design. Since heat rejection from the compressor/aftercooler occurs at approximately 300 K, the thermal subsystem design must include the ability to reject the required amount of heat from the high end of the cycle to enable the cryocooler to cool down to desired operational temperatures on the low side. Finally, since vibrational loads due to operation of the compressor may be present, the mechanical subsystem should be included in deciding where the cryocooler compressor is located on the spacecraft, as well as viable vibration isolation techniques for use. Spaceflight cryocooler manufacturers include Ball Aerospace Corporation, Northrup Grumman Aerospace Systems, Lockheed Martin Corporation, Raytheon Corporation, CREARE, Sunpower Corporation and Thales Corporation. Reviews of cryocooler products can be found in references [76–80] as well as at the company websites. Additional reading material on cryocooler technology may be found in Barron [1], Flynn [2], the *Spacecraft Thermal Control Handbook Vol. II: Cryogenics,* the JPL cryocooler publications website [81], the NIST cryocoolers website [82], the Goddard Space Flight Center Cryogenics and Fluids Branch website [83] and the Proceedings of the International Cryocooler Conference [84].

References

[1] Barron, R., 1985, *Cryogenic Systems*, 2nd edn, Oxford University Press Inc., New York, NY
[2] Flynn, T.M., *Cryogenic Engineering*, 2nd edn, CRC Press, New York, NY
[3] Pobell, F., 2007, *Matter and Methods at Low Temperatures*, 3rd edn, Springer-Verlag, Berlin, Germany
[4] Van Sciver, S.T., 1986, *Helium Cryogenics*, Plenum Press, New York, NY
[5] Ekin, J.W., 2006, *Experimental Techniques for Low Temperature Measurements*, Oxford University Press, Oxford, United Kingdom
[6] Barron, R.F., and Nellis, G.F., 2016, *Cryogenic Heat Transfer*, 2nd edn, CRC Press, New York, NY
[7] The Merriam-Webster Dictionary, 2018, *Cryogenics*, Merriam-Webster, Inc., Springfield, MA
[8] Prigogine, I., 1989, "What Is Entropy," *Naturwissenschaften*, Vol. 76, pp. 1–8
[9] Serway, R.A., 1986, *Physics for Scientists & Engineers*, Saunders College Publishing, Philadelphia, PA
[10] Moran, M.J., Shapiro, H.N., Boettner, D.D., and Bailey, M.B., 2014, *Fundamentals of Engineering Thermodynamics*, John Wiley & Sons, Hoboken, NJ
[11] Collaudin, B., and Rando, N., 2000, "Cryogenics In Space: A Review of the Missions and of the Technologies," *Cryogenics*, Vol. 40, pp. 797–819
[12] Ross, R., 2007, "Aerospace Coolers: A 50-Year Quest for Long-Life Cryogenic Cooling in Space," in *Cryogenic Engineering*, edited by Timmerhaus, K.D. and Reed, R.P., International Cryogenics Monograph Series, Springer, New York, NY
[13] Shirron, P., 2014, "50 Years of Progress In Space Cryogenics," *Cold Facts-Cryogenic Society of America*, Vol. 30, No. 1
[14] Linder, M., Rando, N., Peacock, A., and Collaudin, B., 2001, "Cryogenics In Space: A Review of the Missions and technologies," European Space Agency, *ESA Bulletin* 107
[15] "Properties of solid materials from cryogenic to room temperatures," NIST (National Institute of Standards and Technology), https://trc.nist.gov/cryogenics/materials/materialproperties.htm
[16] Young, W.C., Budynas, R.G., and Sadegh, A.M., 2012, *Roark's Formulas for Stress and Strain*, 8th edn, McGraw-Hill, New York, NY
[17] Nagai, K., Yuri, T., Ogata, T., Umezawa, O., Ishikawa, K., Nishimura, T., Mizoguchi, T., and Ito, Y., 1991, "Cryogenic Mechanical Properties of Ti-6Al-4V Alloys with Three Levels of Oxygen Content," *ISIJ International*, Vol. 31, No. 8, pp. 882–889
[18] Weiss, W., 1982, "Low temperature properties of carbon fiber reinforced epoxide resins," in *Nonmetallic Materials and Composites at Low Temperatures* 2, edited by Hartwig, G., and Evans, D., pp. 293–309, Cryogenic Materials Series, Springer, Boston, MA
[19] Tuttle, J., Canavan, E., and Jahromi, A., 2017, "Cryogenic Thermal Conductivity Measurements on Candidate Materials for Space Missions," *Cryogenics*, Vol. 88, pp. 36–43
[20] Kasen, M.B., MacDonald, G.R., Beekman, D.H., and Schramm, R.E., 1980, "Mechanical, Electrical, and Thermal Characterization of G-10CR and G-11CR Glass-Cloth/Epoxy Laminates between Room Temperature and 4K," *Advances in Cryogenic Engineering*, Vol. 26, pp. 235–244
[21] McKeen, L.W., 2014, *The Effect of Temperature and Other Factors on Plastics and Elastomers*, 3rd edn, William Andrew Inc., Norwich, NY
[22] Hartman, D., Greenwood, M.E., and Miller, D.M., 1996, *High Strength Glass Fibers,* AGY, Aiken, SC
[23] Drude, P., 1890, "Bestimmung der optischen Constanten des Kobalts," *Annalen der Physik*, Vol. 39, pp. 530

[24] Born, M., and Wolf, E., 1965, *Principles of Optics*, 3rd edn, Pergamon Press, New York, NY
[25] Toscano, W.M., and Cravalho, E.G., 1976, "Thermal Radiative Properties of the Noble Metals at Cryogenic Temperatures," *Journal of Heat Transfer*, Vol. 98, No. 3, pp. 438–445
[26] Parker, W.J., and Abbott, G.L., 1964, "Theoretical and Experimental Studies of the Total Emittance of Metals," NASA SP-55, Thermal Radiation of Solids Symposium, March 4–6, San Francisco, CA
[27] Modest, M., 2013, *Radiative Heat Transfer*, 3rd edn, Academic Press, Oxford, United Kingdom
[28] Reuter, G., and Sondheimer, R., 1948, "The Theory of the Anomalous Skin Effect in Metals," *Proceedings of the Royal Society A*, Vol. 195, pp. 336–364
[29] Domoto, G.A., Boehm, R.F., and Tien, C.L., 1969, "Predictions of the Total Emissivity of Metals at Cryogenic Temperatures," *Advances in Cryogenic Engineering*, Vol. 14, pp. 230–239
[30] White, F.M., 1986, *Fluid Mechanics*, 2nd edn, McGraw-Hill Inc., New York, NY
[31] Bardeen, J., Cooper, L.N., and Schrieffer, J.R., 1957, "Theory of Superconductivity," *Physical Review*, Vol. 108, No. 5, pp. 1175–1204
[32] Meissner, W., and Ochsenfeld, R., 1933, "Ein Neuer Effekt Bei Eintritt der Supraleitfahigkeit," *Naturwissenschaften*, Vol. 21, Issue 44, pp. 787–788
[33] Roberts, B.W., 1966, *Superconductive Materials and Some of Their Properties, National Bureau of Standards,* Washington, DC
[34] Bejan, A., 1988, *Advanced Engineering Thermodynamics*, John Wiley & Sons, Hoboken, NJ
[35] Langevin, M.P., 1905, "Magnetisme et théorie des électrons," *Annales de Chimie et de Physique*, Vol. 5, pp. 70–127
[36] Weiss, P., and Piccard, A., 1917, "Le phénomène magnetocalorique," *Journal de physique théorique et appliquée*, Vol. 7, pp. 103–109
[37] Debye, P., 1926, "Einige Bemerkungen zur Magnetisierung bei Tiefer Temperatur," *Annalen der Physik*, Vol. 386, pp. 1154–1160
[38] Giaque, W.F., 1927, "A Thermodynamic Treatment of Certain Magnetic Effects. A Proposed Method of Producing Temperatures Considerably Below 1° Absolute," *Journal of American Chemical Society,* Vol. 49, No. 8, pp. 1864–1870
[39] Shirron, P.J., 2014, "Applications of the Magnetocaloric Effect in Single-stage, Multi-stage and Continuous Adiabatic Demagnetization Refrigerators," *Cryogenics*, Vol. 62, pp. 130–139
[40] Tishin, A.M., and Spichkin, Y.I., 2003, *The Magnetocaloric Effect and Its Applications,* CRC Press, Boca Raton, FL
[41] Weisend II, J.G., 2016, "Principles of Cryostat Design," in *Cryostat Design: Cases Studies, Principles and Engineering*, edited by Weisend II, J.G., Springer International Publishing, Switzerland, Chap. 1, pp. 1–46
[42] Derking, J.H., Holland, H.J., Tirolien, T., and ter Brake, H.J.M., 2012, "A Miniature Joule-Thomson Cooler for Optical Detectors in Space," *Review of Scientific Instruments*, Vol. 83
[43] Salerno, L.J., and Kittel, P., 1997, *Thermal Contact Conductance, NASA Technical Memorandum 110429*, Ames Research Center, Mountain View, CA
[44] Jewell, C.I., 1997, "Overview of Cryogenic Developments in ESA," Sixth European Symposium on Space Environmental Control Systems, Noordwijk, The Netherlands, May 20–22
[45] Anderson, N.T., 1973, "Space Test Program," Space Congress Proceedings, Embry-Riddle University
[46] "NIMBUS Program History," 2004, Goddard Space Flight Center, National Aeronautics and Space Administration
[47] "Mission To Universe: Infrared Astronomical Satellite," NASA Jet Propulsion Laboratory/ California Institute of Technology, November 21, 1983, www.jpl.nasa.gov/missions/infrared-astronomical-satellite-iras/

[48] IRAS Lee, J.H., 1990, "Thermal Performance of a Five Year Lifetime Superfluid Helium Dewar for SIRTF," *Cryogenics*, Vol. 30, Issue 30, pp. 166–172
[49] "COBE," National Aeronautics and Space Administration, May 22, 2016, https://science.nasa.gov/missions/cobe/
[50] Dipirro, M., 2016, "The Superfluid Helium On-Orbit Transfer (SHOOT) Flight Demonstration," in *Cryostat Design: Cases Principles and Engineering*, edited by Weisend II, J.G., Springer International Publishing, Switzerland, Chap. 4, pp. 95–116
[51] "ISO: Infrared Space Observatory," National Aeronautics and Space Administration, May 22, 2016, https://science.nasa.gov/missions/iso
[52] Kessler, M.F., Steinz, J.A., Anderegg, M.E., Clavel, J., Drechsel, G., and Estaria, P., 1996, "The Infrared Space Observatory (ISO) Mission," *Astronomy and Astrophysics*, Vol. 315, pp. L27–L31
[53] Schick, S.H., and Bell, G.A., 1997, "Performance of the Spirit III cryogenic system," Proc. SPIE 3122, Infrared Spaceborne Remote Sensing V, October 23
[54] "Hubble Space Telescope: NICMOS History," STScI (Space Telescope Science Institute), www.stsci.edu/hst/nicmos/design/history/
[55] "Spitzer Space Telescope: Fast Facts," NASA Jet Propulsion Laboratory/California Institute of Technology, www.spitzer.caltech.edu/info/277-Fast-Facts
[56] "Gravity Probe B: Testing Einstein's Universe," National Aeronautics and Space Administration, February 2015, https://einstein.stanford.edu/content/fact_sheet/GPB_FactSheet-0405.pdf
[57] "WISE: Mapping The Infrared Sky," National Aeronautics and Space Administration, September 9, 2009, www.jpl.nasa.gov/news/press_kits/wise-factsheet.pdf
[58] Naes, L., Lloyd, B., and Schick, S., 2008, "WISE Cryogenic Support System Design Overview and Build Status," *Advances in Cryogenic Engineering: Transactions of the Cryogenic Engineering Conference*, Vol. 53, pp. 815–822
[59] Naes, L., Wu, S., Cannon, J., Lloyd, B., and Schick, S., 2005, "WISE Solid Hydrogen Cryostat Design Overview," SPIE Conference on Optics and Photonics 2005, San Diego, CA, July 31–August 4
[60] HERSCHEL "Boiling Into Space," European Space Agency, May 7, 2009, www.esa.int/Our_Activities/Space_Engineering_Technology/Boiling_into_space
[61] "Unveiling Hidden Details of Star and Galaxy Formation and Evolution: HERSCHEL," European Space Agency, https://esamultimedia.esa.int/docs/herschel/HERSCHEL262-LOW-complete.pdf
[62] Nast, T., Frank, D., and Burns, K., 2011, "Cryogenic Propellant Boil-off Reduction Approaches," 49th AIAA Aerospace Sciences Meeting, January 4–7, Orlando, FL
[63] Plachta, D.W., and Guzik, M., 2014, "Cryogenic Boil-off Reduction System," *Cryogenics*, Vol. 60, pp. 62–67
[64] Plachta, D.W., Johnson, W.L., and Feller, J.R., 2016, "Zero Boil-off System Testing," *Cryogenics*, Vol. 74, pp. 88–94
[65] Haberbusch, M.S., Nguyen, C.T., Stochl, R.J., and Hui, T.Y., 2010, "Development of No-Vent™ Liquid Hydrogen Storage System for Space Applications," *Cryogenics*, Vol. 50, pp. 541–548
[66] DeLee, C.H., Barfknecht, P., Breon, S., Boyle, R., DiPirro, M., Francis, J., Huynh, J., Li, X., McGuire, J., Mustafi, S., Tuttle, J., and Wegel, D., 2014, "Techniques for On-Orbit Cryogenic Servicing," *Cryogenics*, Vol. 64, pp. 289–294
[67] Kim, S.Y., and Kang, B.H., 2000, "Thermal Design Analysis of a Liquid Hydrogen Vessel," *International Journal of Hydrogen Energy*, Vol. 25, pp. 133–141

[68] Stautner, S., 2016, "Special Topics in Cryostat Design," in *Cryostat Design: Cases Studies, Principles and Engineering*, edited by Weisend II, J.G., Springer International Publishing, Switzerland, Chap. 7, pp. 195–218

[69] Birur, G.C., and Tsuyuki, G.T., 1992, "A Simplified Generic Cryostat Thermal Model for Predicting Cryogen Mass and Lifetime," *Cryogenics*, Vol. 32, No. 2

[70] Burt, W., 2002, "Regenerative Systems: Stirling and Pulse Tube Cryocoolers," in *Spacecraft Thermal Control Handbook Vol. II: Cryogenics*, edited by Donabedian, M., The Aerospace Press and The American Institute of Aeronautics and Astronautics, Inc., Chap. 8, pp. 135–173

[71] Klein, S., and Nellis, G., 2012, *Thermodynamics*, Cambridge University Press, Cambridge, United Kingdom

[72] Radebaugh, R., 2000, "Pulse Tube Cryocoolers for Cooling Infrared Sensors," Proceedings of SPIE: Infrared Technology and Applications XXVI, Vol. 4130, pp. 363–379

[73] Bowman, R.C., Kiehl, B., and Marquadt, E., 2002, "Closed-Cycle Joule-Thomson Cryocoolers," in *Spacecraft Thermal Control Handbook Vol. II: Cryogenics*, edited by Donabedian, M., The Aerospace Press and The American Institute of Aeronautics and Astronautics, Inc., Chap. 10, pp. 187–216

[74] Swift, W.L., 2002, "Turbo-Brayton Cryocoolers," in *Spacecraft Thermal Control Handbook Vol. II: Cryogenics*, edited by Donabedian, M., The Aerospace Press and The American Institute of Aeronautics and Astronautics, Inc., Chap. 9, pp. 175–186

[75] Ross, R.G., 2002, "Cryocooler Performance Characterization," in *Spacecraft Thermal Control Handbook Vol. II: Cryogenics,* edited by Donabedian, M., The Aerospace Press and The American Institute of Aeronautics and Astronautics, Inc., Chap. 11, pp. 217–262

[76] Marquadt, J.S., Marquadt, E.D., and Boyle, R., 2012, "An Overview of Ball Aerospace Cryocoolers," Proceedings of the International Cryocooler Conference: Cryocoolers 17, pp. 1–8

[77] Raab, J., and Tward, E., 2010, "Northrup Grumman Aerospace Systems Cryocooler Overview," *Cryogenics*, Vol. 50, pp. 572–581

[78] Nast, T., Olson, J., Champagne, P., Evtimov, B., Frank, D., Roth, E., and Renna, T., 2006, "Overview of Lockheed Martin Cryocoolers," *Cryogenics*, Vol. 46, pp. 164–168

[79] Conrad, T., Schaefer, B., Bellis, L., Yates, R., and Barr, M., 2017, "Raytheon Long Life Cryocoolers for Future Space Missions", *Cryogenics*, Vol. 88, pp. 44–50

[80] Arts, R., Mullie, J., Leenders, H., de Jonge G., and Benschop, T., 2017, "Tactical versus Space Cryocoolers: A Comparison", Proceedings of SPIE: Tri-Technology Device Refrigeration (TTDR) II, Vol. 10180

[81] "JPL Cryocooler Program Summary," Jet Propulsion Laboratory, www2.jpl.nasa.gov/adv_tech/coolers/summary.htm

[82] "Cryocooler Technology Resources," National Institute of Standards and Technology, https://trc.nist.gov/cryogenics/cryocoolers.html

[83] "Cryogenics and Fluids Branch Website," NASA Goddard Spaceflight Center, https://cryo.gsfc.nasa.gov/

[84] "International Cryocooler Conference Past Proceedings," International Cryocooler Organization, https://cryocoolerorg.wildapricot.org/Past-Proceedings

Problems

7.1 A space-based cryostat design consists of a series of 0.5 m long struts that separate it from its mounting platform which is operating at room temperature. Determine the amount of contraction occurring along the primary axis of each strut when it cools from 300 K to 55 K using candidate materials of Ti 6Al-4V and stainless steel 304. (Note: The EES function "totalthermalexp" provides a value for the linear expansion integral given an input temperature and the assumption that the reference temperature is at 293 K.)

7.2 A dewar system flying with a stored cryogen for cooling purposes is separated from the main spacecraft structure via low thermal-conductivity struts. Ti 6Al-4V, stainless steel 316, T300 or S-glass are being considered as materials for the struts. Which material provides the best strength at 77 K? And further, which material will provide the best mitigation of conduction-based heat leak? Using Table 7.3 determine:

i) The STC for each of the candidate materials.

ii) Which material is most suitable for the project and why.

7.3 Copper comes in many different grades with varying RRR values. Determine the cryogenic temperature at which the thermal conductivity value begins to vary based on its RRR value. Candidate grades of copper to be considered include those with RRR values of 50, 100, 150, 300 and 500. Provide a comprehensive plot for each of these, showing the thermal conductivity performance between 4 K and 300 K.

7.4 A cryostat design requires electrical wiring that has a length of 1 m (endpoint to endpoint) and a wire size of 32 AWG. A wire material must be selected that will conduct the least amount of heat between endpoint temperatures of 250 K and 30 K.

i) For a single wire, determine the conductive heat load if the wire material is made of copper (RRR = 50), nichrome, platinum and tungsten.

7.5 A 32 AWG copper wire (RRR = 100) with a length of 1 m is carrying 0.1 A of current. The wire is heat sunk in three locations (see Figure P7.5) and is at a temperature of 300 K on the warm end. Determine:

i) The joule heating occurring between each of the temperature stages of the wire.

ii) The comprehensive heat load across the full 1 m length of the wire.

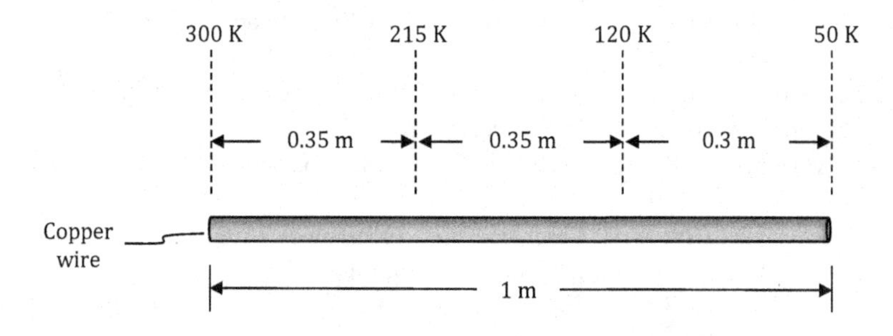

Figure P7.5 Copper wire with temperature sinks at length span locations

7.6 A hollow cubic-shaped storage container (Figure P7.6) made of polystyrene has an internal volume of 0.125 m^3 and a uniform wall thickness of 3.81 cm. The container is filled with 112.5 liters of argon at a saturation pressure of 101 kPa. The exterior wall of the box is at a constant uniform temperature of 300 K. Assuming the interior wall of the polystyrene box has a superheat of 1 K relative to the saturation temperature of the stored argon, how long does it take for the argon to fully evaporate? (Note: Constant pressure throughout the evaporation process can be assumed.)

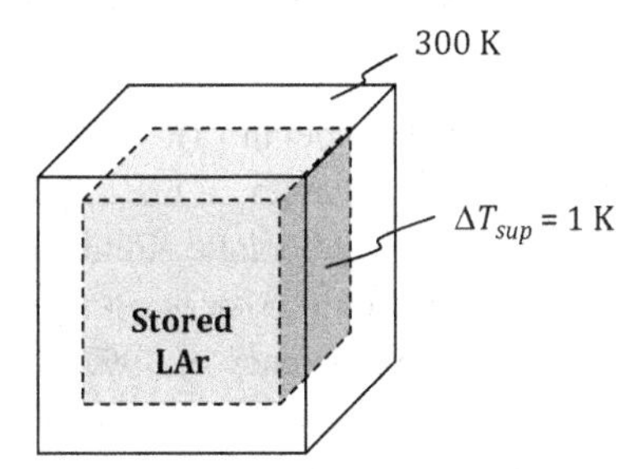

Figure P7.6 Polystyrene storage container

7.7 A Joule–Thomson cryocooler cycle (Figure P7.7) is using neon (R720) as a refrigerant. The mass flow rate within the cycle is 3.5×10^{-4} kg/s. The refrigerant has a state point condition of 150 K at 126.3 kPa at the compressor inlet. The heat rejection occurring at the aftercooler is 15 W. The fluid exits the low-side heat exchanger at saturated vapor conditions. For a cycle compression ratio of 150, determine the recuperative heat-exchanger effectiveness, power input to the compressor, specific power and cooling load if the heat exchanger inlet temperature on the high-pressure side of the cycle is 154 K.

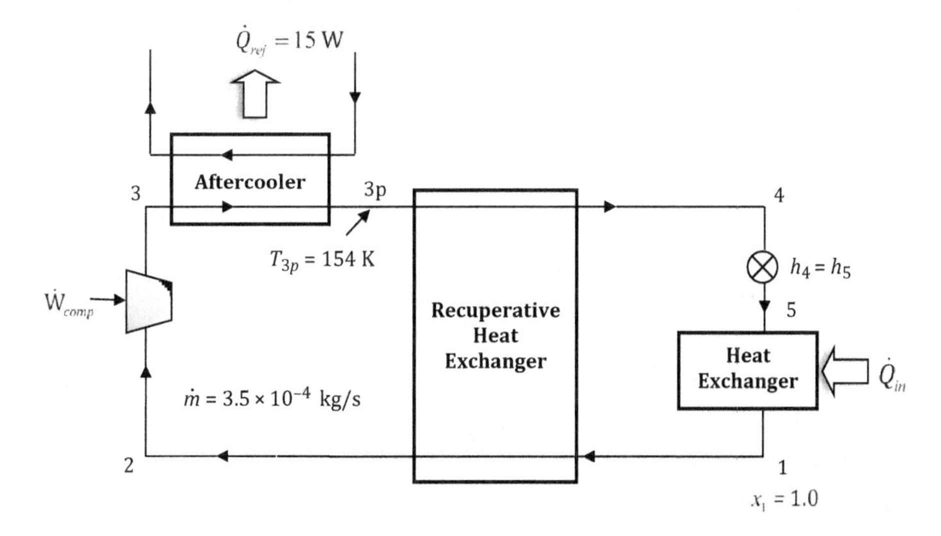

Figure P7.7 Joule–Thomson cryocooler cycle

7.8 A Reverse-Brayton cryocooler cycle charged with helium (R704) has a mass flow rate of 1.3×10^{-3} kg/s. The refrigerant has a state point condition of 300 K at 101 kPa at the compressor inlet. At a cycle compression ratio of 1.6, the compressor is operating with an efficiency of 99%. The recuperator has an inlet temperature of 302.8 K on the high-pressure side and a recuperative heat exchange efficiency of 99%. If the inlet temperature on the low-pressure side heat exchanger is 22.8 K, determine the work input to the cycle, the cooling load, the specific power and the % Carnot for the cycle.

7.9 A Reverse-Brayton cryocooler cycle charged with helium (R704) is providing 300 mW of cooling on its low side while operating with a mass flow rate of 1.5×10^{-3} kg/s. The refrigerant has a compressor inlet state point condition of 291.2 K at 101 kPa and a cycle compression ratio of 1.51. The compressor is operating with an efficiency of 99%. The recuperator has a heat exchange efficiency of 99%. Aftercooling is provided by an ammonia heat pipe operating at a saturation temperature of 319 K. The heat pipe has a total length of 0.75 m (evaporator and condenser lengths of 0.25 m each). The outer encasement has a diameter of 2.54 cm with a wall thickness of 1.5 mm. The wick structure is made of 304 stainless steel with an effective pore diameter of 250 μm, porosity of 0.59 and a radial thickness of 5.1 mm. Assuming that the heat pipe is operating at 97% of its capillary limit, determine the power input, specific power, average temperature cooled to on the low-side heat exchanger and the percent Carnot for the cycle. Please note from Chapter 5 that $d_s = r_{eff}/0.41$.

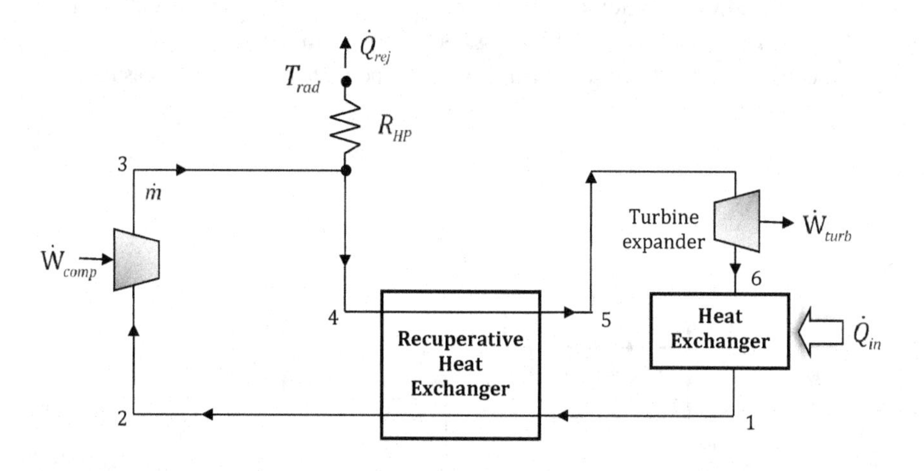

Figure P7.9 Reverse-Brayton cryocooler cycle with heat pipe aftercooling

7.10 A Joule–Thomson cryocooler is providing refrigeration cooling to a dewar via a thermal strap (Figure P7.10). Heat rejection from the cryocooler is occurring at an ambient temperature of 300 K. The input power is 85 W.

i) If the SP is 300, what is the cooling load?
ii) If the cold end is operating at 23 K, what is the cryocooler efficiency in % of Carnot COP?

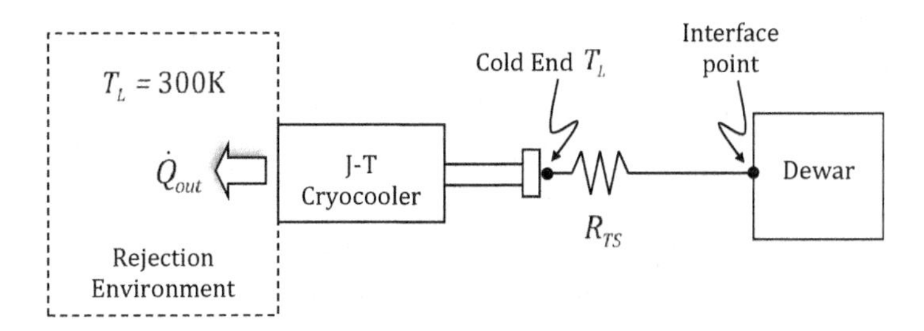

Figure P7.10 Cryocooler configuration

iii) It is anticipated that the thermal strap connecting the cold end of the cryocoooler system to the dewar will be made of multiple foils of OFHC RRR = 150. If 20 foils are used, what is the temperature at the interface point?

Thermal strap assumptions:

- Each foil is 1 mil thick and 1″ in width.
- $\eta_P = 1.0$; $\eta_S = 0.75$; $\eta_E = 1.0$.
- The unfolded length of the thermal strap is 20 cm.
- The temperature value for the thermal conductivity of the foils is T_L.

iv) Assuming the contour surface of the thermal strap (buffed copper) has a full view to the interior side of a cylindrical shroud (buffed aluminum) with a length of 20 cm and a diameter of 5 inches, what would the cooling capacity of the cryocooler need to be to maintain the interface temperature determined in part iii)?

7.11 A dewar tank configuration consists of an inner vessel, an outer vessel and two intermediate radiative shields (Figure P7.11). The cubic inner vessel is filled with 118.8 liters of liquid hydrogen under saturation conditions at 101 kPa. The inner and outer vessels are made of Al-6061. The radiative shields are made of Al-1100. The inner vessel, outer vessel and radiation shields each have a wall thickness of 5 mm and are physically connected to each other through low thermal-conductivity struts. There are a total of 6 A struts, 6 B struts and 6 structure mounts. Both the A and B strut material is Ti 6Al-4V. Both have a hollow cylindrical geometry with a wall thickness of 2.0 mm, an outer diameter of 1.9 cm and a length of 3.5 cm. The structure mounts are also made of Ti 6Al-4V and have a shape factor of 7.121×10^{-3} m. The gaps between the shields, the inner shield and the inner vessel and the outer shield and the outer vessel are under vacuum conditions. Assuming the inner vessel has a uniform temperature that is superheated by 4 K relative to the saturation temperature of the LH_2, 15 layers of aluminized kapton ($\varepsilon = 0.03$) applied to the IV exterior, and the outer vessel is at a uniform temperature of 220 K,

i) Provide a schematic for the thermal circuit.

ii) Determine the total hold time for the LH_2 stored cryogen.

Assume an emissivity of 0.03 for each of the aluminum surfaces. Since the struts and structure mounts have a very small area relative to that of the vessels and the shields,

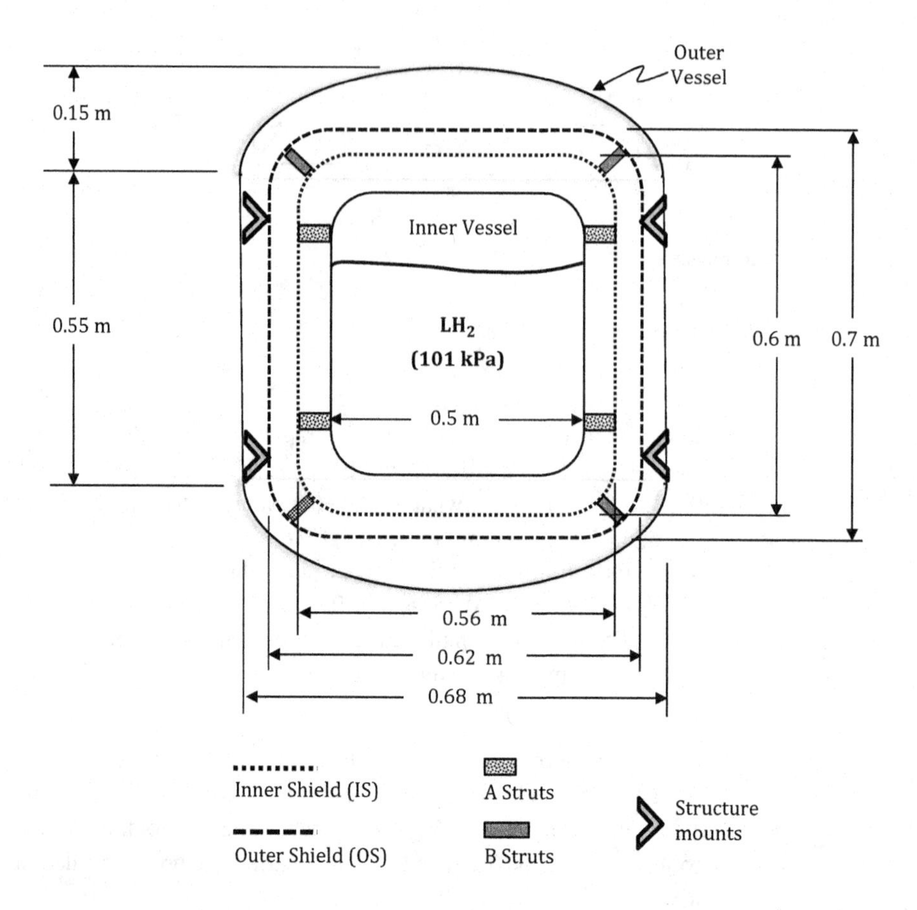

Figure P7.11 Liquid storage dewar configuration with inner and outer radiative shield

radiative thermal exchange between them and the vessels and shields can be disregarded. The shields may be approximated as cylindrical in geometry using the dimensions provided in the figure.

iii) Assuming an RBO hold time technique is applied by extracting 1.26 W of thermal energy from the heat leak reaching the IV, what is the new hold time?"

8 Developmental and Environmental Testing

8.1 Background

In the development of spaceflight hardware systems, items are designed to perform in various operational environments. When the environment for intended use is considered extreme, the hardware in question undergoes ruggedization during the design phase to insure proper functionality once it is placed in the operational environment. The success of the measures taken during the design phase to ruggedize a system is verified by a test regiment imposed at various levels of assembly that expose the hardware to environmental extremes comparable to, and slightly in excess of, what is expected in the operational theater. This is known as design, development and environmental testing. This chapter examines the philosophy behind environmental test practices for spaceflight thermal systems and looks at practical considerations associated with design and ground testing.

8.1.1 Testing for Relevant Environments

The temperature of space is approximately 3 K, which is near the lower end of the cryogenic temperature scale. Environmental loading on high absorptivity surfaces (in either the IR or UV spectrum), however, can easily result in surface temperatures approaching 90°C. The pressure in space is on the order of 10^{-14} Torr. Thus, even though space is not empty, it is considered a vacuum. Objects in space are occasionally subject to bombardment by high-energy particles, such as cosmic rays, that have been known to disrupt the performance of flight electronics while in operation.

All these environmental features mean that space is an extreme environment. However, the design and testing of space systems has an additional level of complexity compared to flight systems designed for use in extreme terrestrial environments: the system still has to be transported there. This is performed with the aid of a launch vehicle. The environmental conditions experienced onboard the launch vehicle during the launch sequence may also be extreme. Mechanical vibrations during launch can have frequency ranges into the 1000s of Hz. Also, acoustic vibrations associated with contemporary launch vehicles easily reach values in the 150–170 dB range [1,2]. Thus, space system design and environmental testing must accommodate both the launch sequence and the operational environment that the hardware will be exposed to.

8.1.2 Significance of Testing in Vacuum

While each of the "extreme" environment features discussed in Section 8.1.1 may be unique to space, demonstrating operational capabilities in a vacuum environment is especially relevant from a heat transfer perspective. As space is a vacuum, the predominant heat transfer mechanisms occurring on space platforms are conduction and radiation. Spaceflight hardware performance must therefore be demonstrated prior to actual launch not only in a relevant vacuum environment, but also using the heat transfer mechanisms that will enable thermal communication with the operational environment during flight.

When testing in air, natural convection currents are often present along the direction of the gravity vector. This can result in temperatures several degrees lower than they would be were convective cooling not present. In regard to MLI, the presence of gas molecules between the interstitial layers of the blanket tends to create thermal shorts through the blanket due to gas conduction. For ground-based testing of spaceflight systems, the best way to replicate the space environment is to conduct tests inside a pressure vessel under vacuum conditions. From a practical perspective, the reason for having vacuum conditions inside a thermal vacuum chamber is to eliminate convective flows along hardware surfaces that promote either heating or cooling, and to eliminate thermal shorts in MLI due to gas conduction between blanket layers. Testing in vacuum also avoids condensation on components operating at temperatures below the dew point. The nominal pressure level in space is a "hard vacuum" relative to earth-based standards. It is extremely difficult to reach space-like vacuum levels in thermal vacuum chambers. Thus, the vacuum conditions used during ground testing only need to be good enough to negate the aforementioned effects. As we saw in Chapter 3, these effects are minimized around pressure levels on the order of 10^{-5} Torr. Thus, thermal vacuum chamber systems used will typically only pull vacuum down to $10^{-6}/10^{-7}$ Torr.

8.2 Spacecraft Systems Test Philosophy

Spacecraft consist of tens of thousands of individual parts. Successful operation of a spacecraft will be dependent upon the performance capability of each of its individual parts. As with any multi-parameter system, there are dominant parameters that strongly influence the system's performance, as well as parameters that have a lesser effect upon the overall system performance. Testing of spaceflight systems to insure proper in-flight performance focuses on screening of parts, engineering design and the final assembled product for features that can lead to marginal or failed performance. Successful screening of system performance during ground-based testing does not guarantee that a system will not experience an in-flight failure [3,4]. Rather, the philosophy behind test programs that screen for performance and/or potential failures is to reduce the risk of in-flight failure. Conversely, the probability of in-flight success increases significantly if all ground-based tests are successfully passed. The military-based approach to the verification of spaceflight systems via ground-based testing is detailed in MIL-STD 1540 (presently revision E) [5]. Within the American aerospace community, this document sets out how to screen for potential design and performance failure in the context of a structured spaceflight test program. While other government and private-sector organizations have their own screening methodologies and guidance documents pertaining to the design and test of spaceflight systems, all such practices and guidance documents are derivatives of MIL-STD 1540. The main difference between the military standard and the approach adopted by other organizations is the rigor of the test program and the amount of risk to performance the organization considers acceptable [3]. Whatever organization is designing and building the space system, the two primary areas of focus for reducing the risk of hardware failure are engineering design and the workmanship at the various assembly levels. Development and testing in these two areas are key activities performed in different phases of the project life cycle.

8.2.1 The Project Life-Cycle

The project life cycle is a "design/build/test/fly" framework. Table 8.1 lists each phase of the life cycle, along with a brief description of activities performed during each phase. The life cycle consists of phases A to F. Phase A establishes a conceptual design to meet the required mission objectives and includes a system requirements review. Phase B consists of activities that further the design accompanied by maturation efforts targeting less mature technologies desired for flight use. During this phase the preliminary design review is held to confirm the thermal control system design that will serve as the basis for iteration and refinement as the project advances. By the beginning of phase C, technology maturation activities should be complete. The final detailed design for the flight model should be established and a critical design review held to assess it. Parts fabrication should have begun by the end of phase C. Nearing the transition to phase D,

Table 8.1 Project life-cycle phases

Phase	Description of activities	Major reviews
A	Subsystem and system architectures refined to establish credibility of the overall design	System Requirements Review
B	Technology maturation of system components further developed along with further refinement of the overall subsystem and system architectures	Preliminary Design Review
C	Subsystem and system designs completed. Fabrication of subsystem/system parts begins.	Critical Design Review System Integration Review
D	Parts fabrication completed. Assembly and test at subsystem, instrument and spacecraft level performed.	Test Readiness Review Pre-ship Review
E	Launch and flight operations	Launch Readiness Review
F	End of mission followed by de-orbit	Disposal Readiness Review

a systems integration review is held to assess the plan for integration and assembly of the hardware. Phase D is dedicated to the completion of outstanding parts fabrication along with integration and test at subsystem, instrument and spacecraft level. Test readiness reviews are held prior to instrument- and spacecraft-level tests to review the status of the hardware and confirm that the project is actually ready to undergo testing at these higher levels of assembly. On completion of the test campaign, a pre-ship review is held to review the campaign results and confirm that the performance requirements have been met and test guidelines adhered to in the process. Phase E consists of launch activities and in-flight operations. A launch readiness review is held prior to launch. Phase F is dedicated to end-of-mission activities, including de-orbit. The disposal readiness review includes an overview of the de-orbit and re-entry plan, as well as an assessment of threats to populated areas of the planet. For further details on the system life cycle, see NASA's *Systems Engineering Handbook* [6].

8.2.2 Design and Development

In spaceflight systems development, test activities are often focused on a particular level of assembly. The type of test performed and the associated requirements to be demonstrated are a function of the level of assembly at which the test is performed. Testing at the component or subsystem level is primarily associated with design and development activities. Within the space community, both components and subsystems are often referred to as units [4,7] or space widgets. Flight projects mandate that all parts and materials used in the system build be certified by their manufacturer as flightworthy prior to purchase. Subassemblies are reviewed to ensure they are designed to meet the desired performance requirements (temperature, stability, heat transfer, etc.). When it has successfully passed design and development tests to meet project performance requirements, the design itself is approved for further screening in the form of qualification and acceptance testing.

An example of a developmental test would be thermal strap conductance testing at specific endpoint temperatures. If the endpoint temperatures are fixed operational values, the overall design of the thermal strap (number of foils and shape factor geometry) will give an effective thermal conductance value that can be measured. This conductance value can in turn be used to calculate the heat flow from the warm end to the cold end. If the net heat flow across the length of the thermal strap and the mounting location interfaces meet the desired amount of heat transfer for the application's temperature difference (mounting surface to mounting surface) then the design is viable for use. If the heat flow at the desired temperature difference does not meet the performance requirement, the thermal strap needs to be redesigned.

8.2.3 Temperature Cycling

Temperature cycling is a technique used to detect flaws in a unit resulting from poor fabrication and/or assembly techniques. Temperature cycling of a unit can be performed either in a pressure vessel with an inert gas or with the chamber under vacuum conditions. The temperature limits that the unit is cycled to (high and low) are dependent on its assembly stage. Component- and subsystem-level testing typically include a greater amount of control to achieve desired temperatures. However, at instrument and spacecraft levels, temperature tests on a particular unit may be limited by nearby units generating heat transfer effects that were not present during standalone testing. As regards the total number of cycles, standard NASA practice is to perform at least eight temperature cycles under thermal vacuum conditions at the component level prior to integration at the next level of assembly (instrument level) and a minimum of four cycles under thermal vacuum conditions at both instrument and spacecraft levels [8].

8.2.4 Qualification and Acceptance

The objective of qualification and acceptance testing is to demonstrate that a given hardware design is robust enough to not merely meet, but exceed in-flight performance requirements. This is demonstrated by temperature cycling the hardware in question to high and low temperatures in excess of levels expected during spaceflight. For military applications, the high and low qualification limits are typically 71°C and −34°C respectively [7]. NASA uses a proto-flight test strategy which includes unit-level temperature cycling to levels lower than those used by the military, yet higher than acceptance limits (proto-qual limits) [4]. Under proto-flight constructs, proto-qual temperature limits include an additional five degrees of temperature margin applied to the high and low acceptance temperature values. Acceptance temperature levels are an additional five degrees beyond the maximum and minimum expected temperatures (based on model predictions). Once the acceptance test requirements have been met, the unit is considered viable for further advancement and integration at higher assembly levels.

Due to the extreme nature of qualification tests, hardware tested to qualification limits is not used for spaceflight. Successful demonstration of a unit's performance capabilities following qualification testing reinforces the ability of the design to operate in a robust manner over the temperature limits tested to. Thus, if a duplicate unit is fabricated to the same design specifications using the same manufacturing practices and tooling, it is very likely to be able to perform to the same temperature levels without being over-stressed thermally during thermal vacuum testing.

8.3 Assembly-Level Testing

The focal points of testing during phase D are the instrument and spacecraft levels. Assembly-level testing may involve thermal communication with other items on the surrounding structure that may affect thermal performance. Prior to assembly-level testing, these thermal inputs are not accounted for. The following subsections look at testing performed at each assembly level.

8.3.1 Instrument

The first level of assembly that unifies multiple subsystems for a specific task or measurement is the instrument level. Throughout the engineering design process leading up to the instrument-level assembly, engineering tools are used to model and predict the thermal performance of components and the subsystem. At the instrument level, thermal balance and thermal cycle tests are performed under vacuum conditions as part of the instrument-level test regiment. Thermal balance testing establishes a baseline against which the thermal model temperature predictions developed using analysis tools can be compared. After correlation of the model with the test results, the refined model is used to approximate the performance of the instrument in the future. Temperature cycles are performed between high and low values for the instrument's units. The extreme temperature values can either be designated acceptance limits or based on model predictions of worst-case hot and worst-case cold with temperature margin added on. During the temperature-cycling part of the thermal vacuum tests, when thermal stabilization (defined in terms of temperature vs. time) of a unit plateaus at the extreme values, testing of the unit is conducted. Tests of the unit's electronics can take the form of aliveness tests (on/off), functional tests that include simplified communications with the unit or comprehensive performance tests.

8.3.2 Spacecraft

The highest level of assembly is the spacecraft level. Spacecraft-level testing is performed under vacuum conditions, and like instrument-level testing it consists of a combination of thermal balance tests, thermal cycles and unit-level functional and/or performance tests. The number of cycles performed will be based on the remaining number of cycles required to fulfill the organization's verification requirements following instrument-level testing. The temperature extremes used for cycles is based on either the high or low extreme temperature values demonstrated during prior testing at the unit level or analysis-based temperature predictions (with margin added) if a correlated thermal model for the system is available. Temperature limits tested to should be kept within the designated red limits for the unit. At each temperature plateau, functional tests are only performed after thermal equilibrium is achieved on the unit of interest. By the end of the successful spacecraft-level test campaign, the whole system must have demonstrated 200 hours [8] of failure-free performance in vacuum before it is considered ready for flight.

8.4 Cryogenic Considerations

Just as special considerations are incorporated into the design of a cryogenic system (see Chapter 7), testing of cryogenic systems must also take into account low-temperature effects on construction materials and unit assemblies. Section 8.2.3 showed that the aim of temperature cycling at both the qualification and acceptance levels is to thermally stress the unit to limits slightly beyond temperature levels expected during in-flight operation. Chapter 7 noted that the cryogenic temperature regime spans all temperatures $\leq$123 K. Cryogenic systems are typically designed to operate at setpoint temperature values that can be as low as the mK scale. For a unit designed to operate at the lower end of the cryogenic temperature regime (e.g., 4 K), cooling down from room temperature to operational temperatures is accompanied by extreme thermal stresses commensurate with the low-temperature operational value. Thus, qualification- and acceptance-level temperature cycling beyond operational temperature levels has little significance and is not performed. Nonetheless, temperature cycling is still relevant for portions of the cryogenic system that may be operating near or well above the cryogenic temperature limit (i.e., 123 K). While the temperature cycling of cryogenic units may not be appropriate at the lower end of the cryogenic temperature regime, performance margin must still be demonstrated. When temperature cycling is involved, margin is applied in the form of a nominal temperature value to the high and low operational values to define the qualification and acceptance test temperature limits. For cryogenic systems, the standard approach is to apply margin to the system in the form of additional heat load. As we saw in Chapter 7, heat leak into a cryogenically cooled volume is always present. Thus, the heat-load margin applied when at the operational temperature accounts for unanticipated heat transfer into the volume. If a cryogenic cooling system can manage heat leak with margin added on and maintain desired operational temperature levels, then the heat-load margin test is deemed successful. This verifies the system's ability to achieve desired in-flight performance. At the design phase, the margin is fixed as 100% on top of the defined heat load at that time. As the design matures, the margin decreases, culminating in a designated 25% margin by the time a project enters phase E. For detailed guidelines pertaining to the test of cryogenic systems, see the *Spacecraft Thermal Control Handbook Vol. II* [7] or NASA Goddard Space Flight Center's *Environmental Verification System Guidelines* [8].

References

[1] *Spacecraft Thermal Control*, 1973, NASA Space Vehicle Design Criteria (Environment), National Aeronautics and Space Administration, Washington DC, NASA SP-8105

[2] Perl, E., 2006, "Test Requirements For Launch, Upper-Stage, and Space Vehicles," Structural Test Subdivision, Aerospace Corporation, Report # TR-2004(8583)-1 Rev. A

[3] Welch, J.W., 2002, "Thermal Testing," in *Spacecraft Thermal Control Handbook Vol. I: Cryogenics*, edited by Gilmore, D., The Aerospace Press and The American Institute of Aeronautics and Astronautics, Inc., Chap. 19, pp. 709–757

[4] Peterson, Max, and Rodberg, E.H., 2005, "Spacecraft Integration and Test," in *Fundamentals of Space Systems*, edited by Pisacane, V.L., Oxford University Press, Inc., 2005, Chap. 14, pp. 725–753

[5] Welch, J.W., 2010, "Flight Unit Qualification Guidelines," Vehicle Systems Division, Aerospace Corporation, Report # TOR-2010(8591)-20

[6] *NASA Systems Engineering Handbook*, 2016, National Aeronautics and Space Administrations, Washington DC, NASA/SP-2016-610S Rev. 2

[7] Donabedian, M., 2002, "Thermal Control Margins, Risk Estimation, and Lessons Learned," in *Spacecraft Thermal Control Handbook Vol. II: Cryogenics,* edited by Gilmore, D., The

Aerospace Press and The American Institute of Aeronautics and Astronautics, Inc., Chap. 19, pp. 469–480

[8] "General Environmental Verification Standard (GEVS) for GSFC Flight Programs and Projects," 2013, GSFC Technical Standards Program, NASA Goddard Space Flight Center, Greenbelt, MD, GSFC-STD-7000A

Appendix A Common View Factor Tables

① [a]Finite Parallel Plates [1,2]

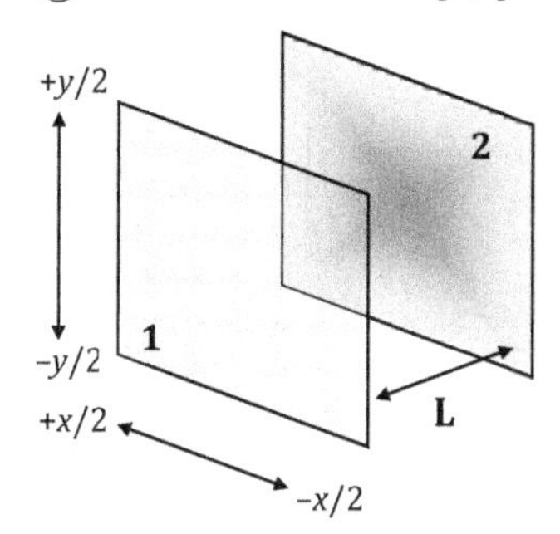

$$A = \frac{x}{L} \quad B = \frac{y}{L}$$

$$F_{1-2} = \frac{2}{\pi AB}\left[\ln\left(\frac{(1+A^2)(1+B^2)}{1+A^2+B^2}\right)^{0.5} + \left(A\sqrt{1+B^2}\right)\cdot\tan^{-1}\left(\frac{A}{\sqrt{1+B^2}}\right) + \left(B\sqrt{1+A^2}\right)\cdot\tan^{-1}\left(\frac{B}{\sqrt{1+A^2}}\right) - A\tan^{1}A - B\tan^{-1}B\right]$$

② [a]Finite rectangles with a common edge at 90° to one another [2]

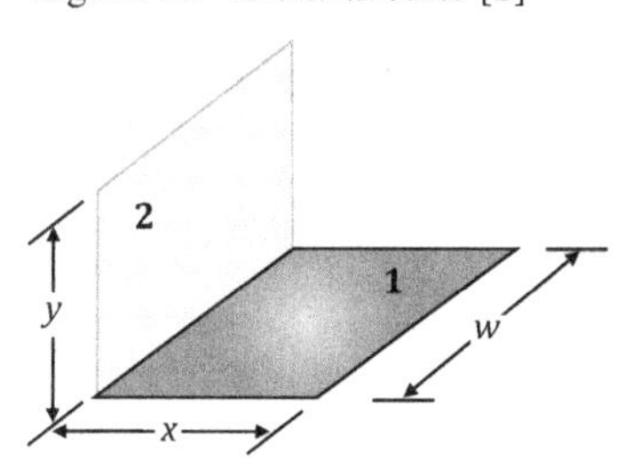

$$A = \frac{y}{w} \quad B = \frac{x}{w}$$

$$F_{1-2} = \frac{1}{\pi B}\left[B\cdot\tan^{-1}\left(\frac{1}{B}\right) + A\cdot\tan^{-1}\left(\frac{1}{A}\right) - \left(\sqrt{A^2+B^2}\right)\cdot\tan^{-1}\left(\frac{1}{\sqrt{A^2+B^2}}\right) + \frac{1}{4}\ln\left(\frac{(1+B^2)(1+A^2)}{1+B^2+A^2}\cdot\left[\frac{W^2(1+B^2+A^2)}{(1+B^2)(B^2+A^2)}\right]^{B^2}\cdot\left[\frac{A^2(1+A^2+B^2)}{(1+A^2)(A^2+B^2)}\right]^{A^2}\right)\right]$$

③ [b]Differential surface element to a circular disk (dA aligned with disk center) [3]

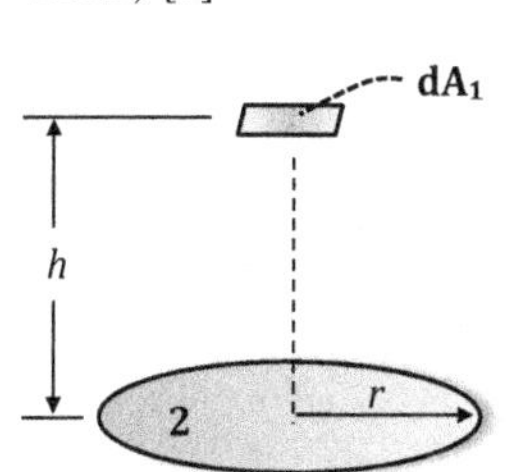

$$F_{1-2} = \frac{r^2}{h^2+r^2}$$

(cont.)

④ [a]Parallel circular disks with aligned centers [1,2,4,5].

$$R_1 = \frac{r_1}{h} \quad R_2 = \frac{r_2}{h}$$

$$A = 1 + \frac{1 + R_2^2}{R_1^2}$$

$$F_{1-2} = \frac{1}{2}\left[A - \sqrt{A^2 - 4\left(\frac{R_2}{R_1}\right)^2}\right]$$

⑤ [c]Sphere to a disk (aligned centers) [6,7]

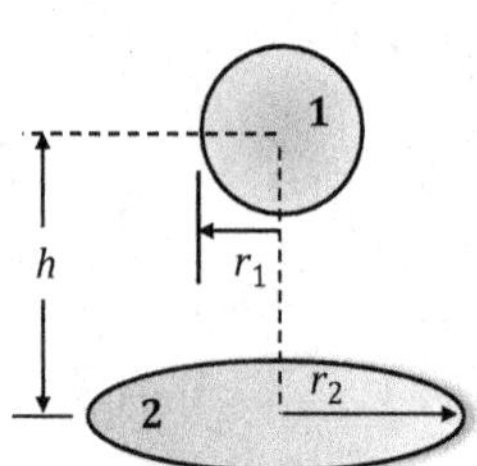

$$R_2 = \frac{r_2}{h}$$

$$F_{1-2} = \frac{1}{2}\left[1 - \frac{1}{\sqrt{1 + R_2^2}}\right]$$

⑥ [a]Differential surface element to disk [1,2]

$$H = \frac{h}{x} \quad R = \frac{r}{x}$$

$$Z = 1 + H^2 + R^2$$

$$F_{1-2} = \frac{H}{2}\left[\frac{Z}{\sqrt{Z^2 - 4R^2}} - 1\right]$$

⑦ [d]Differential ring element on interior of cylinder to disk on cylinder end [8,9].

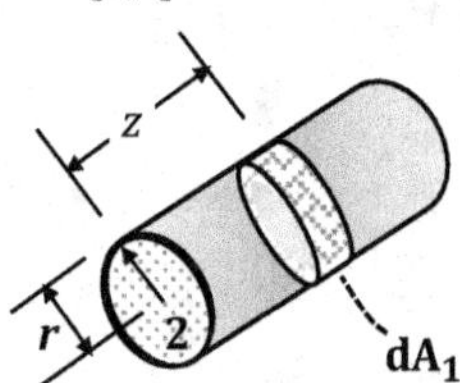

$$Z = \frac{z}{2r}$$

$$F_{1-2} = \frac{Z^2 + 0.5}{\sqrt{Z^2 + 1}} - Z$$

⑧ [b]Point source to a sphere [3]

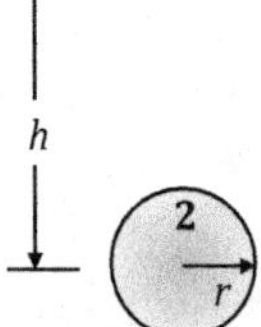

$$R = \frac{r}{h}$$

$$F_{1-2} = 0.5\cdot\left(1 - \sqrt{1 - R^2}\right)$$

(*cont.*)

⑨ [b]Differential element to sphere (element aligned w/sphere normal) [3,10] 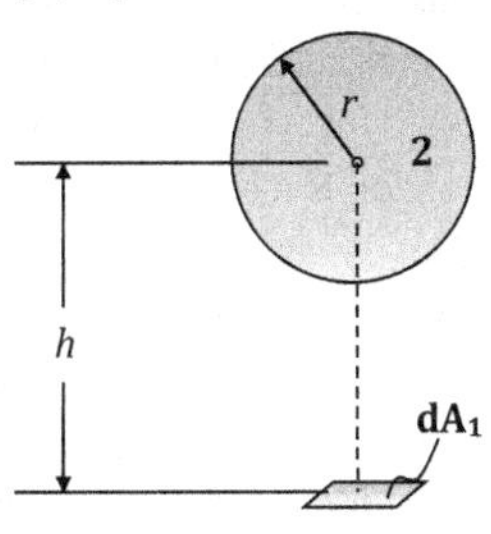	$F_{1-2} = \left(\frac{r}{h}\right)$
⑩ [b]Differential element to sphere (element tangent aligned w/sphere normal) [3,11] 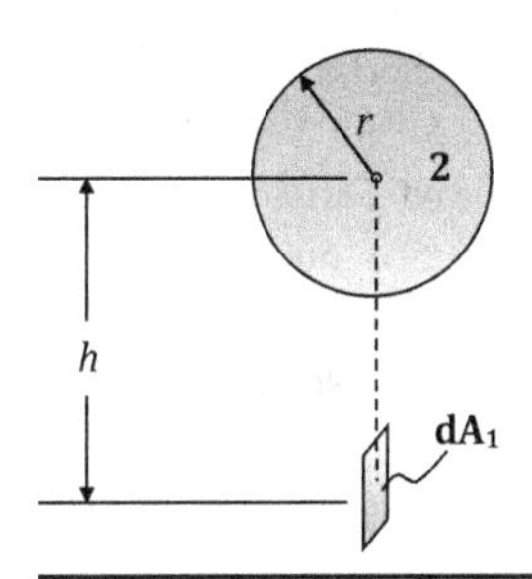	$H = \frac{h}{r}$ $F_{1-2} = \frac{1}{\pi}\left(\tan^{-1}\frac{1}{\sqrt{H^2-1}} - \frac{\sqrt{H^2-1}}{H^2}\right)$

[a] Reprinted from "Radiant-interchange configuration factors," by Hamilton, D.C. and Morgan, W.R., 1952, *NASA TN* 2836.
[b] Reprinted from *Journal of Heat Transfer*, Vol. 94, No. 3, Chung, B.T.F., and Sumitra, P.S., "Radiation Shape Factors from Plane Point Sources," pp. 328–330, 1972, with permission from ASME.
[c] Reprinted from *Journal of Heat Transfer*, Vol. 92, No. 1, Feingold, A., and Gupta, K.G., "New Analytical Approach to the Evaluation of Configuration Factors in Radiation from Spheres and Infinitely Long Cylinders," pp. 69–76, 1970, with permission from ASME.
[d] Reprinted from *International Journal of Heat & Fluid Flow*, Vol. 4, No. 3, Buraczewski, Czeslaw, and Stasiek, "Application of Generalized Pythagoras Theorem to Calculation of Configuration Factors between Surfaces of Channels of Revolution," pp. 157–160, 1983, with permission from Elsevier.

The view factors collected in the Appendix A table are a minor subset of the numerous view factor relations developed over the past century by radiative heat transfer researchers. A comprehensive listing of view factors (compiled by Professor John R. Howell) may be found online at www.thermalradiation.net/indexCat.html.

References

[1] Hamilton, D.C., and Morgan, W.R., 1952, "Radiant-Interchange Configuration Factors," NASA TN 2836.

[2] Hottel, H.C., 1931, "Radiant Heat Transmission between Surfaces Separated by Non-absorbing Media," *Transactions of the ASME*, Vol. 53, FSP-53-196, pp. 265–273.
[3] Chung, B.T.F., and Sumitra, P.S., 1972, "Radiation Shape Factors from Plane Point Sources," *Journal of Heat Transfer*, Vol. 94, No. 3, pp. 328–330.
[4] Keene, H.B., 1913, "Calculation of the Energy Exchange between Two Fully Radiative Coaxial Circular Apertures at Different Temperatures," *Proceedings of the Royal Society*, Vol. LXXXVIII-A, pp. 59–60.
[5] Leuenberger, H., and Person, R.A., 1956, "Compilation of Radiation Shape Factors for Cylindrical Assemblies," Paper no. 56-A-144, ASME, November.
[6] Feingold, A., and Gupta, K.G., 1970, "New Analytical Approach to the Evaluation of Configuration Factors in Radiation from Spheres and Infinitely Long Cylinders," *Journal of Heat Transfer*, Vol. 92, No. 1, pp. 69–76.
[7] Naraghi, M.H.N., and Chung, B.T.F., 1982, "Radiation Configuration between Disks and a Class of Axisymmetric Bodies," *Journal of Heat Transfer*, Vol. 104, No. 3, pp. 426–431.
[8] Hottel, Hoyt C., and Keller, J.D., 1933, "Effect of Reradiation on Heat Transmission in Furnaces and Through Openings," *Transactions of the ASME*, Vol. 55, IS-55-6, pp. 39–49.
[9] Buraczewski, Czeslaw, and Stasiek, Jan, 1983, "Application of Generalized Pythagoras theorem to Calculation of Configuration Factors between Surfaces of Channels of Revolution," *International Journal of Heat & Fluid Flow*, Vol. 4, No. 3, pp. 157–160.
[10] Juul, N.H., 1979, "Diffuse Radiation View Factors from Differential Plane Sources to Spheres," *Journal of Heat Transfer*, Vol. 101, No. 3, pp. 558–560.
[11] Naraghi, M.H.N., 1988b, "Radiation view factors from spherical segments to planar surfaces," *Journal of Thermophysics and Heat Transfer*, Vol. 2, No. 4, pp. 373–375.

Appendix B Thermal Coatings Tables†

White and/or color coatings		
Coating name	α (Solar)	ε (IR)
Astro-Quartz	0.06	–
Astro-Quartz with 900°C bake out	0.03	–
Aeroglaze A971 yellow paint	0.43	0.89
AZ4301	0.19	–
Beta Cloth/VDA Sheldahl	0.38	0.85
Beta Cloth/VDA Chemfab 500 no silicone	0.45	0.90
Beta Cloth Glass/Al	0.29	0.91
Barium Sulfate with Polyvinyl Alcohol	0.06	0.88
Biphenyl-white solid	0.23	0.86
Catalac white paint	0.24	0.90
Dupont Lucite acrylic lacquer	0.35	0.90
Dow Corning white paint	0.19	0.88
Gortex expanded Teflon/VDA (VDA side)	0.54	0.54
Gortex expanded Teflon/VDA (Teflon side)	0.43	0.43
GSFC white paint NS43-C	0.20	0.92
GSFC white paint NS44-B	0.34	0.91
GSFC white paint MS74	0.17	0.92
Aeroglaze A276	0.26	0.88
Aeroglaze A276/K1100 composite 1500ESH	0.53	0.89
Tedlar white plastic	0.39	0.87
S13GLO	0.20	0.90
S13GPLO/Kevlar composite	0.21	0.90
TOR-RC Triton Systems	0.18	0.82
COR-RC Triton Systems	0.28	0.71
YB71 white paint	0.18	0.90
Z93	0.16	0.92
Zinc orthotitanate with potassium silicate	0.13	0.92
Zinc oxide with sodium silicate	0.15	0.92
Zirconium oxide with 650 glass resin	0.23	0.88

† Emmissivity and absorptivity table data republished courtesy of NASA.

Thermally conductive paints		
Coating name	α (Solar)	ε (IR)
Electrodag	0.90	0.68
GSFC NS43-G	0.26	0.90
NS43-G/Hincom	0.18	0.92
GSFC green NS53-B	0.52	0.87
GSFC green NS43-E	0.57	0.89
GSFC green NS55-F	0.57	0.91
GSFC green NS-79	0.57	0.91
Z93SC55	0.14	0.94

Anodized aluminum coatings		
Coating name	α (Solar)	ε (IR)
AnoBlack	0.94	0.89
Black (1)	0.65	0.82
Black (2)	0.86	0.86
Black		
0.0001 mils	0.51	0.75
0.0005 mils	0.60	0.82
0.001 mils	0.67	0.84
Black anodize with sealer		
0.001 mils	0.71	0.84
0.002 mils	0.70	0.86
1.5 mils	0.73	0.86
Blue (1)	0.67	0.87
Blue (2)	0.53	0.82
Brown	0.73	0.86
Chromic	0.44	0.56
Clear (1)	0.27	0.76
Clear (2)	0.35	0.84
Clear/7075 aluminum	0.68	0.82
Clear		
0.0001 mils	0.32	0.75
0.0005 mils	0.37	0.81
0.001 mils	0.44	0.85
Clear type 2 Class 1	0.43	0.67
Clear with sealer		
0.001 mils	0.50	0.84
0.002 mils	0.70	0.86
1.5 mils	0.59	0.87
Green	0.66	0.88
Gold (1)	0.48	0.82
Gold		
0.0001 mils	0.40	0.74
0.0005 mils	0.46	0.81
0.001 mils	0.52	0.82

(cont.)

Anodized aluminum coatings		
Coating name	α (Solar)	ε (IR)
Hard anodize		
Aluminum 7075-T6	0.83	0.87
Aluminum 6061-T6	0.90	0.86
Aluminum 2024-T6	0.76	0.90
Plain	0.26	0.04
Red	0.57	0.88
Sulphuric	0.42	0.87
Yellow	0.47	0.87
Blue anodized titanium foil	0.70	0.13

Metal and conversion coatings		
Coating name	α (Solar)	ε (IR)
Alodyne aluminum 2024	0.42	0.10
Alodyne aluminum 6061-T2	0.44	0.14
Aluminum polished	0.14	0.03
Alzac A-2	0.88	0.88
Alzac A-5	0.18	–
Alodine beryllium	0.68–0.76	0.06–0.15
Beryllium copper oxidized	0.83–0.87	0.09
Beryllium		
Lathe finish	0.51	0.05
Lap finish	0.71	0.12
Wire finish	0.77	0.28
Milled finish	0.52	0.07
Black chrome	0.96	0.62
Black copper	0.98	0.63
Black irridite	0.62	0.17
Black nickel	0.91	0.66
Black nickel/ZK60Mg	0.78–0.80	0.56–0.61
Black nickel/magnesium	0.92	0.74
Black nickel/titanium	0.91	0.77
Buffed aluminum	0.16	0.03
Buffed copper	0.30	0.03
Constantan-metalsStrip	0.37	0.09
Copper foil tape		
Plain	0.32	0.02
Sanded	0.26	0.04
Tarnished	0.55	0.04
Dow 7 on polished magnesium	–	0.49
Dow 7 on sanded magnesium	–	0.65
Dow 9 on magnesium	–	0.87
Dow 23 on magnesium	0.62	0.67

(*cont.*)

Metal and conversion coatings		
Coating name	α (Solar)	ε (IR)
Electroplated gold	0.23	0.03
Electrolytic gold	0.20	0.02
Electroless nickel	0.39	0.07
Gold-plated beryllium mirror	0.21	0.03
Irridite aluminum (1)	–	0.11
Irridite aluminum (2)	0.24	0.04
Inconel X foil	0.52	0.10
Kannigen–nickel alloy	0.45	0.08
Mu metal	–	0.09
Plain beryllium copper	0.31	0.03
Platinum foil	0.33	0.04
Tantalum foil	0.40	0.05
TiN/titanium	0.52	0.11
Titanium anodized	0.84–0.86	0.44–0.48
Tiodize titanium TinFin400	0.74	0.73
Tiodize titanium TinFin200	0.74	0.23
Tungsten polished	0.44	0.03
Stainless steel		
Polished	0.42	0.11
Machined	0.47	0.14
Sandblasted	0.58	0.38
Machine rolled	0.39	0.11
Boom polished	0.44	0.10
1-mil 304 foil	0.40	0.05

Vapor deposited coatings		
Coating name	α (Solar)	ε (IR)
Aluminum	0.08	0.02
Aluminum on black kapton	0.12	0.03
Aluminum on fiberglass	0.15	0.07
Aluminum on stainless steel	0.08	0.02
Aluminum on lacquer	0.08	0.02
Aluminum on abraded lacquer	0.33	0.07
Chromium	0.56	0.17
Chromium on 5-mil kapton	0.57	0.24
Germanium	0.52	0.09
Gold	0.19	0.02
Iron oxide	0.85	0.56
Molybdenum	0.56	0.21
Nickel	0.38	0.04
Rhodium	0.18	0.03
Silver	0.04	0.02
Titanium	0.52	0.12
Tungsten	0.60	0.27

Composite coatings		
Coating name	α (Solar)	ε (IR)
Aluminum composite coating on 3 mil kapton	0.14	0.67
Aluminum composite over aluminum foil	0.09	0.72
Aluminum oxide (Al_2O_3) – (12λ /4) on buffed aluminum		
initial	0.13	0.23
2560 ESH UV + P+	0.13	0.23
Aluminum oxide (Al_2O_3) – (12λ /4) on fused silica	0.12	0.24
Silver beryllium copper (AgBeCu)	0.19	0.03
Kapton overcoating	0.31	0.57
Parylene C overcoating	0.22	0.34
Teflon overcoating	0.12	0.38
GSFC dark mirror coating SiO-Cr-Al	0.86	0.04
GSFC composite SiOx-Al_2O_3-Ag	0.07	0.68
Helios second surface mirror/silver backing		
Initial	0.07	0.79
24 hours at 5 Suns	0.07	0.80
48 hours at 11 Suns + P^+	0.08	0.79
Inconel with teflon overcoating – 1 mil	0.55	0.46
SiOx /VDA/0.5 mil kapton	0.19	0.14
Vespel polyimide SP1	0.89	0.90
M46J/RS-3 composite substrate	0.91	0.73
M55J/954-3	0.92	0.80
M40 graphite composite	0.94	–
GSFC spectrally selective composite coating	0.13	0.68
Silver composite (Al_2O_3/Ag/0.33 mil kapton)	0.73	0.25
Silver aluminum oxide (Ag/Al_2O_3/1 mil Kapton)	0.07	0.32–0.44
Teflon impregnated anodized titanium	0.75	0.26
Upliex-S 0.5 mil with 10,000 Å Al_2O_3	0.08	0.25
Upliex-S 1.0 mil with 10,000 Å Al_2O_3	0.08	0.23

Films and tapes		
Name	α (Solar)	ε (IR)
Aclar film (aluminum backing)		
1 mil	0.12	0.45
2 mil	0.11	0.62
5 mil	0.11	0.73
Black kapton	0.93	0.85
Copper foil tape		
Plain	0.32	0.02
Sanded	0.26	0.04
Tarnished	0.55	0.04
Germanium black kapton	0.55	0.84
Germanium black kapton reinforced	0.50	0.85
Kapton film (aluminum backing)		
0.08 mil	0.23	0.24

(*cont.*)

Films and tapes		
Name	α (Solar)	ε (IR)
0.15 mil	0.25	0.34
0.25 mil	0.31	0.45
0.50 mil	0.34	0.55
1.0 mil	0.38	0.67
1.5 mil	0.40	0.71
2.0 mil	0.41	0.75
3.0 mil	0.45	0.82
5.0 mil	0.46	0.86
Kapton film (Chromium–silicon oxide–aluminum backing (green)		
1.0 mil	0.79	0.78
Kapton film (Aluminum–aluminum oxide overcoating) 1 mil		
Initial	0.12	0.20
1800 ESH UV	0.12	0.20
Kapton film (Aluminum–silicon oxide overcoatinog)		
Initial	0.11	0.33
2400 ESH UV	0.22	0.33
Kapton film (Silver–aluminum oxide overcoating)		
Initial	0.08	0.19
24000 ESH UV	0.08	0.21
Kapton Film (Aluminum–silicon oxide overcoating)		
Initial	0.12	0.18
4000 ESH UV	0.28	0.24
Kapton 3 mil (ITO/VDA/Kapton)	0.09	0.02
Kapton 3 mil (ITO/Kapton/VDA)	0.43	0.82
Kimfoil polycarbonate film (aluminum backing)		
0.08 mil	0.19	0.23
0.20 mil	0.20	0.30
0.24 mil	0.17	0.28
Mylar film		
Aluminum backing		
0.15 mil	0.14	0.28
0.25 mil	0.15	0.34
3.0 mil	0.17	0.76
5.0 mil	0.19	0.77
Skylab sail		
Initial	0.15	0.35
1900 ESH UV	0.19	0.36
Skylab parasol fabric (orange)		
Initial	0.51	0.86
2400 ESH UV	0.65	0.86
Tedlar (gold backing)		
0.5 mil	0.30	0.49
1.0 mil	0.26	0.58
Polyester film yellow tape IP8102YE	0.52	0.82
Sumitomo bakelite	0.17	0.82
Tedlar 1.5 mil overcoat SiOx	0.21	0.63

(cont.)

Films and tapes		
Name	α (Solar)	ε (IR)
Teflon		
Aluminum backing		
2 mil	0.08	0.66
5 mil	0.13	0.81
10 mil	0.13	0.87
ATO/5 mil Teflon/Ag/niobium	0.09	0.80
Gold backing		
0.5 mil	0.24	0.43
1.0 mil	0.22	0.52
5 mil	0.22	0.81
10 mil	0.23	0.82
Silver backing		
2 mil	0.08	0.68
5 mil	0.08	0.81
10 mil	0.09	0.88
Tefzel (gold backing)		
0.05 mil	0.29	0.47
1.0 mil	0.26	0.61
General tapes		
235-3M black	0.20	0.03
425-3M aluminum foil	0.22	0.03
1170-3M electrical tape	0.15	0.59
850-3M Mylar–aluminum backing	0.09	0.03
7361-Mystic aluminized kapton	0.14	0.03
7452-Mystic aluminum foil	0.21	0.03
7800-Mystic aluminum foil	0.19	0.03
Y9360-3M aluminized Mylar	0.23	0.02
3M Gold tape	0.26	0.04
3M Gold tape	0.40	0.89
ITO/FEP type A/Inconel/Y966 Courtaulds	0.11	0.79
Tedlar 3M838	0.87–0.90	0.86

Reference

[1] Kauder, L., 2005, "Spacecraft Thermal Control Coatings References," NASA/TP-2005-212792.

Appendix C Earth IR and Albedo tables

Table C.1 Cold case Earth IR and Albedo factors[§]

Surface sensitivity	Orbit period (hours)	Albedo factor	EIR_{Flux} (W/m^2)
Orbit inclination $\leq 30°$			
UV	1.5	0.11	258
	6.0	0.14	245
	24.0	0.16	240
IR	1.5	0.25	206
	6.0	0.19	224
	24.0	0.18	230
UV and IR	1.5	0.14	228
	6.0	0.16	232
	24.0	0.16	235
$30° <$ Orbit inclination $\leq 60°$			
UV	1.5	0.16	239
	6.0	0.18	238
	24.0	0.19	233
IR	1.5	0.3	200
	6.0	0.31	207
	24.0	0.25	210
UV and IR	1.5	0.19	218
	6.0	0.19	221
	24.0	0.20	223
$60° <$ Orbit inclination $\leq 90°$			
UV	1.5	0.16	231
	6.0	0.18	231
	24.0	0.18	231
IR	1.5	0.26	193
	6.0	0.27	202
	24.0	0.24	205
UV and IR	1.5	0.19	218
	6.0	0.20	224
	24.0	0.20	224

[§] Values republished with permission of NASA and DXC Technology/CSC (Computer Sciences Corporation) from NASA/TM 2001-211221 (reference [1]).

Table C.2 Hot case Earth IR and Albedo factors§

Surface sensitivity	Orbit period (hours)	Albedo factor	EIR_{Flux} (W/m^2)
Orbit inclination ≤30°			
UV	1.5	0.28	237
	6.0	0.23	248
	24.0	0.22	251
IR	1.5	0.20	285
	6.0	0.19	269
	24.0	0.19	262
UV and IR	1.5	0.24	275
	6.0	0.21	264
	24.0	0.20	260
30°< Orbit inclination ≤60°			
UV	1.5	0.31	204
	6.0	0.31	212
	24.0	0.28	224
IR	1.5	0.22	274
	6.0	0.21	249
	24.0	0.21	245
UV and IR	1.5	0.26	257
	6.0	0.24	248
	24.0	0.24	247
60°< Orbit inclination ≤90°			
UV	1.5	0.28	214
	6.0	0.27	218
	24.0	0.24	224
IR	1.5	0.22	250
	6.0	0.22	221†
	24.0	0.20	217†
UV and IR	1.5	0.26	244
	6.0	0.24	233
	24.0	0.23	232

§ Values republished with permission of NASA and DXC Technology/CSC (Computer Sciences Corporation) from NASA/TM 2001-211221 (reference [1]).

Table C.3 Albedo correction factors§

Beta angle	Albedo correction factor
0	0.04
10	0.04
20	0.05
30	0.06
40	0.07

Table C.3 *(cont.)*

Beta angle	Albedo correction factor
50	0.09
60	0.12
70	0.16
80	0.22
90	0.31

§ Values republished with permission of NASA and DXC Technology/CSC (Computer Sciences Corporation) from NASA/TM 2001-211221 (reference [1]).

Reference

[1] Anderson, B.J., Justus, C.G., and Batts, G.W., 2001, "Guidelines for the Selection of Near-Earth Thermal Environment Parameters for Spacecraft Design," National Aeronautics and Space Administration, NASA/TM 2001-211221

Index

Solutions

Chapter 1 Solutions

1.1

$$\dot{E}_{in} - \dot{E}_{out} = \dot{E}_{st}$$

$$40\text{ W} - 25\text{ W} = \dot{E}_{st}$$

$$\dot{E}_{st} = 15\text{ W}$$

1.2

$$\dot{E}_{in} - \dot{E}_{out} = \dot{E}_{st} = 0$$

$$\dot{E}_{in} = \dot{E}_{out}$$

$$\dot{E}_{out} = \left(50\text{ W/m}^2\right)\left(0.25\text{ m}^2\right) + 10\text{ W} + 10\text{ W}$$

$$\dot{E}_{out} = 12.5\text{ W} + 10\text{ W} + 10\text{ W}$$

$$\dot{E}_{out} = 32.5\text{ W}$$

1.3 Fourier's law for conduction applies. The energy balance equation for this heat transfer mechanism is

$$\dot{Q} = \frac{kA(T_o - T_L)}{\Delta x}$$

Solving for the temperature at $x = L$ (i.e., T_L) our relation becomes

$$T_L = T_o - \frac{\dot{Q}\Delta x}{kA}$$

$$T_L = 285\text{ K} - \frac{(100\text{ W})(0.75\text{ m})}{\left(167\frac{\text{W}}{\text{m}\cdot\text{K}}\right)(0.4\text{ m})(0.2\text{ m})}$$

$$T_L = 279.39\text{ K}$$

1.4 i) To solve for the heat transfer from the surface to the stream we must use Newton's law of cooling

$$\dot{Q} = hA\left(T_{Surf} - T_{\mathrm{H2O}}\right)$$

$$\dot{Q} = \left(1.5\ \mathrm{W/m^2 \cdot K}\right)\left(1\ \mathrm{m^2}\right)(353\ \mathrm{K} - 303\ \mathrm{K})$$

$$\dot{Q} = 75\ \mathrm{W}$$

ii) We know that 1.0 W = 1 J/s. To get the number of Joules transferred we must multiply both sides of the equal sign by the time elapsed (i.e., 120 seconds).

$$\text{Joules} = 1.0\ \mathrm{W} \cdot 120\ \mathrm{s}$$

$$\text{Joules} = 120$$

1.5 To determine the heat transfer from the active surface we must apply the Stefan–Boltzmann Law

$$\dot{Q} = \varepsilon \sigma A T^4$$

where σ is 5.67×10^{-8} W/m^2 · K

$$\dot{Q} = (0.89)\left(5.67 \times 10^{-8}\ \mathrm{W/m^2 \cdot K}\right)\left(0.5\ \mathrm{m^2}\right)(180\ \mathrm{K})^4$$

$$\dot{Q} = 2.523\ \mathrm{W}$$

1.6 i) Since the resistivity of the material is given, we can place this in the Wiedemann–Franz relationship to determine the thermal conductivity.

$$k = \frac{L_o}{\rho_{elec}} T$$

where $L_o = 2.45 \times 10^{-8}$ W · Ω/K^2

$$k = \frac{\left(2.45 \times 10^{-8}\ \mathrm{W \cdot \Omega/K^2}\right)}{\left(11 \times 10^{-8}\ \Omega \cdot \mathrm{m}\right)}(293\mathrm{K})$$

$$k = 63.26\ \mathrm{W/m \cdot K}$$

ii) The thermal conductivity value from the matweb tables is 69.1 W/m · K. The quote from matweb.com is slightly greater than the value determined in part a).

1.7 i) To determine the mean free path we use equation 1.7

$$\lambda_{MFP} = \frac{k_B T}{2^{5/2} \pi R_{mP}^2 p}$$

Both the temperature and the pressure of the system are provided, as well as the diameter of an air molecule. From this diameter we can determine the radius. Also, we know that Boltzmann's constant is 1.381×10^{-23} J/K. Placing values in the mean free path relation we have

$$\lambda_{MFP} = \frac{(1.381\times10^{-23}\ \mathrm{J/K})(263\ \mathrm{K})}{2^{5/2}\pi\left[(2\times10^{-10}\ \mathrm{m}^2)^2\right](20{,}000\ \mathrm{N/m}^2)}$$

$$\lambda_{MFP} = 2.55\times10^{-7}\ \mathrm{m}$$

ii) Assuming the molecules in part A are contained in a cubic box of dimensions 50 cm × 50 cm × 50 cm, the characteristic length for travel of the molecules is significantly greater than the mean free path. The continuum hypothesis is therefore satisfied for convective heat transfer internal to the box.

Chapter 2 Solutions

2.1 i)

$\dot{q}_x$ → T_1 — $\frac{2L}{kA}$ — T_2 — $\frac{L}{kA}$ — T_3 — $\frac{0.5L}{kA}$ — T_4 → $\dot{q}_x$

ii)

$$\dot{q}_x = \frac{(T_1 - T_4)}{\left[\frac{2L}{k_1A} + \frac{L}{k_2A} + \frac{0.5L}{k_3A}\right]}$$

iii) Assuming $k_1 = k_2 = k_3 = k$ then the resistances become $2L/kA$, L/kA and $0.5L/kA$.The limiting resistance is the largest resistance along the heat flow path.

2.2 i)

$$R_{1-2} = \frac{0.5L}{k_1A} + \frac{0.5\Delta x}{k_2A} + \frac{0.5\Delta y}{k_2A} = \frac{0.5L}{k_1(th\cdot\Delta y)} + \frac{0.5\Delta x}{k_2(th\cdot\Delta y)} + \frac{0.5\Delta y}{k_2(th\cdot\Delta x)}$$

$$R_{2-3} = \frac{0.5\Delta y}{k_2A} + \frac{0.5\Delta x}{k_2A} + \frac{0.5L}{k_3A} = \frac{0.5\Delta y}{k_2(th\cdot\Delta x)} + \frac{0.5\Delta x}{k_2(th\cdot\Delta y)} + \frac{0.5L}{k_3(th\cdot\Delta y)}$$

$$\frac{1}{R_{1-3}} = \frac{1}{R_{1-2} + R_{2-3}}$$

$$\frac{1}{R_{1-3}} = \frac{1}{\left[\frac{0.5L}{k_1(th\cdot\Delta y)} + \frac{0.5\Delta x}{k_2(th\cdot\Delta y)} + \frac{0.5\Delta y}{k_2(th\cdot\Delta x)} + \frac{0.5\Delta y}{k_2(th\cdot\Delta x)} + \frac{0.5\Delta x}{k_2(th\cdot\Delta y)} + \frac{0.5L}{k_3(th\cdot\Delta y)}\right]}$$

ii) Assuming $T_1 > T_2 > T_3$

$$\dot{q}_x = \frac{(T_1 - T_3)}{[R_{1-2} + R_{2-3}]}$$

$$\dot{q}_x = \frac{(T_1 - T_3)}{\left[\dfrac{0.5L}{k_1(th \cdot \Delta y)} + \dfrac{0.5\Delta x}{k_2(th \cdot \Delta y)} + \dfrac{0.5\Delta y}{k_2(th \cdot \Delta x)} + \dfrac{0.5\Delta y}{k_2(th \cdot \Delta x)} + \dfrac{0.5\Delta x}{k_2(th \cdot \Delta y)} + \dfrac{0.5L}{k_3(th \cdot \Delta y)}\right]}$$

$$\dot{q}_x = \frac{(T_1 - T_3)}{\left[\dfrac{0.5L}{th \cdot \Delta y}\left(\dfrac{1}{k_1} + \dfrac{1}{k_3}\right) + \dfrac{\Delta x}{k_2(th \cdot \Delta y)} + \dfrac{\Delta y}{k_2(th \cdot \Delta x)}\right]}$$

2.3 i)

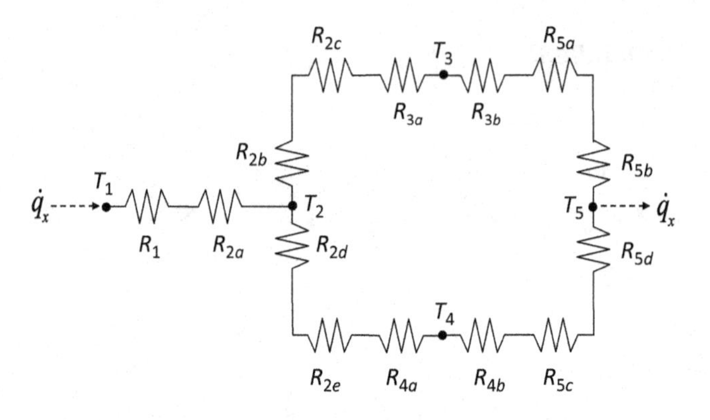

where

$$R_{1-2} = R_1 + R_{2a} = \frac{0.25L}{k_1\left(\dfrac{\Delta y}{3}\right) \cdot th} + \frac{0.25L}{k_2 \Delta y \cdot th} = \frac{0.75L}{k_1(\Delta y \cdot th)} + \frac{0.25L}{k_2(\Delta y \cdot th)}$$

$$R_{2-3} = R_{2b} + R_{2c} + R_{3a} = \frac{0.5\Delta y - 0.1\Delta y}{k_2(0.5L) \cdot th} + \frac{1.25L}{k_2(\Delta y \cdot th)} + \frac{2.5L}{k_3(\Delta y \cdot th)}$$

$$= \frac{0.8\Delta y}{k_2(L \cdot th)} + \frac{1.25L}{k_2(\Delta y \cdot th)} + \frac{2.5L}{k_3(\Delta y \cdot th)}$$

By symmetry, we know that

$$R_{2-4} = R_{2d} + R_{2e} + R_{4a} = \frac{0.5\Delta y - 0.1\Delta y}{k_2(0.5L) \cdot th} + \frac{1.25L}{k_2(\Delta y \cdot th)} + \frac{2.5L}{k_4(\Delta y \cdot th)}$$

$$= \frac{0.8\Delta y}{k_2(L \cdot th)} + \frac{1.25L}{k_2(\Delta y \cdot th)} + \frac{2.5L}{k_4(\Delta y \cdot th)}$$

As for the thermal couplings between members 3 to 5 and 4 to 5, we know that

$$R_{3-5} = R_{3b} + R_{5a} + R_{5b} = \frac{2.5L}{k_3(\Delta y \cdot th)} + \frac{0.125L}{k_5\left(\dfrac{\Delta y}{5} \cdot th\right)} + \frac{0.4\Delta y}{k_5(0.25L \cdot th)}$$

$$= \frac{2.5L}{k_3(\Delta y \cdot th)} + \frac{0.625L}{k_5(\Delta y \cdot th)} + \frac{1.6\Delta y}{k_5(L \cdot th)}$$

By symmetry, we know

$$R_{4-5} = R_{4b} + R_{5c} + R_{5d} = \frac{2.5L}{k_4(\Delta y \cdot th)} + \frac{0.125L}{k_5\left(\frac{\Delta y}{5} \cdot th\right)} + \frac{0.4\Delta y}{k_5(0.25L \cdot th)}$$

$$= \frac{2.5L}{k_4(\Delta y \cdot th)} + \frac{0.625L}{k_5(\Delta y \cdot th)} + \frac{1.6\Delta y}{k_5(L \cdot th)}$$

ii) Paths 2–3–5 and 2–4–5 are parallel. Thus

$$\frac{1}{R_{2-5,\text{II}}} = \frac{1}{R_{2-3-5}} + \frac{1}{R_{2-4-5}}$$

$$R_{2-5,\text{II}} = \left[\frac{1}{R_{2-3-5}} + \frac{1}{R_{2-4-5}}\right]^{-1}$$

$$R_{Tot} = R_{1-2} + R_{2-5,\text{II}} = R_{1-2} + \left[\frac{1}{R_{2-3-5}} + \frac{1}{R_{2-4-5}}\right]^{-1}$$

$$\frac{1}{R_{Tot}} = \frac{1}{R_{1-2} + \left[\frac{1}{R_{2-3-5}} + \frac{1}{R_{2-4-5}}\right]^{-1}}$$

where $R_{2-3-5} = R_{2-3} + R_{3-5}$ and $R_{2-4-5} = R_{2-4} + R_{4-5}$

iii) Using the resistances developed in part i), we can substitute in actual values for Δy and the thermal conductivity and compare. Alternatively, we know the geometry is the same along both paths. Also, the materials along the paths are the same, with the exception of material 3 and material 4. Thus, we can simply compare the resistances across these two elements.

$$R_3 = \frac{L}{k_3\left(\frac{\Delta y}{5} \cdot th\right)} = \frac{5L}{k_3(\Delta y \cdot th)}$$

$$R_4 = \frac{L}{k_4\left(\frac{\Delta y}{5} \cdot th\right)} = \frac{5L}{k_4(\Delta y \cdot th)}$$

$k_3 = k_{CU} = 390$ W/m · K; $k_4 = k_{\text{Al}} = 167$ W/m · K

Since copper has a higher k value than aluminum, by inspection one can determine that the path of least resistance is through material 3.

2.4 i)

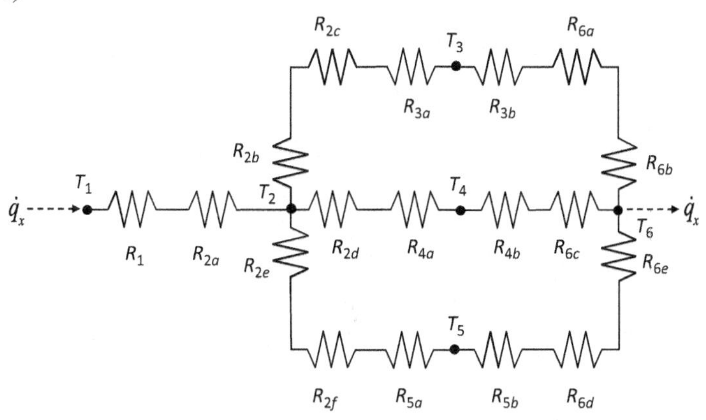

where

$$R_{1-2} = R_1 + R_{2a} = \frac{0.25L}{k_1\left(\frac{\Delta y}{3}\right)\cdot th} + \frac{0.25L}{k_2\Delta y \cdot th} = \frac{0.75L}{k_1(\Delta y \cdot th)} + \frac{0.25L}{k_2(\Delta y \cdot th)}$$

$$R_{2-3} = R_{2b} + R_{2c} + R_{3a} = \frac{0.5\Delta y - 0.0625\Delta y}{k_2(0.5L)\cdot th} + \frac{2L}{k_2(\Delta y \cdot th)} + \frac{4L}{k_3(\Delta y \cdot th)}$$

$$= \frac{0.875\Delta y}{k_2(L \cdot th)} + \frac{2L}{k_2(\Delta y \cdot th)} + \frac{4L}{k_3(\Delta y \cdot th)}$$

$$R_{2-4} = R_{2d} + R_{4a} = \frac{0.25L}{k_2(\Delta y \cdot th)} + \frac{0.5L}{k_4\left(\frac{\Delta y}{3}\cdot th\right)} = \frac{0.25L}{k_2(\Delta y \cdot th)} + \frac{1.5L}{k_4(\Delta y \cdot th)}$$

$$R_{2-5} = R_{2e} + R_{2f} + R_{5a} = \frac{0.5\Delta y - 0.125\Delta y}{k_2(0.5L)\cdot th} + \frac{0.25L}{k_2(0.25\Delta y \cdot th)} + \frac{0.5L}{k_5(0.25\Delta y \cdot th)}$$

$$= \frac{0.75\Delta y}{k_2(L \cdot th)} + \frac{L}{k_2(\Delta y \cdot th)} + \frac{2L}{k_5(\Delta y \cdot th)}$$

$$R_{3-6} = R_{3b} + R_{6a} + R_{6b} = \frac{0.5L}{k_3(0.125\Delta y \cdot th)} + \frac{0.125L}{k_6(0.125\Delta y \cdot th)} + \frac{0.5\Delta y - 0.0625\Delta y}{k_6(0.25L)\cdot th}$$

$$= \frac{4L}{k_3(\Delta y \cdot th)} + \frac{L}{k_6(\Delta y \cdot th)} + \frac{1.75\Delta y}{k_6(L \cdot th)}$$

$$R_{4-6} = R_{4b} + R_{6c} = \frac{0.5L}{k_4\left(\frac{\Delta y}{3}\cdot th\right)} + \frac{0.125L}{k_6(\Delta y \cdot th)} = \frac{1.5L}{k_4(\Delta y \cdot th)} + \frac{0.125L}{k_6(\Delta y \cdot th)}$$

$$R_{5-6} = R_{5b} + R_{6d} + R_{6e} = \frac{0.5L}{k_5(0.25\Delta y \cdot th)} + \frac{0.125L}{k_6(0.25\Delta y \cdot th)} + \frac{0.5\Delta y - 0.125\Delta y}{k_6(0.25L)\cdot th}$$

$$= \frac{2L}{k_5(\Delta y \cdot th)} + \frac{0.5L}{k_6(\Delta y \cdot th)} + \frac{1.5\Delta y}{k_6(L \cdot th)}$$

ii)

$$\frac{1}{R_{Tot}} = \frac{1}{R_{1-2} + R_{2-6,\mathrm{II}}}$$

where

$$\frac{1}{R_{2-6,\mathrm{II}}} = \frac{1}{R_{2-3} + R_{3-6}} + \frac{1}{R_{2-4} + R_{4-6}} + \frac{1}{R_{2-5} + R_{5-6}}$$

thus

$$\frac{1}{R_{Tot}} = \frac{1}{R_{1-2} + \left[\frac{1}{R_{2-3} + R_{3-6}} + \frac{1}{R_{2-4} + R_{4-6}} + \frac{1}{R_{2-5} + R_{5-6}}\right]^{-1}}$$

iii) Let $k_{OFHC} = 390$ W/m · K; $k_{\mathrm{Al},2024\text{-}T6} = 177$ W/m · K; $k_{304-SS} = 14.9$ W/m · K; We also know that

$$k_1 = k_2 = k_5 = k_6 = k_{\mathrm{Al},2024\text{-}T6}$$

$$k_3 = k_{OFHC}$$

$$k_4 = k_{SS}$$

Since the heat flow through the thermal circuit splits into three distinct paths, we must determine the resistance value along each path for comparison purposes. These three distinct paths are

Path 1: $R_{2-3} + R_{3-6}$
Path 2: $R_{2-4} + R_{4-6}$
Path 3: $R_{2-5} + R_{5-6}$

Using the resistances determined in part i) and given that $\Delta y = 2L$, the resistances of interest are

$$R_{2-3} = \frac{0.875 \cdot 2L}{k_2(L \cdot th)} + \frac{2L}{k_2(2L \cdot th)} + \frac{4L}{k_3(2L \cdot th)} = \frac{1.75}{k_2 \cdot th} + \frac{1}{k_2 \cdot th} + \frac{2}{k_3 \cdot th} = \frac{0.021}{th}\ \mathrm{m \cdot K/W}$$

$$R_{3-6} = \frac{0.5L}{k_3(0.125 \cdot 2L \cdot th)} + \frac{0.125L}{k_6(0.125 \cdot 2L \cdot th)} + \frac{2L(0.5 - 0.0625)}{k_6(0.25L) \cdot th}$$
$$= \frac{2}{k_3 \cdot th} + \frac{0.5}{k_6 \cdot th} + \frac{3.5}{k_6 \cdot th} = \frac{0.028}{th}\ \mathrm{m \cdot K/W}$$

Thus, the total resistance for path 1 is $\frac{0.049}{th}$ m · K/W

$$R_{2-4} = \frac{0.25L}{k_2(2L \cdot th)} + \frac{0.5L}{k_4\left(\frac{2L}{3} \cdot th\right)} = \frac{0.125L}{k_2 \cdot th} + \frac{0.75L}{k_4 \cdot th} = \frac{0.051}{th}\ \mathrm{m \cdot K/W}$$

$$R_{4-6} = R_{4b} + R_{6c} = \frac{0.5L}{k_4\left(\frac{2L}{3} \cdot th\right)} + \frac{0.125L}{k_6(2L \cdot th)} = \frac{0.75}{k_4 \cdot th} + \frac{0.0625}{k_6 \cdot th} = \frac{0.0503}{th}\ \mathrm{m \cdot K/W}$$

Thus, the total resistance for path 2 is $\frac{0.101}{th}$ m · K/W

$$R_{2-5} = \frac{2L(0.5 - 0.125)}{k_2(0.5L) \cdot th} + \frac{0.25L}{k_2(0.25 \cdot 2L \cdot th)} + \frac{0.5L}{k_5(0.25 \cdot 2L \cdot th)} = \frac{1.5}{k_2 \cdot th} + \frac{0.5}{k_2 \cdot th} + \frac{1}{k_5 \cdot th}$$
$$= \frac{0.017}{th}\ \mathrm{m \cdot K/W}$$

$$R_{5-6} = \frac{0.5L}{k_5(0.25 \cdot 2L \cdot th)} + \frac{0.125L}{k_6(0.25 \cdot 2L \cdot th)} + \frac{2L(0.5 - 0.125)}{k_6(0.25L) \cdot th} = \frac{1}{k_5 \cdot th} + \frac{0.25}{k_6 \cdot th} + \frac{3}{k_6 \cdot th}$$
$$= \frac{0.024}{th}\ \mathrm{m \cdot K/W}$$

Thus, the total resistance for path 3 is $\frac{0.041}{th}$ m · K/W

Comparison of the three different paths shows that path 3 is the path of least resistance.

2.5 i)

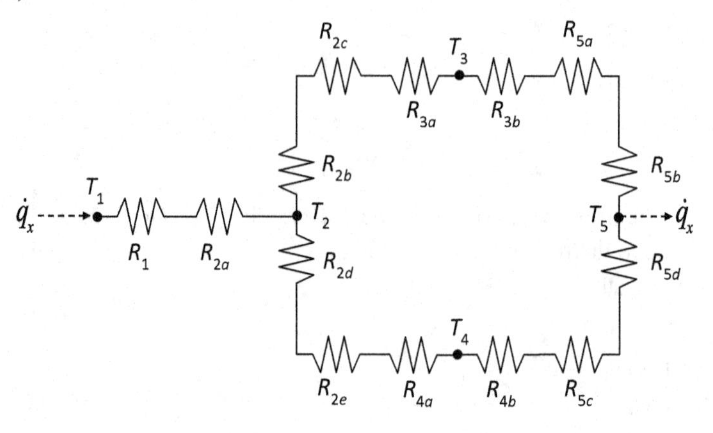

$$R_{1-2} = R_1 + R_{2a} = \frac{0.5L}{k_1\left(\frac{\Delta x}{3}\right)\cdot th} + \frac{0.25L}{k_2(\Delta x \cdot th)} = \frac{1.5L}{k_1(\Delta x \cdot th)} + \frac{0.25L}{k_2(\Delta x \cdot th)}$$

$$R_{2-3} = R_{2b} + R_{2c} + R_{3a} = \frac{0.375\Delta x}{k_2\left(\frac{L}{2}\right)\cdot th} + \frac{0.25L}{k_2\left(\frac{\Delta x}{4}\right)\cdot th} + \frac{0.5L}{k_3\left(\frac{\Delta x}{4}\right)\cdot th}$$

$$R_{3-5} = R_{3b} + R_{5a} + R_{5b} = \frac{0.5L}{k_3\left(\frac{\Delta x}{4}\right)\cdot th} + \frac{0.166L}{k_5\left(\frac{\Delta x}{4}\right)\cdot th} + \frac{0.375\Delta x}{k_5\left(\frac{L}{3}\right)\cdot th}$$

$$R_{2-4} = R_{2d} + R_{2e} + R_{4a} = \frac{0.25\Delta x}{k_2\left(\frac{L}{2}\right)\cdot th} + \frac{0.25L}{k_2\left(\frac{\Delta x}{2}\right)\cdot th} + \frac{0.5L}{k_4\left(\frac{\Delta x}{2}\right)}$$

$$R_{4-5} = R_{4b} + R_{5c} + R_{5d} = \frac{0.5L}{k_4\left(\frac{\Delta x}{2}\right)\cdot th} + \frac{0.166L}{k_5\left(\frac{\Delta x}{2}\right)\cdot th} + \frac{0.25L}{k_5\left(\frac{L}{3}\right)\cdot th}$$

ii)

$$\frac{1}{R_{Tot}} = \frac{1}{R_{1-2} + R_{2-5}}$$

where

$$\frac{1}{R_{2-5}} = \frac{1}{R_{2-3} + R_{3-5}} + \frac{1}{R_{2-4} + R_{4-5}}$$

so

$$R_{2-5} = \left[\frac{1}{R_{2-3} + R_{3-5}} + \frac{1}{R_{2-4} + R_{4-5}}\right]^{-1}$$

$$\frac{1}{R_{Tot}} = \frac{1}{R_{1-2} + \left[\frac{1}{R_{2-3} + R_{3-5}} + \frac{1}{R_{2-4} + R_{4-5}}\right]^{-1}}$$

$$\dot{q}_{1-5} = \frac{(T_1 - T_5)}{\left[R_{1-2} + \left[\frac{1}{R_{2-3} + R_{3-5}} + \frac{1}{R_{2-4} + R_{4-5}}\right]^{-1}\right]}$$

iii) To determine the path of least resistance between material elements 2 and 3 we must determine and compare R_{2-3-5} and R_{2-4-5}

$$R_{2-3-5} = \frac{0.375\Delta x}{k_2\left(\frac{L}{2}\right)\cdot th} + \frac{0.25L}{k_2\left(\frac{\Delta x}{4}\right)\cdot th} + \frac{0.5L}{k_3\left(\frac{\Delta x}{4}\right)\cdot th} + \frac{0.166L}{k_5\left(\frac{\Delta x}{4}\right)\cdot th} + \frac{0.375\Delta x}{k_5\left(\frac{L}{3}\right)\cdot th}$$

$$R_{2-3-5} = \frac{0.75\Delta x}{k_2(L\cdot th)} + \frac{L}{k_2(\Delta x\cdot th)} + \frac{4L}{k_3(\Delta x\cdot th)} + \frac{0.664L}{k_5(\Delta x\cdot th)} + \frac{1.125\Delta x}{k_5(L\cdot th)}$$

we know that $k_2 = k_3 = k_4 = k$
so

$$R_{2-3-5} = \frac{0.75\Delta x}{k(L\cdot th)} + \frac{L}{k(\Delta x\cdot th)} + \frac{4L}{k(\Delta x\cdot th)} + \frac{0.664L}{k(\Delta x\cdot th)} + \frac{1.125\Delta x}{k(L\cdot th)}$$

$$R_{2-3-5} = \frac{1.875\Delta x}{k(L\cdot th)} + \frac{5.664L}{k(\Delta x\cdot th)}$$

and

$$R_{2-4-5} = \frac{0.25\Delta x}{k_2\left(\frac{L}{2}\right)\cdot th} + \frac{0.25L}{k_2\left(\frac{\Delta x}{2}\right)\cdot th} + \frac{0.5L}{k_4\left(\frac{\Delta x}{2}\right)\cdot th} + \frac{0.5L}{k_4\left(\frac{\Delta x}{2}\right)\cdot th} + \frac{0.166L}{k_5\left(\frac{\Delta x}{2}\right)\cdot th} + \frac{0.25\Delta x}{k_5\left(\frac{L}{3}\right)\cdot th}$$

$$R_{2-4-5} = \frac{0.5\Delta x}{k_2(L\cdot th)} + \frac{0.5L}{k_2(\Delta x\cdot th)} + \frac{L}{k_4(\Delta x\cdot th)} + \frac{L}{k_4(\Delta x\cdot th)} + \frac{0.332L}{k_5(\Delta x\cdot th)} + \frac{0.75\Delta x}{k_5(L\cdot th)}$$

since $k_2 = k_4 = k_5 = k$ we have

$$R_{2-4-5} = \frac{1.25\Delta x}{k(L\cdot th)} + \frac{2.882L}{k(\Delta x\cdot th)}$$

By inspection of values in the numerators for the common terms, we can determine that R_{2-4-5} has a lower resistance. Thus, the path of least resistance passes through material 4.

iv) Since $k_2 \neq k_3 \neq k_4$ in this case, we must return to the definitions of R_{2-3-5} and R_{2-4-5} and place in conductivity values assuming

$k_{Al} = 177$ W/m·K
$k_{SS} = 15$ W/m·K
$k_{CU} = 390$ W/m·K

we get

$$R_{2-3-5} = \frac{0.00478\Delta x}{(L \cdot th)} + \frac{0.02676L}{(\Delta x \cdot th)}; \quad R_{2-4-5} = \frac{0.003\Delta x}{(L \cdot th)} + \frac{0.13513L}{(\Delta x \cdot th)}$$

The shape factor variable in this case (i.e., $\Delta x/L$) inverts between the first and second term in each resistance. This suggests that there will be an inflection point where R_{2-3-5} is greater than or less than R_{2-4-5}. To find this, one must take a variation of $\Delta x/L$.

for $\Delta x/L < 7.8$; R_{2-3-5} has the least resistance
for $\Delta x/L \geq 7.8$; R_{2-4-5} has the least resistance

Please note that if the shape factors were the same for materials 3 and 4 we would simply compare R_3 and R_4 to determine the path of least resistance.

2.6 i)

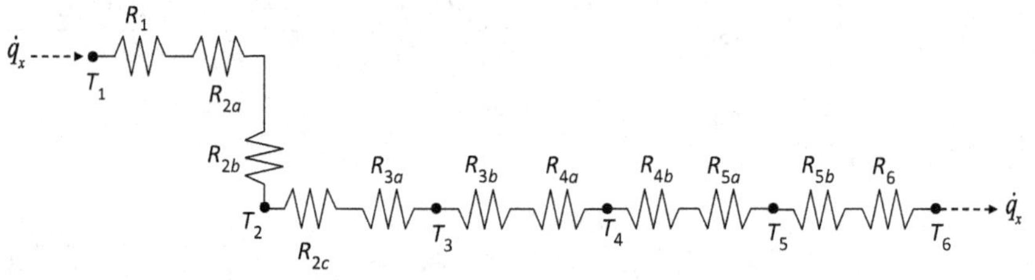

where

$$R_{1-2} = R_1 + R_{2a} + R_{2b} = \frac{0.5L}{k_1\left(\frac{\Delta y}{3}\right) \cdot th} + \frac{0.5L}{k_2\left(\frac{\Delta y}{3}\right) \cdot th} + \frac{\Delta y/3}{k_2(L \cdot th)}$$

$$R_{2-3} = R_{2c} + R_{3a} = \frac{0.5L}{k_2(\Delta y \cdot th)} + \frac{0.25L}{k_3\left(\frac{\Delta y}{3}\right) \cdot th}$$

$$R_{3-4} = R_{3b} + R_{4a} = \frac{0.25L}{k_3\left(\frac{\Delta y}{3}\right) \cdot th} + \frac{0.75L}{k_4(\Delta y \cdot th)}$$

$$R_{4-5} = R_{4b} + R_{5a} = \frac{0.75L}{k_4(\Delta y \cdot th)} + \frac{0.25L}{k_5(\Delta y \cdot th)}$$

$$R_{5-6} = R_{5b} + R_6 = \frac{0.25L}{k_5(\Delta y \cdot th)} + \frac{0.5L}{k_6\left(\frac{\Delta y}{3}\right) \cdot th}$$

ii) Using the resistances determined in part i), the given that $L = \Delta y/3$ and the thermal conductivity assumptions of $k_3 = 2k$; $k_6 = 3k$; $k_2 = k_4 = k_5 = k_1 = k$, the resistances of interest are

$$R_1 = \frac{3L}{k_1(\Delta y \cdot th)} = \frac{3L}{k \cdot (3L) \cdot th} = \frac{1}{k \cdot th}$$

$$R_2 = R_{2a} + R_{2b} + R_{2c} = \frac{0.5L}{k_2\left(\frac{\Delta y}{3}\right) \cdot th} + \frac{\frac{\Delta y}{3}}{k_2(L \cdot th)} + \frac{0.5L}{k_2(\Delta y \cdot th)} = \frac{0.5}{k \cdot th} + \frac{1}{k \cdot th} + \frac{0.5L}{k(3L \cdot th)}$$

$$= \frac{1.666}{k \cdot th}$$

$$R_3 = \frac{0.5L}{k_3\left(\frac{\Delta y}{3} \cdot th\right)} = \frac{0.5L}{2k \cdot (L \cdot th)} = \frac{0.25}{k \cdot th}$$

$$R_4 = \frac{1.5L}{k_4(\Delta y \cdot th)} = \frac{0.5}{k \cdot th}$$

$$R_5 = \frac{0.5L}{k_5(\Delta y \cdot th)} = \frac{0.1666}{k \cdot th}$$

$$R_6 = \frac{L}{k_6(\Delta y \cdot th)} = \frac{0.333}{3k \cdot th} = \frac{0.111}{k \cdot th}$$

R_2 is the largest, thus making it the limiting resistance for heat flow.

2.7 i)

$\dot{q}_r$ ---→ T_0 —R_{0-1}— T_1 —R_{1-2}— T_2 —R_{2-3}— T_3 ---→ $\dot{q}_r$

where $R_{0-1} = \frac{\ln(R/R_0)}{2\pi \cdot th \cdot k_1}$; $R_{1-2} = \frac{\ln(2.5R/R)}{2\pi \cdot th \cdot k_2}$; $R_{2-3} = \frac{\ln(3R/2.5R)}{2\pi \cdot th \cdot k_3}$

ii)

$$\dot{q}_r = \frac{(T_0 - T_3)}{\left[\frac{\ln(R/R_0)}{2\pi \cdot th \cdot k_1} + \frac{\ln(2.5R/R)}{2\pi \cdot th \cdot k_2} + \frac{\ln(3R/2.5R)}{2\pi \cdot th \cdot k_3}\right]}$$

$$\dot{q}_r = \frac{(T_0 - T_3)}{\frac{1}{2\pi \cdot th} \cdot \left[\frac{\ln(R/R_0)}{k_1} + \frac{\ln(2.5R/R)}{k_2} + \frac{\ln(3R/2.5R)}{k_3}\right]}$$

2.8 i) Using the polynomial relations for the thermal conductivity of copper, (OFHC)/RRR = 150, 304 stainless steel, aluminum 1100, aluminum 6061-T6 and titanium 15-3-3-3 from the NIST Cryogenics website, the following plots are produced.

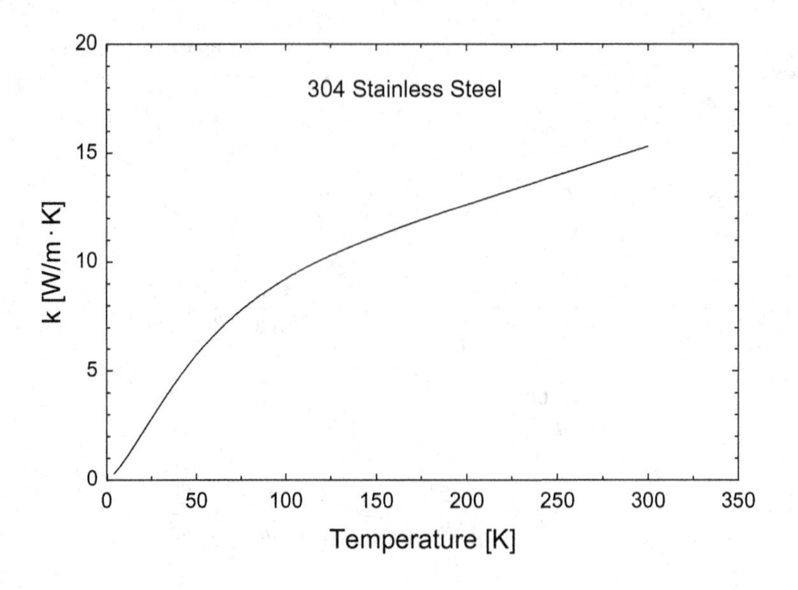
304 Stainless Steel
k [W/m·K]
Temperature [K]
0
5
10
15
20
0
50
100
150
200
250
300
350

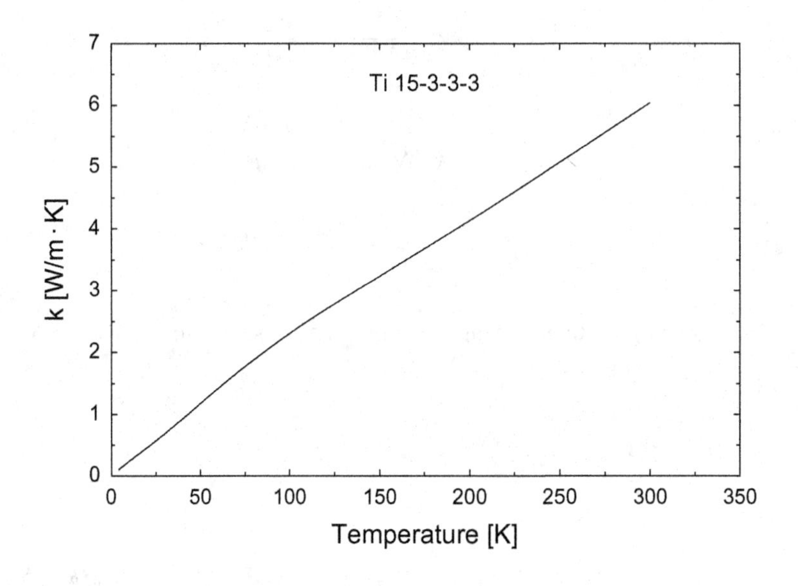
Ti 15-3-3-3
k [W/m·K]
Temperature [K]
0
1
2
3
4
5
6
7
0
50
100
150
200
250
300
350

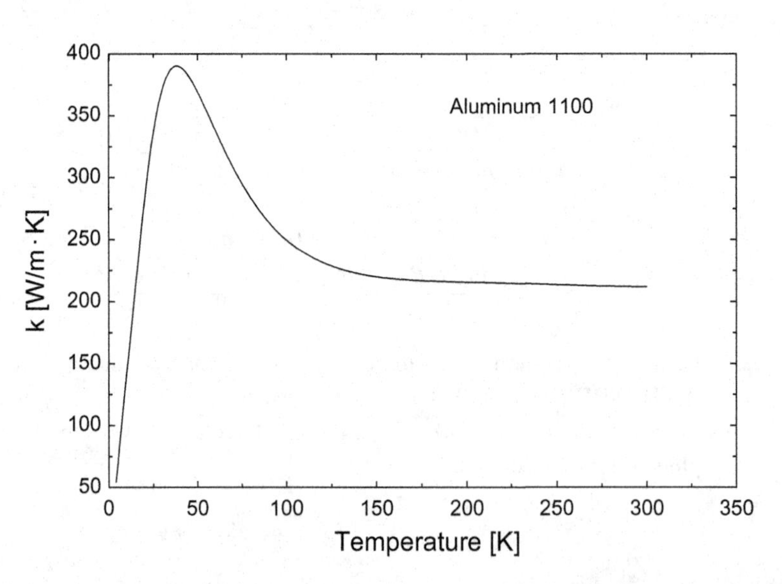
Aluminum 1100
k [W/m·K]
Temperature [K]
50
100
150
200
250
300
350
400
0
50
100
150
200
250
300
350

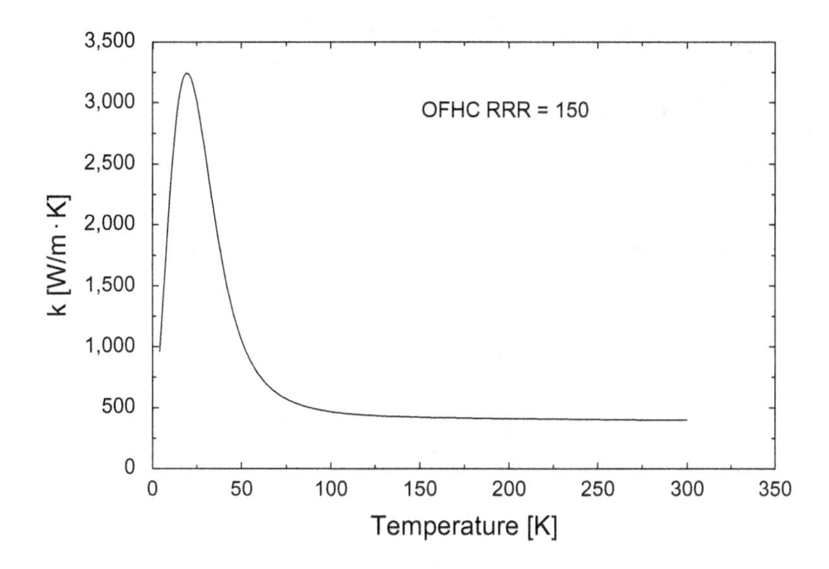

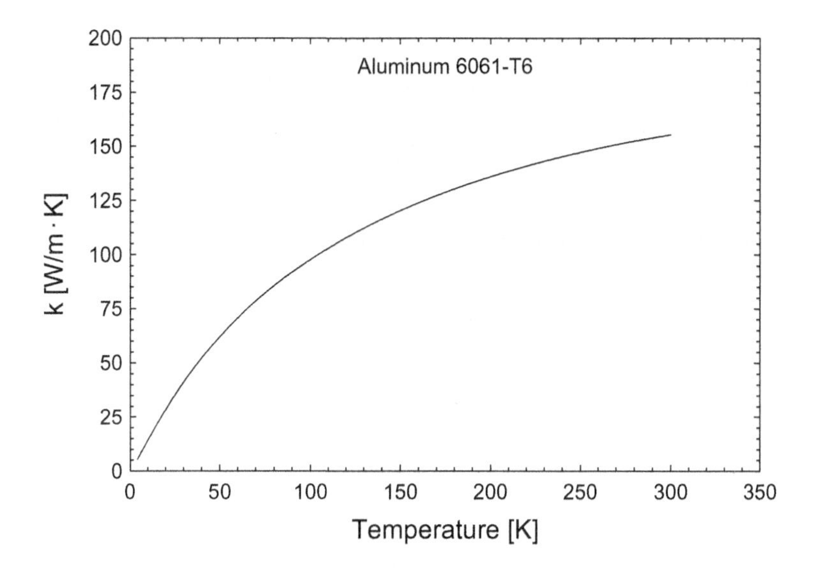

ii) Below 50 K, OFHC/RRR = 150 has the largest variation in thermal conductivity (i.e., delta).

iv) Between 300 K and 100 K, OFHC/RRR = 150 has little variation. The thermal conductivity is actually increasing. As the temperature reduces to the 100 K value (and ever lower, down to 30 K), the thermal conductance and heat transfer increase given the same source and sink temperatures.

Material	k at 300 K	k at 100 K	Delta
Copper OFHC/RRR = 150	397.6	466.1	68.5
304 Stainless steel	15.31	9.367	5.943
Aluminum 1100	211.8	249.7	37.9
Aluminum 6061-T6	155.3	97.7	57.6
Titanium 15-3-3-3	6.041	2.3	3.741

2.9 Using the equation for sensible heating

$$\dot{Q} = \frac{mc_p}{\Delta t} \cdot \left(T_{final} - T_{init}\right)$$

Solving for T_{final}

$$T_{final} = T_{init} + \frac{\dot{Q} \cdot \Delta t}{mc_p}$$

We know that $\rho = \dfrac{m}{Vol} \Rightarrow m = \rho \cdot Vol$

$$T_{final} = T_{init} + \frac{\dot{Q} \cdot \Delta t}{\rho \cdot Vol \cdot c_p}$$

Now the specific heat, density and volume for the cylinder must be determined. We know

$$Vol = \pi\left[r_o^2 - r_i^2\right] \cdot l$$

$$Vol = \pi[(0.16 \text{ m})^2 - (0.14 \text{ m})^2] \cdot (0.15 \text{ m})$$

$$Vol = 0.00282 \text{ m}^3$$

Referencing tables either in EES or at the NIST website, the specific heat and density at room temperature is

$$c_p|_{T = 293\text{K}} = 875 \text{ J/kg} \cdot \text{K}; \ \rho|_{T = 293\text{K}} = 2{,}770 \text{ kg/m}^3$$

Note: $\Delta t = 20$ min $= 1{,}200$ s. Placing values in the main equation we will have

$$T_{final} = 293 \text{ K} + \frac{(50 \text{ W})(1{,}200 \text{ s})}{(2{,}770 \text{ kg/m}^3)(0.00282 \text{ m}^3)(875 \text{ J/kg} \cdot \text{K})}$$

$$T_{final} = 293 \text{ K} + 8.77 \text{ K}$$

$$T_{final} = 301.77 \text{ K or } 28.7°\text{C}$$

2.10 i)

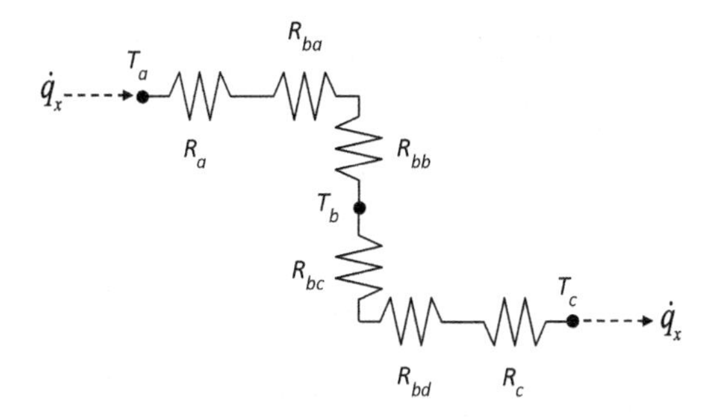

$$R_{a-b} = R_a + R_{ba} + R_{bb} = \frac{0.5\Delta x}{k_a(\Delta y \cdot th)} + \frac{0.5\Delta x}{k_b(\Delta y \cdot th)} + \frac{\Delta y}{k_b(\Delta x \cdot th)}$$

$$R_{b-c} = R_{bc} + R_{bd} + R_c = \frac{\Delta y}{k_b(\Delta x \cdot th)} + \frac{0.5\Delta x}{k_b(\Delta y \cdot th)} + \frac{0.5\Delta x}{k_c(\Delta y \cdot th)}$$

ii) $\dot{q}_{in} = \frac{(T_a - T_b)}{R_{a-b}} = \frac{(T_b - T_c)}{R_{b-c}}$ or $\dot{q}_{in} = \frac{(T_a - T_c)}{[R_{a-b} + R_{b-c}]}$

Solving for R_{a-b} and R_{b-c}

$$R_{a-b} = \frac{0.5\ \text{m}/2}{(177\ \text{W/m}\cdot\text{K})(0.05\ \text{m}\cdot 0.003\ \text{m})} + \frac{0.5\ \text{m}/2}{(177\ \text{W/m}\cdot\text{K})(0.05\ \text{m}\cdot 0.003\ \text{m})} + \frac{0.05\ \text{m}}{(177\ \text{W/m}\cdot\text{K})(0.05\ \text{m}\cdot 0.003\ \text{m})}$$

$$R_{a-b} = 0.942\ \text{K/W} + 0.942\ \text{K/W} + 1.883\ \text{K/W}$$

$$R_{a-b} = 3.767\ \text{K/W}$$

By symmetry we know that $R_{a-b} = R_{b-c}$. Thus $R_{b-c} = 3.767$ K/W. Using the heat transfer relation, we can solve for T_a

$$T_a = T_c + \dot{q}_{in}(R_{a-b} + R_{b-c})$$

$$T_a = 278\ \text{K} + 8\ \text{W}(3.767\ \text{K/W} + 3.767\ \text{K/W})$$

$$T_a = 278\ \text{K} + 8\ \text{W}(7.534\ \text{K/W})$$

$$T_a = 278\ \text{K} + 60.3\ \text{K}$$

$$T_a = 338.3\ \text{K or } 65.3°\text{C}$$

For T_b we can use

$$\dot{q}_{in} = \frac{(T_a - T_b)}{R_{a-b}}$$

$$T_b = T_a - \dot{q}_{in}(R_{a-b})$$

$$T_b = 338\text{ K} - 8\text{ W}(3.767\text{ K/W})$$

$$T_b = 338\text{ K} - 30.14\text{ K}$$

$$T_b = 307.86\text{ K} \quad \text{or } 34.9°\text{C}$$

iii)

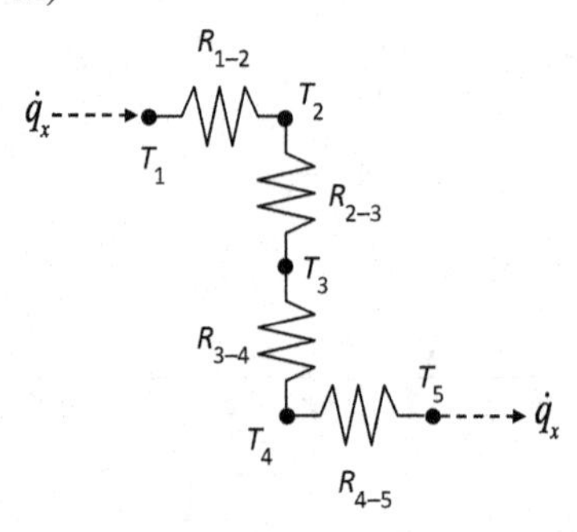

Due to the fact that $\Delta x = \Delta y$, $k_1 = k_2 = k_3 = k_4 = k_5 = k$ and $R_{1-2} = R_{2-3} = R_{3-4} = R_{4-5}$ then

$$R_{1-2} = \frac{0.5\Delta x}{k_1(\Delta y \cdot th)} + \frac{0.5\Delta x}{k_2(\Delta y \cdot th)} = \frac{\Delta x}{k(\Delta y \cdot th)}$$

iv) We know that $\dot{q}_{in} = \frac{(T_1 - T_5)}{R_{Tot}}$ where

$$\frac{1}{R_{Tot}} = \frac{1}{[R_{1-2} + R_{2-3} + R_{3-4} + R_{4-5}]}$$

$$\frac{1}{R_{Tot}} = \frac{1}{4 \cdot R_{1-2}} \text{ so } \dot{q}_{in} = \frac{(T_1 - T_5)}{4 \cdot R_{1-2}}$$

Thus,

$$R_{1-2} = \frac{\Delta x}{k(\Delta y \cdot th)} = \frac{0.05\text{ m}}{(177\text{ W/m} \cdot \text{K})(0.05\text{ m} \cdot 0.003\text{ m})} = 1.883\text{ K/W}$$

$$T_1 = T_5 + \dot{q}_{in}(4 \cdot R_{1-2})$$

$$T_1 = 278\text{ K} + 8\text{ W}(4 \cdot 1.833\text{ K/W})$$

$$T_1 = 278\text{ K} + 60.26\text{ K}$$

$$T_1 = 338.3\text{ K} \quad \text{or } 65.3°\text{C}$$

For T_2 we know that $\dot{q}_{in} = \frac{(T_1 - T_2)}{R_{1-2}}$

Thus,

$$T_2 = T_1 - \dot{q}_{in}(R_{1-2})$$

$$T_2 = 338\ \text{K} - 8\ \text{W}(1.833\ \text{K/W})$$

$$T_2 = 323.6\ \text{K} \quad \text{or } 50.6°\text{C}$$

Also for T_3 we know that $\dot{q}_{in} = \dfrac{(T_2 - T_3)}{R_{2-3}}$

Thus,

$$T_3 = T_2 - \dot{q}_{in}(R_{2-3})$$

$$T_3 = 323.6\ \text{K} - 8\ \text{W}(1.833\ \text{K/W})$$

$$T_3 = 309\ \text{K} \quad \text{or } 36°\text{C}$$

Last, for T_4 we know that $\dot{q}_{in} = \dfrac{(T_3 - T_4)}{R_{3-4}}$

Thus,

$$T_4 = T_3 - \dot{q}_{in}(R_{3-4})$$

$$T_4 = 309\ \text{K} - 8\ \text{W}(1.833\ \text{K/W})$$

$$T_4 = 294.34\ \text{K} \quad \text{or } 21.3°\text{C}$$

** T_5 was given as $T_5 = 5°\text{C}$

v) Case B

$$T_{avg} = \frac{T_2 + T_3 + T_4}{3} = \frac{50.6°\text{C} + 36°\text{C} + 21.3°\text{C}}{3} = 36°\text{C}$$

T_b for Case A was $\approx 35°\text{C}$, thus they are within 1°C.

2.11 i) We know the bellows tube can be fully extended or fully contracted. In the fully extended configuration the inner and outer radius of the corrugated section is approximated as being the same as the non-corrugated endpoint segments. Thus, in our resistance relation

$$R_{bellow} = \frac{L}{kA}$$

one can assume the path length for heat transfer is the sum of the path along the wall for the six corrugations and the non-corrugated segments.

$$L = 10\ \text{mm} + 12 \cdot 5\ \text{mm} + 10\ \text{mm} = 80\ \text{mm}$$

We know $r_i = 1\ cm$. Since the tube wall thickness is $t = 0.02$ cm, $r_o = 1.02$ cm

$$A = \pi\left(r_o^2 - r_i^2\right)$$

$$A = \pi((0.0102\ \text{m})^2 - (0.01\ \text{m})^2)$$

$$A = 1.269 \times 10^{-5}\ \text{m}^2$$

$$R_{bellow} = \frac{0.08\ \text{m}}{(14.9\ \text{W/m} \cdot \text{K})(1.269 \times 10^{-5}\ \text{m}^2)}$$

$$R_{bellow} = 423\ \text{K/W or } C_{bellow} = 0.0023\ \text{W/K}$$

ii)

$$Ratio = \frac{C_{bellow}}{C_{Tube}}$$

$$Ratio = \frac{\dfrac{k_{SS}A}{L_{bellow}}}{\dfrac{k_{SS}A}{L_{Tube}}} = \frac{L_{Tube}}{L_{bellow}}$$

$$Ratio = \frac{44\ \text{mm}}{80\ \text{mm}} = 0.55$$

2.12 i)

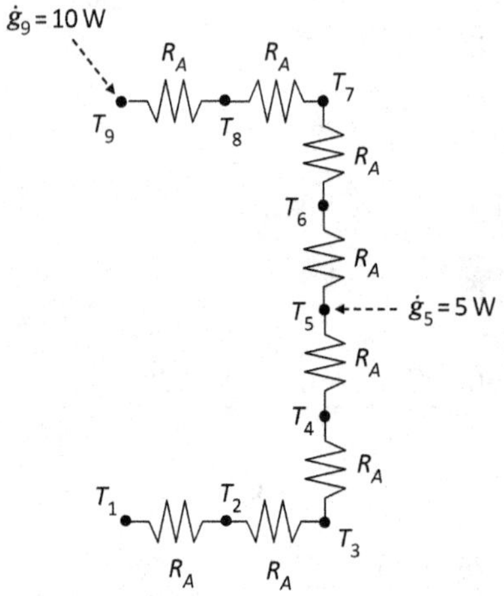

where $R_{1-2} = R_{2-3} = R_{3-4} = R_{4-5} = R_{5-6} = R_{6-7} = R_{7-8} = R_{8-9} = R_A$

$$R_A = \frac{L}{kA} = \frac{\Delta x}{k_{Al}(\Delta x \cdot th)} = \frac{1}{k_{Al} \cdot th} = \frac{1}{(177\ \text{W/m} \cdot \text{K})(0.01\ \text{m})} = 0.56497 \frac{\text{K}}{\text{W}}$$

ii) Taking an energy balance at node 5 gives

$$\dot{g}_9 + \dot{g}_5 = \frac{T_5 - T_1}{4R_A}$$

$$4R_A \cdot (\dot{g}_9 + \dot{g}_5) = T_5 - T_1$$

$$T_1 = T_5 - 4R_A \cdot (\dot{g}_9 + \dot{g}_5)$$

$$T_1 = 293\text{ K} - 4(0.056497\text{ K/W})(10\text{ W} + 5\text{ W})$$

$$T_1 = 259.1\text{ K} \approx -13.9°\text{C}$$

iii) Taking the energy balance at element 9 we have

$$\dot{g}_9 = \frac{T_9 - T_5}{4R_A}$$

$$4R_A \cdot \dot{g}_9 = T_9 - T_5$$

$$T_9 = 4R_A \cdot \dot{g}_9 + T_5$$

$$T_9 = 4(0.056497\text{ K/W})(10\text{ W}) + 293\text{ K}$$

$$T_9 = 315.6\text{ K} = 42.6°\text{C}$$

2.13

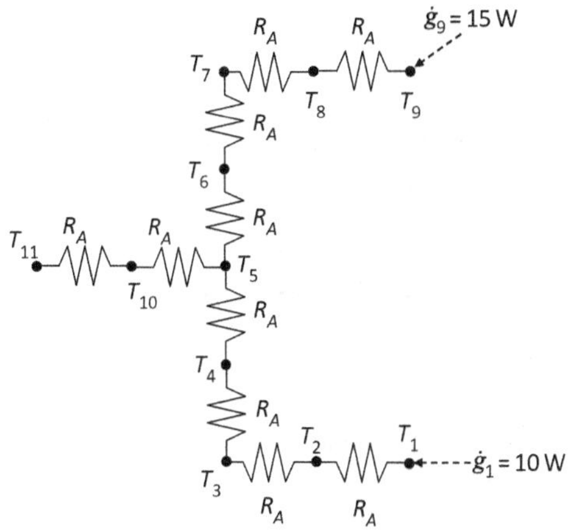

where $R_A = \dfrac{\Delta x}{k_{Al}(\Delta y \cdot t)} = \dfrac{1}{k_{Al} \cdot t} = \dfrac{1}{(177\text{ W/m}\cdot\text{K})(0.01\text{ m})} = 0.565\text{ K/W}$

ii) Let's assume element 9 has a temperature of 20°C. The central location for heat flow from either side of the wishbone configuration is element 5. Determination of the temperature at element 5 will be a necessary intermediate step in the process to determining the temperature of element 11. An energy balance at element 9 that includes element 5 would be

$$\frac{T_9 - T_5}{4 \cdot R_A} = \dot{g}_9 \text{ or } T_9 - T_5 = 4 \cdot R_A \cdot \dot{g}_9$$

Solving for the temperature at element 5 we would have

$$T_5 = T_9 - 4 \cdot R_A \cdot \dot{g}_9$$

$$T_5 = 293\text{ K} - 4 \cdot (0.56497\text{ K/W}) \cdot (15\text{ W})$$

$$T_5 = 259.1\text{ K} = -13.9°\text{C}$$

Now, taking an energy balance at element 11, we would have

$$\frac{T_5 - T_{11}}{2 \cdot R_A} = \dot{g}_9 + \dot{g}_1$$

Solving for the temperature at element 11

$$T_5 - T_{11} = 2 \cdot R_A \cdot (\dot{g}_9 + \dot{g}_1)$$

$$T_{11} = 259.1\text{ K} - 2 \cdot R_A \cdot (\dot{g}_9 + \dot{g}_1)$$

$$T_{11} = -13.9°\text{C} - 2 \cdot (0.565\text{ K/W}) \cdot (25\text{ W})$$

$$T_{11} = 230.9\text{ K} = -42.2°\text{C}$$

iii) Using the temperature value determined in part ii) for element 5, we can perform an energy balance from element 1 to element 5. This would be

$$\frac{T_1 - T_5}{4 \cdot R_A} = \dot{g}_1 \text{ or } T_1 - T_5 = 4 \cdot R_A \cdot \dot{g}_1$$

Solving for the temperature at element 1 we would have

$$T_1 = T_5 + 4 \cdot R_A \cdot \dot{g}_1$$

$$T_1 = 259.1\text{ K} + 4 \cdot (0.56497\text{ K/W}) \cdot (10\text{ W})$$

$$T_5 = 281.7\text{ K} = 8.7°\text{C}$$

2.14 i) For the honeycomb conductance, we can use

$$C_T = \left(\frac{8k\delta}{3S}\right)\left(\frac{LW}{T}\right)$$

where

$L = 1$ m; $W = 1$ m; $\sigma = 4/3$; $\delta = 0.001$ m; $S = 0.01$ m; $T = 0.0254$ m; $k = 167$ W/m · K

Placing values in the honeycomb through conductance equation we have

$$C_T = \left(\frac{8 \cdot (167\text{ W/m} \cdot \text{K})(0.001\text{ m})}{3(0.01\text{ m})}\right)\left(\frac{1\text{ m} \cdot 1\text{ m}}{0.0254\text{ m}}\right)$$

$$C_T = 1{,}753.3\text{ W/K}$$

Next, for the slab we know

$$C_{slab} = \frac{kA}{L} = \frac{(167\text{ W/m} \cdot \text{K})(1\text{ m} \cdot 1\text{ m})}{(0.0254\text{ m})}$$

$$C_{slab} = 6{,}574.8\text{ W/K}$$

The ratio of the honeycomb through conductance to the solid material conductance is

$$\frac{C_T}{C_{slab}} = 0.2666$$

ii) The density can be defined as $\rho = m \cdot Vol$. For 6061-T6 Al we know that

$$\rho|_{6061-T6,\ Al} = 2{,}700 \text{ kg/m}^3$$

Calculating for the mass we have

$$m|_{Honeycomb} = \frac{\rho\sigma\delta LW \cdot th}{S} = \frac{2 \cdot (4/3)(0.001 \text{ m})(1 \text{ m})(1 \text{ m})(0.0254 \text{ m})}{(0.01 \text{ m})}$$

$$m|_{Honeycomb} = 18.29 \text{ kg}$$

For the slab we would have

$$m|_{slab} = (2{,}700 \text{ kg/m}^3)(1 \text{ m})(1 \text{ m})(0.0254 \text{ m})$$

$$m|_{slab} = 68.5\ kg$$

The conductance to mass ratio for each would then be

Honeycomb

$$\frac{C_T}{m} = \frac{1{,}753.3 \text{ W/K}}{18.29 \text{ kg}}$$

$$\frac{C_T}{m} = 95.86 \frac{\text{W}}{\text{K} \cdot \text{kg}}$$

Slab

$$\frac{C_T}{m} = \frac{6{,}574.8 \text{ W/K}}{68.5 \text{ kg}}$$

$$\frac{C_T}{m} = 95.98 \frac{\text{W}}{\text{K} \cdot \text{kg}}$$

The ratio between the two then becomes

$$ratio = \frac{Honeycomb}{Slab} = \frac{95.86 \text{ W/K} \cdot \text{kg}}{95.98 \text{ W/K} \cdot \text{kg}} = 0.998$$

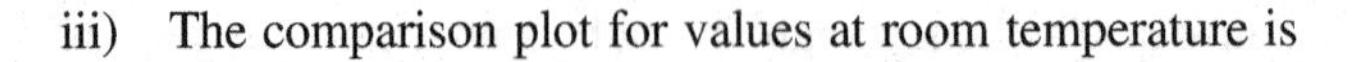

iii) The comparison plot for values at room temperature is

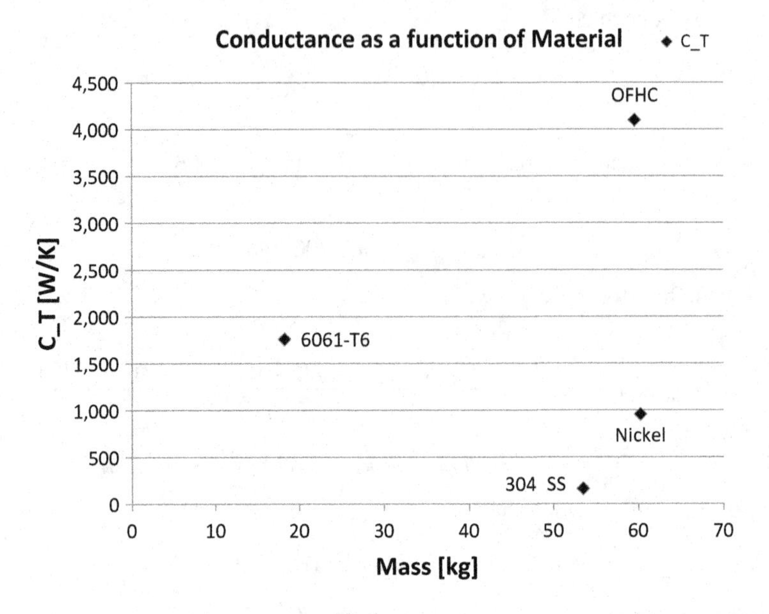

Material	Conductivity	Density	Volume	Mass	C_T
304 SS	14.9	7900	0.006773	53.51	156.43
Nickel	90.7	8900	0.006773	60.28	952.23
6061-T6	167	2700	0.006773	18.29	1753.28
OFHC	390	8800	0.006773	59.61	4094.48

Clearly, aluminum 6061-T6 honeycomb has a significant mass saving compared with the other materials. However, the aluminum does have a modest conductance relative to the best performer (conductance-wise) OFHC.

iv) The comparison plot for values at room temperature is

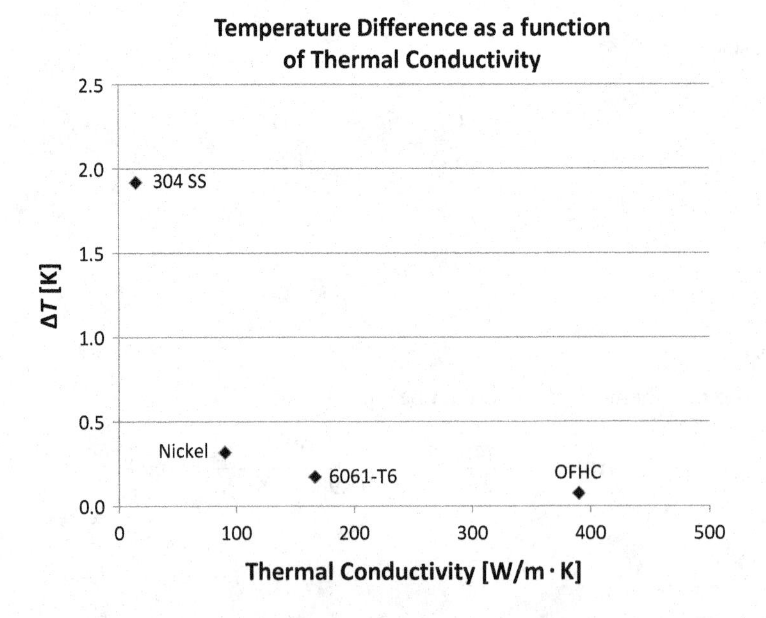

Material	Conductivity	Density	Volume	Mass	C_T	Delta T
304 SS	14.9	7900	0.006773	53.51	156.43	1.92
Nickel	90.7	8900	0.006773	60.28	952.23	0.32
6061-T6	167	2700	0.006773	18.29	1753.28	0.17
OFHC	390	8800	0.006773	59.61	4094.49	0.07

The lowest temperature difference for the materials selected is observed in aluminum 6061-T6 and OFHC. In light of the disparity in through conductance (shown in part iii) between the OFHC and 6061-T6 aluminum, the ΔTs only have a difference of a tenth of a degree. When factoring in the mass savings elucidated in part iii), aluminum becomes a primary candidate material for the honeycomb radiator.

2.15 Using the EES code below, the temperature of member 5 was determined to be 274.7 K.

```
{Problem 2.15 E-boxes attached to inner walls of 6-sided enclosure}
k_Al=167              [W/m-K]              {6061-Al}

{Lower cube dimensions}
th=0.006 [m]          {Side wall thickness}
s_lower=0.5       [m]       {Side wall length}

{E-boxes side length}
s_EBox=0.15 [m]

{Dissipations}
g_dot_3=25 [W]
g_dot_4=30 [W]

{Boundary temperatures}
T_Ebox_4=293

{Lower box inter-resistances}
R_lower=s_lower/(k_Al*s_lower*th)

{Resistances from E-boxes to lower members}
R_couple=(1/k_Al)*((0.5*s_EBox)/(s_EBox^2)+(0.5*th)/(s_lower^2))

{Energy balances}
T_1=(T_6+T_2+T_5+T_4)/4

T_2=(T_6+T_3+T_5+T_1)/4

T_3=(g_dot_3*R_lower+T_2+T_4+T_5+T_6)/4

T_4=T_EBox_4-g_dot_4*R_couple
```

```
T_5=((T_1+T_2+T_3+T_4)-(g_dot_3+g_dot_4)*R_lower)/4

T_6=(T_1+T_2+T_3+T_4)/4

T_EBox_3=T_3+g_dot_3*R_couple
```

2.16 i) The diagram for the thermal circuit is shown below

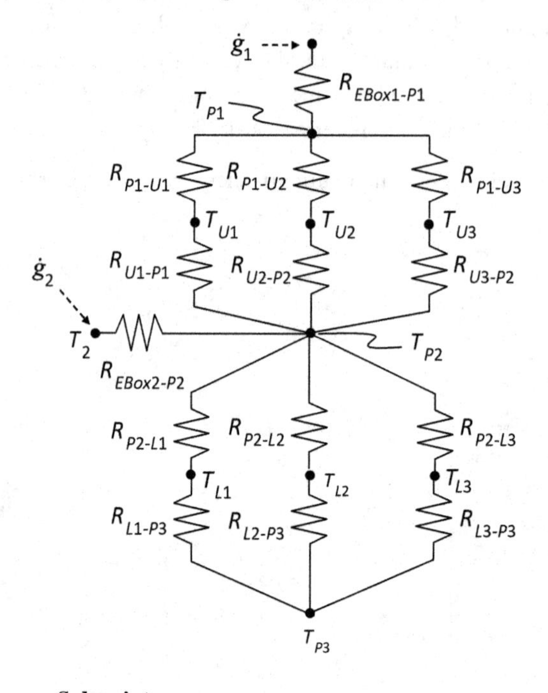

Subscripts

U: Upper stand-off
L: Lower stand-off
P: Plate

ii) To determine the required temperature on plate 3 we will use the following EES code

```
{Problem 2.16 Mulitplate problem, part ii}

{Boundary temperature}
T_Ebox2=273+25

{Thermal conductivities}
k_6061Al=167 [W/m-K]
k_Ni=90.5 [W/m-K]
k_Ti=21.9 [W/m-K]
k_OFHC=390 [W/m-K]
```

```
{Dissipations}
g_dot_1=15 [W]
g_dot_2=7.5 [W]

{Plates}
t_plate=0.25*2.54/100 [m]
Diam_plate=35/100 [m]

{Stand-off rods}
Diam_rod=1.5/100 [m]
h=25/100 [m]

{Area}
A_Ebox=0.01 [m^2]
A_plate=pi*(Diam_plate/2)^2 [m^2]
A_rod=pi*(Diam_rod/2)^2 [m^2]

{Resistances}
R_plate=0.5*t_plate/(k_6061Al*A_plate)
R_Ebox=.05/(k_6061Al*A_Ebox)
R_upper=h/(k_Ni*A_rod)
R_lower=h/(k_Ni*A_rod)

1/R_3upper=3/R_upper
1/R_3lower=3/R_lower

g_dot_2=(T_Ebox2-T_plate2)/(R_Ebox+R_plate)

g_dot_1+g_dot_2=(T_plate2-T_plate3)/(R_plate+R_3lower+R_plate)

g_dot_1=(T_Ebox1-T_plate2)/(R_Ebox+2*R_plate+R_3upper+R_plate)
```

Execution of the code predicts plate 3 will need to have a temperature of 180.5 K.

iii) Using a combination of aluminum and copper stand-offs, the temperature can be maintained between the designated temperatures. The corresponding EES code is shown below.

```
{Problem 2.16 Mulitplate problem, part iii}

{Boundary temperature}
T_Ebox2=273+25

{Thermal conductivities}
k_6061Al=167 [W/m-K]
k_Ni=90.5 [W/m-K]
```

```
k_Ti=21.9 [W/m-K]
k_OFHC=390 [W/m-K]

{Dissipations}
g_dot_1=15 [W]
g_dot_2=7.5 [W]

{Plates}
t_plate=0.25*2.54/100 [m]
Diam_plate=35/100 [m]

{Stand-off rods}
Diam_rod=1.5/100 [m]
h=25/100 [m]

{Area}
A_Ebox=0.01 [ m^2]
A_plate=pi*(Diam_plate/2)^2 [m^2]
A_rod=pi*(Diam_rod/2)^2 [m^2]

{Resistances}
R_plate=0.5*t_plate/(k_6061Al*A_plate)
R_Ebox=.05/(k_6061Al*A_Ebox)
R_upper=h/(k_OFHC*A_rod)
R_lower=h/(k_OFHC*A_rod)

1/R_3upper=3/R_upper
1/R_3lower=3/R_lower

g_dot_2=(T_Ebox2-T_plate2)/(R_Ebox+R_plate)

g_dot_1+g_dot_2=(T_plate2-T_plate3)/(R_plate+R_3lower+R_plate)

g_dot_1=(T_Ebox1-T_plate2)/(R_Ebox+2*R_plate+R_3upper+R_plate)
```

2.17 i) Using the general equations for length- and width-based honeycomb panel thermal conductance, we know that

$$C_L = \left(\frac{2k\delta}{\sigma S}\right)\left(\frac{WT}{L}\right) \text{ and } C_W = \left(\frac{k\delta\sigma}{2S}\right)\left(\frac{LT}{W}\right)$$

Thus, we can take the ratio

$$\frac{C_L}{C_W} = \frac{\left(\frac{2k\delta}{\sigma S}\right)\left(\frac{WT}{L}\right)}{\left(\frac{k\delta\sigma}{2S}\right)\left(\frac{LT}{W}\right)} = \frac{\left(\frac{2W}{\sigma L}\right)}{\left(\frac{\sigma L}{2W}\right)} = \left(\frac{2W}{\sigma L}\right)^2 = \left(\frac{2}{\sigma}\right)^2 = \left(\frac{2}{1.172}\right)^2 = 1.706$$

ii) Taking the ratio, we will have

$$\frac{C_T}{C_L} = \frac{\left(\frac{2k\sigma\delta}{S}\right)\left(\frac{LW}{T}\right)}{\left(\frac{2k\delta}{\sigma S}\right)\left(\frac{WT}{L}\right)} = \frac{\left(\frac{\sigma L}{T}\right)}{\left(\frac{T}{\sigma L}\right)} = \left(\frac{\sigma L}{T}\right)^2 = \left(\frac{1.172 \cdot 0.5\ \text{m}}{2 \cdot 0.03\ \text{m}}\right)^2 = 381.6$$

2.18 i)

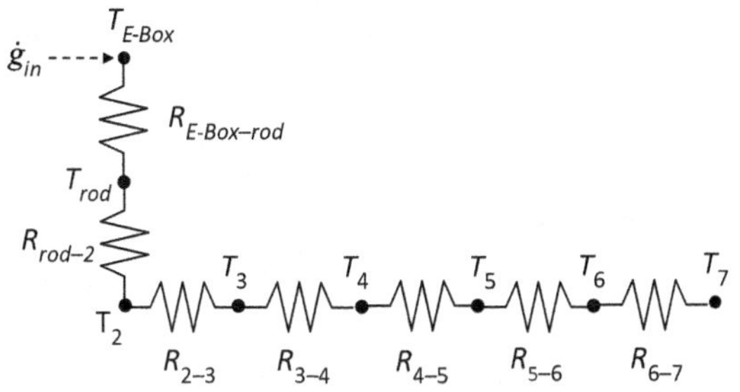

ii)

$$R_{e\text{-}rod} = \frac{0.06\ \text{m}/2}{k_{E\text{-}Box}(A_{E\text{-}Box})} + \frac{0.18\ \text{m}/2}{k_{rod}(A_{rod})}$$

$$R_{e\text{-}rod} = \frac{0.03\ \text{m}}{k_{E\text{-}Box}(0.06\ \text{m})^2} + \frac{0.09\ \text{m}}{k_{rod}\left(\pi \cdot (0.015\ \text{m})^2\right)}$$

$$R_{e\text{-}rod} = \frac{8.333\ \text{m}^{-1}}{k_{E\text{-}Box}} + \frac{127.33\ \text{m}^{-1}}{k_{rod}}$$

$$R_{rod-2} = \frac{0.09\ \text{m}}{k_{rod}\left(\pi \cdot (0.0015\ \text{m})^2\right)} + \frac{0.003\ \text{m}/2}{k_2(\Delta x \cdot \Delta y)}$$

$$R_{rod-2} = \frac{0.09\ \text{m}}{k_{rod}(7.068 \times 10^{-4}\ \text{m}^2)} + \frac{0.0015\ \text{m}}{k_2(0.09)(0.09)}$$

$$R_{rod-2} = \frac{127.33\ \text{m}^{-1}}{k_{rod}} + \frac{0.18519\ \text{m}^{-1}}{k_2}$$

$$R_{2-3} = \frac{0.5\Delta x}{k_2(\Delta y \cdot th)} + \frac{0.5\Delta x}{k_3(\Delta y \cdot th)}$$

However, since $k_1 = k_2 = k_3 = k_4 = k_5 = k_6 = k_7$ we can rewrite this as $R_{2-3} = \Delta x / k_2(\Delta y \cdot th)$ and by symmetry and use of like materials $R_{2-3} = R_{3-4} = R_{4-5} = R_{5-6} = R_{6-7}$

iii) We know that

$$\dot{g}_{in} = \frac{(T_{E\text{-}Box} - T_7)}{R_{Tot}}$$

In this case

$$\frac{1}{R_{Tot}} = \frac{1}{R_{E\text{-}Box-rod} + R_{rod-2} + R_{2-3} + R_{3-4} + R_{4-5} + R_{5-6} + R_{6-7}}$$

This is actually equal to

$$\frac{1}{R_{Tot}} = \frac{1}{R_{E\text{-}Box-rod} + R_{rod-2} + 5R_{2-3}}$$

Thus,

$$\dot{q} = \frac{T_{E\text{-}Box} - T_7}{R_{E\text{-}Box-rod} + R_{rod-2} + 5R_{2-3}}$$

Solving for the temperature at element 7

$$T_7 = T_{E\text{-}Box} - \dot{q}(R_{E\text{-}Box-rod} + R_{rod-2} + 5R_{2-3})$$

we know that

$$R_{2-3} = \frac{\Delta x}{k_2(\Delta y \cdot th)}$$

$$R_{2-3} = \frac{0.09 \text{ m}}{(200 \text{ W/m} \cdot \text{K})(0.09 \text{ m})(0.003 \text{ m})}$$

$$R_{2-3} = 1.667 \text{ K/W}$$

$$R_{e\text{-}rod} = \frac{8.333 \text{ m}^{-1}}{200 \text{ W/m} \cdot \text{K}} + \frac{127.33 \text{ m}^{-1}}{16.2 \text{ W/m} \cdot \text{K}} = 0.0416 \text{ K/W} + 7.86 \text{ K/W} = 7.9015 \text{ K/W}$$

$$R_{rod-2} = \frac{127.33 \text{ m}^{-1}}{16.2 \text{ W/m} \cdot \text{K}} + \frac{0.18519 \text{ m}^{-1}}{200 \text{ W/m} \cdot \text{K}} = 7.86 \text{ K/W} = 9.26 \times 10^{-4} \text{ K/W} = 7.8609 \text{ K/W}$$

Solving for T_7

$$T_7 = T_{E\text{-}Box} - \dot{q}(R_{E\text{-}Box-rod} + R_{rod-2} + 5R_{2-3})$$

$$T_7 = 308 \text{ K} - 10 \text{ W}(7.9015 \text{ K/W} + 7.8609 \text{ K/W} + 5 \cdot 1.667 \text{ K/W})$$

$$T_7 = 308 \text{ K} - 10 \text{ W}(24.1 \text{ K/W})$$

$$T_7 = 67 \text{ K}$$

** This is fairly cold

iv) Substituting in a thermal conductivity value of 390 W/m · K for OFHC, the resistance calculations now become

$$R_{e\text{-}rod} = \frac{8.333 \text{ m}^{-1}}{200 \text{ W/m} \cdot \text{K}} + \frac{127.33 \text{ m}^{-1}}{390 \text{ W/m} \cdot \text{K}} = 0.0416 \text{ K/W} + 0.326 \text{ K/W} = 0.368 \text{ K/W}$$

$$R_{rod-2} = \frac{127.33 \text{ m}^{-1}}{390 \text{ W/m} \cdot \text{K}} + \frac{0.18519 \text{ m}^{-1}}{200 \text{ W/m} \cdot \text{K}} = 0.326 \text{ K/W} = 9.26 \times 10^{-4} \text{ K/W} = 0.3269 \text{ K/W}$$

Solving for T_7

$$T_7 = T_{E\text{-}Box} - \dot{q}\left(R_{E\text{-}Box-rod} + R_{rod-2} + 5R_{2-3}\right)$$

$$T_7 = 308\text{ K} - 10\text{ W}(0.368\text{ K/W} + 0.3269\text{ K/W} + 5 \cdot 1.667\text{ K/W})$$

$$T_7 = 308\text{ K} - 10\text{ W}(9.0284\text{ K/W})$$

$$T_7 = 217.7\text{ K or } -55.3°\text{C}$$

v) Additional measures can be taken to reduce the temperature difference between the E-box and element 7. These options include:

- Increasing the cross-sectional area of the rod
- Decreasing the length of the rod
- Decreasing the path length from element 2 to element 7
- Changing the materials used for elements 2 through 7 to a material with higher thermal conductivity.

2.19 i) The thermal circuit can be diagramed as

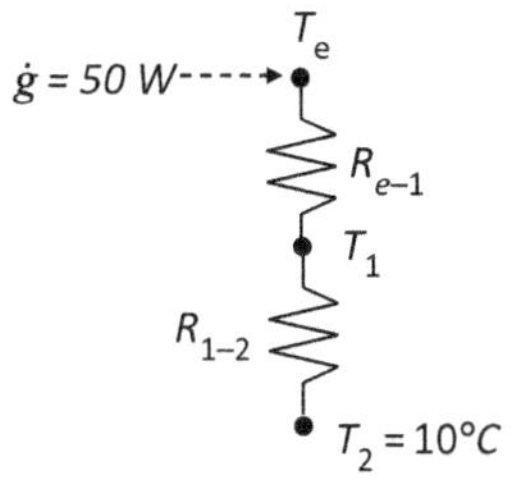

$\dot{g} = 50$ W, T_e, R_{e-1}, T_1, R_{1-2}, $T_2 = 10°C$

$k_{Al}|_{2024\text{-}T6} = 177$ W/m · K and $T_2 = 283$ K

By examining the circuit we know that $R_{Tot} = R_{e-1} + R_{1-2}$

$$R_{e-1} = \frac{0.1\text{ m}}{\left(177\,\dfrac{\text{W}}{\text{m}\cdot\text{K}}\right)(0.3\text{ m})(0.2\text{ m})} + \frac{0.005\text{ m}}{\left(177\,\dfrac{\text{W}}{\text{m}\cdot\text{K}}\right)(0.3\text{ m})(0.2\text{ m})} + \frac{0.15\text{ m}}{\left(177\,\dfrac{\text{W}}{\text{m}\cdot\text{K}}\right)(0.01\text{ m})(0.3\text{ m})}$$

$$R_{e-1} = 0.0094\text{ K/W} + 4.7\times10^{-4}\text{ K/W} + 0.282\text{ K/W}$$

$$R_{e-1} = 0.2923\text{ K/W}$$

$$R_{1-2} = \frac{0.25\text{ m}}{\left(177\,\dfrac{\text{W}}{\text{m}\cdot\text{K}}\right)(0.01\text{ m})(0.3\text{ m})} + \frac{0.005\text{ m}}{\left(177\,\dfrac{\text{W}}{\text{m}\cdot\text{K}}\right)(0.3\text{ m})(0.5\text{ m})}$$

$$R_{1-2} = 0.47\text{ K/W} + 1.88\times10^{-4}\text{ K/W}$$

$$R_{1-2} = 0.470188 \text{ K/W}$$

Also, we know that $\dot{g}_e = (T_e - T_2)/R_{Tot}$. Solving for T_e we have

$$T_e = \dot{g}_e \cdot R_{Tot} + T_2$$

$$T_e = 50 \text{ W} \cdot 0.7624 \text{ K/W} + 283 \text{ K}$$

$$T_e = 38.124 \text{ K} + 283 \text{ K}$$

$$T_e = 321.12 \text{ K} = 48.12°\text{C}$$

ii) The thermal circuit diagram now becomes

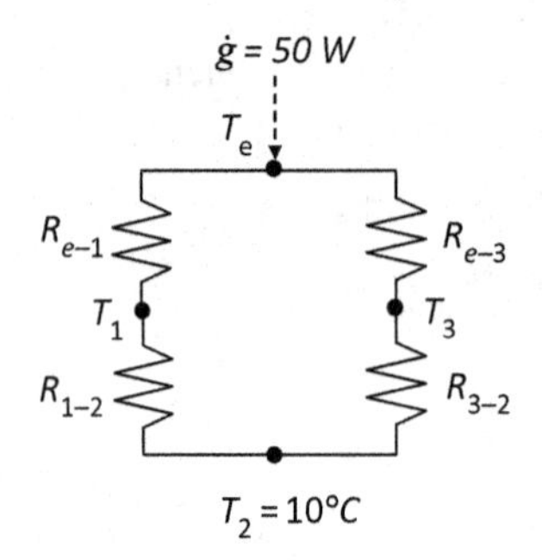

$R_{e-1} + R_{e-3}$ and $R_{1-2} + R_{3-2}$, thus $R_{e-1} + R_{e-3} = 0.2923$ K/W. R_{1-2} must be recalculated.

$$R_{1-2} = \frac{0.25 \text{ m}}{\left(177 \dfrac{\text{W}}{\text{m} \cdot \text{K}}\right)(0.01 \text{ m})(0.3 \text{ m})} + \frac{0.005 \text{ m}}{\left(177 \dfrac{\text{W}}{\text{m} \cdot \text{K}}\right)(0.3 \text{ m})(0.5 \text{ m})} + \frac{0.105 \text{ m}}{\left(177 \dfrac{\text{W}}{\text{m} \cdot \text{K}}\right)(0.01 \text{ m})(0.3 \text{ m})}$$

$$R_{1-2} = 0.4708 \text{ K/W} + 1.883 \times 10^{-4} \text{ K/W} + 0.1977 \text{ K/W}$$

$$R_{1-2} = 0.6687 \text{ K/W}$$

We must now determine the total resistance

$$\frac{1}{R_{Tot}} = \frac{1}{R_{e-1} + R_{1-2}} + \frac{1}{R_{e-3} + R_{3-2}}$$

$$\frac{1}{R_{Tot}} = \frac{1}{0.961 \text{ K/W}} + \frac{1}{0.961 \text{ K/W}}$$

$$\frac{1}{R_{Tot}} = 2.0811 \text{ W/K or } R_{Tot} = 0.4805 \text{ K/W}$$

Calculating the temperature of the electronics box now we have

$$T_e = \dot{g}_e \cdot R_{Tot} + T_2$$

$$T_e = 50\ \text{W} \cdot 0.4805\ \text{K/W} + 283\ \text{K}$$

$$T_e = 24.02\ \text{K} + 283\ \text{K}$$

$$T_e = 307.02\ \text{K} = 34°\text{C}$$

2.20 To determine the outer wall temperature required we will create an EES program to perform the energy balance calculations.

```
{Problem 2.20 Hollow Cubic with 3 E-box dissipations temperature
determination using shape factors}

{Energy dissipations}
g_dot_1=25 [W]
g_dot_2=30 [W]
g_dot_3=40 [W]

g_dot_total=g_dot_1+g_dot_2+g_dot_3

{Temperatures}
{T_avg=25}
T_avg=(T_i+T_o)/2  {Average temperature of the hollow cubic box}
T_3=20             {Temperature of E-box 3}

{Thermal conductivity}
k_Al=conductivity(Aluminum_6061, T=T_avg)

{E-Box 3 thermal resistance to interface with lower box}
side=12.5/100 [m]
R_EBox3_Cubic=(0.5*side)/(k_Al*side^2)

{Cubic shape factor determination}
x_o=50/100          [m]          {Outer cubic side length.....m}
x_i=(50-1)/100      [m]          {Inner cubic side length......m}

V=x_o^2-x_i^2      {Volume difference between outer and inner cube}

A_i=6*x_i^2        {Total surface area of inner cube}

L=A_i^0.5

s_inf=3.391

ratio=x_o/x_i
```

```
{Non-dimensional shape factor}
s_star= (2*(pi^0.5))/((1+((pi^0.5)/6)*(ratio^3-1))^(1/3)-1)+s_inf

S=s_star*A_i/L

{Energy balance relations}
T_i=T_3-g_dot_3*R_EBox3_Cubic

T_o=T_i-g_dot_total/(S*k_Al)
```

On execution of the code, the outer wall temperature requirement is approximately 19°C.

2.21 We will use EES to solve the problem. The code is shown below.

```
{Problem 2.21 Honeycomb panel problem}

x=0.75        [m]
y=0.75        [m]

N=3
M=3

delta_x=x/N
delta_y=y/M

Area=delta_x*delta_y

{Facesheet specifications}
thick=0.5/1000     {Thickness of facesheet in mm}

k_6061=177         [W/m-K]
k_Diamond=1500     [W/m-K]
k_value=k_6061

R_face=delta_x/(k_value*delta_y*thick)

{Honeycomb specifications}
del=0.5/1000      {Ribbon thickness}
S=1.5/100         {Honeycomb cell width}
sigma=4/3
z=2.54/100        {Thickness of honeycomb, cm}

R_rdir=(2*S*y)/(3*k_6061*del*x*z)
R_lat=(S*x)/(k_6061*del*x*z)
R_z=(3*S*z)/(8*k_6061*del*x*y)

R_totz=R_z+(thick)/(k_value*delta_x*delta_y)
```

```
{Interior facesheet}

Q_dot[1] =5
Q_dot[2] =0
Q_dot[3] =10
Q_dot[4] =0
Q_dot[5] =30
Q_dot[6] =0
Q_dot[7] =15
Q_dot[8] =0
Q_dot[9] =8

T[5] =30

{Inner facesheet conductances w/coupling to outer facesheet}
Q_dot[1] = ((T[1] -T[2] ) /R_face) + ((T[1] -T[4] ) /R_face) + ((T[1] -&
T[11] ) /R_totz)

Q_dot[2] = ((T[2] -T[1] ) /R_face) + ((T[2] -T[3] ) /R_face) + ((T[2] -&
T[5] ) /R_face) + ((T[2] -T[12] ) /R_totz)

Q_dot[3] = ((T[3] -T[2] ) /R_face) + ((T[3] -T[6] ) /R_face) +&
((T[3] -T[13] ) /R_totz)

Q_dot[4] = ((T[4] -T[1] ) /R_face) + ((T[4] -T[7] ) /R_face) +&
((T[4] -T[5] ) /R_face) + ((T[4] -T[14] ) /R_totz)

Q_dot[5] = ((T[5] -T[2] ) /R_face) + ((T[5] -T[8] ) /R_face) + ((T[5] -&
T[4] ) /R_face) + ((T[5] -T[6] ) /R_face) + ((T[5] -T[15] ) /R_totz)

Q_dot[6] = ((T[6] -T[3] ) /R_face) + ((T[6] -T[9] ) /R_face) +&
((T[6] -T[5] ) /R_face) + ((T[6] -T[16] ) /R_totz)

Q_dot[7] = ((T[7] -T[4] ) /R_face) + ((T[7] -T[8] ) /R_face) + ((T[7] -&
T[17] ) /R_totz)

Q_dot[8] = ((T[8] -T[7] ) /R_face) + ((T[8] -T[9] ) /R_face) + ((T[8] -&
T[5] ) /R_face) + ((T[8] -T[18] ) /R_totz)

Q_dot[9] = ((T[9] -T[6] ) /R_face) + ((T[8] -T[6] ) /R_face) + ((T[9] -&
T[19] ) /R_totz)

{Outer facesheet conductances w/coupling to inner facesheet}
Q_flux[11] =-100
Q_flux[12] =-100
Q_flux[13] =-100
```

```
Q_flux[14] =-100
{ Q_flux[15] =0}
Q_flux[16] =-100
Q_flux[17] =-100
Q_flux[18] =-100
Q_flux[19] =-100

Q_flux[ 11]*Area=((T[11] -T[12] )/R_face)+((T[11] -T[14] )/&
R_face)+((T[11] -T[1] )/R_totz)

Q_flux[ 12]*Area=((T[12] -T[11] )/R_face)+((T[12] -T[13] )/&
R_face)+((T[12] -T[15] )/R_face)+((T[12] -T[2] )/R_totz)

Q_flux[ 13]*Area=((T[13] -T[12] )/R_face)+((T[13] -T[16] )/&
R_face)+((T[13] -T[3] )/R_totz)

Q_flux[ 14]*Area=((T[14] -T[11] )/R_face)+((T[14] -T[17] )/&
R_face)+((T[14] -T[15] )/R_face)+((T[14] -T[4] )/R_totz)

Q_flux[ 15]*Area=((T[15] -T[12] )/R_face)+((T[15] -T[18] )/&
R_face)+((T[15] -T[14] )/R_face)+((T[15] -T[16] )/R_face)+&
((T[15] -T[5] )/R_totz)

Q_flux[ 16]*Area=((T[16] -T[13] )/R_face)+((T[16] -T[19] )/&
R_face)+((T[16] -T[15] )/R_face)+((T[16] -T[6] )/R_totz)

Q_flux[ 17]*Area=((T[17] -T[14] )/R_face)+((T[17] -T[18] )/&
R_face)+((T[17] -T[7] )/R_totz)

Q_flux[ 18]*Area=((T[18] -T[17] )/R_face)+((T[18] -T[19] )/&
R_face)+((T[18] -T[15] )/R_face)+((T[18] -T[8] )/R_totz)

Q_flux[ 19]*Area=((T[19] -T[16] )/R_face)+((T[19] -T[16] )/&
R_face)+((T[19] -T[9] )/R_totz)
```

i) -264 W/m^2

ii) 32.9 K

iii) 3.9 K

2.22 i)

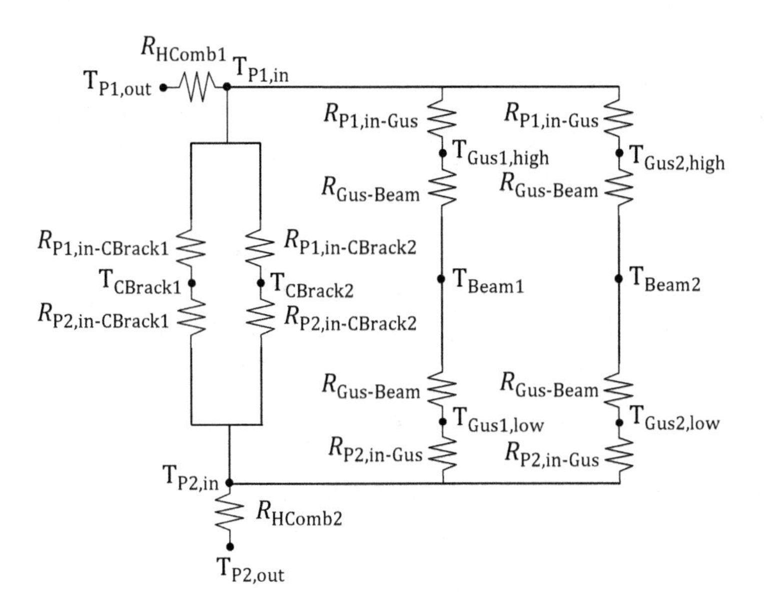

ii) We will use EES to solve this problem. The EES code is

```
{Problem 2.22 Honeycomb space truss with 80/20}

{Facesheet specifications}
thick=1.0/1000          { Thickness of facesheet in mm}
x=0.75             [m]
y=0.75             [m]

k_6061=177                                   [W/m-K]
k_6105=193                                   [W/m-K]
k_Diamond=1500                               [W/m-K]
k_value=k_6061

R_facesheet=(0.5*thick)/(k_6061*x*y)

{Honeycomb specifications}
del=0.5/1000     {Ribbon thickness}          [m]
S=1.5/100        {Honeycomb cell width}      [m]
sigma=4/3
z=2.54/100       {Thickness of honeycomb}

R_rdir=(2*S*y)/(3*k_6061*del*x*z)
R_lat=(S*x)/(k_6061*del*x*z)
R_z=(3*S*z)/(8*k_6061*del*x*y)
```

```
{Single panel calculations}
R_panel=2*R_facesheet+R_z

{1515 Eight-hole corner bracket}
A_CB_Bracket=(3.0*2.810)*(2.54*2.54)*(1e-4)-4*pi*&
((0.5*0.328*2.54/100)^2)
th_CB_Bracket=0.25*2.54/100

{Panel resistance through corner brackets}
R_CB_Bracket=(3.0*2.54/100)/(k_6105*(2.810*2.54/100)&
*th_CB_Bracket)+2*0.5*th_CB_Bracket/(k_6105*A_CB_Bracket)

{1515 Cross T-slot cross-beams}
R_Beam_Mid=(1.06-2*(0.5*4.243*2.54/100))/&
(k_6105*1.279*2.52*2.54*(1e-4))
R_Beam_Ends=(2*0.75*2.54/100)/(k_6105*1.5*1.312*2.54*&
2.54*(1e-4))

{ 1515 Corner gusset resistances}
th_Gusset=0.250*2.54/100
A_G_z1=4.506013*2.54*2.54*(1e-4)
A_G_z2=3.767*2.54*2.54*(1e-4)
L_Short=(0.5*4.243+0.5*3.0)*2.54/100
L_Long=(0.5*4.243+4.5)*2.54/100
A_Gusset_th=(0.25*1.312)*2.54*2.54*(1e-4)
1/R_Gusset=k_6105*A_Gusset_th/L_Short+k_6105*A_Gusset_th/L_Long

{Panel resistance through T-slot beam}
R_P1_G_P2=R_Beam_Mid+R_Beam_Ends+0.5*R_Gusset+0.5*&
th_Gusset/(k_6105*A_G_z1)+0.5*th_Gusset/(k_6105*A_G_z2)

{Parallel resistances between honeycomb panels}
1/R_parallel=2/R_CB_Bracket+2/R_P1_G_P2

{Panel facesheet to panel facesheet resistance}
R_Tot=2*R_panel+2*R_facesheet+R_parallel

{Panel facesheet temperature on loading side}
Q_dot=80
T_P2_out=260 [ K]

T_P1_out=Q_dot*R_Tot+T_P2_out
```

The temperature is 291.6 K

Chapter 3 Solutions

3.1 i) The spectral emissive power for a blackbody is defined as

$$E_{\lambda b}(\lambda, T) = \frac{2\pi c_1}{\lambda^5\left[\exp^{c_2/\lambda T} - 1\right]}$$

where $c_1 = hc_o^2$. The variables h and c_o are Planck's constant and the speed of light respectively. They have values of h = Planck's constant, 6.6256×10^{-34} J · s and c_o = speed of light 3.0×10^8 m/s. In addition, $c_2 = hc_o/k_B$ where k_B is Boltzmann's constant. This has a value of k_B = Boltzmann's constant, 1.3805×10^{-23} J/K. Constants c_1 and c_2 can be calculated as

$$c_1 = (6.6256 \times 10^{-34}\ \text{J} \cdot \text{s}) \cdot (3.0 \times 10^8\ \text{m/s})^2 = 5.963 \times 10^{-17}\ \text{W} \cdot \text{m}^2 \text{ and}$$

$$c_2 = \frac{\left(6.6256 \times 10^{-34}\ \text{J} \cdot \text{s}\right) \cdot \left(3.0 \times 10^8\ \text{m/s}\right)}{\left(1.3805 \times 10^{-23}\ \text{J/K}\right)} = 0.014398\ \text{m} \cdot \text{K or } 1.439 \times 10^4\ \mu\text{m} \cdot \text{K}$$

The value for the power on the exponent in the denominator must also be determined. This is calculated as follows

$$\frac{c_2}{\lambda T} = \frac{\left(1.439 \times 10^4\ \mu\text{m} \cdot \text{K}\right)}{(100\ \mu\text{m}) \cdot (800\ \text{K})} = 0.1799$$

Now we can place all of these values into our original relation for blackbody spectral emissive power. That will be

$$E_{\lambda b} = \frac{2\pi \cdot \left(5.963 \times 10^{-17}\ \text{W} \cdot \text{m}^2\right)}{(100\ \mu\text{m})^5\left[\exp^{0.1799} - 1\right]} = \frac{3.75 \times 10^{-16}\ \text{W} \cdot \text{m}^2}{1.971 \times 10^9\ \mu\text{m}^5}$$

$$E_{\lambda b} = \frac{3.75 \times 10^{-4}\ \text{W} \cdot \mu\text{m}^2}{1.971 \times 10^9\ \mu\text{m}^5}$$

$$E_{\lambda b} = 1.903 \times 10^{-13}\ \text{W}/\mu\text{m}^3 \text{ or } 0.1903\ \text{W/m}^2 \cdot \mu\text{m}$$

ii) The total blackbody emissive power is defined as $E_b = \sigma T^4$. Given that the temperature is known, we can solve for the blackbody emissive power as

$$E_b = (5.67 \times 10^{-8}\ \text{W/m}^2 \cdot \text{K}^4)(800\ \text{K})^4$$

$$E_b = 23{,}224.3\ \text{W/m}^2 \text{ or } E_b = 23.2\ \text{kW/m}^2$$

iii) By definition, Wien's Displacement Law is $\lambda_{max} \cdot T = c_3$ where $c_3 = 2{,}897.8\ \mu\text{m} \cdot \text{K}$. Solving for temperature we will have

$$T = \frac{c_3}{\lambda_{max}} = \frac{2{,}897.8\ \mu\text{m} \cdot \text{K}}{100\ \mu\text{m}} \approx 29\ \text{K}$$

3.2 To obtain Wien's Displacement Law one must differentiate Planck's Law for spectral distribution of emissive power w.r.t. wavelength (λ) and set the final form equal to a value of 0. In doing such we can solve for the maximum λ value (i.e., λ_{max}). We know Planck's Law is

$$E_{\lambda b}(\lambda, T) = \frac{2\pi c_1}{\lambda^5[\exp^{c_2/\lambda T} - 1]}$$

where

$c_1 = hc_o^2$; c_o = speed of light 3.0×10^8 m/s; h = Planck's constant, 6.6256×10^{-34} J · s

$c_2 = hc_o/k_B$; k_B = Boltzmann's constant, 1.3805×10^{-23} J/K

Taking the derivate of Planck's Law w.r.t. wavelength and setting that equal to zero we will have

$$\frac{d}{d\lambda}[E_{\lambda b}(\lambda, T)] = \frac{d}{d\lambda}\left[\frac{2\pi c_1}{\lambda^5[\exp^{c_2/\lambda T} - 1]}\right] = 0$$

For purposes of simplifying the procedure leading to our solution we will do a variable change. As such, let $\eta = \lambda T$. Thus

$$d\eta = Td\lambda \text{ or } d\lambda = \frac{d\eta}{T}$$

Our equality now becomes

$$T \cdot \frac{d}{d\eta}\left[\frac{2\pi c_1 T^5}{\eta^5[\exp^{c_2/\eta} - 1]}\right] = 0$$

which can be rewritten as

$$2\pi c_1 T^6 \cdot \frac{d}{d\eta}\left[\eta^{-5}\left(\exp^{c_2/\eta} - 1\right)^{-1}\right] = 0$$

Dividing both sides of the equal sign by $2\pi c_1 T^6$ we will have

$$\frac{d}{d\eta}\left[\eta^{-5}\left(\exp^{c_2/\eta} - 1\right)^{-1}\right] = 0$$

Distributing the derivative, we will have

$$(-5) \cdot \eta^{-6} \cdot \left(\exp^{c_2/\eta} - 1\right)^{-1} + (-1) \cdot \eta^{-5} \cdot \left(\exp^{c_2/\eta} - 1\right)^{-2} \cdot \exp^{c_2/\eta} \cdot \frac{c_2}{\eta^2} \cdot (-1) = 0$$

Divide through by $(\exp^{c_2/\eta} - 1)^{-1}$ and our equation now becomes

$$(-5) \cdot \eta^{-6} + \eta^{-5} \cdot \left(\exp^{c_2/\eta} - 1\right)^{-1} \cdot \exp^{c_2/\eta} \cdot \frac{c_2}{\eta^2} = 0$$

which can be rewritten as

$$\frac{5}{\eta^6} = \frac{c_2}{\eta^7} \cdot \frac{\exp^{c_2/\eta}}{[\exp^{c_2/\eta} - 1]}$$

divide both sides by 5 and obtain

$$\frac{1}{\eta^6} = \frac{1}{5} \cdot \frac{c_2}{\eta^7} \cdot \frac{\exp^{c_2/\eta}}{[\exp^{c_2/\eta} - 1]}$$

multiply both side by η^7 and the equation now becomes

$$\eta = \frac{c_2}{5} \cdot \frac{\exp^{c_2/\eta}}{[\exp^{c_2/\eta} - 1]}$$

We can change the form of the fraction or the RHS by multiplying the numerator and denominator by a fraction that is equal to unity. If we take that fraction as $1/\exp^{c_2/\eta}$ our equation will be

$$\eta = \frac{c_2}{5} \cdot \frac{\exp^{c_2/\eta}}{[\exp^{c_2/\eta} - 1]} \cdot \frac{1/\exp^{c_2/\eta}}{1/\exp^{c_2/\eta}}$$

$$\eta = \frac{c_2}{5} \cdot \frac{1}{\left[1 - \frac{1}{\exp^{c_2/\eta}}\right]}$$

which can also be written as

$$\eta = \frac{c_2}{5} \cdot \frac{1}{[1 - \exp^{-c_2/\eta}]}$$

As a last step we will substitute the original variables back in for η to obtain

$$\lambda T = \frac{c_2}{5} \cdot \frac{1}{[1 - \exp^{-c_2/\lambda T}]}$$

where λ as shown actually is λ_{max}. Solving for λT in EES will reveal that the product λT equals a constant value of c_3 = 2,897.8 μm · K

3.3 i) The geometry for this problem most closely matches view factor 2 in the Appendix A table. To match the form of view factor 2 focus will initially be placed on the view of composite surface 1,2 by composite surface 3,4. If composite surface 1,2 were subdivided it would have the form

$$A_{3,4}\, F_{3,4-1,2} = A_{3,4}\, F_{3,4-1} + A_{3,4}\, F_{3,4-2}$$

By reciprocity the following holds

$$A_{3,4}\, F_{3,4-1,2} = A_{1,2}\, F_{1,2-3,4}$$

$$A_{3,4}\, F_{3,4-1} = A_1 F_{1-3,4}$$

$$A_{3,4}\, F_{3,4-2} = A_2 F_{2-3,4}$$

These can be substituted back into the main subdivision equation to attain

$$A_{1,2} F_{1,2-3,4} = A_1 F_{1-3,4} + A_2 F_{2-3,4}$$

The view factor F_{1-4} needs to be isolated. This can be accomplished by subdividing $F_{1-3,4}$

$$A_{1,2} F_{1,2-3,4} = A_1 (F_{1-3} + F_{1-4}) + A_2 F_{2-3,4}$$

Solving for F_{1-4}

$$F_{1-4} = \frac{A_{1,2}F_{1,2-3,4} - A_2 F_{2-3,4}}{A_1} - F_{1-3}$$

With the aid of view factor 2, $F_{1,2-3,4}$, $F_{2-3,4}$, $F_{3-1,2}$ and F_{3-2} can be determined using the following EES code

```
"Example view factor for perpendicular surfaces sharing a
common edge"
{Calculation for surface 3 to surface 2}
x=0.4
y=0.3
w=0.75
A=y/w
B=x/w

Term1=B*arctan(1/B)+A*arctan(1/A)-((A^2+B^2)^0.5)*&
arctan(1/((A^2+B^2)^0.5))

Term2=(1+B^2)*(1+A^2)/(1+B^2+A^2)
Term3=((B^2)*(1+B^2+A^2)/((1+B^2)*(B^2+A^2)))^(B^2)

Term4=((A^2)*(1+A^2+B^2)/((1+A^2)*(A^2+B^2)))^(A^2)

F_A_2=(1/(pi*B))*(Term1+0.25*ln(Term2*Term3*Term4))
```

It should be noted that to solve for $F_{1,2-3,4}$ and $F_{2-3,4}$ the additional step of invoking the reciprocity relation for these view factors is required. View factor F_{1-3} needs to be determined. When viewing composite surfaces 1,2 from surface 3 subdivision can be invoked to obtain

$$F_{3-1,2} = F_{3-1} + F_{3-2}$$

Solving for F_{3-1} would give

$$F_{3-1} = F_{3-1,2} - F_{3-2}$$

by reciprocity $A_3F_{3-1} = A_1F_{1-3}$. Solving for F_{1-3} would give

$$F_{1-3} = \frac{A_3}{A_1}F_{3-1}$$

The following can be determined using view factor 2 and reciprocity rules

$F_{1,2-3,4} = 0.1626$; $F_{2-3,4} = 0.3192$; $F_{3-1,2} = 0.2969$; $F_{3-2} = 0.2087$

Thus

$$F_{3-1} = 0.2969 - 0.2087 = 0.0882$$

Given that the geometric dimensions of the surfaces are known, the areas can be calculated

$$A_1 = 0.525\ \text{m}^2;\ A_2 = 0.225\ \text{m}^2;\ A_3 = 0.3\ \text{m}^2;\ A_{1,2} = 0.75\ \text{m}^2$$

As such

$$F_{1-3} = \frac{0.3\ \text{m}^2}{0.525\ \text{m}^2}(0.0882) = 0.050$$

Returning to the relation for the view factor from surface 1 to surface 3

$$F_{1-4} = \frac{(0.75\ \text{m}^2)(0.1626) - (0.225\ \text{m}^2)(0.3192)}{0.525\ \text{m}^2} - 0.50$$

$$F_{1-3} = 0.045$$

ii) $\dot{q}_{1-4} = A_1 F_{1-4} \sigma \left[T_1^4 - T_4^4\right]$

$$\dot{q}_{1-4} = \left(0.525\ \text{m}^2\right) \cdot (0.045) \cdot \left(5.67 \times 10^{-8}\ \text{W/m}^2 \cdot \text{K}^4\right) \cdot \left[(300\ \text{K})^4 - (250\ \text{K})^4\right]$$

$$\dot{q}_{1-4} = 5.62\ \text{W}$$

3.4 i) The geometry for this problem most closely matches view factor 2 in the Appendix A table. However, for the actual view factor equation to apply the geometry in question must be the same. Thus, a surface needs to be added within the plane of surface 3 extending from it to the edge of surface 2. The new geometry will look like

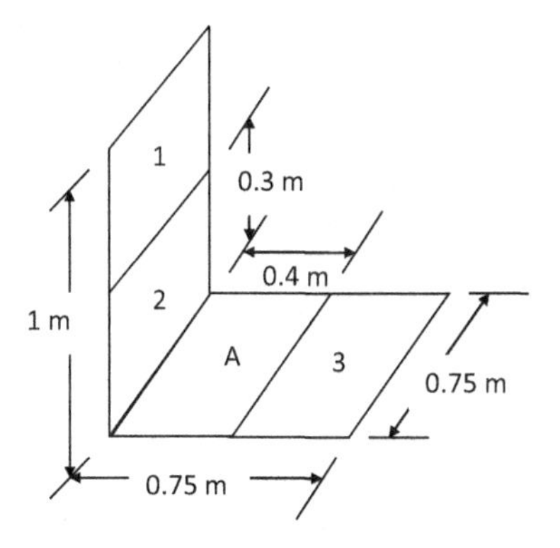

where A is an imaginary surface to be used only for solution purposes. To match the form of view factor 2 focus will initially be placed on the view of composite surface 1,2 by composite surface A,3. If composite surface 1,2 were subdivided it would have the form

$$A_{A,3}F_{A,3-1,2} = A_{A,3}F_{A,3-1} + A_{A,3}F_{A,3-2}$$

where the RHS of the equality consists of views to composite surface 1,2 being subdivided. By reciprocity, the following holds

$$A_{A,3}F_{A,3-1,2} = A_{1,2}F_{1,2-A,3}$$

$$A_{A,3}F_{A,3-1} = A_1F_{1-A,3}$$

$$A_{A,3}F_{A,3-2} = A_2F_{2-A,3}$$

The reciprocity relation can then be substituted back into the main view factor equation established for subdivision of composite surface 1,2. The new form becomes

$$A_{1,2}F_{1,2-A,3} = A_1F_{1-A,3} + A_2F_{2-A,3}$$

The $F_{1-A,3}$ can be further reduced by rewriting the equation with composite surface 1,3 subdivided

$$A_{1,2}F_{1,2-A,3} = A_1(F_{1-A} + F_{1-3}) + A_2F_{2-A,3}$$

The F_{1-3} term can be solved for by isolating it as

$$F_{1-3} = \frac{A_{1,2}F_{1,2-A,3} - A_2F_{2-A,3}}{A_1} - F_{1-A}$$

With the aid of view factor 2, $F_{1,2-A,3}$, $F_{2-A,3}$, $F_{A-1,2}$ and F_{A-2} can be determined using the following EES code

```
"Example view factor for perpendicular surfaces sharing a
common edge"
{Calculation for surface A to surface 2}
x=0.4
y=0.3
w=0.75
A=y/w
B=x/w

Term1=B*arctan(1/B)+A*arctan(1/A)-((A^2+B^2)^0.5)*&
arctan(1/((A^2+B^2)^0.5))

Term2=(1+B^2)*(1+A^2)/(1+B^2+A^2)

Term3=((B^2)*(1+B^2+A^2)/((1+B^2)*(B^2+A^2)))^(B^2)

Term4=((A^2)*(1+A^2+B^2)/((1+A^2)*(A^2+B^2)))^(A^2)

F_A_2=(1/(pi*B))*(Term1+0.25*ln(Term2*Term3*Term4))
```

It should be noted that to solve for $F_{1,2-A,3}$ and $F_{2-A,3}$ the additional step of invoking the reciprocity relation for these view factors is required. View factor F_{1-A} needs to be determined. When viewing composite surfaces 1,2 from surface A subdivision can be invoked to obtain

$$F_{A-1,2} = F_{A-1} + F_{A-2}$$

Solving for F_{A-1} would give

$$F_{A-1} = F_{A-1,2} - F_{A-2}$$

by reciprocity $A_A F_{A-1} = A_1 F_{1-A}$. Solving for F_{1-A} would give

$$F_{1-A} = \frac{A_A}{A_1} F_{A-1}$$

The following can be determined using view factor 2 and reciprocity rules

$$F_{1,2-A,3} = 0.1626;\ F_{2-A,3} = 0.3192;\ F_{A-1,2} = 0.2969;\ F_{A-2} = 0.2087$$

Thus

$$F_{A-1} = 0.2969 - 0.2087 = 0.0882$$

Given that the geometric dimensions of the surfaces are known, the areas can be calculated

$$A_1 = 0.525\ \text{m}^2;\ A_2 = 0.225\ \text{m}^2;\ A_A = 0.3\ \text{m}^2;\ A_{1,2} = 0.75\ \text{m}^2$$

As such

$$F_{1-A} = \frac{0.3\ \text{m}^2}{0.525\ \text{m}^2}(0.0882) = 0.050$$

Returning to the relation for the view factor from surface 1 to surface 3

$$F_{1-3} = \frac{(0.75\ \text{m}^2)(0.1626) - (0.225\ \text{m}^2)(0.3192)}{0.525\ \text{m}^2} - 0.50$$

$$F_{1-3} = 0.045$$

ii) $\dot{q}_{1-3} = A_1 F_{1-3} \sigma \left[T_1^4 - T_3^4\right]$

$$\dot{q}_{1-3} = \left(0.525\ \text{m}^2\right) \cdot (0.045) \cdot \left(5.67 \times 10^{-8}\ \text{W/m}^2 \cdot \text{K}^4\right) \cdot \left[(330\ \text{K})^4 - (263\ \text{K})^4\right]$$

$$\dot{q}_{1-3} = 9.48\ \text{W}$$

3.5 i) Find F_{1-4}
we can subdivide the originating surface as

$$A_{3,4} F_{3,4-1,2} = A_{3,4} F_{3,4-1} + A_{3,4} F_{3,4-2}$$

By reciprocity, we know that

$$A_{3,4} F_{3,4-1,2} = A_{1,2} F_{1,2-3,4}$$

$$A_{3,4} F_{3,4-1} = A_1 F_{1-3,4}$$

$$A_{3,4} F_{3,4-2} = A_1 F_{2-3,4}$$

So we can rewrite the main view factor equation as

$$A_{1,2}F_{1,2-3,4} = A_1F_{1-3,4} + A_2F_{2-3,4}$$

$$A_{1,2}F_{1,2-3,4} = A_1F_{1-3} + A_1F_{1-4} + A_2F_{2-3} + A_2F_{2-4}$$

By the reciprocity rule for diagonal members of rectangular surfaces joined by a common boundary, we also know that

$$A_1F_{1-4} = A_2F_{2-3}$$

Substituting into our previously expanded equation, we will get

$$A_{1,2}F_{1,2-3,4} = A_1F_{1-3} + A_1F_{1-4} + A_1F_{1-4} + A_2F_{2-4}$$

or

$$A_{1,2}F_{1,2-3,4} = A_1F_{1-3} + 2A_1F_{1-4} + A_2F_{2-4}$$

Solving for F_{1-4} we have

$$F_{1-4} = \frac{A_{1,2}F_{1,2-3,4} - A_1F_{1-3} - A_2F_{2-4}}{2A_1}$$

Based on geometry, we know that

$$A_{1,\,2} = 0.375\ \text{m}^2;\ A_1 = 0.225\ \text{m}^2;\ A_2 = 0.15\ \text{m}^2;$$

Also, using either the view factor ② (Appendix A) or the View Factor Library in EES, we can determine that

$$F_{1,2-3,4} = 0.226;\ F_{1-3} = 0.192;\ F_{2-4} = 0.162$$

Placing these values in the relation for F_{1-4} and solving will lead to $F_{1,4} = 0.038$

ii)

$$\dot{q}_{1-4} = A_1F_{1-4}\sigma\left[T_1^4 - T_4^4\right]$$

$$\dot{q}_{1-4} = \left(0.225\ \text{m}^2\right)\cdot(0.038)\cdot\left(5.67\times10^{-8}\ \text{W/m}^2\cdot\text{K}^4\right)\cdot\left[(320\ \text{K})^4 - (280\ \text{K})^4\right]$$

$$\dot{q}_{1-4} = 2.04\ \text{W}$$

iii)

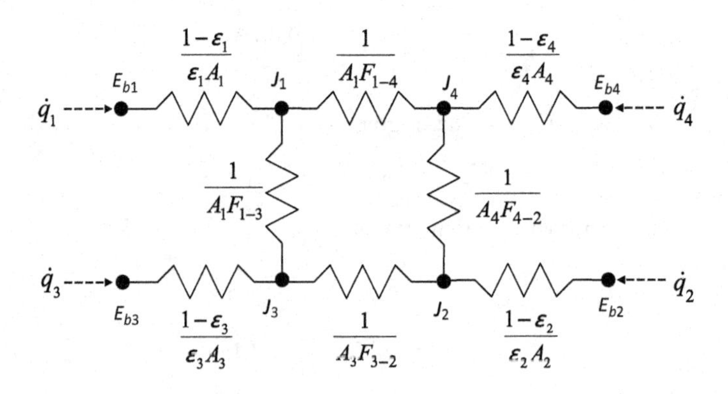

iv) Since we are dealing with gray bodies, in order to solve for the net heat transfer we must incorporate the $\hat{F}$ technique. We will use EES to develop relations leading to a solution.

```
{Problem 3.5, Part iv) Gray body net heat transfer between surface
1 and 4}

A_1=0.5*0.75            [m^2]
A_2=0.5*0.3             [m^2]
A_3=0.5*0.75            [m^2]
A_4=0.5*0.3             [m^2]

A_5=A_1+A_2+A_3+A_4+2*0.5*0.5 [m^2] {Effective area of surroudings}

{View factors}
F[1,1]=0; F[1,2]=0; F[1,3]=0.192; F[1,4]=0.038

F[2,1]=0; F[2,2]=0; F[2,3]=(A_1/A_2)*F[1,4]; F[2,4]=0.162

F[3,1]=(A_1/A_3)*F[1,3]; F[3,2]=(A_2/A_3)*F[2,3]; F[3,3]=0; F[3,4]=0

F[4,1]=(A_1/A_4)*F[1,4]; F[4,2]=(A_2/A_4)*F[2,4]; F[4,3]=0; F[4,4]=0

F[1,5]=1-(F[1,1]+F[1,2]+F[1,3]+F[1,4])
F[2,5]=1-(F[2,1]+F[2,2]+F[2,3]+F[2,4])
F[3,5]=1-(F[3,1]+F[3,2]+F[3,3]+F[3,4])
F[4,5]=1-(F[4,1]+F[4,2]+F[4,3]+F[4,4])

F[5,1]=A_1*F[1,5]/A_5
F[5,2]=A_2*F[2,5]/A_5
F[5,3]=A_3*F[3,5]/A_5
F[5,4]=A_4*F[4,5]/A_5
F[5,5]=1-(F[4,1]+F[4,2]+F[4,3])

{emissivities}
eps_1=0.88; eps_2=0.88; eps_3=0.88; eps_4=0.88; eps_5=1.0

{reflectivities}
rho[1]=1-eps_1; rho[2]=1-eps_2; rho[3]=1-eps_3; rho[4]=1-eps_4
rho[5]=0

{Temperatures}
T_1=320
T_4=280
```

```
{F-hat values}
Duplicate M=1,5
  Duplicate N=1,5
    F_hat[M,N]=F[M,N]+SUM(rho[K]*F[M,K]*F_hat[K,N],K=1,5)
  End
End

Q_dot_14=eps_1*eps_4*sigma#*A_1*F_hat[1,4]*(T_1^4-T_4^4)
```

The heat transfer between surface 1 and surface 2 is 2.719 W.

3.6 i) and iii) To determine the total emissive power for our surface we would incorporate the use of the Stefan–Boltzmann Law

$$E = \varepsilon \sigma A T^4$$

The emissivity was specificied as unity. We know the Stefan–Boltzmann constant to be 5.67×10^{-8} W/m · K^4. The area will have to be converted into metric units. Given that 1 in = 0.0254 m, the area in question turns out to be 0.06032 m^2. Using EES we can create a table to vary the temperature between 3 K and 300 K. An overlay of the corresponding plots for the emissive power as a function of temperature for parts i) and iii) will be

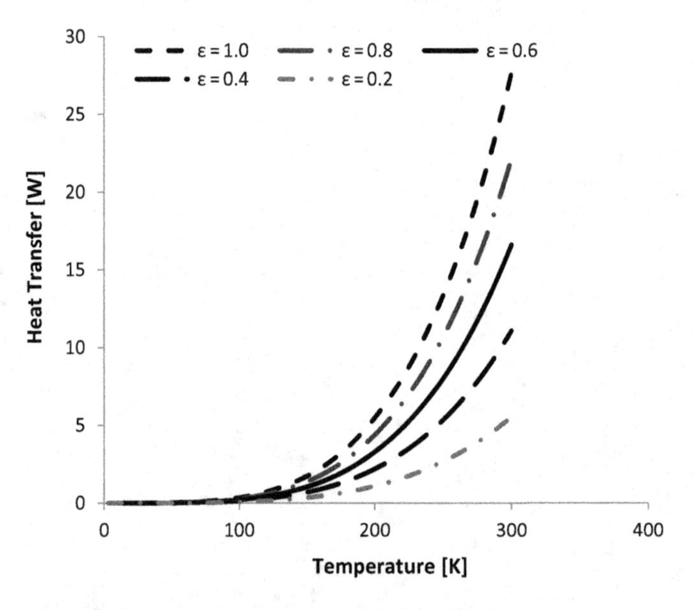

Plot for full temperature range spanning 3 K–300 K

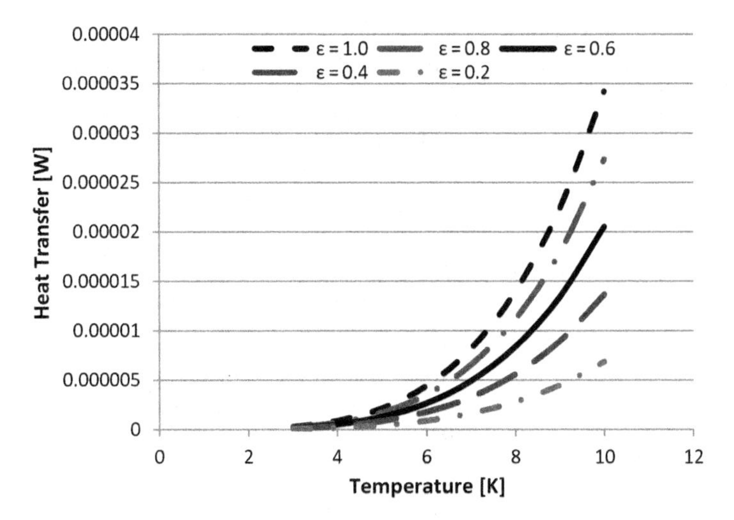

Heat transfer performance ≤10 K

Regardless of the surfaces' emissivity, the heat transfer performance falls within the single-digit μW range by 7 K. These values are approximately negligible for single-digit Kelvin thermal control systems.

ii) You can either use the Stefan–Boltzmann Law for radiation to calculate the temperature value at a given heat rejection or determine a value based on inspection of the plots. Either way the surface begins to emit <50 μW at 11 K. What this tells us is that when designing a cryogenic volume that is going to the 11 K or less temperature range, the net radiative thermal energy exchange between objects having surface areas less than the size of a sheet of paper will be on the order of tens of microwatts. This is low enough for the performance of this thermal design to be approximated under certain conditions at temperature using conduction-based analysis alone.

3.7 For this we will use the equation for effective emissivity

$$\varepsilon^* = \frac{1}{(N+1)\left[\dfrac{1}{\varepsilon_1} + \dfrac{1}{\varepsilon_2} - 1\right]}$$

Using values from Appendix B we can create the following table of values

Number of layers	Case i	Case ii	Case iii
1	0.210	0.010	0.385
2	0.140	0.007	0.257
3	0.105	0.005	0.192
4	0.084	0.004	0.154
5	0.070	0.003	0.128
6	0.060	0.003	0.110
7	0.052	0.002	0.096
8	0.047	0.002	0.086

(*cont.*)

Number of layers	Case i	Case ii	Case iii
9	0.042	0.002	0.077
10	0.038	0.002	0.070
11	0.035	0.002	0.064
12	0.032	0.002	0.059
13	0.030	0.001	0.055
14	0.028	0.001	0.051
15	0.026	0.001	0.048
16	0.025	0.001	0.045
17	0.023	0.001	0.043
18	0.022	0.001	0.041
19	0.021	0.001	0.038
20	0.020	0.001	0.037
21	0.019	0.001	0.035
22	0.018	0.001	0.033
23	0.017	0.001	0.032
24	0.017	0.001	0.031
25	0.016	0.001	0.030
26	0.016	0.001	0.029
27	0.015	0.001	0.027
28	0.014	0.001	0.027
29	0.014	0.001	0.026
30	0.014	0.001	0.025

Plotting these values we obtain the following

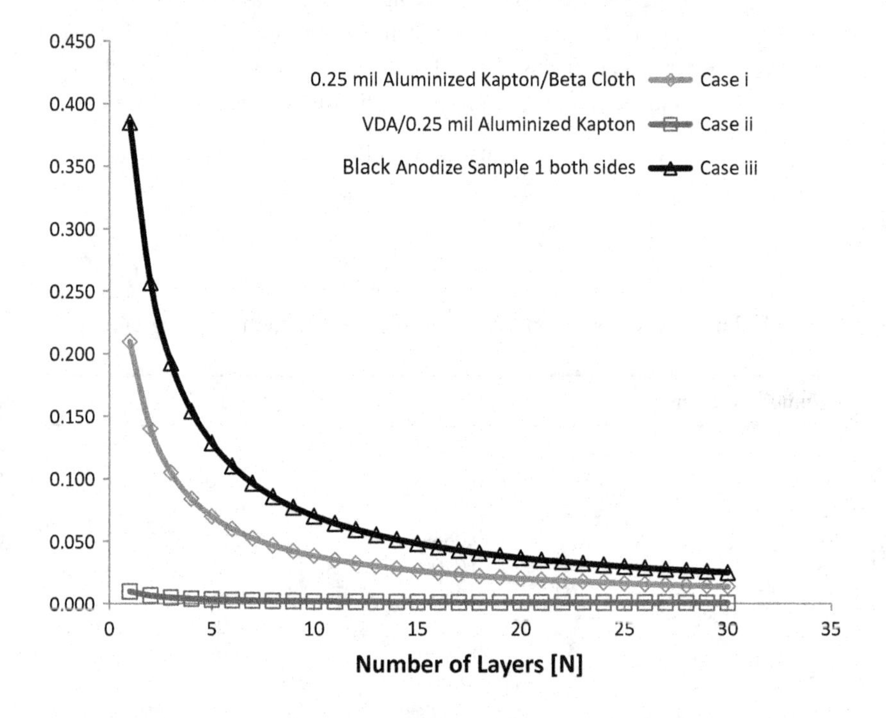

The lowest effective emissivity is achieved by using the combination of coatings having the lowest emissivities. For the problem here that would be case (ii). This case then creates the best radiative insulation of the three options listed. Note that each of the options shows very little variation beyond 25 layers.

3.8 Assuming diffuse gray body radiation:

i)

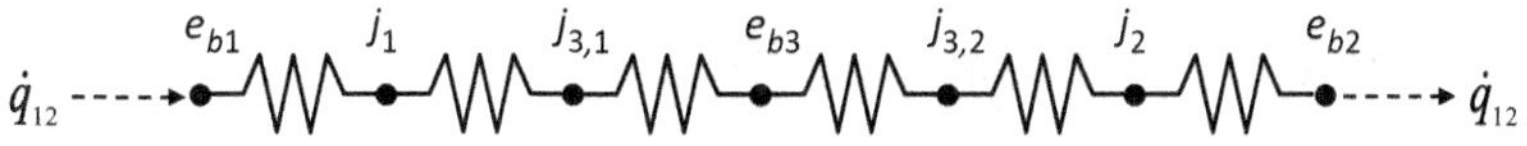

$\dot{q}_{12}$ → e_{b1} — j_1 — $j_{3,1}$ — e_{b3} — $j_{3,2}$ — j_2 — e_{b2} → $\dot{q}_{12}$

$$\frac{1-\varepsilon_1}{\varepsilon_1 A_1} \quad \frac{1}{A_1 F_{1-3,1}} \quad \frac{1-\varepsilon_{3,1}}{\varepsilon_{3,1} A_3} \quad \frac{1-\varepsilon_{3,2}}{\varepsilon_{3,2} A_3} \quad \frac{1}{A_3 F_{3,2-2}} \quad \frac{1-\varepsilon_2}{\varepsilon_2 A_2}$$

ii) For determination of the total resistance we must first find the view factors for $F_{1-3,1}$ and $F_{3,2-2}$. We know by summation rules for an enclosure that

$$F_{1-1} + F_{1-3,1} = 1$$

Since $F_{1-1} = 0$ here, $F_{1-3,1} = 1$
If we do a summation for the surface 3,2 then we have

$$F_{3,2-3,2} + F_{3,2-2} = 1$$

We know that $F_{3,2-3,2} = 0$ thus $F_{3,2-2} = 1$. The individual resistances are tallied as follows

$$\varepsilon_1 = 0.05,\ \varepsilon_2 = 0.05 \ \text{and}\ \varepsilon_{3,1} = \varepsilon_{3,2} = 0.02$$

iii) Given that $T_1 = 200$ K, $T_2 = 20$ K and $A_2 = 0.75\ \text{m}^2$

$$\dot{q}_{1-2} = \frac{\sigma(T_1^4 - T_2^4)}{\left(\dfrac{482.31}{A_2}\right)}$$

$$\dot{q}_{1-2} = \frac{\left(5.67 \times 10^{-8} \dfrac{\text{W}}{\text{m}^2 \cdot \text{K}^4}\right)\left[(200\ \text{K})^4 - (20\ \text{K})^4\right]}{\left(\dfrac{482}{0.75\ \text{m}^2}\right)}$$

$$\dot{q}_{1-2} = 0.141\ \text{W}$$

3.9 i) The calculation for effective emissivity is

$$\varepsilon^* = \frac{1}{(N+1)\left[\dfrac{1}{\varepsilon_A} + \dfrac{1}{\varepsilon_B} - 1\right]}$$

where A and B denote the individual sides of a thermal shield layer. The value for emissivity is provided in the problem statement. We also know from

Kirchoff's Law that $\varepsilon = \alpha$. However, we still need the value for the emissivity on the other side of the shield. For an opaque surface $\tau = 0.0$. Thus, the standard summation of irradiation components

$$\alpha + \rho + \tau = 1.0$$

reduces to

$$\alpha + \rho = 1.0$$

One can easily extrapolate from there that $\alpha = 1.0 - \rho$. Thus, for the shields being used

$$\alpha = 1.0 - 0.3 = 0.7 = \varepsilon$$

Having attained the value of the emissivity for the surface on the left of the shield, the effective emissivity can now be calculated as

$$\varepsilon^* = \frac{1}{(5+1)\left[\dfrac{1}{0.7} + \dfrac{1}{0.4} - 1\right]}$$

$$\varepsilon^* = \frac{1}{17.57} = 0.0569$$

ii) The net heat transfer occurs from the warmer surface to the colder surface (i.e., the 300 K surface to the 150 K surface). Assuming the warmer surface is on the inboard side of the MLI, the equation for heat transfer through the MLI will be

$$\dot{q}_{MLI} = \varepsilon^* A\sigma\left(T_i^4 - T_o^4\right)$$

where subscripts i and o denote inner and outer. The calculation is

$$\dot{q}_{MLI} = (0.0569)\cdot\left(5.67\times10^{-8}\ \mathrm{W/m^2\cdot K^4}\right)\cdot\left(0.75\ \mathrm{m^2}\right)\left[(300\ \mathrm{K})^4 - (150\ \mathrm{K})^4\right]$$

$$\dot{q}_{MLI} = 18.38\ \mathrm{W}$$

3.10 i)

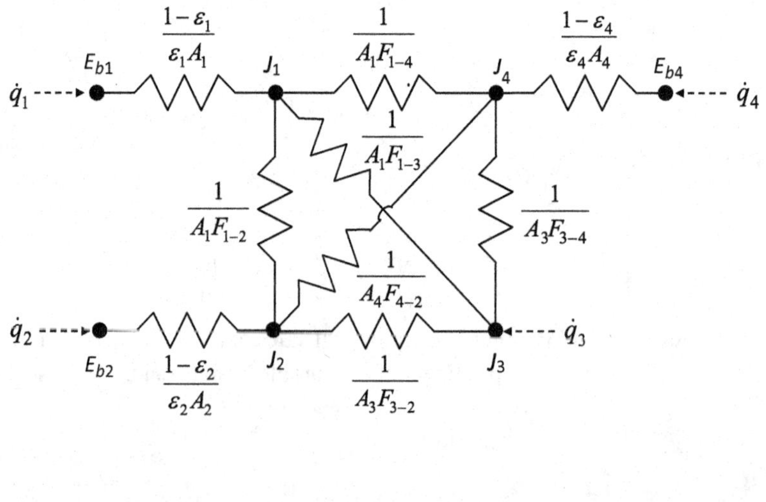

$\frac{1-\varepsilon_1}{\varepsilon_1 A_1}$ $\frac{1}{A_1F_{1-4}}$ $\frac{1-\varepsilon_4}{\varepsilon_4 A_4}$

E_{b1} J_1 J_4 E_{b4}

$\dot{q}_1$ $\dot{q}_4$

$\frac{1}{A_1F_{1-3}}$

$\frac{1}{A_1F_{1-2}}$ $\frac{1}{A_3F_{3-4}}$

$\frac{1}{A_4F_{4-2}}$

$\dot{q}_2$ $\dot{q}_3$

E_{b2} $\frac{1-\varepsilon_2}{\varepsilon_2 A_2}$ J_2 $\frac{1}{A_3F_{3-2}}$ J_3

ii) $$\frac{E_{b4}-J_4}{\dfrac{1-\varepsilon_4}{\varepsilon_4 A_4}}=\frac{J_4-J_1}{(A_4\mathrm{F}_{4-1})^{-1}}+\frac{J_4-J_2}{(A_4\mathrm{F}_{4-2})^{-1}}+\frac{J_4-J_3}{(A_4\mathrm{F}_{4-3})^{-1}}$$

iii) By the blackbody assumption, $\varepsilon = 1$ and there is no reflection (i.e., the radiosity includes no reflection). Another way to interpret this is that for a blackbody surface $J = E_b$. Thus, the radiosity can be recast as

$$J_3 = E_{b3} = A\sigma T_3^4$$

$$J_3 = E_{b3} = (0.25\ \mathrm{m}^2)(5.67\times10^{-8}\ \mathrm{W/m}^2\cdot\mathrm{K}^4)[(320\ \mathrm{K})^4]$$

$$J_3 = 148.6\ \mathrm{W}$$

3.11

i)

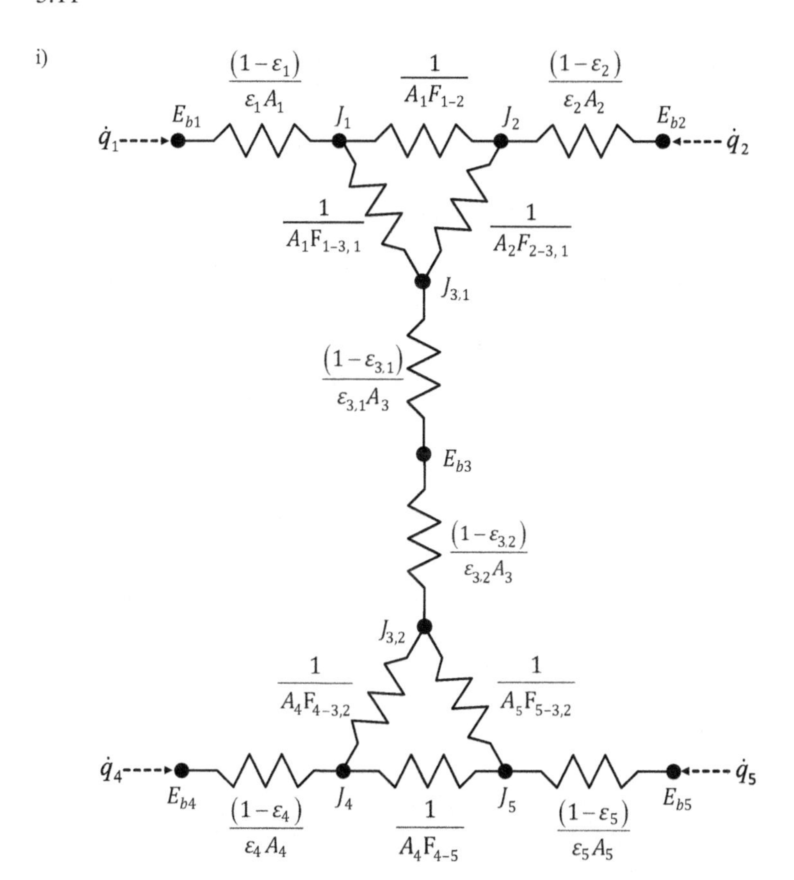

ii) $$\frac{E_{b4}-J_4}{\dfrac{1-\varepsilon_4}{\varepsilon_4 A_4}}=\frac{J_4-J_5}{(A_4\mathrm{F}_{4-5})^{-1}}+\frac{J_4-J_{3,2}}{(A_4\mathrm{F}_{4-3,2})^{-1}}$$

iii) We know $\rho_2 = 0.3$, $A_2 = 0.5\ \mathrm{m}^2$ and $T_2 = 300$ K

We must apply the Stefan–Boltzmann Law

$$\dot{Q} = A\sigma T^4$$

We know that $\alpha + \rho = 1$, thus

$$\alpha = 1 - \rho$$

$$\alpha = 1 - 0.3 = 0.7$$

Also, by Kirchhoff's Law we know $\varepsilon = \alpha = 0.7$
Thus,

$$\dot{Q} = (0.7)\left(0.5\ \text{m}^2\right)\left(5.67 \times 10^{-8}\ \text{W/m}^2 \cdot \text{K}^4\right)\left[(300\ \text{K})^4\right]$$

$$\dot{Q} = 160.7\ \text{W}$$

3.12 i)

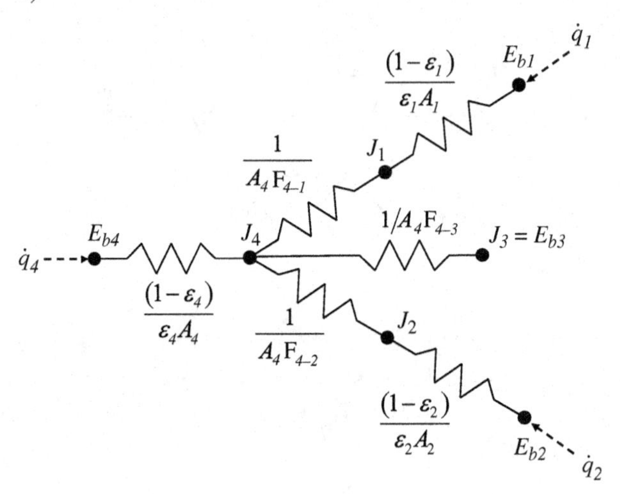

ii)

$$\frac{E_{b4} - J_4}{\frac{1 - \varepsilon_4}{\varepsilon_4 A_4}} = \frac{J_4 - J_1}{\left(A_4 F_{4-1}\right)^{-1}} + \frac{J_4 - J_2}{\left(A_4 F_{4-2}\right)^{-1}} + \frac{J_4 - J_3}{\left(A_4 F_{4-3}\right)^{-1}}$$

iii) We know that for a blackbody surface $J = E_b$. Thus, the radiosity can be recast as

$$J_3 = E_{b3} = A\sigma T_3^4$$

$$J_3 = E_{b3} = (0.5\ \text{m}^2)(5.67 \times 10^{-8}\ \text{W/m}^2 \cdot \text{K}^4)[(295\ \text{K})^4]$$

$$J_3 = 214.7\ \text{W}$$

3.13 Let the base be surface 1. The sphere is surface 2 and the box is surface 3. The view factor summation for the base surface is

$$F_{1-1} + F_{1-2} + F_{1-3} = 1.0$$

Since the base does not view itself $F_{1-1} = 0$. The summation relation now becomes

$$F_{1-2} + F_{1-3} = 1.0$$

which can be rearranged to solve for the view of the box from the base as

$$F_{1-3} = 1.0 - F_{1-2}$$

However, before this can be solved the view factor from the base to the sphere must be determined. This can be calculated with the aid of view factor 5 from the Appendix A table. Based on the surfaces defined, this view factor is

$$F_{2-1} = \frac{1}{2} \cdot \left[1 - \frac{1}{\sqrt{1 + (R_1)^2}} \right]$$

where

$$R_1 = \frac{r_1}{h} = \frac{0.25\ \text{m}}{0.74\ \text{m}} = 0.333$$

Placing this in the view factor relation gives

$$F_{2-1} = \frac{1}{2} \cdot \left[1 - \frac{1}{\sqrt{1 + (0.333)^2}} \right] = \frac{1}{2} \cdot [1 - 0.948] = 0.025$$

By reciprocity $A_2 F_{2-1} = A_1 F_{1-2}$ thus

$$F_{1-2} = \frac{A_2}{A_1} F_{2-1}$$

Placing values in this becomes

$$F_{1-2} = \left[\frac{4\pi \cdot (0.15\ \text{m})^2}{4\pi \cdot (0.25\ \text{m})^2} \right] (0.025) = 0.036$$

Placing this back into the view factor summation relation for the base to the box gives

$$F_{1-3} = 1.0 - F_{1-2}$$

$$F_{1-3} = 1.0 - 0.036$$

$$F_{1-3} = 0.964$$

3.14 i) The summation rule applies for the sphere in the enclosure. It should be noted that the Teflon string is roughly the thickness of a strand of human hair. As such, the view to it can be neglected. Surface 1 is the sphere. The summation rule as applied to it is

$$F_{1-1} + F_{1-2} + F_{1-3} + F_{1-4} = 1.0$$

However, since the surface is convex, the view upon itself will be zero. Thus, the summation for the sphere reduces to

$$F_{1-2} + F_{1-3} + F_{1-4} = 1.0$$

The equation for calculating both F_{1-2} and F_{1-3} can be obtained with the aid of View Factor 5 from the Appendix A table.

View to Surface 2

$$R_2 = \frac{r_2}{h} = \frac{0.2\ \text{m}}{0.3\ \text{m}} = 0.666;\ F_{1-2} = \frac{1}{2} \cdot \left[1 - \frac{1}{\sqrt{1 + (0.666)^2}}\right] = 0.0838$$

View to Surface 3

$$R_3 = \frac{r_3}{h} = \frac{0.2\ \text{m}}{0.7\ \text{m}} = 0.2857;\ F_{1-3} = \frac{1}{2} \cdot \left[1 - \frac{1}{\sqrt{1 + (0.2857)^2}}\right] = 0.0192$$

From the view factor summation as applied to surface 1, it is known that

$$F_{1-4} = 1 - F_{1-2} - F_{1-3}$$

Thus, F_{1-4} can be determined

$$F_{1-4} = 1 - 0.0838 - 0.0192$$

$$F_{1-4} = 0.897$$

In order to find the view from the inner wall of the cylinder to the sphere (i.e., F_{4-1}) we will have to apply the reciprocity rule. By reciprocity, we know

$$A_1 F_{1-4} = A_4 F_{4-1}$$

Thus $F_{4-1} = \frac{A_1}{A_4} F_{1-4}$. However, before calculating this view factor, the areas for both surfaces must be determined. The areas for surfaces 1 and 4 are

$$A_1 = 4\pi r^2 = 4\pi \cdot \left(\frac{0.25\ \text{m}}{2}\right)^2 = 0.1963\ \text{m}^2$$

$$A_4 = 2\pi r l = 2\pi \cdot (0.2\ \text{m}) \cdot (1.0\ \text{m}) = 1.2566\ \text{m}^2$$

Placing these values in the reciprocity relation gives

$$F_{4-1} = \left(\frac{0.1963\ \text{m}^2}{1.2566\ \text{m}^2}\right) \cdot (0.897) = 0.14$$

3.15 Based on geometry, we know that the following applies

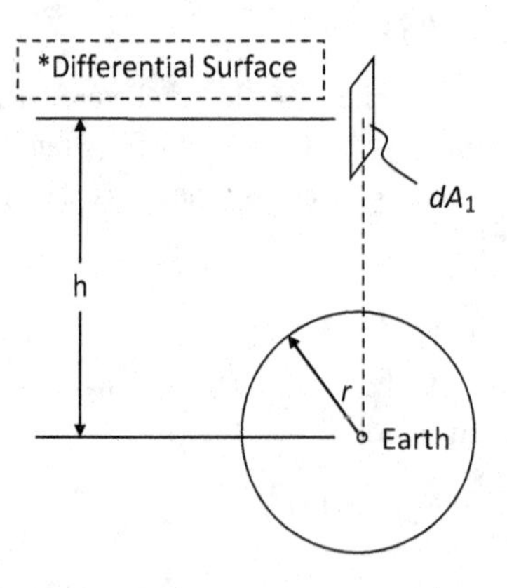

Let's use VF10 from Appendix A. The equation for it is below.

$$H = \left(\frac{h}{r}\right)$$

$$F_{d1-2} = \frac{1}{\pi}\left(\tan^{-1}\frac{1}{\sqrt{H^2-1}} - \frac{\sqrt{H^2-1}}{H^2}\right)$$

Known:

1) Radius of the Earth is 6,380 km
2) Altitude of space cube is 350 km
3) $dA_1 = 1\ m^2$

$h = r_{earth} + \text{altitude}$
$h = 6{,}380\ \text{km} + 350\ \text{km}$
$h = 6{,}730\ \text{km}$

$$H = \left(\frac{h}{r}\right) = \frac{6{,}730\ \text{km}}{6{,}380\ \text{km}} = 1.0548$$

$$F_{d1-2} = \frac{1}{\pi}\left(\tan^{-1}\frac{1}{\sqrt{(1.0548)^2-1}} - \frac{\sqrt{(1.0548)^2-1}}{(1.0548)^2}\right)$$

$$F_{d1-2} = 0.3008$$

3.16 If the SpaceCube is divided into six individual sides then the nadir-facing side's view of Earth can be determined using view factor 9 in Appendix A. For the surfaces on the side of the SpaceCube (i.e., facing deep space) their view factor to Earth can be calculated using view factor 10 from the Appendix A table. In order to account for the individual view factors between each side of the SpaceCube and the Earth, the surfaces of the SpaceCube that are within the FOV of the Earth need to be subdivided. As such, the solution methodology will begin by taking the view factor from the Earth to the SpaceCube and subdividing each of the SpaceCube's sides. In equation form this is

$$F_{E-SCube} = F_{E-SCube1} + F_{E-SCube2} + F_{E-SCube3} + F_{E-SCube4} + F_{E-SCube5} + F_{E-SCube6}$$

Let's assign surface 1 as the nadir-facing side. Surfaces 3, 4, 5 and 6 are the deep space facing sides of the SpaceCube. Surface 2 is the anti-nadir-facing side of the SpaceCube. Since surface 2 does not view the Earth, it has a view factor of $F_{E-SCube2} = 0$. In addition, due to geometry considerations and commonality of the Earth within their FOV, it can be assumed that

$$F_{E-SCube,Side} = F_{E-SCube3} = F_{E-SCube4} = F_{E-SCube5} = F_{E-SCube6}$$

As such, the initial equation established for subdivision of the SpaceCube surfaces can be recast as

$$F_{E-SCube} = F_{E-SCube1} + 4 \cdot F_{E-SCube,Side}$$

By reciprocity, the following equalities apply

$$A_E F_{E-SCube} = A_{SCube} F_{SCube-E}$$

$$A_E F_{E-SCube1} = A_{SCube1} F_{SCube1-E}$$

$$A_E F_{E-SCube,Side} = A_{SCube,Side} F_{SCube,Side-E}$$

If the subdivision equation is multiplied by A_E then it becomes

$$A_E F_{E-SCube} = A_E F_{E-SCube1} + 4 \cdot A_E \cdot F_{E-SCube,Side}$$

Using the reciprocity relations previously defined, this can be rewritten as

$$A_{SCube} F_{SCube-E} = A_{SCube1} F_{SCube1-E} + 4 \cdot A_{SCube,Side} \cdot F_{SCube,Side-E}$$

Solving for $F_{SCube-E}$

$$F_{SCube-E} = \frac{A_{SCube1} F_{SCube1-E} + 4 \cdot A_{SCube,\,Side} \cdot F_{SCube,\,Side-E}}{A_{SCube}}$$

Given that the length of the SpaceCube is 1.0 m

$$A_{SCube1} = A_{SCube,\ Side} = 1.0\ \text{m}^2;\ A_{Scube} = 6.0\ \text{m}^2$$

For $F_{SCube1-E}$, it is defined as $F_{SCube1-E} = \left(\frac{r}{h}\right)^2$
where r is the radius of the earth and $h = r_{earth} + altitude$. Placing in values and solving for h
$h = 6{,}380\ \text{km} + 250\ \text{km}$
which sums to $h = 6{,}630$ km. Placing in the view factor calculation gives

$$F_{SCube1-E} = \left(\frac{r}{h}\right)^2 = \left(\frac{6{,}380\ \text{km}}{6{,}630\ \text{km}}\right)^2 = 0.962^2$$

$$F_{SCube1-E} = 0.9254$$

For the SpaceCube side surfaces, view factor 10 will be relied upon.
$H = \left(\frac{h}{r}\right)$ where $h = 6{,}380\ \text{km} + 250\ \text{km} = 6{,}630$ km. The view factor relation then becomes

$$H = \left(\frac{h}{r}\right) = \frac{6{,}630\ \text{km}}{6{,}380\ \text{km}} = 1.039$$

$$F_{SCube,\,Side-E} = \frac{1}{\pi}\left(\tan^{-1}\frac{1}{\sqrt{(1.039)^2 - 1}} - \frac{\sqrt{(1.039)^2 - 1}}{(1.039)^2}\right)$$

$$F_{SCube,Side-E} = 0.3294$$

Placing both of these view factors into the overall view factor for the SpaceCube to Earth

$$F_{SCube-E} = \frac{(1.0\ \text{m}^2) \cdot (0.9254) + 4 \cdot (1.0\ \text{m}^2) \cdot (0.3294)}{(6.0\ \text{m}^2)}$$

$$F_{SCube-E} = 0.3738$$

3.17 The boundary temperatures (i.e., $T_1 = 348$ K and $T_2 = 248$ K) will be assigned to both the conductive and radiative representation of the heat transfer. The effective conductance can be calculated as

$$\dot{q}_{cond} = \dot{q}_{rad}$$

$$\frac{kA}{l} \cdot (T_1 - T_2) = \varepsilon^* \sigma A \left(T_1^4 - T_2^4\right)$$

$$C_{eff} = \frac{kA}{l} = \frac{\varepsilon^* \sigma A \left(T_1^4 - T_2^4\right)}{(T_1 - T_2)}$$

$$C_{eff} = \frac{(0.015)\left(5.67 \times 10^{-8} \dfrac{\text{W}}{\text{m}^2 \cdot \text{K}^4}\right)(0.5\ \text{m}^2)\left((348\ \text{K})^4 - (248\ \text{K})^4\right)}{(348\ \text{K} - 248\ \text{K})}$$

$$C_{eff} = 0.0463\ \text{K/W or } R_{Total} = 21.6\ \text{K/W}$$

3.18 i) From the MLI equation, we know

$$\varepsilon^* = \frac{1}{(N+1)\left[\dfrac{1}{\varepsilon_A} + \dfrac{1}{\varepsilon_B} - 1\right]}$$

Since the effective emissivity is known, we can isolate the $(N + 1)$ on the LHS as

$$(N+1) = \frac{1}{(\varepsilon^*)\left[\dfrac{1}{\varepsilon_A} + \dfrac{1}{\varepsilon_B} - 1\right]}$$

Solving for N would give

$$N = \frac{1}{(\varepsilon^*)\left[\dfrac{1}{\varepsilon_A} + \dfrac{1}{\varepsilon_B} - 1\right]} - 1$$

Substituting in values we have

$$N = \frac{1}{(0.01753)\left[\dfrac{1}{0.26} + \dfrac{1}{0.35} - 1\right]} - 1$$

$$N = \frac{1}{(0.01753)(5.70329)}$$

$$N = 10.002 - 1$$

$$N = 10.002 - 1$$

$$N = 9.002$$

Rounding down, nine shields are required to achieve the effective emissivity quoted in the problem statement.

ii) The heat loss can be modeled as

$$\dot{Q}_{LOSS} = \varepsilon^* A \sigma \left(T_i^4 - T_o^4\right)$$

Isolating the fourth-order temperature difference on the RHS we would have

$$\frac{\dot{Q}_{LOSS}}{\varepsilon^* A\sigma} = T_i^4 - T_o^4$$

Solving for the inner temperature term gives

$$T_i^4 = \frac{\dot{Q}_{LOSS}}{\varepsilon^* A\sigma} + T_o^4$$

Taking the fourth root gives

$$T_i = \left[\frac{\dot{Q}_{LOSS}}{\varepsilon^* A\sigma} + T_o^4\right]^{\frac{1}{4}}$$

Before calculating the temperature, the area must be determined.

$$A = 4\pi r^2; \qquad r = d/2$$

$$A = 4\pi\left(\frac{d}{2}\right)^2$$

$$A = \cancel{4}\pi\frac{d^2}{\cancel{4}}$$

$$A = \pi d^2 = \pi(0.5\text{ m})^2$$

$$A = 0.7854\text{ m}^2$$

Placing the known values in the equation for the temperature, we have

$$T_i = \left[\frac{6\text{ W}}{(0.01753)(0.7854\text{ m}^2)(5.67\times10^{-8}\text{ W/m}^2\cdot\text{K}^4)} + (142\text{ K})^4\right]^{\frac{1}{4}}$$

$$T_i = 299.9\text{ K} \approx 300\text{ K}$$

3.19 i) For the inner radius we will have $r_i = \dfrac{1.27\text{ cm}}{2} = 0.635\text{ cm}$

For the outer radius we will have $r_o = \dfrac{3.81\text{ cm}}{2} = 1.905\text{ cm}$

Also, we know that $k|_{304,SS} = 14.9\text{ W/m}\cdot\text{K}$. Note that this is taken at room temperature.

The resistance relation for radial heat transfer in cylindrical coordinates applies. That would be

$$R = \frac{\ln(r_o/r_i)}{2\pi kl}$$

We can define the total resistance as

$$R_{Total} = R_{tube} + R_{Polystyrene} + R_{sleeve}$$

The resistance through the polystyrene will be

$$R_{Polystyrene} = \frac{\ln\left(\frac{18.05\text{ mm}}{7.35\text{ mm}}\right)}{2\pi \cdot \left(330 \times 10^{-6}\text{ W/m}\cdot\text{K}\right) \cdot 1\text{ m}}$$

$$R_{Polystyrene} = 4.333 \text{ K/W or } 0.23078 \text{ W/K}$$

The resistance through the inner tube is

$$R_{tube} = \frac{\ln\left(\frac{6.35\text{ mm}}{7.35\text{ mm}}\right)}{2\pi \cdot (14.9\text{ W/m}\cdot\text{K}) \cdot 1\text{ m}} = 0.00183\frac{\text{K}}{\text{W}}$$

$$R_{sleeve} = \frac{\ln\left(\frac{19.05\text{ mm}}{18.05\text{ mm}}\right)}{2\pi \cdot (14.9\text{ W/m}\cdot\text{K}) \cdot 1\text{ m}} = 0.01127\frac{\text{K}}{\text{W}}$$

$$R_{Total} = 0.00183\frac{\text{K}}{\text{W}} + 4.333\frac{\text{K}}{\text{W}} + 0.01127\frac{\text{K}}{\text{W}}$$

$$R_{Total} = 4.3461\frac{\text{K}}{\text{W}} \text{ or } C_{Total} = 0.2300\frac{\text{W}}{\text{K}}$$

ii) Here we shall recast the radiation occurring in the vacuum gap space between the tube and the outer sleeve as conduction using the equality

$$\dot{Q}_{cond} = \dot{Q}_{rad,net}$$

where

$$\dot{Q}_{rad,net} = \frac{\sigma\left(T_o^4 - T_i^4\right)}{\left[\frac{1-\varepsilon_i}{\varepsilon_i A_i} + \frac{1}{A_i F_{i-o}} + \frac{1-\varepsilon_o}{\varepsilon_o A_o}\right]}$$

We can further expand out the conduction/radiation equality as

$$\frac{k_{eff} A}{l} \cdot (T_o - T_i) = \frac{\sigma\left(T_o^4 - T_i^4\right)}{\left[\frac{1-\varepsilon_i}{\varepsilon_i A_i} + \frac{1}{A_i F_{i-o}} + \frac{1-\varepsilon_o}{\varepsilon_o A_o}\right]}$$

which can also be expressed as

$$C_{eff} \cdot (T_o - T_i) = \frac{\sigma\left(T_o^4 - T_i^4\right)}{\left[\frac{1-\varepsilon_i}{\varepsilon_i A_i} + \frac{1}{A_i F_{i-o}} + \frac{1-\varepsilon_o}{\varepsilon_o A_o}\right]}$$

Solving for the effective conductance we will have

$$C_{eff} = \frac{\sigma\left(T_o^4 - T_i^4\right)}{\left[\frac{1-\varepsilon_i}{\varepsilon_i A_i} + \frac{1}{A_i F_{i-o}} + \frac{1-\varepsilon_o}{\varepsilon_o A_o}\right] \cdot (T_o - T_i)}$$

which can also be redefined as

$$C_{eff} = \frac{\sigma(T_o^4 - T_i^4)}{\frac{1}{A_i} \cdot \left[\frac{1-\varepsilon_i}{\varepsilon_i} + \frac{1}{F_{i-o}} + \left(\frac{A_i}{A_o}\right)\left(\frac{1-\varepsilon_o}{\varepsilon_o}\right)\right] \cdot (T_o - T_i)}$$

By definition of the area we know that
$A_i/A_o = r_i/r_o$ and the assumption that the inner tube has full view of the outer tube's inner wall we will let $F_{i-o} = 1.0$. The overall effective conductance relation now becomes

$$C_{eff} = \frac{\sigma(T_o^4 - T_i^4)}{\frac{1}{A_i} \cdot \left[\frac{1-\varepsilon_i}{\varepsilon_i A_i} + \left(\frac{r_i}{r_o}\right)\left(\frac{1-\varepsilon_o}{\varepsilon_o}\right) + 1\right] \cdot (T_o - T_i)}$$

Also, based on the assumption of polished SS for the inner and outer surfaces, we know that $\varepsilon_i = \varepsilon_o = 0.11$. Next we must determine the inner surface area. This is

$$A_{inner} = 2\pi r l = 2\pi \cdot (7.35\text{ mm}) \cdot (1\text{ m})$$

$$A_{inner} = 2\pi \cdot (0.00735\text{ mm}) \cdot (1\text{ m})$$

$$A_{inner} = 0.0462\text{ m}^2$$

Placing all these values in the relation for effective conductance we will have

$$C_{eff} = \frac{\left(5.67 \times 10^{-8} \frac{\text{W}}{\text{m}^2 \cdot \text{K}^4}\right)\left[(300\text{ K})^4 - (77\text{ K})^4\right]}{\left(\frac{1}{0.0462\text{ m}^2}\right) \cdot \left[\frac{1-0.11}{0.11} + \left(\frac{18.05\text{ mm}}{7.35\text{ mm}}\right)\left(\frac{1-0.11}{0.11}\right) + 1\right] \cdot (300\text{ K} - 77\text{ K})}$$

$$C_{eff} = \frac{\left(5.67 \times 10^{-8} \frac{\text{W}}{\text{m}^2 \cdot \text{K}^4}\right)\left[(300\text{ K})^4 - (77\text{ K})^4\right]}{(21.645\text{ m}^{-2}) \cdot [8.09 + (2.46)(8.09) + 1] \cdot (223\text{ K})}$$

$$C_{eff} = 0.00327\text{ W/K}$$

3.20 This problem requires the use of EES.

i) Assuming the following configuration for surfaces

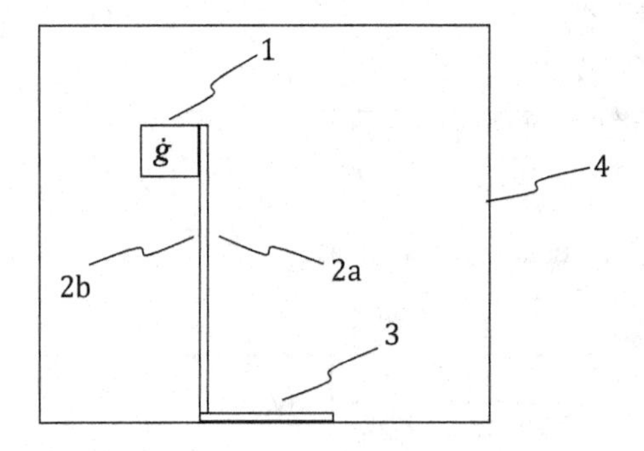

The thermal circuit can be diagramed as

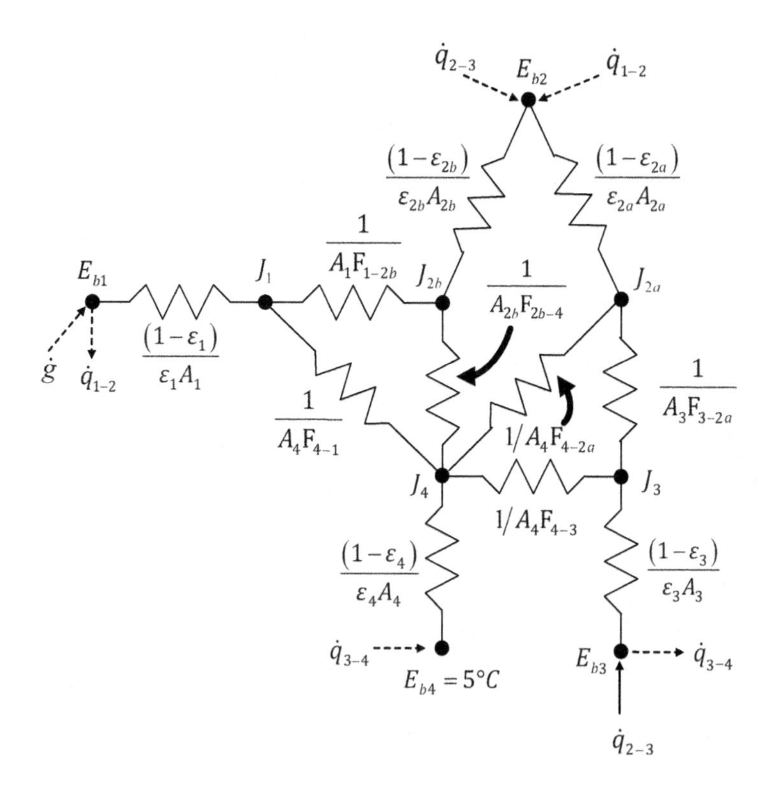

ii) The solution for the view factors incorporates the surface areas for each of the members in the circuit. Beginning with the surface areas, in EES these are defined as

```
"Areas including sub-divided surfaces"
A_1a=0.10*0.35
A_1b=0.10*0.35
A_1c=0.10*0.35
A_1d=0.10*0.10
A_1e=0.10*0.10
A_1=A_1a+A_1b+A_1c+A_1d+A_1e
A_2a=0.35*0.5
A_2b=0.35*0.5-0.35*0.1
A_2=A_2a+A_2b
A_3=0.35*0.35
A_4=6*0.70*0.70-0.35*0.35
```

The view factors for each of the surfaces can then be defined as

```
"View factors"
F_1b_4=1.0
F_1c_4=1.0
F_1d_4=1.0
F_1e_4=1.0
```

```
A_1a*F_1a_2b=A_2b*F_2b_1a
F_1_1=0.0
F_1_1+F_1_2b+F_1_4=1.0

F_2b_1=F_2b_1a+F_2b_1b+F_2b_1c+F_2b_1d+F_2b_e
F_2b_1a=0.08982
F_2b_1b=0.0
F_2b_1c=0.0
F_2b_1d=0.0
F_2b_1e=0.0
A_2b*F_2b_1=A_1*F_1_2b
A_2*F_2_3=A_2a*F_2a_3
F_2a_3=0.2201
F_2a_1=0.0
F_2a_2a=0.0
F_2b_1=F_2b_1a
F_2b_2b=0.0
F_2b_3=0.0
F_2_2=0.0
F_2a_1+F_2a_2a+F_2a_3+F_2a_4=1.0
A_2a*F_2a_3=A_3*F_3_2a
F_2b_1+F_2b_2b+F_2b_3+F_2b_4=1.0
F_3_1=0.0
A_3*F_3_2=A_3*F_3_2a+A_3*F_3_2b
F_3_2b=0.0
F_3_3=0.0
F_3_1+F_3_2a+F_3_3+F_3_4=1.0
A_3*F_3_4=A_4*F_4_3
A_2a*F_2a_4=A_4*F_4_2a
A_2b*F_2b_4=A_4*F_4_2b
A_1a*F_1a_4=A_4*F_4_1a
A_1b*F_1b_4=A_4*F_4_1b
A_1c*F_1c_4=A_4*F_4_1c
A_1d*F_1d_4=A_4*F_4_1d
A_1e*F_1e_4=A_4*F_4_1e
F_4_1=F_4_1a+F_4_1b+F_4_1c+F_4_1d+F_4_1e
A_4*F_4_1=A_1*F_1_4
F_4_1+F_4_2a+F_4_2b+F_4_3+F_4_4=1.0
```

iii) The energy balance relations for the circuit are defined in EES as the following

```
"Heat flows"
g=15

"E_b1 Energy balance"
g=(E_b1-J1)/((1-eps_1)/(eps_1*A_1))+(T1-T2)/R_1_2
```

```
"J_1 Energy balance"
(E_b1-J1)/((1-eps_1)/(eps_1*A_1))=(J1-J2b)/(1/&
(A_1*F_1_2b))+(J1-J4)/(1/(A_1*F_1_4))

"E_b2 Energy balance"
(T1-T2)/R_1_2=(E_b2-J2a)/((1-eps_2a)/(eps_2a*A_2a))+&
(E_b2-J2b)/((1-eps_2b)/(eps_2b*A_2b))+(T2-T3)/R_2_3

"J_2b Energy balance"
(E_b2-J2b)/(((1-eps_2b)/(eps_2b*A_2b)))=(J2b-J1)/(1/&
(A_2b*F_2b_1))+(J2b-J4)/(1/(A_2b*F_2b_4))

"J_2a Energy balance"
(E_b2-J2a)/(((1-eps_2a)/(eps_2a*A_2a)))=(J2a-J3)/(1/&
(A_2a*F_2a_3))+(J2a-J4)/(1/(A_2a*F_2a_4))

"E_b3 Energy balance"
(T2-T3)/R_2_3=(E_b3-J3)/(((1-eps_3)/(eps_3*A_3)))+(T3-T4)&
/R_3_4

"J_3 Energy balance"
(E_b3-J3)/((1-eps_3)/(eps_3*A_3))=(J3-J2a)/(1/&
(A_3*F_3_2a))+(J3-J4)/(1/(A_3*F_3_4))

"E_b4 Energy balance"
(T3-T4)/R_3_4=(E_b4-J4)/((1-eps_4)/(eps_4*A_4))
```

iv) Before we solve for the temperatures throughout the circuit (including the E-box temperature), we must establish the rest of the program. This will include constants, surface coatings, conductive resistances, blackbody energy fluxes and the boundary temperature definition. The remaining code for each of these sections is shown below.

```
"Chapter 3: Problem 3.20, Oven bracket solution"

sigma=5.67e-8

"Coatings by surface"
eps_1=0.88
eps_1a=0.88
eps_1b=0.88
eps_1c=0.88
eps_1d=0.88
eps_1e=0.88
eps_2=0.34
eps_2a=0.34
```

```
eps_2b=0.34
eps_3=0.34
eps_4=0.82

"Resistances"
R_1_2=0.05/(210*0.10*0.10)+0.005/(210*0.1*0.1)+0.2/&
(210*0.005*0.35)
R_2_3=(0.25/(210*0.005*0.35))+(0.175/(210*0.005*0.35))
R_3_4=0.005/(210*0.35*0.35)+0.005/(210*6*0.7*0.7)

T4=278

"Blackbody energy fluxes"
E_b1=sigma*(T1^4)
E_b2=sigma*(T2^4)
E_b3=sigma*(T3^4)
E_b4=sigma*(T4^4)

"Kelvin to Celsius temperature conversion"
T1C=T1-273
T2C=T2-273
T3C=T3-273
T4C=T4-273
```

Upon execution of the program, the temperature values throughout the circuit are

$T_1 = 16.3°C$
$T_2 = 11.3°C$
$T_3 = 5°C$
$T_4 = 5°C$

3.21 Solutions to parts i), ii) and iii) are as follows

i) Placing an imaginary surface at the location of the aperture would result in a full enclosure. As such, summation rules apply for each surface facing the interior of the enclosure. With regard to the sensor, the summation would be

$$1 = F_{sensor-sensor} + F_{sensor-cavity} + F_{sensor-aperture}$$

Since the sensor surface is flat and it cannot see itself, this reduces to

$$1 = F_{sensor-cavity} + F_{sensor-aperture}$$

If the view factor for the sensor to the aperture was known then the view factor for the sensor to the cavity wall could be determined.

$$F_{sensor-cavity} = 1 - F_{sensor-aperture}$$

Given the size of the sensor surface (i.e., 5 mm × 5 mm), it may be considered a differential surface area relative to the aperture surface area. Using the view factor for a differential surface area to a circle where the circle's center is aligned with the differential element, we can determine a value.

$$F_{d1-2} = \frac{r^2}{h^2 + r^2} \text{ where } r = 0.225 \text{ m}$$

However, h must be determined in order to apply the view factor relation. Based on geometry, it is known that the distance from the center of the cavity to the contact line between the aperture and sphere is half of the diameter (i.e., 0.5 m). A right triangle can then be defined that includes the center of the aperture, the edge of the aperture and the center of the spherical cavity.

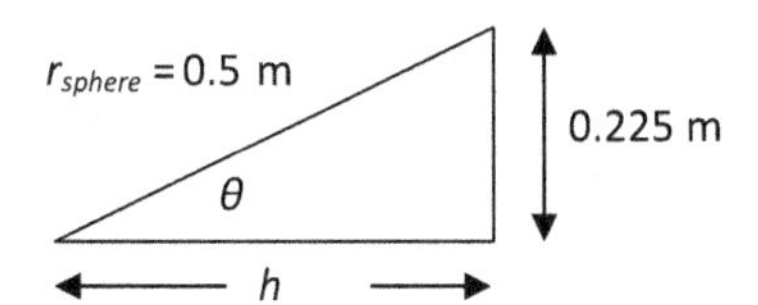

Using this interpretation

$$\theta = \sin^{-1}\left[\frac{y}{r_{sphere}}\right] = \sin^{-1}[0.45] = 26.7°$$

The length h can be solved for by invoking the definition of the cosine of an angle.

$$\cos\theta = \left[\frac{h}{r_{sphere}}\right]$$

Solving for h, we would have

$$h = r_{sphere}(\cos\theta + 1)$$

$$h = (0.5 \text{ m})[\cos(26.7°) + 1] = 0.9466 \text{ m}$$

As such,

$$F_{sensor-aperture} = \frac{(0.225 \text{ m})^2}{(0.9466 \text{ m})^2 + (0.225 \text{ m})^2} = 0.0534$$

Finally,

$$F_{sensor-cavity} = 1 - 0.0534 = 0.947$$

ii) Taking a summation on the fictitious aperture surface

$$1 = F_{aperture-aperture} + F_{aperture-sensor} + F_{aperture-cavity}$$

Since the aperture cannot see itself, this reduces to

$$1 = F_{aperture-sensor} + F_{aperture-cavity} \text{ or}$$

$$F_{aperture-cavity} = 1 - F_{aperture-sensor}$$

Since the view factor for the sensor to the aperture is known, reciprocity can be applied to determine the view factor for the aperture to the sensor

$$A_{aperture}F_{aperture-sensor} = A_{sensor}F_{sensor-aperture}$$

$$F_{aperture-sensor} = \left(\frac{A_{sensor}}{A_{aperture}}\right)F_{sensor-aperture}$$

$$F_{aperture-sensor} = \left(\frac{0.005 \text{ m} \cdot 0.005 \text{ m}}{\pi(0.225 \text{ m})^2}\right)(0.0534 \text{ m}) = 8.39 \times 10^{-6}$$

Thus,

$$F_{aperture-cavity} = 1 - 8.39 \times 10^{-6} = 0.99999$$

iii)

$$\dot{q}_{out} = \varepsilon\sigma A_{sensor}T^4_{sensor}$$

$$\dot{q}_{out} = (5.67\times10^{-8} \text{ W/m}^2\cdot\text{K}^4)(0.005 \text{ m}\cdot0.005 \text{ m})(0.0534)(220 \text{ K})^4 = 1.773\times10^{-4} \text{ W}$$

3.22 i) The view factor between E-box 1 and E-box 2, as well as E-box 2 and E-box 3, can be determined using either the relations for common view factors in Appendix A or the function calls in EES. For the view factor between E-boxes 1 and 3, Hottel's method can be used since the geometry is provided. The view factor for each of these to the enclosure structure can be determined using the summation rule. Values determined are as follows.

F_1_1 = 0
F_1_2 = 0.1553
F_1_3 = 0.08928
F_1_4 = 0.7554
F_2_2 = 0
F_2_1 = 0.1553
F_2_3 = 0.07229
F_2_4 = 0.7724
F_3_1 = 0.08928
F_3_2 = 0.07229
F_3_3 = 0
F_3_4 = 0.8384
F_4_1 = 0.04125
F_4_2 = 0.04218
F_4_3 = 0.04579
F_4_4 = 0.8708

ii) For the determination of the E-box temperatures an EES code was developed to model the conductive and radiative heat flows between each of the four elements in the thermal circuit. The code is shown below.

```
"Problem 3_22: Three E-boxes in a temperature-controlled box"
sigma=5.67e-8

"Thermal conductivities"
k_SS=14.9
k_Al_6061=167
k_Box=k_SS

eps_1=0.91               {NASA/GSFC NS-37 white paint}
eps_2=0.75               {Paladin black lacquer}
eps_3=0.02               {Vapor-deposited aluminum}
eps_4=0.88               {S13GLO white paint}

A_1=5*(0.25^2)                              "Ebox_1"
A_2=5*(0.25^2)                              "Ebox_2"
A_3=5*(0.25^2)                              "Ebox_3"
A_4=6.0*(1-0.015)-3*(0.25^2)                "Structure"

F_1_1=0.0
F_2_2=0.0
F_3_3=0.0
F_2_3=0.07229            {Parallel plate view factor}
F_3_2=F_2_3
F_1_2=0.1553            {Parallel plate view factor}

{Using Hottel's method}
F_1_3=(0.25)*(2*1.0296-2*0.918)/(2*5*0.25*0.25)

A_2*F_2_1=A_1*F_1_2
F_3_1=(A_1/A_3)*F_1_3

F_1_1+F_1_2+F_1_3+F_1_4=1.0
F_2_1+F_2_2+F_2_3+F_2_4=1.0
F_3_1+F_3_2+F_3_3+F_3_4=1.0

F_4_1=(A_1/A_4)*F_1_4
F_4_2=(A_2/A_4)*F_2_4
F_4_3=(A_3/A_4)*F_3_4
F_4_1+F_4_2+F_4_3+F_4_4=1.0

"Energy dissipations"

g1=55
g2=60
g3=75
{g4=0.0}
```

```
"Boundaries"
T3=310

"Balances"
R_1_4=(0.125/(k_Al_6061*0.25*0.25))+(0.005/&
(k_Box*0.25*0.25))
R_4_1=R_1_4

R_2_4=(0.125/(k_Al_6061*0.25*0.25))+(0.005/&
(k_Box*0.25*0.25))
R_4_2=R_2_4

R_3_4=(0.125/(k_Al_6061*0.25*0.25))+(0.005/&
(k_Box*0.25*0.25))
R_4_3=R_3_4

"Heat flows"
g1=(E_b1-J1)/((1-eps_1)/(eps_1*A_1))+(T1-T4)/R_1_4

(E_b1-J1)/((1-eps_1)/(eps_1*A_1))=(J1-J4)/(1/(A_1*F_1_4))&
+(J1-J2)/(1/(A_1*F_1_2))+(J1-J3)/(1/(A_1*F_1_3))

g2=(E_b2-J2)/((1-eps_2)/(eps_2*A_2))+(T2-T4)/R_2_4

(E_b2-J2)/(((1-eps_2)/(eps_2*A_2)))=(J2-J4)/(1/&
(A_2*F_2_4))+(J2-J1)/(1/(A_2*F_2_1))+(J2-J3)/(1/&
(A_2*F_2_3))

g3=(E_b2-J3)/((1-eps_3)/(eps_3*A_3))+(T3-T4)/R_3_4

(E_b3-J3)/((1-eps_3)/(eps_3*A_3))=(J3-J4)/(1/(A_3*F_3_4))&
+(J3-J2)/(1/(A_3*F_3_2))+(J3-J1)/(1/(A_3*F_3_1))

g4=(E_b4-J4)/((1-eps_4)/(eps_4*A_4))+(T4-T1)/R_1_4+(T4-&
T2)/R_2_4+(T4-T3)/R_3_4

(E_b4-J4)/((1-eps_4)/(eps_4*A_4))=(J4-J3)/(1/(A_3*F_3_4))&
+(J4-J2)/(1/(A_2*F_2_4))+(J4-J1)/(1/(A_1*F_1_4))

"Temperatures"
E_b1=sigma*(T1^4)
E_b2=sigma*(T2^4)
E_b3=sigma*(T3^4)
E_b4=sigma*(T4^4)
```

```
TC1=T1-273
TC2=T2-273
TC3=T3-273
TC4=T4-273
```

Upon execution of the code, the temperature values are

T1 = 309.6 K
T2 = 309.7 K
T3 = 310 K
T4 = 308.7 K

Chapter 4 Solutions

4.1 For a surface which is experiencing full frontal solar loading, we know that the energy balance will be

$$\alpha_S SA = \varepsilon A \sigma T^4$$

If we solve for the temperature we can obtain a form of

$$T = \left[\left(\frac{S}{\sigma}\right) \cdot \left(\frac{\alpha_S}{\varepsilon}\right)\right]^{1/4}$$

The ratio of the solar constant to the Stefan Boltzmann constant (i.e., S/σ) results in a number that also is constant. Thus, the only independent variables on the RHS of our equation that are affecting the temperature are the absorptivity and the emissivity. In assuming the value of α_S/ε varies between 0.1 and 1.5 we can obtain corresponding temperatures for each value of this ratio. The resultant plot for the variation is shown below.

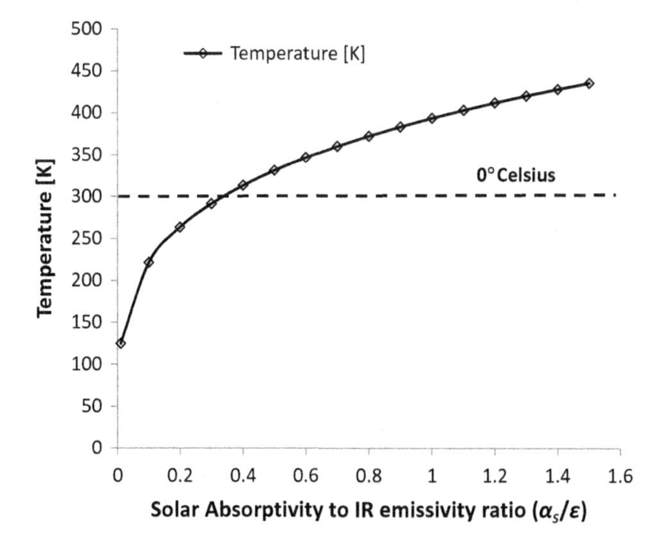

4.2 For the surface in question we will need to take into consideration the projected surface area along the solar constant. The energy balance for the surface would be

$$\alpha_S S A_p = \varepsilon A_{surf} \sigma T_{surf}^4$$

or

$$\alpha_S S A_{surf} \cos\gamma = \varepsilon A_{surf} \sigma T_{surf}^4$$

Solving for the temperature we would have

$$T_{surf} = \left[\cos\gamma \cdot \left(\frac{S}{\sigma}\right) \cdot \left(\frac{\alpha_S}{\varepsilon}\right)\right]^{1/4}$$

Given that the surface area, absorptivity and emissivity for Z306 Black Polyurethane paint (referencing Appendix B, $\alpha_S = 0.95$ and $\varepsilon = 0.87$) and the constants to the equation (i.e., the solar constant and the Stefan–Boltzmann constant) are known we can calculate the temperature at each angle. A table listing the values is shown below.

α	ε	S [W/m²]	Angle	T_surf [K]	T_surf [°C]
0.95	0.87	1361	0	402.4	129.4
0.95	0.87	1361	15	398.9	125.9
0.95	0.87	1361	30	388.2	115.2
0.95	0.87	1361	45	369.0	96.0
0.95	0.87	1361	60	338.3	65.3
0.95	0.87	1361	75	287.0	14.0

4.3 The relations for the view factors can be referenced in Appendix A. The EES code used to calculate the view factors in each case is

```
{Problem 4.3: View factor determination for a surface with a normal
aligned with the nadir direction}
r_Earth=6380

a[1] =200
h[1] =r_Earth+a[1]

{Surface normal aligned with nadir}
F_plate_E[1] =(r_Earth/h[1] )^2

{Surface normal at 90 degrees relative to nadir}
H_bar[1] =h[1] /r_Earth

F_tplate_E[1] =(1/pi)* (arctan(1/sqrt((H_bar[1] ^2-1)))-&
sqrt((H_bar[1] ^2-1))/(H_bar[1] ^2))

FULL=696
```

```
Duplicate K=1,FULL

  a[ K+1] =a[ K] +50
  h[ K+1] =r_Earth+a[ K+1]
  F_plate_E[ K+1] =(r_Earth/h[ K+1] )^2 { Surface normal aligned with nadir}

H_bar[ K+1] =h[ K+1] /r_earth { Surface normal at 90 degrees relative to nadir}

F_tplate_E[ K+1] =(1/pi)* (arctan(1/sqrt((H_bar[ K+1] ^2-1)))-&
sqrt((H_bar[ K+1] ^2-1))/(H_bar[ K+1] ^2))

End
```

The plots for both cases is

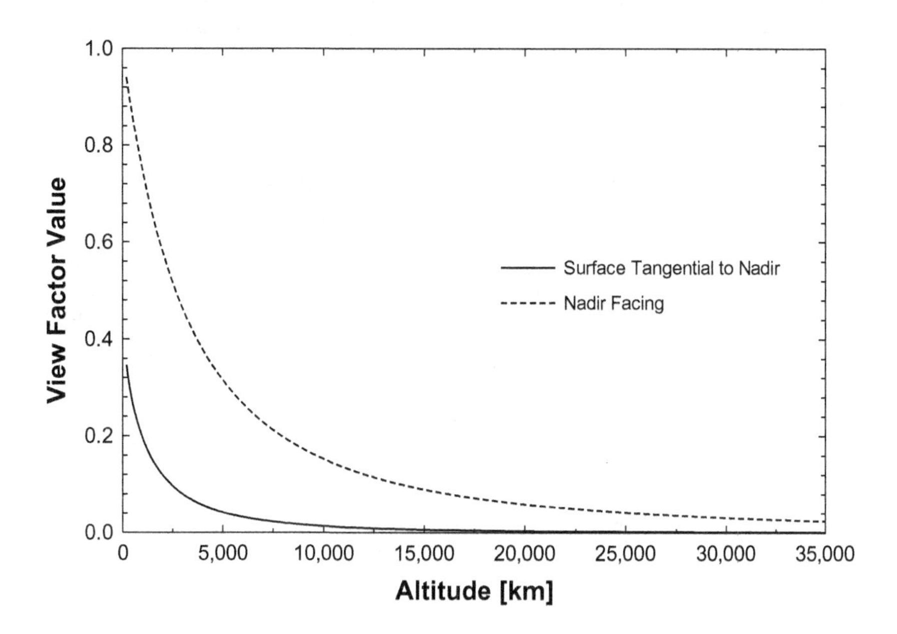

4.4 **Max and Min On-Orbit Temperature Analysis for the Nadir-Facing Flat Plate**

A 1.0 m^2 flat plate has one active surface and the opposite surface is covered with MLI. The active surface's coating is Z93 white paint ($\alpha_S = 0.19$, $\varepsilon = 0.89$) and it is nadir facing. The non-active side of the plate is covered with 10 layers of 0.15 mil double-sided aluminized kapton ($\alpha_S = 0.25$, $\varepsilon = 0.34$) with an outer layer of Beta cloth. Assuming an energy dissipation of 30 W on the plate, what are the Max and Min temperatures for the plate if it has an orbit period of 90 minutes?

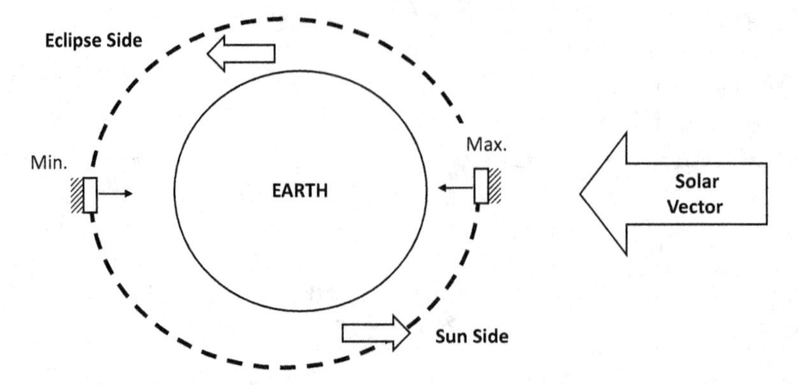

Figure 4P.4 Example Schematic for Maximum and Minimum orbit positions (shown with orbit inclination of 90° relative to the ecliptic plane and $\beta = 0°$).

The orbit for this problem will resemble that shown in the schematic in Figure 4P.4. At the Max orbit location the plate's energy balance relation is

$$\dot{g} + \varepsilon_{surf} A_{surf} EIR_{flux} F_{surf-earth} + Alb \alpha_S SA_{surf} F_{surf-earth} = \varepsilon_{surf} A_{surf} \sigma T_{surf}^4 + \varepsilon^* A_{surf} \sigma \left(T_{surf}^4 - T_o^4\right)$$

Upon substitution of actual values in (with exception of the temperatures) the balance relation becomes

$$30\ \text{W} + (0.89)(1.0\ \text{m}^2)\left(250\ \frac{\text{W}}{\text{m}^2}\right)(0.92) + (0.26)(0.19)\left(1{,}361\ \frac{\text{W}}{\text{m}^2}\right)(1.0\ \text{m}^2) = (0.89)(1.0\ \text{m}^2)\sigma T_{surf}^4 + \varepsilon^* \cdot (1.0\ \text{m}^2)\sigma\left(T_{surf}^4 - T_o^4\right)$$

As for the MLI surface, the energy balance relation is

$$\dot{Q}_{Env} + \dot{Q}_{plate-MLI} = \dot{Q}_{out,rej}$$

Upon expanding out the $\dot{Q}$ terms we now have

$$\alpha_S SA_{surf} + \varepsilon^* A_{surf} \sigma \left(T_{surf}^4 - T_o^4\right) = \varepsilon_{MLI} A_{surf} \sigma T_o^4$$

When we substitute in the values for variables other than the temperatures, our balance equation becomes

$$(0.34)\left(1{,}361 \frac{\text{W}}{\text{m}^2}\right)(1.0\ \text{m}^2) + \varepsilon^* (1.0\ \text{m}^2)\sigma\left(T_{surf}^4 - T_o^4\right) = (0.34)(1.0\ \text{m}^2)\sigma T_o^4$$

Referencing Table C.2, the Earth-IR value is 250 W/m^2 and the corresponding albedo factor is 0.22. These equations can be solved simultaneously in EES using the following code

```
{Problem 4.4, max case, nadir-facing flat plate with 0.15
double-sided aluminized kapton}
sigma=5.67e-8 [W/m^2-K^4]
g_dot=30 [W]
```

```
S=1361
EIR_H=250              {Earth IR Hot Case}
EIR_C=193              {Earth IR Cold Case}
A=1.0 [ m^2]
Alb_H=0.26
N=10
F=0.92

eps_Beta=0.86          {Beta cloth}
alpha_Beta=0.4         {Beta cloth}
eps_1=0.34             {0.15 mil aluminized kapton}
eps_2=0.34             {0.15 mil aluminized kapton}
eps_Z93=0.89           {Z93 white paint}
alpha_Z93=0.19         {Z93 white paint}
eps_surf2=0.84         {Polyurethane white paint}
estar=1/((N+1)*(1/eps_1+1/eps_2-1))

T_surf=T_i

{Energy balance at the plate}
eps_Z93*A*EIR_H*F+Alb_H*alpha_Z93*S*A*F+g_dot=estar*A*&
sigma*(T_i^4-T_o^4)+eps_Z93*A*sigma*T_i^4

{Energy balance on the outer MLI surface}
alpha_Beta*S*A+estar*A*sigma*(T_i^4-T_o^4)=eps_Beta*A*&
sigma*T_o^4
```

we will get $T_{surf} = 278.1\text{ K} \approx 5.1°\text{C}$ and $T_o = 324.3\text{ K} \approx 51.3°\text{C}$. At the min orbit location, the plate's energy balance relation is

$$\dot{g} + \varepsilon_{surf} A_{surf} EIR_{flux} F_{surf-earth} = \varepsilon_{surf} A_{surf} \sigma T_{surf}^4 + \varepsilon^* A_{surf} \sigma \left(T_{surf}^4 - T_o^4\right)$$

Substituting in values for variables other than the temperature the relation now becomes

$$30\text{W} + (0.89)\left(1.0\text{m}^2\right)\left(193\frac{\text{W}}{\text{m}^2}\right)(0.92) = (0.89)\left(1.0\text{m}^2\right)\sigma T_{surf}^4 + \varepsilon^* \cdot \left(1.0\text{m}^2\right)\sigma\left(T_{surf}^4 - T_o^4\right)$$

where the EIR_{flux} value (i.e., 193 W/m^2) has been referenced from Appendix Table C.1. For the MLI surface which is facing deep space, the energy balance relation is

$$\dot{Q}_{Env} + \dot{Q}_{plate-MLI} = \dot{Q}_{out,rej}$$

Expanding out the $\dot{Q}$ terms and substituting in for variables other than the temperatures, the relation becomes

$$\varepsilon^* \left(1.0\text{ m}^2\right)\sigma\left(T_{surf}^4 - T_o^4\right) = (0.34)\left(1.0\text{ m}^2\right)\sigma T_o^4$$

We can solve these equations simultaneously. Relative to our Max case EES code we would only need to change the energy balance relations for the plate and the MLI outer surface. In EES this would be

```
{Energy Balance at the plate}
eps_Z93*A*EIR_C*F+g_dot=estar*A*sigma*(T_i^4-T_o^4)+&
eps_Z93*A*sigma*T_i^4

{Energy Balance on the outer MLI surface}
estar*A*sigma*(T_i^4-T_o^4)=eps_Beta*A*sigma*T_o^4
```

Upon solving for the temperatures we obtain T_{surf} = 245.8 K ≈ –27.2°C and T_o = 93.8 K ≈ –179.2°C

4.5 Given an altitude of 250 km, we must first determine the view factor of the earth from the active surface area. Using view factor 9 from Appendix A, the equation and schematic for it are below.

$$F_{d1-2} = \left(\frac{r}{h}\right)^2$$

Known:

- Radius of earth is 6,380 km
- Altitude of space cube is 250 km
- dA_1 = 1 m^2

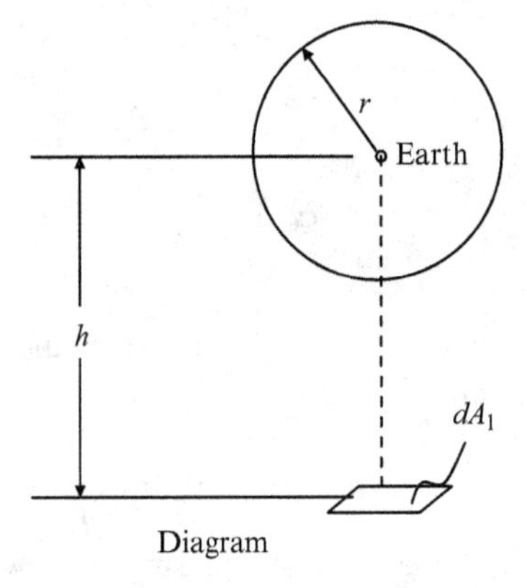

Diagram

$h = r_{earth} + altitude$
h = 6,380 km + 250 km
h = 6,580 km

$$F_{d1-2} = \left(\frac{r}{h}\right)^2 = \left(\frac{6{,}380\text{ km}}{6{,}630\text{ km}}\right)^2 = 0.962^2$$

$$= 0.926$$

Thus $F_{d1-2} = 0.926 = F_{surf-Earth}$

i) Min case

The surface is coated with Martin Black Velvet paint ($\alpha_S = 0.91; \varepsilon = 0.94$)

The plate will be facing Earth in eclipse. For the IR-sensitive surface, the IR flux will then be EIR_{flux} = 193 W/m^2. We know

$$\dot{Q}_{rej} = \varepsilon\sigma A_{surf} T^4_{surf}$$

We also know

$$\dot{Q}_{env} = \varepsilon EIR_{flux} A F_{surf-Earth}$$

Thus, an energy balance on the plate would be

$$\varepsilon EIR_{flux} A F_{surf-Earth} = \varepsilon\sigma A T^4_{surf}$$

$$EIR_{flux} F_{surf-Earth} = \sigma T^4_{surf}$$

Solving for T_{surf} we get

$$T_{surf} = \left[\frac{EIR_{flux} F_{surf-Earth}}{\sigma}\right]^{\frac{1}{4}}$$

Placing values in the relation

$$T_{surf} = \left[\frac{(193\ \mathrm{W/m^2})(0.926)}{5.67\times10^{-8}\ \mathrm{W/m^2\cdot K^4}}\right]^{\frac{1}{4}}$$

$$T_{surf} = 236.9\ \mathrm{K} = -36°\mathrm{C}$$

Max case

The solar constant is 1,361 W/m^2
For the Albedo & IR-sensitive surface, the IR flux is EIR_{flux} = 244 W/m^2
The albedo factor (including the orbit average correction) is Alb = 0.3

We know

$$\dot{Q}_{rej} = \varepsilon\sigma A_{surf} T^4_{surf}$$

In addition, the environmental load may be captured in equation form as

$$\dot{Q}_{env} = \varepsilon EIR_{flux} A F_{surf-Earth} + Alb\alpha_S SAF_{surf-Earth}$$

Performing an energy balance on the active plate gives

$$\varepsilon EIR_{flux} A F_{surf-Earth} + Alb\alpha_S SAF_{surf-Earth} = \varepsilon\sigma A T^4_{surf}$$

Solving for T_{surf} we get

$$T_{surf} = \left[\frac{\varepsilon EIR_{flux} F_{surf-Earth} + Alb\alpha_S SF_{surf-Earth}}{\varepsilon\sigma}\right]^{\frac{1}{4}}$$

Placing values in

$$T_{surf} = \left[\frac{(0.94)(244\ \mathrm{W/m^2})(0.926) + (0.3)(0.91)(1{,}361\ \mathrm{W/m^2})(0.926)}{(0.94)\left(5.67\times10^{-8}\ \mathrm{W/m^2\cdot K^4}\right)}\right]^{\frac{1}{4}}$$

$$T_{surf} = 319.7\ \mathrm{K} = 46.7°\mathrm{C}$$

ii) Min case

The surface is coated with Polyurethane white paint ($\alpha_S = 0.27; \varepsilon = 0.84$)
For the IR-sensitive surface, the IR flux will then be $EIR_{flux} = 193$ W/m^2.
Based on the energy balance derived in part i) we know
Solving for T_{surf} we get

$$T_{surf} = \left[\frac{EIR_{flux}F_{surf-Earth}}{\sigma}\right]^{\frac{1}{4}}$$

Placing values in the relation

$$T_{surf} = \left[\frac{(193\ \text{W/m}^2)(0.926)}{5.67 \times 10^{-8}\ \text{W/m}^2 \cdot \text{K}^4}\right]^{\frac{1}{4}}$$

$$T_{surf} = 236.9\ \text{K} = -36°\text{C}$$

Max case

The solar constant is $Solar = 1{,}361$ W/m^2
For the IR-sensitive surface, the IR flux is $EIR_{flux} = 250$ W/m^2
The albedo factor (including the orbit average correction) is $Alb = 0.26$

Based on the derivation in part i), we know

$$T_{surf} = \left[\frac{\varepsilon EIR_{flux}F_{surf-Earth} + Alb\alpha_S SF_{surf-Earth}}{\varepsilon\sigma}\right]^{\frac{1}{4}}$$

$$T_{surf} = \left[\frac{(0.84)(250\ \text{W/m}^2)(0.926) + (0.26)(0.27)(1{,}361\ \text{W/m}^2)(0.926)}{(0.84)\left(5.67 \times 10^{-8}\ \text{W/m}^2 \cdot \text{K}^4\right)}\right]^{\frac{1}{4}}$$

$$T_{surf} = 277.6\ \text{K} = 4.6°\text{C}$$

4.6 Given the dimension for the length of a single side of the cubic structure, the surface area of one side will be

$$A_{side} = 0.5\ \text{m} \times 0.5\ \text{m} = 0.25\ \text{m}^2$$

and the total surface area is

$$A_{total} = 6 \cdot A_{side} = 6 \cdot 0.25\ \text{m}^2 = 1.5\ \text{m}^2$$

Next we must determine the projected surface area for the structure. Assuming the solar vector is aligned with the 0° position, a full rotation of 360° will expose each of the cubic's lateral sides that are aligned with the structure's rotational axis to solar loading. In each case, the loading will take place between 90° and 270°. From projected surface area theory, we know that

$$A_{p,average} = 4 \cdot \left[\frac{1}{2\pi} \int_{0}^{2\pi} A_{side} \cos \gamma d\gamma \right]$$

$$A_{p,average} = 4A_{side} \cdot \left[\frac{1}{2\pi} \int_{3\pi/2}^{\pi/2} \cos \gamma d\gamma \right]$$

$$A_{p,average} = \frac{4A_{side}}{\pi}$$

Placing this new average area into an energy balance will give us

$$\varepsilon A_{surf} \sigma T_{surf}^4 = \alpha_S S \cdot \left[\frac{4A_{side}}{\pi} \right]$$

Solving for T_{surf} we have

$$T_{surf} = \left[\left(\frac{\alpha_S S}{\varepsilon \sigma} \right) \cdot \left[\frac{4A_{side}}{6\pi A_{side}} \right] \right]^{\frac{1}{4}}$$

$$T_{surf} = \left[\left(\frac{(0.95)(1{,}361\ \mathrm{W/m^2})}{(0.75)(5.67 \times 10^{-8}\ \mathrm{W/m^2 \cdot K^4})} \right) \cdot \left[\frac{4}{6\pi} \right] \right]^{\frac{1}{4}}$$

$$T_{surf} = 283.4\ \mathrm{K} = 10.4°\mathrm{C}$$

4.7 i) The EES code for the problem is

```
{Problem 4.7 Transient analysis of nadir-facing flat plate
in orbit with ideal MLI on back side}

{A is iter, B is Earth_in, C is g_dot, D is Alb_in and Y is
Q_in}
Function test(A,B,C,D)
  If (A=19) or (A=53) Then
    Y:=B+22.6+C
  Endif

  If (A=20) or (A=52) Then
    Y:=B+45+C
  Endif

  If (A=21) or (A=51) Then
    Y:=B+67.12+C
  Endif
```

```
   If (A>21) and (A<51) Then
    Y:=B+C
   Endif

   If (A<19) or (A>53) Then
    Y:=B+C+D
   Endif

   test:=Y
End

{Main program}
sigma=5.67e-8              {Boltzmann's Constant}

{Radiator features}
l_rad=1.0
w_rad=1.0
thick_rad=0.0254
A_surf=l_rad*w_rad
Vol_rad=l_rad*w_rad*thick_rad
rho=2700.0
c_p=896.0
k_6061=167

{Orbit statistics}
Orbit_Periodm=90           {Orbit period in minutes}
Orbit_Periods=90*60        {Orbit period in seconds}
Orbit_Degrees=360          {Degrees in orbit}
Delta_Degrees=5
Del_T=Delta_Degrees*Orbit_Periods/Orbit_Degrees
INK=Orbit_Degrees/5
Revs=5.0                    {#Revolutions in transient analysis}
Full=Revs*INK

Fo=(k_6061*1.25*60)/((rho*c_p)*(thick_rad^2))

{Surface coatings}
alpha_S=0.19;
eps=0.89;

{Environmental loading}
EIR=193
Solar=1361
Albedo=0.26
Earth_in=eps*A_surf*EIR
Alb_in=alpha_S*Albedo*A_surf*Solar
```

```
{Plate dissipation}
g_dot=0.0

Ang[1]=0
T[1]=273
Time[1]=0.0
Time_min[1]=0.0

Q_in[1]=Earth_in+Alb_in+g_dot

Duplicate K=2,FULL
  Ang[K]=Ang[K-1]+5
  Time[K]=Time[K-1]+75.0
  Time_min[K]=Time[K]/60
  {T[K]=0}

  z[K]=floor(K/INK)
  iter[K]=K-z[K]*INK

  Q_in[K]=test(iter[K],Earth_in,g_dot,Alb_in)

  T[K]=(Q_in[K-1]-eps*sigma*A_surf*((T[K-1])^4))*(Del_T/&
  (rho*Vol_rad*c_p))+T[K-1]
```

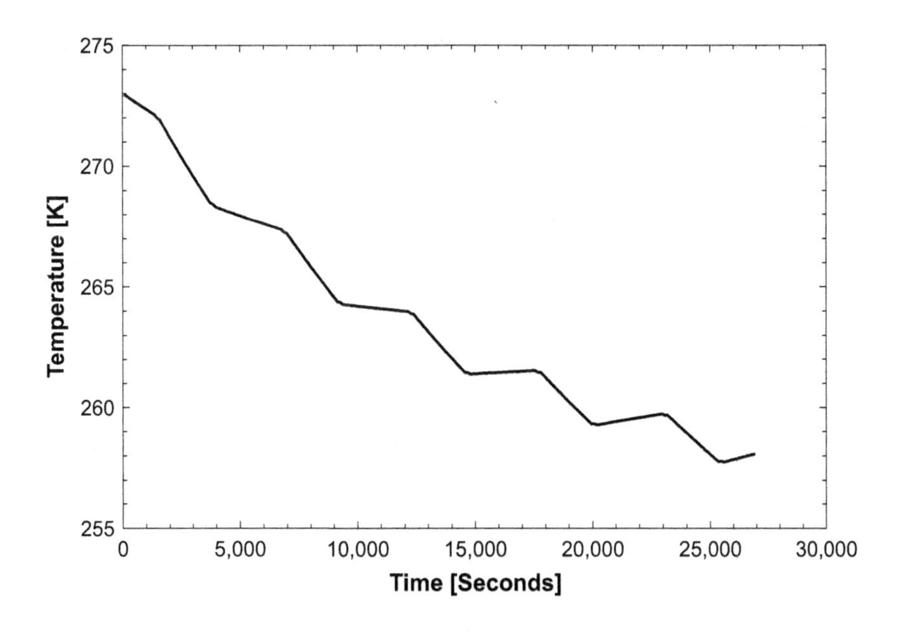

In the no heating case there is a steady decrease in temperature from 273K with each revolution in the orbit. However, after five orbits the temperature has not yet reached harmonic steady-state oscillations.

ii)

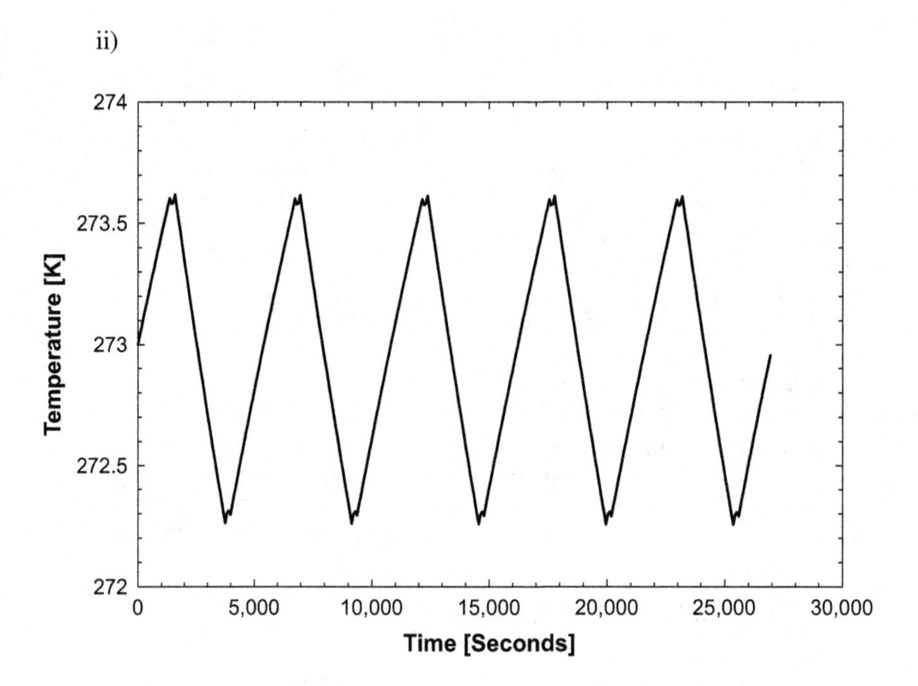

With the application of a continuous 69 W during these five orbits we can attain harmonic steady-state temperatures on the plate. Notice the peak-to-peak temperature difference is within 2°C

4.8 i) To determine the net heat transfer from shield 2 to shield 3, as well as the temperature of shield 3, we will use EES. The following code predicts both of these values

```
{Problem 4.8 Gray body heat transfer between radiation
shields In space}

{Constants}
S=1361    [ W/m^2]                      {Solar constant}

{Geometry}
a=1.5        [m]
b=1.5        [m]
c=0.5        [m]

A_1=a*b
A_2=a*b
A_3=a*b
A_4=8*c*a

{View factors}
F[ 1,1] =0
F[ 2,2] =0
F[3,3] =0
```

```
F[1,3]=0
F[3,1]=0

F[1,2]=f3d_1(a,b,c)
F[2,1]=A_1*F[1,2]/(2*A_2)

F[3,2]=f3d_1(a,b,c)
F[2,3]=A_3*F[3,2]/(2*A_2)

F[1,4]=1-(F[1,1]+F[1,2]+F[1,3])
F[2,4]=1-(F[2,1]+F[2,2]+F[2,3])
F[3,4]=1-(F[3,1]+F[3,2]+F[3,3])

F[4,1]=A_1*F[1,4]/A_4
F[4,2]=2*A_2*F[2,4]/A_4
F[4,3]=A_3*F[3,4]/A_4
F[4,4]=1-(F[4,1]+F[4,2]+F[4,3])

{Emissivities}
eps_1=0.45; eps_2=0.45; eps_3=0.45;    {0.25 mil aluminized kapton}

{Emissivity of space node}
eps_4=1.0

{Solar absorptivity}
abs_1=0.31; abs_2=0.31; abs_3=0.31;    {0.25 mil aluminized kapton}

{Absorptivity of space node}
abs_4=1.0

{Reflectivities}
rho[1]=1-eps_1; rho[2]=1-eps_2; rho[3]=1-eps_3; rho[4]=0

{Temperatures}
T_4=3

{F-hat values}
Duplicate M=1,4
  Duplicate N=1,4
    F_hat[M,N]=F[M,N]+SUM(rho[K]*F[M,K]*F_hat[K,N],K=1,4)
  End
End

{Energy balance on shield 1}
Q_dot_1_Solar=abs_1*S*A_1

Q_dot_1_2=eps_1*eps_2*sigma#*A_1*F_hat[1,2]*(T_1^4-T_2^4)
```

```
T_1=(Q_dot_1_rej_a/(eps_1*sigma#*A_1))^0.25

Q_dot_1_rej_b=eps_1*eps_4*sigma#*A_1*F_hat[1,4]*(T_1^4-T_4^4)

Q_dot_1_rej_a=Q_dot_1_Solar-Q_dot_1_2-Q_dot_1_rej_b

{Energy balance on shield 2}
Q_dot_2_rej=eps_2*eps_4*sigma#*2*A_2*F_hat[2,4]*(T_2^4-T_4^4)

Q_dot_2_3=eps_2*eps_3*sigma#*A_2*F_hat[2,3]*(T_2^4-T_3^4)

Q_dot_1_2=Q_dot_2_rej+Q_dot_2_3

{Energy balance on shield 3}
Q_dot_3_rej_a=eps_2*eps_3*sigma#*A_3*F_hat[3,4]*(T_3^4-T_3^4)

Q_dot_3_rej_b=Q_dot_2_3-Q_dot_3_rej_a

T_3=(Q_dot_3_rej_b/(eps_3*sigma#*A_3))^0.25
```

Execution of the code shows that the heat transfer from shield 2 to shield 3 is 9.7 W. The temperature of shield 3 is 114 K.

ii) Creating a table in EES that varies the separation distance from 0.5 m to 1.5 m, a plot of the temperatures on shield 1 through shield 3 can be created.

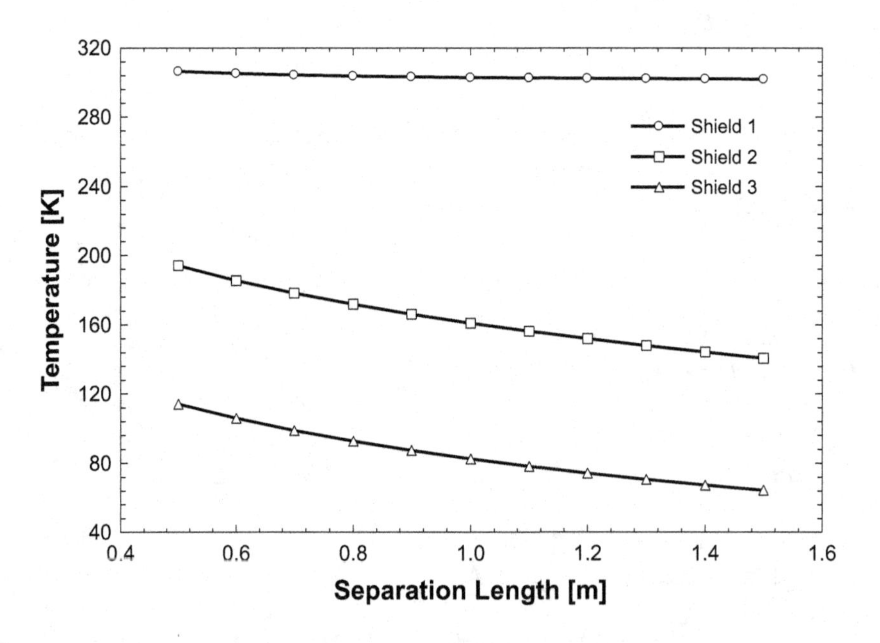

4.9 We still start by setting our energy generation term to the definition for the Stefan–Boltzmann law

$$\dot{Q}_{gen} = \varepsilon\sigma A_{surf} T_{E\text{-}box}^4$$

Given that the energy dissipation and the surface area rejecting the waste heat is known, we only need to find the emissivity in order to solve for the E-box temperature. Referencing Appendix B, we know that for S13 GLO white paint $\alpha_S = 0.19$ and $\varepsilon = 0.89$. If we solve our primary equation for the E-box temperature we will have

$$T_{E\text{-}box} = \left[\frac{\dot{Q}_{gen}}{\varepsilon\sigma A_{surf}}\right]^{1/4}$$

Placing in actual values gives

$$T_{E\text{-}box} = \left[\frac{40\text{ W}}{(0.89)(5.67\times10^{-8}\text{ W/m}^2\cdot\text{K}^4)(0.5\text{ m}^2)}\right]^{1/4}$$

$$T_{E\text{-}box} = 199.5\text{ K } or = -73.5°\text{C}$$

The E-box therefore needs additional heating in order to maintain a temperature of at least –20°C.

ii) If we are to use heater power to increase the temperature of the E-box, an energy balance equation for the heat flows would then be

$$\dot{Q}_{gen} + \dot{Q}_{Htr} = \varepsilon\sigma A_{surf} T_{E\text{-}box}^4$$

We can find the heater power needed by setting the E-box temperature to its lowest acceptable limit (i.e., –20°C) and solving for $\dot{Q}_{Htr}$ in our energy balance relation. This will be

$$\dot{Q}_{Htr} = \varepsilon\sigma A_{surf} T_{E\text{-}box}^4 - \dot{Q}_{gen}$$

$$\dot{Q}_{Htr} = (0.89)(5.67\times10^{-8}\text{ W/m}^2\cdot\text{K}^4)(0.5\text{ m}^2)\left[(253\text{ K})^4\right] - 40\text{ W}$$

$$\dot{Q}_{Htr} = 103.4\text{ W} - 40\text{ W}$$

$$\dot{Q}_{Htr} = 63.4\text{ W}$$

iii) Options for additional/alternate sources of heat input include
- ➢ Insulating a portion of the presently exposed active surface;
- ➢ Changing the active surface's coating to one of a lower emissivity;
- ➢ Orienting the spacecraft to make use of environmental heating from the Earth.

4.10 Referencing the coatings table, we know Dow Corning white paint has coatings values of $\alpha_S = 0.19$; $\varepsilon = 0.88$

i) From the energy balance on the SpaceCube we know

$$\alpha_S S A_p = \varepsilon\sigma A_{surf} T_{surf}^4$$

where the projected surface area (A_p) of the sphere onto a plane that is perpendicular to the Sun's rays equals a circle (i.e., $A_p = \pi r^2$). From basic geometry, we can then determine that

$$A_p = \pi r^2 = \frac{1}{4} A_{surf}$$

Placing this in the energy balance equation, we then have

$$0.25 \alpha_S S A_p = \varepsilon \sigma A_{surf} T_{surf}^4$$

Solving for T_{surf} we would then have

$$T_{surf} = \left[\frac{0.25 \alpha_S S}{\varepsilon \sigma}\right]^{\frac{1}{4}}$$

$$T_{surf} = \left[\frac{0.25(0.19)(1{,}361\ \text{W/m}^2)}{(0.88)(5.67 \times 10^{-8}\ \text{W/m}^2 \cdot \text{K}^4)}\right]^{\frac{1}{4}}$$

$$T_{surf} = 189.7\ \text{K}$$

ii) With one side of the cube receiving solar loading and all six rejecting thermal energy, the energy balance becomes

$$\frac{\alpha_S S A_{surf}}{6} = \varepsilon \sigma A_{surf} T_{surf}^4$$

$$\frac{\alpha_S S}{6} = \varepsilon \sigma T_{surf}^4$$

Solving for T_{surf} we would then have

$$T_{surf} = \left[\frac{\alpha_S S}{6 \varepsilon \sigma}\right]^{\frac{1}{4}}$$

$$T_{surf} = \left[\frac{(0.19)(1{,}361\ \text{W/m}^2)}{6(0.88)(5.67 \times 10^{-8}\ \text{W/m}^2 \cdot \text{K}^4)}\right]^{\frac{1}{4}}$$

$$T_{surf} = 171.4\ \text{K}$$

4.11 The total environmental load in full eclipse is from Earth-IR. Each side of the SpaceCube has a surface area of 1.0 m^2. For the surface coatings we can reference Appendix B. The emissivity and absorptivity of Martin Black Velvet is $\alpha_S = 0.94$, $\varepsilon = 0.91$. Using Table C.1 (cold case), we will need to know the period of the orbit. From equation 4.6 we know the orbit period is defined as

$$OP = 2\pi \left(\frac{a^3}{\mu}\right)^{1/2}$$

where μ is the universal gravitational constant and a is the semi-major axis. In our case,

$$a = 250\text{ km} + 6{,}380\text{ km} = 6{,}630\text{ km}$$

where we have summed the altitude and the radius of the Earth (6,380 km). Returning to our relation for the orbit period we then have

$$Period = 2\pi\left[\frac{(6{,}630\text{ km})^3}{3.98603\times10^{14}\frac{\text{m}^3}{\text{s}^2}}\right]^{1/2}$$

$$Period = 5,\ 372.54\text{ s} = 89.54\text{ min}$$

We will approximate to a 90-min orbit basis on our calculation. For the inclination angle provided we will source data from the 0 to 30 degree category with IR sensitivity. The Earth-IR flux then becomes $EIR_{Flux} = 206\text{ W/m}^2$. Using view factors 9 and 10 from Appendix A we will have

$$F_{Nadir-Earth} = 0.926 \text{ and } F_{Side-Earth} = 0.329$$

Thus our total environmental loading becomes

$$\dot{Q}_{tot} = \dot{Q}_{Earth-IR/Nadir} + \dot{Q}_{Earth-IR/Side}$$

$$\dot{Q}_{tot} = \varepsilon AF_{Nadir-Earth}EIR_{Flux} + 4\cdot\varepsilon AF_{Side-Earth}EIR_{Flux}$$

$$\dot{Q}_{tot} = (0.91)(1.0\text{ m}^2)(0.926)(206\ {}^{\text{W}}/_{\text{m}^2}) + 4\cdot(0.91)(1.0\text{ m}^2)(0.329)(206\ {}^{\text{W}}/_{\text{m}^2})$$

$$\dot{Q}_{tot} = 173.6\text{ W} + 246.7\text{ W}$$

$$\dot{Q}_{tot} = 420.3\text{ W}$$

4.12 For each part we must calculate the heat flow through the MLI in both the hot and cold cases.

Part i) Using the Lockheed MLI equation we will rely upon EES to perform our calculation. In the hot case the code is

```
{Problem 4.12, Part i, hot case flat plate w/double-sided
aluminized Mylar, Lockheed Equation}
sigma=5.67e-8       [W/m^2-K^4]
g_dot=30     [W]
Solar=1361
EIR_Hot=250
EIR_Cold=193
A_surf=1.0  [m^2]
Albedo_H=0.22
N_S=10

eps_Beta=0.86                                   {Beta cloth}
alpha_Beta=0.4                                  {Beta cloth}
```

```
eps_1=0.28                         {0.15 mil aluminized Mylar}
eps_2=0.28                         {0.15 mil aluminized Mylar}
eps_surf1=0.89                     {Z93 white paint}
alpha_surf1=0.19                   {Z93 white paint}
eps_surf2=0.84                     {Polyurethane white paint}

N_bar=28
eps_tr=0.031
C_S=2.11e-9
C_r=5.39e-10
T_m=(T_surf+T_o)/2

{Lockheed Equation}
q_MLI_Flux=((C_S*(N_bar^3.56)*T_m)/(N_S+1)*(T_surf-T_o))&
+((C_r*eps_tr)/N_S)*((T_surf^4.67)-(T_o^4.67))

q_dot=q_MLI_Flux*A_surf

{Energy balance at the plate}
eps_surf1*A_surf*EIR_Hot+Albedo_H*alpha_surf1*Solar*&
A_surf+g_dot=q_dot&+eps_surf1*A_surf*sigma*T_surf^4

{Energy balance on the outer MLI surface}
alpha_Beta*Solar*A_surf+q_dot=eps_Beta*A_surf*sigma*T_o^4
```

In the Cold case the code is

```
{Problem 4.12, Part ii, cold case flat plate w/double-sided
aluminized Mylar, Lockheed Equation}
sigma=5.67e-8      [W/m^2-K^4]
g_dot=30    [W]
Solar=1361
EIR_Hot=250
EIR_Cold=193
A_surf=1.0  [m^2]
Albedo_H=0.22
N_S=10

eps_Beta=0.86                      {Beta cloth}
alpha_Beta=0.4                     {Beta cloth}
eps_1=0.28                         {0.15 mil aluminized Mylar}
eps_2=0.28                         {0.15 mil aluminized Mylar}
eps_surf1=0.89                     {Z93 white paint}
alpha_surf1=0.19                   { Z93 white paint}
eps_surf2=0.84                     { Polyurethane white paint}
```

```
N_bar=28
eps_tr=0.031
C_S=2.11e-9
C_r=5.39e-10
T_m=(T_surf+T_o)/2

{Lockheed Equation}
q_MLI_Flux=((C_S*(N_bar^3.56)*T_m)/(N_S+1))*(T_surf-T_o)&
+((C_r*eps_tr)/N_S)*((T_surf^4.67)-(T_o^4.67))

q_dot=q_MLI_Flux*A_surf

{Energy balance at the plate}
g_dot+eps_surf1*A_surf*EIR_Cold=q_dot+eps_surf1*A_surf*&
sigma*T_surf^4

{Energy balance on the outer MLI surface}
q_dot=eps_Beta*A_surf*sigma*T_o^4
```

The hot and cold case temperatures are 280 K and 251.1 K respectively.

Part ii) The code for the hot case will be

```
{Problem 4.12, Part ii, hot case flat plate w/double-sided
aluminized Mylar}
sigma=5.67e-8      [W/m^2-K^4]
g_dot=30    [W]
Solar=1361
EIR_Hot=250
EIR_Cold=193
A_surf=1.0  [m^2]
Albedo_H=0.22
N_S=10

eps_Beta=0.86                      {Beta cloth}
alpha_Beta=0.4                     {Beta cloth}
eps_1=0.28                         {0.15 mil aluminized Mylar}
eps_2=0.28                         {0.15 mil aluminized Mylar}
eps_surf1=0.89                     {Z93 white paint}
alpha_surf1=0.19                   {Z93 white paint}
eps_surf2=0.84                     {Polyurethane white paint}
estar=1/((N_S+1)*(1/eps_1+1/eps_2-1))
```

```
{Energy balance at the plate}
eps_surf1*A_surf*EIR_Hot+Albedo_H*alpha_surf1*Solar*&
A_surf+g_dot=estar*A_surf*sigma*(T_surf^4-T_o^4)&
+eps_surf1*A_surf*sigma*T_surf^4

{Energy balance on the outer MLI surface}
alpha_Beta*Solar*A_surf+estar*A_surf*sigma*(T_surf^4-&
T_o^4)=eps_Beta*A_surf*sigma*T_o^4

Q_MLI=estar*A_surf*sigma*(T_surf^4-T_o^4)
```

The code for the Cold case is

```
{Problem 4.12, Part ii, cold case flat plate w/double-sided
aluminized Mylar}
sigma=5.67e-8     [W/m^2-K^4]
g_dot=30    [W]
Solar=1361
EIR_Hot=250
EIR_Cold=193
A_surf=1.0  [m^2]
Albedo_H=0.22
N_S=10

eps_Beta=0.86                   {Beta cloth}
alpha_Beta=0.4                  {Beta cloth}
eps_1=0.28                      {0.15 mil aluminized Mylar}
eps_2=0.28                      {0.15 mil aluminized Mylar}
eps_surf1=0.89                  {Z93 white paint}
alpha_surf1=0.19                {Z93 white paint}
eps_surf2=0.84                  {Polyurethane white paint}
estar=1/((N_S+1)*(1/eps_1+1/eps_2-1))

{Energy balance at the plate}
g_dot+eps_surf1*A_surf*EIR_Cold=estar*A_surf*sigma*&
(T_surf^4-T_o^4)+eps_surf1*A_surf*sigma*T_surf^4

{Energy balance on the outer MLI surface}
estar*A_surf*sigma*(T_surf^4-T_o^4)=eps_Beta*A_surf*&
sigma*T_o^4
```

The hot and cold case temperatures are 280.7 K and 250.4 K respectively. As such, for the orbit case and simplified geometry. There is little change in the MLI performance between the modeling cases and the subsequent temperature of the plate.

4.13 Problem solutions for parts i) and ii) are shown below.

i) Referencing Appendix B we know the following

Catalac black paint $\alpha_S = 0.96$, $\varepsilon = 0.88$
NASA/GSFC NS-53-B green paint $\alpha_S = 0.52$, $\varepsilon = 0.87$
Hughson A-276 white paint $\alpha_S = 0.26$, $\varepsilon = 0.88$
Magnesium oxide white paint $\alpha_S = 0.09$, $\varepsilon = 0.90$

Based on contributions from environmental thermal energy sources and onboard dissipations, we know the energy balance is

$$\dot{g} + \dot{Q}_{Solar} + \dot{Q}_{Albedo} + \dot{Q}_{Earth-IR} = \dot{Q}_{out}$$

Expanding this out would give

$$\dot{g} + \alpha_S S A_{sphere} + 0.5 \cdot Alb \alpha_S S A_{sphere} F_{sphere-earth} + 0.5 \cdot \varepsilon EIR_{flux} A_{sphere} F_{sphere-earth} = \varepsilon A_{sphere} \sigma T^4_{sphere}$$

Solving for T_{sphere} we would have

$$T_{sphere} = \left(\frac{\dot{g} + \alpha_S S A_{p,sphere} + 0.5 \cdot Alb \alpha_S S A_{sphere} F_{sphere-earth} + 0.5 \cdot \varepsilon EIR_{flux} A_{sphere} F_{sphere-earth}}{\varepsilon \sigma A_{sphere}} \right)^{\frac{1}{4}}$$

Based on the equation for T_{sphere}, the solar constant (S) and the view factor for the sphere to earth ($F_{sphere\text{-}earth}$) must be determined for the sphere's location. Using view factor 8 from Appendix A, we know

$$F_{sphere-earth} = 0.5 \cdot \left(1 - \sqrt{1 - R^2}\right) \text{ where } R = \frac{r}{h}$$

Since the radius of Earth is 6,380 km, we also know that

$$h = 1.488M \text{ km} + 6{,}380 \text{ km} = 1{,}494{,}380 \text{ km}$$

$$\text{Thus } R = \frac{6{,}380 \text{ km}}{1{,}494{,}380 \text{ km}} = 0.004269$$

Upon calculation, $F_{sphere-earth} = 0.0000045561$. In light of this extremely small view factor, the albedo and Earth-IR contributions go to zero. The temperature equation now becomes

$$T_{sphere} = \left(\frac{\dot{g} + \alpha_S S A_{p,sphere}}{\varepsilon \sigma A_{sphere}} \right)^{\frac{1}{4}}$$

As for the intensity of the Sun's rays 1/100th the distance between the Earth and the Sun, we know an identity can be drawn between the intensity at Earth and the intensity at the sphere based on the Sun's average power

$$I_{earth} \cdot 4\pi \cdot \bar{r}^2_{earth} = P_{avg} = I_{sphere} \cdot 4\pi \cdot \bar{r}^2_{sphere}$$

Solving for the intensity at the sphere we would have

$$I_{sphere} = I_{earth}\frac{\bar{r}^2_{earth}}{\bar{r}^2_{sphere}} = \left(1{,}361\ \frac{\mathrm{W}}{\mathrm{m}^2}\right)\left[\frac{(93M\ \mathrm{miles})^2}{(92.07M\ \mathrm{miles})^2}\right] = 1{,}388.6\ \frac{\mathrm{W}}{\mathrm{m}^2}$$

Placing values in the following EES equation

```
{Problem 4.13 Temperature of a 1.0 m diameter sphere at L1, part i)}
sigma=5.67e-8
S=1361                     [W/m^2]

{Coatings}
alpha_S1=0.96; eps_1=0.88    {Catalac black paint}
alpha_S2=0.52; eps_2=0.82    {NASA/GSFC NS-53-B green paint}
alpha_S3=0.26; eps_3=0.88    {Hughson A-276 white paint}
alpha_S4=0.09; eps_4=0.90    {Magnesium oxide white paint}

{Intensity of Sun's rays at sphere}
i_earth=S
r_bar_earth=93*1000000*1.6                              [km]
r_bar_sphere=r_bar_earth*(1-1/100)                      [km]
i_sphere=i_earth*((r_bar_earth^2)/(r_bar_sphere^2))

{View factors}
r_earth=6380                                            [km]
r_sphere=(1/100)*r_bar_earth
h=r_sphere+r_earth
R=r_earth/h
F_sphere_Earth=0.5*((1-(1-R^2)^0.5))

{Sphere surface area}
D_sphere=1.0                                            [m]
A_psphere=pi*(D_sphere^2)/4; A_sphere=4*pi*
(D_sphere^2)/4                                      [m^2]

{Environmental loads}
alpha_S=alpha_S4
eps=eps_4

Q_dot_solar=alpha_S*i_sphere*A_psphere

{Energy balance relations}
g_dot=50

T_sphere=((g_dot+Q_dot_solar)/(eps*sigma*A_sphere))^0.25
```

T_{sphere} values for each coating are

Catalac black paint T_{sphere} = 289.2 K
NASA/GSFC NS-53-B green paint T_{sphere} = 255 K
Hughson A-276 white paint T_{sphere} = 214.8 K
Magnesium oxide white paint T_{sphere} = 174.4 K

ii) At L3, the sphere is the same distance from the Sun as the Earth. Similar to part i), environmental contributions in the form of Earth albedo and Earth-IR are negligible. The corresponding EES code for solution is

```
{Problem 4.13 Temperature of a 1.0 m diameter sphere at L3,
part ii)}
sigma=5.67e-8S=1361                          [W/m^2]

{Coatings}
alpha_S1=0.96; eps_1=0.88   {Catalac black paint}
alpha_S2=0.52; eps_2=0.82   {NASA/GSFC NS-53-B green paint}
alpha_S3=0.26; eps_3=0.88   {Hughson A-276 white paint}
alpha_S4=0.09; eps_4=0.90   {Magnesium oxide white paint}

{Intensity of Sun's rays at sphere}
i_earth=S
r_bar_earth=93*1000000*1.6
r_bar_sphere=r_bar_earth
i_sphere=i_earth*((r_bar_earth^2)/(r_bar_sphere^2))

{View factors}
r_earth=6380                                            [km]
r_sphere=(1/100)*r_bar_earth
h=r_sphere+r_earth
R=r_earth/h
F_sphere_Earth=0.5*((1-(1-R^2)^0.5))

{Sphere surface area}
D_sphere=1.0                                            [m]
A_psphere=pi*(D_sphere^2)/4; A_sphere=4*pi*
(D_sphere^2)/4                                          [m^2]

{Environmental Loads}
alpha_S=alpha_S1
eps=eps_1

Q_dot_solar=alpha_S*i_earth*A_psphere

{Energy balance relations}
g_dot=50

T_sphere=((g_dot+Q_dot_solar)/(eps*sigma*A_sphere))^0.25
```

T_{sphere} values for each coating are

Catalac black paint T_{sphere} = 287.9 K
NASA/GSFC NS-53-B green paint T_{sphere} = 253.8 K
Hughson A-276 white paint T_{sphere} = 213.9 K
Magnesium oxide white paint T_{sphere} = 173.8 K

4.14 Given the orbit period is known, the orbit altitude can be determined and then used to calculate the view factor for the radiator surface to the Earth. The orbit altitude calculated is 272.6 km. This gives a corresponding view factor to Earth of 0.92. The following EES code can be used to determine the Max case temperature predictions.

```
{Problem 4.14: Max temperature determination for nadir-
facing SpaceCube dissipating 60W}

{Altitude determination}
P_min=90
P_sec=P_min*60

mu=3.98603e14

a=(mu*(P_sec/(2*pi))^2)^(1/3)            [km]

r_Earth=6380 [km]

altitude=a/1000-r_Earth

{Constants}
sigma=5.67e-8                {Boltzmann's constant: W/m^2-K^4}
S=1361                       {Solar constant}
Alb_H=0.26
EIR_Hot=250
EIR_Cold=193

th=0.015                     {SpaceCube wall thickness}
g_dot=60                     {E-box dissipation}

{Radiator surface area}
A_1b=0.8
F_rad_E=0.92

{Coatings}
eps_1a=0.02                  {VDA}
eps_2=0.87                   {Z306 black polyurethane paint}
eps_1b=0.89                  {Z93 white paint}
alpha_1b=0.19                {Z93 white paint solar absorptance}
```

```
{E-Box (Node 2) Energy balance}
g_dot=(T_2-T_1)/R_2_1+sigma*(T_2^4-T_1^4)/Rad_Res

Rad_Res=(1-eps_1a)/(eps_1a*A_1a)+1/(A_2*F_2_1a)+(1-eps_2)&
/(eps_2*A_2)

EBox_length=0.4                    [m]
EBox_side=EBox_length^2
A_2=5*(EBox_side)
s_length=0.8              [m]   {SpaceCube exterior wall length}
in_length=s_length-2*th
A_1a=(in_length^2)-EBox_side+5*(in_length^2)

k_6061=155.3                       [W/m-K]
k_ins=0.596

{View Factors}
F_2_1a+F_2_2=1.0
F_2_2=0.0
F_1a_2=(A_2/A_1a)*F_2_1a

l_space=0.01                       [m]              {Spacer length}

R_1=0.5*th/(k_6061*6*in_length)
R_2=0.5*EBox_length/(k_6061*EBox_side)
R_spacer=l_space/(k_ins*EBox_side)

R_2_1=R_1+R_2+R_spacer

{MLI Exterior surface thermal loads}
A_MLI=0.8                                [m^2]
alpha_S_MLI=0.25          {Absorptivity of aluminized kapton}
F_MLI_Earth=0.322         {Edgewise MLI to Earth view factor}

Q_MLI1=alpha_S_MLI*S*A_MLI    {Sunviewing Surface}

Q_MLI_side=eps_MLI*A_MLI*F_MLI_Earth*EIR_Hot+Alb_H*&
alpha_S_MLI*S*F_MLI_Earth*A_MLI

Q_MLI2=Q_MLI_side
Q_MLI3=Q_MLI_side
Q_MLI4=Q_MLI_side
Q_MLI5=Q_MLI_side

{Non-ideal MLI definition}
N=12                              {Number of MLI shields}
```

```
eps_MLI=0.34                              {Aluminized kapton}
estar=1/((N+1)*(1/eps_MLI+1/eps_MLI-1)) {Effective emissivity}

{MLI surface energy balances}
T_MLI1^4=Q_MLI1+estar*A_MLI*sigma*(T_1^4-T_MLI1^4)/&
(eps_MLI*A_MLI*sigma)

T_MLI2^4=Q_MLI2+estar*A_MLI*sigma*(T_1^4-T_MLI2^4)/&
(eps_MLI*A_MLI*sigma)

T_MLI3^4=Q_MLI3+estar*A_MLI*sigma*(T_1^4-T_MLI3^4)/&
(eps_MLI*A_MLI*sigma)

T_MLI4^4=Q_MLI4+estar*A_MLI*sigma*(T_1^4-T_MLI4^4)/&
(eps_MLI*A_MLI*sigma)

T_MLI5^4=Q_MLI5+estar*A_MLI*sigma*(T_1^4-T_MLI5^4)/&
(eps_MLI*A_MLI*sigma)

{Energy balance on SpaceCube}
g_dot+Alb_H*F_rad_E*alpha_1b*S*A_1b+eps_1b*F_rad_E*A_1b&
*EIR_Hot=estar*A_MLI*sigma*(T_1^4-T_MLI1^4)+estar&
*A_MLI*sigma*(T_1^4-T_MLI2^4)+estar*A_MLI*sigma*&
(T_1^4-T_MLI3^4)+estar*A_MLI*sigma*(T_1^4-T_MLI4^4)&
+estar*A_MLI*sigma*(T_1^4-T_MLI5^4)&
+eps_1b*A_1b*sigma*T_1^4
```

The EES code for the Min case is as follows

```
{Problem 4.14: Min Temperature determination for nadir-
facing SpaceCube dissipating 60W}

{Altitude determination}
P_min=90
P_sec=P_min*60

mu=3.98603e14

a=(mu*(P_sec/(2*pi))^2)^(1/3)            [km]

r Earth=6380 [km]

altitude=a/1000-r_Earth
```

```
{Constants}
sigma=5.67e-8              {Boltzmann's constant: W/m^2-K^4}
S=1361                     {Solar constant}
Alb_H=0.26
EIR_Hot=250
EIR_Cold=193

th=0.015                   {SpaceCube wall thickness}
g_dot=60                   {E-box dissipation}

{Radiator surface area}
A_1b=0.8
F_rad_E=0.92

{Coatings}
eps_1a=0.02                {VDA}
eps_2=0.87                 {Z306 black polyurethane paint}
eps_1b=0.89                {Z93 white paint}
alpha_1b=0.19              {Z93 white paint solar absorptance}

{E-Box (Node 2) Energy balance}
g_dot=(T_2-T_1)/R_2_1+sigma*(T_2^4-T_1^4)/Rad_Res

Rad_Res=(1-eps_1a)/(eps_1a*A_1a)+1/(A_2*F_2_1a)+(1-eps_2)&
/(eps_2*A_2)

EBox_length=0.4            [m]
EBox_side=EBox_length^2
A_2=5*(EBox_side)
s_length=0.8               [m]    { SpaceCube exterior wall length}
in_length=s_length-2*th
A_1a=(in_length^2)-EBox_side+5*(in_length^2)

k_6061=155.3               [W/m-K]
k_ins=0.596

{View factors}
F_2_1a+F_2_2=1.0
F_2_2=0.0
F_1a_2=(A_2/A_1a)*F_2_1a

l_space=0.01               [m]    {Spacer length}

R_1=0.5*th/(k_6061*6*in_length)
R_2=0.5*EBox_length/(k_6061*EBox_side)
R_spacer=l_space/(k_ins*EBox_side)
```

```
R_2_1=R_1+R_2+R_spacer

{MLI exterior surface thermal loads}
A_MLI=0.8                                [m^2]
alpha_S_MLI=0.25          {Absorptivity of aluminized kapton}
F_MLI_Earth=0.322         {Edgewise MLI to Earth view factor}

Q_MLI1=0.0               {Space viewing surface}

Q_MLI_side=eps_MLI*A_MLI*F_MLI_Earth*EIR_Cold

Q_MLI2=Q_MLI_side
Q_MLI3=Q_MLI_side
Q_MLI4=Q_MLI_side
Q_MLI5=Q_MLI_side

{Non-ideal MLI definition}
N=12                                     {Number of MLI shields}
eps_MLI=0.34                             {Aluminized kapton}
estar=1/((N+1)*(1/eps_MLI+1/eps_MLI-1))  {Effective emissivity}

{MLI surface energy balances}
T_MLI1^4=Q_MLI1+estar*A_MLI*sigma*(T_1^4-T_MLI1^4)/&
(eps_MLI*A_MLI*sigma)

T_MLI2^4=Q_MLI2+estar*A_MLI*sigma*(T_1^4-T_MLI2^4)/&
(eps_MLI*A_MLI*sigma)

T_MLI3^4=Q_MLI3+estar*A_MLI*sigma*(T_1^4-T_MLI3^4)/&
(eps_MLI*A_MLI*sigma)

T_MLI4^4=Q_MLI4+estar*A_MLI*sigma*(T_1^4-T_MLI4^4)/&
(eps_MLI*A_MLI*sigma)

T_MLI5^4=Q_MLI5+estar*A_MLI*sigma*(T_1^4-T_MLI5^4)/&
(eps_MLI*A_MLI*sigma)

{Energy balance on SpaceCube}
g_dot+eps_1b*F_rad_E*A_1b*EIR_Cold=estar*A_MLI*sigma*&
(T_1^4-T_MLI1^4)+estar*A_MLI*sigma*(T_1^4-T_MLI2^4)&
+estar*A_MLI*sigma*(T_1^4-T_MLI3^4)&
+estar*A_MLI*sigma*(T_1^4-T_MLI4^4)&
+estar*A_MLI*sigma*(T_1^4-T_MLI5^4)&
+eps_1b*A_1b*sigma*T_1^4
```

Temperature predictions for both cases are shown in the table below.

	Max (hot) case		Min (cold) case	
Solution	T_1 [K]	T_2 [K]	T_1 [K]	T_2 [K]
Non-ideal MLI	281.1	287.6	255.4	262

4.15 Assuming IR-sensitive surfaces for the radiator and the outer layer of the MLI, the EES program for the hot case is

```
{Problem 4.15  MEO hot case orbit temperature determinations
for SpaceCube}

{Altitude determination}
r_Earth=6380000                                   [m]

altitude=10385000                                 [m]

a=r_Earth+altitude

mu_ugc=3.98603e14                                   {m^3/s^2}

OP_seconds=2*pi*((a^3/mu_ugc)^(1/2))

OP_hours=OP_seconds/3600

angle11=(pi/180)*23.4
angle12=(pi/180)*66.6

{Constants}
sigma=5.67e-8                 {Boltzmann's constant: W/m^2-K^4}
S=1361                        {Solar constant}

Alb_H=0.19+0.04               {Inclination of 0 degrees}
EIR_Hot=269                   {Inclination of 0 degrees}
EIR_Cold=224                  {Inclination of 0 degrees}

th=0.015                      {SpaceCube wall thickness}
g_dot=75                      {E-box dissipation}

{Active surface area}
A_1b=0.8

{Coatings}
eps_1a=0.02                   {VDA}
eps_2=0.87                    {Z306 Black Polyurethane paint}
```

```
eps_1b=0.89                {Z93 white paint}
alpha_1b=0.19              {Z93 white paint solar absorptance}


{E-Box (Node 2) Energy balance}
g_dot=(T_2-T_1)/R_2_1+sigma*(T_2^4-T_1^4)/Rad_Res

Rad_Res=(1-eps_1a)/(eps_1a*A_1a)+1/(A_2*F_2_1a)+(1-eps_2)&
/(eps_2*A_2)

EBox_length=0.3         [m]
EBox_side=EBox_length^2
A_2=5*(EBox_side)
s_length=0.75           [m]  {SpaceCube exterior wall length}
in_length=s_length-2*th
A_1a=(in_length^2)-EBox_side+5*(in_length^2)

k_6061=155.3                     [W/m-K]
k_ins=0.596

{ View factors}
F_2_1a+F_2_2=1.0
F_2_2=0.0
F_1a_2=(A_2/A_1a)*F_2_1a

{View factor: surface normal aligned with nadir}
F_rad_E=(r_Earth/a)^2

{View factor: surface normal tangential to nadir}
H_bar=a/r_Earth
F_MLI_Earth=(1/pi)*(arctan(1/sqrt((H_bar^2-1)))&
-sqrt((H_bar^2-1))/(H_bar^2))

{Resistances}
l_space=0.01            [m]            {Spacer length}

R_1=0.5*th/(k_6061*6*in_length)
R_2=0.5*EBox_length/(k_6061*EBox_side)
R_spacer=l_space/(k_ins*EBox_side)

R_2_1=R_1+R_2+R_spacer

{MLI exterior surface thermal loads}
A_MLI=0.75             [m^2]
alpha_S_MLI=0.25          {Absorptivity of aluminized kapton}
```

```
Q_MLI1=alpha_S_MLI*S*cos(angle11)*A_MLI    { Sunviewing surface}

Q_MLI_side=eps_MLI*A_MLI*F_MLI_Earth*EIR_Hot+Alb_H*&
alpha_S_MLI*S*F_MLI_Earth*A_MLI

Q_MLI2=eps_MLI*A_MLI*F_MLI_Earth*EIR_Hot+Alb_H*alpha_S_&
MLI*S*F_MLI_Earth*A_MLI+alpha_S_MLI*S*cos(angle12)*A_MLI
Q_MLI3=Q_MLI_side
Q_MLI4=Q_MLI_side
Q_MLI5=Q_MLI_side

{Non-ideal MLI definition}
N=15                                       {Number of MLI shields}
eps_MLI=0.34                               {Aluminized kapton}
estar=1/((N+1)*(1/eps_MLI+1/eps_MLI-1)) {Effective emissivity}

{MLI Surface energy balances}
T_MLI1^4=Q_MLI1+estar*A_MLI*sigma*(T_1^4-T_MLI1^4)/&
(eps_MLI*A_MLI*sigma)

T_MLI2^4=Q_MLI2+estar*A_MLI*sigma*(T_1^4-T_MLI2^4)/&
(eps_MLI*A_MLI*sigma)

T_MLI3^4=Q_MLI3+estar*A_MLI*sigma*(T_1^4-T_MLI3^4)/&
(eps_MLI*A_MLI*sigma)

T_MLI4^4=Q_MLI4+estar*A_MLI*sigma*(T_1^4-T_MLI4^4)/&
(eps_MLI*A_MLI*sigma)

T_MLI5^4=Q_MLI5+estar*A_MLI*sigma*(T_1^4-T_MLI5^4)/&
(eps_MLI*A_MLI*sigma)

{Energy balance on SpaceCube}
g_dot+Alb_H*F_rad_E*alpha_1b*S*A_1b+eps_1b*F_rad_E*&
A_1b*EIR_Hot=estar*A_MLI*sigma*(T_1^4-T_MLI1^4)&
+estar*A_MLI*sigma*(T_1^4-T_MLI2^4)+estar*A_MLI*sigma*&
(T_1^4-T_MLI3^4)+estar*A_MLI*sigma*(T_1^4-T_MLI4^4)&
+estar*A_MLI*sigma*(T_1^4-T_MLI5^4)&
+eps_1b*A_1b*sigma*T_1^4
```

The Cold case EES code is

```
{Problem 4.15  MEO cold case orbit temperature determina-
tions for SpaceCube}

{Altitude determination}
r_Earth=6380000                        [m]
```

```
altitude=10385000                                  [m]

a=r_Earth+altitude

mu_ugc=3.98603e14                                    {m^3/s^2}

OP_seconds=2*pi*((a^3/mu_ugc)^(1/2))

OP_hours=OP_seconds/3600

angle11=(pi/180)*23.4
angle12=(pi/180)*66.6

{Constants}
sigma=5.67e-8             {Boltzmann's constant: W/m^2-K^4}
S=1361                    {Solar constant}

Alb_H=0.19+0.04           {Inclination of 0 degrees}
EIR_Hot=269               {Inclination of 0 degrees}
EIR_Cold=224              {Inclination of 0 degrees}

th=0.015                  {SpaceCube wall thickness}
g_dot=75                  {E-box dissipation}

{ Active Surface Area}
A_1b=0.8

{Coatings}
eps_1a=0.02                {VDA}
eps_2=0.87                 {Z306 black polyurethane paint}
eps_1b=0.89                {Z93 white paint}
alpha_1b=0.19              {Z93 white paint solar absorptance}

{E-box (node 2) energy balance}
g_dot=(T_2-T_1)/R_2_1+sigma*(T_2^4-T_1^4)/Rad_Res

Rad_Res=(1-eps_1a)/(eps_1a*A_1a)+1/(A_2*F_2_1a)+(1-eps_2)&
/(eps_2*A_2)

EBox_length=0.3        [m]
EBox_side=EBox_length^2
A_2=5*(EBox_side)
s_length=0.75          [m]    {SpaceCube exterior wall length}
in_length=s_length-2*th
A_1a=(in_length^2)-EBox_side+5*(in_length^2)
```

```
k_6061=155.3                    [W/m-K]
k_ins=0.596

{View factors}
F_2_1a+F_2_2=1.0
F_2_2=0.0
F_1a_2=(A_2/A_1a)*F_2_1a

{View factor: surface normal aligned with nadir}
F_rad_E=(r_Earth/a)^2

{View factor: surface normal tangential to nadir}
H_bar=a/r_Earth
F_MLI_Earth=(1/pi)*(arctan(1/sqrt((H_bar^2-1)))-&
sqrt((H_bar^2-1))/(H_bar^2))

{Resistances}
l_space=0.01       [m]    {Spacer length}

R_1=0.5*th/(k_6061*6*in_length)
R_2=0.5*EBox_length/(k_6061*EBox_side)
R_spacer=l_space/(k_ins*EBox_side)

R_2_1=R_1+R_2+R_spacer

{MLI exterior surface thermal loads}
A_MLI=0.75          [m^2]
alpha_S_MLI=0.25          {Absorptivity of aluminized kapton}

Q_MLI1=0.0                {Space viewing surface}

Q_MLI_side=eps_MLI*A_MLI*F_MLI_Earth*EIR_Cold

Q_MLI2=Q_MLI_side
Q_MLI3=Q_MLI_side
Q_MLI4=Q_MLI_side
Q_MLI5=Q_MLI_side

{Non-ideal MLI definition}
N=15                                     {Number of MLI shields}
eps_MLI=0.34                             {Aluminized kapton}
estar=1/((N+1)*(1/eps_MLI+1/eps_MLI-1)) {Effective emissivity}

{MLI surface energy balances}
T_MLI1^4=Q_MLI1+estar*A_MLI*sigma*(T_1^4-T_MLI1^4)/&
(eps_MLI*A_MLI*sigma)
```

```
T_MLI2^4=Q_MLI2+estar*A_MLI*sigma*(T_1^4-T_MLI2^4)/&
(eps_MLI*A_MLI*sigma)

T_MLI3^4=Q_MLI3+estar*A_MLI*sigma*(T_1^4-T_MLI3^4)/&
(eps_MLI*A_MLI*sigma)

T_MLI4^4=Q_MLI4+estar*A_MLI*sigma*(T_1^4-T_MLI4^4)/&
(eps_MLI*A_MLI*sigma)

T_MLI5^4=Q_MLI5+estar*A_MLI*sigma*(T_1^4-T_MLI5^4)/&
(eps_MLI*A_MLI*sigma)

{Energy balance on SpaceCube}
g_dot+eps_1b*F_rad_E*A_1b*EIR_Cold=estar*A_MLI*sigma*&
(T_1^4-T_MLI1^4)+estar*A_MLI*sigma*(T_1^4-T_MLI2^4)&
+estar*A_MLI*sigma*(T_1^4-T_MLI3^4)&
+estar*A_MLI*sigma*(T_1^4-T_MLI4^4)+estar*&
A_MLI*sigma*(T_1^4-T_MLI5^4)&
+eps_1b*A_1b*sigma*T_1^4
```

The temperature predictions for each case are shown in the table below.

	Max (hot) case		Min (cold) case	
Inclination	T_1 [K]	T_2 [K]	T_1 [K]	T_2 [K]
0°	224.7	239.1	218.6	232.9

4.16 Assuming IR-sensitive surfaces for the radiator and the outer layer of the MLI, as well as the offset between the ecliptic plane and the Earth's equatorial plane:

i) The EES program for the hot case is

```
{Problem 4.16  Hot case orbit SpaceCube temperature deter-
minations}

{Altitude determination}
r_Earth=6380000                              [m]

altitude=273000                            [m]

a=r_Earth+altitude

mu_ugc=3.98603e14                                {m^3/s^2}

OP_seconds=2*pi*((a^3/mu_ugc)^(1/2))

OP_minutes=OP_seconds/60

OP_hours=OP_seconds/3600
```

```
angle=(pi/180)*(30-23.4)

{Constants}
sigma=5.67e-8                {Boltzmann's constant: W/m^2-K^4}
S=1361                       {Solar constant}

Alb_H=0.2+0.04               {Inclination of 0 to 30 degrees}
EIR_Hot=285                  {Inclination of 0 to 30 degrees}
EIR_Cold=206                 {Inclination of 0 to 30 degrees}

th=0.015                     {SpaceCube wall thickness}
g_dot=65                     {E-box dissipation}
g_enc=0                      {Enclosure heating}

{ Active surface area}
A_1b=0.8

{Coatings}
eps_1a=0.02                  {VDA}
eps_2=0.91                   {Black Velvet paint, BOL}
eps_1b=0.89                  {Z93 white paint}
alpha_1b=0.19                {Z93 white Paint Solar Absorptance}

{ E-Box (Node 2) Energy Balance}
g_dot=(T_2-T_1)/R_2_1+sigma*(T_2^4-T_1^4)/Rad_Res

Rad_Res=(1-eps_1a)/(eps_1a*A_1a)+1/(A_2*F_2_1a)+(1-eps_2)&
/(eps_2*A_2)

EBox_length=0.4     [m]
EBox_side=EBox_length^2
A_2=5*(EBox_side)
s_length=0.75       [m]       {SpaceCube exterior wall length}
in_length=s_length-2*th
A_1a=(in_length^2)-EBox_side+5*(in_length^2)

k_6061=155.3                  [W/m-K]
k_ins=0.596

{View factors}
F_2_1a+F_2_2=1.0
F_2_2=0.0
F_1a_2=(A_2/A_1a)*F_2_1a

{View factor: surface normal aligned with nadir}
F_rad_E=(r_Earth/a)^2
```

```
{View factor: surface normal tangential to nadir}
H_bar=a/r_Earth
F_MLI_Earth=(1/pi)*(arctan(1/sqrt((H_bar^2-1)))-&
sqrt((H_bar^2-1))/(H_bar^2))

{Resistances}
l_space=0.01            [m]             {Spacer length}

R_1=0.5*th/(k_6061*6*in_length)
R_2=0.5*EBox_length/(k_6061*EBox_side)
R_spacer=l_space/(k_ins*EBox_side)

R_2_1=R_1+R_2+R_spacer

{MLI exterior surface thermal loads}
A_MLI=0.75                       [m^2]
alpha_S_MLI=0.25         {Absorptivity of aluminized kapton}

Q_MLI1=alpha_S_MLI*S*cos(angle)*A_MLI      { Sunviewing surface}

Q_MLI_side=eps_MLI*A_MLI*F_MLI_Earth*EIR_Hot+Alb_H*&
alpha_S_MLI*S*F_MLI_Earth*A_MLI

Q_MLI2=eps_MLI*A_MLI*F_MLI_Earth*EIR_Hot+Alb_H*alpha_S_&
MLI*S*F_MLI_Earth*A_MLI+alpha_S_MLI*S*cos(angle)*A_MLI
Q_MLI3=Q_MLI_side
Q_MLI4=Q_MLI_side
Q_MLI5=Q_MLI_side

{Non-ideal MLI definition}
N=11                                  {Number of MLI shields}
eps_MLI=0.34                            {Aluminized kapton}
estar=1/((N+1)*(1/eps_MLI+1/eps_MLI-1)) {Effective emissivity}

{MLI surface energy balances}
T_MLI1^4=Q_MLI1+estar*A_MLI*sigma*(T_1^4-T_MLI1^4)/&
(eps_MLI*A_MLI*sigma)

T_MLI2^4=Q_MLI2+estar*A_MLI*sigma*(T_1^4-T_MLI2^4)/&
(eps_MLI*A_MLI*sigma)

T_MLI3^4=Q_MLI3+estar*A_MLI*sigma*(T_1^4-T_MLI3^4)/&
(eps_MLI*A_MLI*sigma)

T_MLI4^4=Q_MLI4+estar*A_MLI*sigma*(T_1^4-T_MLI4^4)/&
(eps_MLI*A_MLI*sigma)
```

```
T_MLI5^4=Q_MLI5+estar*A_MLI*sigma*(T_1^4-T_MLI5^4)/&
(eps_MLI*A_MLI*sigma)

{Energy balance on SpaceCube}
g_dot+g_enc+Alb_H*F_rad_E*alpha_1b*S*A_1b+eps_1b*&
F_rad_E*A_1b*EIR_Hot=estar*A_MLI*sigma&
*(T_1^4-T_MLI1^4)+estar*A_MLI*sigma*(T_1^4-&
T_MLI2^4)+estar*A_MLI*sigma*(T_1^4-T_MLI3^4)&
+estar*A_MLI*sigma*(T_1^4-T_MLI4^4)+estar*A_MLI*sigma*&
(T_1^4-T_MLI5^4)+eps_1b*A_1b*sigma*T_1^4
```

The temperature predictions for the structure and the E-Box are 287.0 K and 294.1 K respectively.

ii) The EES code for the Cold case is

```
{Problem 4.16  cold case orbit SpaceCube temperature deter-
minations}

{Altitude determination}
r_Earth=6380000                                [m]

altitude=273000                              [m]

a=r_Earth+altitude

mu_ugc=3.98603e14                                 { m^3/s^2}

OP_seconds=2*pi*((a^3/mu_ugc)^(1/2))

OP_minutes=OP_seconds/60

OP_hours=OP_seconds/3600

angle=(pi/180)*(30-23.4)

{Constants}
sigma=5.67e-8               {Boltzmann's constant: W/m^2-K^4}
S=1361                      {Solar constant}

Alb_H=0.2+0.04                   {Inclination of 0 degrees}
EIR_Hot=285                      {Inclination of 0 degrees}
EIR_Cold=206                     {Inclination of 0 degrees}
```

```
th=0.015                              {SpaceCube wall thickness}
g_dot=65                              {E-box dissipation}
g_enc=0                               {Enclosure heating}

{Active surface area}
A_1b=0.8

{Coatings}
eps_1a=0.02              {VDA}
eps_2=0.91               {Black Velvet paint, BOL}
eps_1b=0.89              {Z93 white paint}
alpha_1b=0.19            {Z93 white paint solar absorptance}

{E-box (node 2) energy balance}
g_dot=(T_2-T_1)/R_2_1+sigma*(T_2^4-T_1^4)/Rad_Res

Rad_Res=(1-eps_1a)/(eps_1a*A_1a)+1/(A_2*F_2_1a)+(1-eps_2)&
/(eps_2*A_2)

EBox_length=0.4     [m]
EBox_side=EBox_length^2
A_2=5*(EBox_side)
s_length=0.75       [m]    {SpaceCube exterior wall length}
in_length=s_length-2*th
A_1a=(in_length^2)-EBox_side+5*(in_length^2)

k_6061=155.3                         [W/m-K]
k_ins=0.596

{View factors}
F_2_1a+F_2_2=1.0
F_2_2=0.0
F_1a_2=(A_2/A_1a)*F_2_1a

{View factor: surface normal aligned with nadir}
F_rad_E=(r_Earth/a)^2

{View factor: surface normal tangential to nadir}
H_bar=a/r_Earth
F_MLI_Earth=(1/pi)*(arctan(1/sqrt((H_bar^2-1)))-&
sqrt((H_bar^2-1))/(H_bar^2))

{Resistances}
l_space=0.01                [m]                {Spacer length}
```

```
R_1=0.5*th/(k_6061*6*in_length)
R_2=0.5*EBox_length/(k_6061*EBox_side)
R_spacer=l_space/(k_ins*EBox_side)

R_2_1=R_1+R_2+R_spacer

{MLI exterior surface thermal loads}
A_MLI=0.75            [m^2]
alpha_S_MLI=0.25           {Absorptivity of aluminized kapton}

Q_MLI1=0.0                 {Space viewing surface}

Q_MLI_side=eps_MLI*A_MLI*F_MLI_Earth*EIR_Cold

Q_MLI2=Q_MLI_side
Q_MLI3=Q_MLI_side
Q_MLI4=Q_MLI_side
Q_MLI5=Q_MLI_side

{Non-ideal MLI definition}
N=11                                    {Number of MLI shields}
eps_MLI=0.34                            {Aluminized kapton}
estar=1/((N+1)*(1/eps_MLI+1/eps_MLI-1)) {Effective emissivity}

{MLI surface energy balances}
T_MLI1^4=Q_MLI1+estar*A_MLI*sigma*(T_1^4-T_MLI1^4)/&
(eps_MLI*A_MLI*sigma)

T_MLI2^4=Q_MLI2+estar*A_MLI*sigma*(T_1^4-T_MLI2^4)/&
(eps_MLI*A_MLI*sigma)

T_MLI3^4=Q_MLI3+estar*A_MLI*sigma*(T_1^4-T_MLI3^4)/&
(eps_MLI*A_MLI*sigma)

T_MLI4^4=Q_MLI4+estar*A_MLI*sigma*(T_1^4-T_MLI4^4)/&
(eps_MLI*A_MLI*sigma)

T_MLI5^4=Q_MLI5+estar*A_MLI*sigma*(T_1^4-T_MLI5^4)/&
(eps_MLI*A_MLI*sigma)

{ Energy balance on SpaceCube}
g_dot+g_enc+eps_1b*F_rad_E*A_1b*EIR_Cold=estar*A_MLI*&
sigma*(T_1^4-T_MLI1^4)+estar*A_MLI*sigma*(T_1^4&
-T_MLI2^4)+estar*A_MLI*sigma*(T_1^4-T_MLI3^4)&
+estar*A_MLI*sigma*(T_1^4-T_MLI4^4)+estar*A_MLI*sigma*&
(T_1^4-T_MLI5^4)+eps_1b*A_1b*sigma*T_1^4
```

The temperature predictions for the structure and the E-box are 259.9 K and 267 K respectively.

iii) With an additional heat input of approximately 19 W the E-box temperature will operate at 273 K in the Min case.

4.17 The same EES code can be used to calculate the values at L1 and L3. However, the first step is to determine the environmental loading. View factor calculations for the nadir-facing surface and the surfaces that are tangential to the Earth nadir reveal that the view factors are on order of 1.0e–5 or less. As such, environmental loading from the Earth is negligible. The only loading to be accounted for is solar when in Sun view. In addition, the solar constant must be recalculated when at L1 to account for the closer radial distance to the source.

i) The EES code for the L1 case is

```
{Problem 4.17: L1 and L3 position temperature predictions}

{Constants}
sigma=5.67e-8
S=1361*((1/0.99)^2)                              {Solar constant}

Alb_H=0.19+0.04 {Within ecliptic plane, arbitrarily selected value}
EIR_Hot=220     {Within ecliptic plane, arbitrarily selected value}
EIR_Cold=220    {Within ecliptic plane, arbitrarily selected value}

th=0.015        {SpaceCube wall thickness}
g_dot=60        {E-box dissipation}

{Lagrange point distances}
Distance_L1=93000000*1.6*0.01                          [km]

Distance_L1_m=Distance_L1*1000                         [m]

Distance_L2_m=Distance_L1_m

Distance_L3=93000000*1.6*1000                          [m]

{Active surface view factors for L1 and L2}
r_Earth=6380000                                        [m]

altitude=Distance_L1_m

a=r_Earth+altitude

{View factor: surface normal aligned with nadir}
F_rad_E=(r_Earth/a)^2
```

```
{View factor: surface normal tangential to nadir}
H_bar=a/r_Earth
F_MLI_Earth=(1/pi)*(arctan(1/sqrt((H_bar^2-1)))-&
sqrt((H_bar^2-1))/(H_bar^2))

{Active surface area}
A_1b=0.8

{Coatings}
eps_1a=0.88                {Black anodize (sample 3)}
eps_2=0.88                 {Black anodize (sample3)}
eps_1b=0.89                {Z93 white paint}
alpha_1b=0.19              {Z93 white paint solar absorptance}

{E-box (node 2) energy balance}
g_dot=(T_2-T_1)/R_2_1+sigma*(T_2^4-T_1^4)/Rad_Res

Rad_Res=(1-eps_1a)/(eps_1a*A_1a)+1/(A_2*F_2_1a)+(1-eps_2)&
/(eps_2*A_2)

EBox_length=0.4                 [m]
EBox_side=EBox_length^2
A_2=5*(EBox_side)
s_length=0.8          [m]     {SpaceCube exterior wall length}
in_length=s_length-2*th
A_1a=(in_length^2)-EBox_side+5*(in_length^2)

k_6061=155.3                    [W/m-K]
k_ins=0.35                         {Vespel}

{View factors}
F_2_1a+F_2_2=1.0
F_2_2=0.0
F_1a_2=(A_2/A_1a)*F_2_1a

{Resistances}
l_space=0.01          [m]       { Spacer length}

R_1=0.5*th/(k_6061*6*in_length)
R_2=0.5*EBox_length/(k_6061*EBox_side)
R_spacer=l_space/(k_ins*EBox_side)

R_2_1=R_1+R_2+R_spacer
```

```
{MLI exterior surface thermal loads}
A_MLI=0.8                                       [m^2]
alpha_S_MLI=0.25           {Absorptivity of aluminized kapton}

Q_MLI1=alpha_S_MLI*S*A_MLI               { Sunviewing surface}

Q_MLI_side=eps_MLI*A_MLI*F_MLI_Earth*EIR_Hot+Alb_H*&
alpha_S_MLI*S*F_MLI_Earth*A_MLI

Q_MLI2=Q_MLI_side
Q_MLI3=Q_MLI_side
Q_MLI4=Q_MLI_side
Q_MLI5=Q_MLI_side

{Non-ideal MLI definition}
N=20                                     {Number of MLI shields}
eps_MLI=0.34                             {Aluminized kapton}
estar=1/((N+1)*(1/eps_MLI+1/eps_MLI-1))  {Effective emissivity}

{MLI surface energy balances}
T_MLI1^4=Q_MLI1+estar*A_MLI*sigma*(T_1^4-T_MLI1^4)/&
(eps_MLI*A_MLI*sigma)

T_MLI2^4=Q_MLI2+estar*A_MLI*sigma*(T_1^4-T_MLI2^4)/&
(eps_MLI*A_MLI*sigma)

T_MLI3^4=Q_MLI3+estar*A_MLI*sigma*(T_1^4-T_MLI3^4)/&
(eps_MLI*A_MLI*sigma)

T_MLI4^4=Q_MLI4+estar*A_MLI*sigma*(T_1^4-T_MLI4^4)/&
(eps_MLI*A_MLI*sigma)

T_MLI5^4=Q_MLI5+estar*A_MLI*sigma*(T_1^4-T_MLI5^4)/&
(eps_MLI*A_MLI*sigma)

{Energy balance on SpaceCube}
g_dot+Alb_H*F_rad_E*alpha_1b*S*A_1b+eps_1b*F_rad_E*A_1b*&
EIR_Hot=estar*A_MLI*sigma*(T_1^4-T_MLI1^4)+estar*A_MLI&
*sigma*(T_1^4-T_MLI2^4)+estar*A_MLI*sigma*(T_1^4-&
T_MLI3^4)+estar*A_MLI*sigma*(T_1^4-T_MLI4^4)+estar*A_MLI&
*sigma*(T_1^4-T_MLI5^4)+eps_1b*A_1b*sigma*T_1^4
```

The temperature predictions for the structure and the E-box are 193.8 K and 202.9 K respectively.

ii) At L3 we will use a solar constant of 1,361 W/m^2. Since the active surface is Sun viewing, the EES code must be modified to reflect solar loading on the active surface. The corresponding sections of code that need to be changed are

```
Q_MLI1=0.0                                     {Sunviewing surface}

Q_MLI_side=0.0

Q_MLI2=Q_MLI_side
Q_MLI3=Q_MLI_side
Q_MLI4=Q_MLI_side
Q_MLI5=Q_MLI_side
```

and

```
{Energy balance on SpaceCube}
g_dot+alpha_1b*A_1b*S=estar*A_MLI*sigma*(T_1^4-T_MLI1^4)&
+estar*A_MLI*sigma*(T_1^4-T_MLI2^4)+estar*A_MLI*sigma*&
(T_1^4-T_MLI3^4)+estar*A_MLI*sigma*(T_1^4&
-T_MLI4^4)+estar*A_MLI*sigma*(T_1^4-T_MLI5^4)&
+eps_1b*A_1b*sigma*T_1^4
```

The temperature predictions for the structure and the E-box are 281.4 K and 288.2 K respectively.

4.18 Assuming IR-sensitive surfaces for the radiator and the outer layer of the MLI, as well as a 6.6° offset between the Earth's equatorial plane and the orbit plane, the EES code in the hot case will be.

```
{Problem 4.18: L-bracket in a SpaceCube problem: hot case}

{Constants}
sigma=5.67e-8
S=1361                                 {Solar constant}
k_6061=155.3                          [W/m-K]

{Environmental loading}
Alb_H=0.2+0.04                 {Inclination of 0 to 30 degrees}
EIR_Hot=285                    {Inclination of 0 to 30 degrees}
EIR_Cold=206                   {Inclination of 0 to 30 degrees}

{Altitude determination}
r_Earth=6380000                                  [m]

altitude=273000                                  [m]
```

```
a=r_Earth+altitude

mu_ugc=3.98603e14                              { m^3/s^2}

OP_seconds=2*pi*((a^3/mu_ugc)^(1/2))

OP_minutes=OP_seconds/60

OP_hours=OP_seconds/3600

angle1=(pi/180)*(30)
angle2=(pi/180)*(60)

"Areas including sub-divided surfaces"
A_1a=0.10*0.35
A_1b=0.10*0.35
A_1c=0.10*0.35
A_1d=0.10*0.10
A_1e=0.10*0.10
A_1=A_1a+A_1b+A_1c+A_1d+A_1e
A_2a=0.35*0.5
A_2b=0.35*0.5-0.35*0.1
A_2=A_2a+A_2b
A_3=0.35*0.35
A_4a=6*0.70*0.70-0.35*0.35

{Active surface area}
A_4b=0.7

"View factors"
F_1b_4a=1.0
F_1c_4a=1.0
F_1d_4a=1.0
F_1e_4a=1.0
A_1a*F_1a_2b=A_2b*F_2b_1a
F_1_1=0.0
F_1_1+F_1_2b+F_1_4a=1.0

F_2b_1=F_2b_1a+F_2b_1b+F_2b_1c+F_2b_1d+F_2b_e
F_2b_1a=0.08982
F_2b_1b=0.0
F_2b_1c=0.0
F_2b_1d=0.0
F_2b_1e=0.0
A_2b*F_2b_1=A_1*F_1_2b
A_2*F_2_3=A_2a*F_2a_3
```

```
F_2a_3=0.2201
F_2a_1=0.0
F_2a_2a=0.0
F_2b_1=F_2b_1a
F_2b_2b=0.0
F_2b_3=0.0
F_2_2=0.0
F_2a_1+F_2a_2a+F_2a_3+F_2a_4a=1.0
A_2a*F_2a_3=A_3*F_3_2a
F_2b_1+F_2b_2b+F_2b_3+F_2b_4a=1.0
F_3_1=0.0
A_3*F_3_2=A_3*F_3_2a+A_3*F_3_2b
F_3_2b=0.0
F_3_3=0.0
F_3_1+F_3_2a+F_3_3+F_3_4a=1.0
A_3*F_3_4a=A_4a*F_4a_3
A_2a*F_2a_4a=A_4a*F_4a_2a
A_2b*F_2b_4a=A_4a*F_4a_2b
A_1a*F_1a_4a=A_4a*F_4a_1a
A_1b*F_1b_4a=A_4a*F_4a_1b
A_1c*F_1c_4a=A_4a*F_4a_1c
A_1d*F_1d_4a=A_4a*F_4a_1d
A_1e*F_1e_4a=A_4a*F_4a_1e
F_4a_1=F_4a_1a+F_4a_1b+F_4a_1c+F_4a_1d+F_4a_1e
A_4a*F_4a_1=A_1*F_1_4a
F_4a_1+F_4a_2a+F_4a_2b+F_4a_3+F_4a_4a=1.0

{View factor: surface normal aligned with nadir}
F_rad_E=(r_Earth/a)^2

{View factor: surface normal tangential to nadir}
H_bar=a/r_Earth
F_MLI_Earth=(1/pi)*(arctan(1/sqrt((H_bar^2-1)))-&
sqrt((H_bar^2-1))/(H_bar^2))

"Heat flows"
g_dot=35
g_enc=0

"Resistances"
th=0.015                              {SpaceCube wall thickness}

R_1_2=0.05/(k_6061*0.10*0.10)+0.005/(k_6061*0.1*0.1)+0.2/&
(k_6061*0.005*0.35)
```

```
R_2_3=(0.25/(k_6061*0.005*0.35))+(0.175/&
(k_6061*0.005*0.35))
R_3_4=0.005/(k_6061*0.35*0.35)+th/(k_6061*6*0.7*0.7)

"Coatings by surface"
eps_1=0.88              {Black anodize, Sample 3}
eps_1a=0.88             {Black anodize, Sample 3}
eps_1b=0.88             {Black anodize, Sample 3}
eps_1c=0.88             {Black anodize, Sample 3}
eps_1d=0.88             {Black anodize, Sample 3}
eps_1e=0.88             {Black anodize, Sample 3}
eps_2=0.88              {Black anodize, Sample 3}
eps_2a=0.88             {Black anodize, Sample 3}
eps_2b=0.88             {Black anodize, Sample 3}
eps_3=0.88              {Black anodize, Sample 3}
eps_4a=0.88             {Black anodize, Sample 3}
eps_4b=0.89             {Z93 white paint}
alpha_4b=0.19           {Z93 white paint solar absorptance}

{Non-ideal MLI definition}
N=12                                         {Number of MLI shields}
eps_MLI=0.34                                 {Aluminized kapton}
estar=1/((N+1)*(1/eps_MLI+1/eps_MLI-1))      {Effective emissivity}

"E_b1 Energy balance"
g_dot=(E_b1-J1)/((1-eps_1)/(eps_1*A_1))+(T_1-T_2)/R_1_2

"J_1 Energy balance"
(E_b1-J1)/((1-eps_1)/(eps_1*A_1))=(J1-J2b)/(1/&
(A_1*F_1_2b))+(J1-J4a)/(1/(A_1*F_1_4a))

"E_b2 Energy balance"
(T_1-T_2)/R_1_2=(E_b2-J2a)/((1-eps_2a)/(eps_2a*A_2a))&
+(E_b2-J2b)/((1-eps_2b)/(eps_2b*A_2b))+(T_2-T_3)/R_2_3

"J_2b Energy balance"
(E_b2-J2b)/(((1-eps_2b)/(eps_2b*A_2b)))=(J2b-J1)/(1/&
(A_2b*F_2b_1))+(J2b-J4a)/(1/(A_2b*F_2b_4a))

"J_2a Energy balance"
(E_b2-J2a)/(((1-eps_2a)/(eps_2a*A_2a)))=(J2a-J3)/(1/&
(A_2a*F_2a_3))+(J2a-J4a)/(1/(A_2a*F_2a_4a))
```

```
"E_b3 Energy balance"
(T_2-T_3)/R_2_3=(E_b3-J3)/(((1-eps_3)/(eps_3*A_3)))+&
(T_3-T_4)/R_3_4

"J_3 Energy balance"
(E_b3-J3)/((1-eps_3)/(eps_3*A_3))=(J3-J2a)/(1/&
(A_3*F_3_2a))+(J3-J4a)/(1/(A_3*F_3_4a))

"J_4a Energy balance"
(E_b4-J4a)/((1-eps_4a)/(eps_4a*A_4a))=(J4a-J1)/(1/&
(A_1*F_1_4a))+(J4a-J2a)/(1/(A_2a*F_2a_4a))+(J4a-J2b)&
/(1/(A_2b*F_2b_4a))+(J4a-J3)/(1/(A_3*F_3_4a))

{MLI Exterior surface thermal loads}
A_MLI=0.75                             [m^2]
alpha_S_MLI=0.25           {Absorptivity of aluminized kapton}

Q_MLI1=alpha_S_MLI*S*cos(angle1)*A_MLI      {Sunviewing surface}

Q_MLI_side=eps_MLI*A_MLI*F_MLI_Earth*EIR_Hot+Alb_H*&
alpha_S_MLI*S*F_MLI_Earth*A_MLI

Q_MLI2=eps_MLI*A_MLI*F_MLI_Earth*EIR_Hot+Alb_H*alpha_S_&
MLI*S*F_MLI_Earth*A_MLI+alpha_S_MLI*S*cos(angle2)*A_MLI
Q_MLI3=Q_MLI_side
Q_MLI4=Q_MLI_side
Q_MLI5=Q_MLI_side

{MLI surface energy balances}
T_MLI1^4=Q_MLI1+estar*A_MLI*sigma*(T_4^4-T_MLI1^4)/&
(eps_MLI*A_MLI*sigma)

T_MLI2^4=Q_MLI2+estar*A_MLI*sigma*(T_4^4-T_MLI2^4)/&
(eps_MLI*A_MLI*sigma)

T_MLI3^4=Q_MLI3+estar*A_MLI*sigma*(T_4^4-T_MLI3^4)/&
(eps_MLI*A_MLI*sigma)

T_MLI4^4=Q_MLI4+estar*A_MLI*sigma*(T_4^4-T_MLI4^4)/&
(eps_MLI*A_MLI*sigma)

T_MLI5^4=Q_MLI5+estar*A_MLI*sigma*(T_4^4-T_MLI5^4)/&
(eps_MLI*A_MLI*sigma)
```

```
{Energy balance on SpaceCube}
g_dot+g_enc+Alb_H*F_rad_E*alpha_4b*S*A_4b+eps_4b*F_rad_E&
*A_4b*EIR_Hot=estar*A_MLI*sigma&
*(T_4^4-T_MLI1^4)+estar*A_MLI*sigma*(T_4^4-&
T_MLI2^4)+estar*A_MLI*sigma*(T_4^4-T_MLI3^4)&
+estar*A_MLI*sigma*(T_4^4-T_MLI4^4)+estar*A_MLI*&
sigma*(T_4^4-T_MLI5^4)+eps_4b*A_4b*sigma*T_4^4

"Blackbody energy fluxes"
E_b1=sigma*(T_1^4)
E_b2=sigma*(T_2^4)
E_b3=sigma*(T_3^4)
E_b4=sigma*(T_4^4)
```

The temperature predictions for the structure and the E-box are 280.4 K and 305.8 K respectively.

In the cold case the EES code would be

```
{Problem 4.18: L-bracket in a SpaceCube problem: cold case}

{Constants}
sigma=5.67e-8
S=1361                              {Solar constant}
k_6061=155.3                        [W/m-K]

{Environmental loading}
Alb_H=0.2+0.04               {Inclination of 0 to 30 degrees}
EIR_Hot=285                  {Inclination of 0 to 30 degrees}
EIR_Cold=206                 {Inclination of 0 to 30 degrees}

{Altitude determination}
r_Earth=6380000                                 [m]

altitude=273000                               [m]

a=r_Earth+altitude

mu_ugc=3.98603e14                                  {m^3/s^2}

OP_seconds=2*pi*((a^3/mu_ugc)^(1/2))

OP_minutes=OP_seconds/60

OP_hours=OP_seconds/3600
```

```
angle1=(pi/180)*(30)
angle2=(pi/180)*(60)

"Areas including sub-divided surfaces"
A_1a=0.10*0.35
A_1b=0.10*0.35
A_1c=0.10*0.35
A_1d=0.10*0.10
A_1e=0.10*0.10
A_1=A_1a+A_1b+A_1c+A_1d+A_1e
A_2a=0.35*0.5
A_2b=0.35*0.5-0.35*0.1
A_2=A_2a+A_2b
A_3=0.35*0.35
A_4a=6*0.70*0.70-0.35*0.35

{Active surface area}
A_4b=0.7

"View factors"
F_1b_4a=1.0
F_1c_4a=1.0
F_1d_4a=1.0
F_1e_4a=1.0
A_1a*F_1a_2b=A_2b*F_2b_1a
F_1_1=0.0
F_1_1+F_1_2b+F_1_4a=1.0

F_2b_1=F_2b_1a+F_2b_1b+F_2b_1c+F_2b_1d+F_2b_e
F_2b_1a=0.08982
F_2b_1b=0.0
F_2b_1c=0.0
F_2b_1d=0.0
F_2b_1e=0.0
A_2b*F_2b_1=A_1*F_1_2b
A_2*F_2_3=A_2a*F_2a_3
F_2a_3=0.2201
F_2a_1=0.0
F_2a_2a=0.0
F_2b_1=F_2b_1a
F_2b_2b=0.0
F_2b_3=0.0
F_2_2=0.0
F_2a_1+F_2a_2a+F_2a_3+F_2a_4a=1.0
A_2a*F_2a_3=A_3*F_3_2a
```

```
F_2b_1+F_2b_2b+F_2b_3+F_2b_4a=1.0
F_3_1=0.0
A_3*F_3_2=A_3*F_3_2a+A_3*F_3_2b
F_3_2b=0.0
F_3_3=0.0
F_3_1+F_3_2a+F_3_3+F_3_4a=1.0
A_3*F_3_4a=A_4a*F_4a_3
A_2a*F_2a_4a=A_4a*F_4a_2a
A_2b*F_2b_4a=A_4a*F_4a_2b
A_1a*F_1a_4a=A_4a*F_4a_1a
A_1b*F_1b_4a=A_4a*F_4a_1b
A_1c*F_1c_4a=A_4a*F_4a_1c
A_1d*F_1d_4a=A_4a*F_4a_1d
A_1e*F_1e_4a=A_4a*F_4a_1e
F_4a_1=F_4a_1a+F_4a_1b+F_4a_1c+F_4a_1d+F_4a_1e
A_4a*F_4a_1=A_1*F_1_4a
F_4a_1+F_4a_2a+F_4a_2b+F_4a_3+F_4a_4a=1.0

{View factor: surface normal aligned with nadir}
F_rad_E=(r_Earth/a)^2

{View factor: surface normal tangential to nadir}
H_bar=a/r_Earth
F_MLI_Earth=(1/pi)*(arctan(1/sqrt((H_bar^2-1)))-&
sqrt((H_bar^2-1))/(H_bar^2))

"Heat flows"
g_dot=35
g_enc=0

"Resistances"
th=0.015                              {SpaceCube wall thickness}

R_1_2=0.05/(k_6061*0.10*0.10)+0.005/(k_6061*0.1*0.1)+0.2/&
(k_6061*0.005*0.35)
R_2_3=(0.25/(k_6061*0.005*0.35))+(0.175/&
(k_6061*0.005*0.35))
R_3_4=0.005/(k_6061*0.35*0.35)+th/(k_6061*6*0.7*0.7)

"Coatings by surface"
eps_1=0.88              {Black anodize, Sample 3}
eps_1a=0.88             {Black anodize, Sample 3}
eps_1b=0.88             {Black anodize, Sample 3}
eps_1c=0.88             {Black anodize, Sample 3}
eps_1d=0.88             {Black anodize, Sample 3}
eps_1e=0.88             {Black anodize, Sample 3}
eps_2=0.88              {Black anodize, Sample 3}
```

```
eps_2a=0.88                  {Black anodize, Sample 3}
eps_2b=0.88                  {Black anodize, Sample 3}
eps_3=0.88                   {Black anodize, Sample 3}
eps_4a=0.88                  {Black anodize, Sample 3}
eps_4b=0.89                  {Z93 white paint}
alpha_4b=0.19                {Z93 White paint solar absorptance}

{Non-ideal MLI definition}
N=12                                      {Number of MLI shields}
eps_MLI=0.34                              {Aluminized kapton}
estar=1/((N+1)*(1/eps_MLI+1/eps_MLI-1))   {Effectiveemissivity}

"E_b1 Energy balance"
g_dot=(E_b1-J1)/((1-eps_1)/(eps_1*A_1))+(T_1-T_2)/R_1_2

"J_1 Energy balance"
(E_b1-J1)/((1-eps_1)/(eps_1*A_1))=(J1-J2b)/(1/&
(A_1*F_1_2b))+(J1-J4a)/(1/(A_1*F_1_4a))

"E_b2 Energy balance"
(T_1-T_2)/R_1_2=(E_b2-J2a)/((1-eps_2a)/(eps_2a*A_2a))&
+(E_b2-J2b)/((1-eps_2b)/(eps_2b*A_2b))+(T_2-T_3)/R_2_3

"J_2b Energy balance"
(E_b2-J2b)/(((1-eps_2b)/(eps_2b*A_2b)))=(J2b-J1)/(1/&
(A_2b*F_2b_1))+(J2b-J4a)/(1/(A_2b*F_2b_4a))

"J_2a Energy balance"
(E_b2-J2a)/(((1-eps_2a)/(eps_2a*A_2a)))=(J2a-J3)/(1/&
(A_2a*F_2a_3))+(J2a-J4a)/(1/(A_2a*F_2a_4a))

"E_b3 Energy balance"
(T_2-T_3)/R_2_3=(E_b3-J3)/(((1-eps_3)/(eps_3*A_3)))+&
(T_3-T_4)/R_3_4

"J_3 Energy balance"
(E_b3-J3)/((1-eps_3)/(eps_3*A_3))=(J3-J2a)/(1/&
(A_3*F_3_2a))+(J3-J4a)/(1/(A_3*F_3_4a))

"J_4a Energy balance"
(E_b4-J4a)/((1-eps_4a)/(eps_4a*A_4a))=(J4a-J1)/(1/&
(A_1*F_1_4a))+(J4a-J2a)/(1/(A_2a*F_2a_4a))+(J4a-J2b)&
/(1/(A_2b*F_2b_4a))+(J4a-J3)/(1/(A_3*F_3_4a))

{MLI exterior surface thermal loads}
A_MLI=0.75                                [m^2]
alpha_S_MLI=0.25             {Absorptivity of aluminized kapton}
```

```
Q_MLI1=0.0                    {Space viewing surface}

Q_MLI_side=eps_MLI*A_MLI*F_MLI_Earth*EIR_Cold

Q_MLI2=Q_MLI_side
Q_MLI3=Q_MLI_side
Q_MLI4=Q_MLI_side
Q_MLI5=Q_MLI_side

{MLI surface energy balances}
T_MLI1^4=Q_MLI1+estar*A_MLI*sigma*(T_4^4-T_MLI1^4)/&
(eps_MLI*A_MLI*sigma)

T_MLI2^4=Q_MLI2+estar*A_MLI*sigma*(T_4^4-T_MLI2^4)/&
(eps_MLI*A_MLI*sigma)

T_MLI3^4=Q_MLI3+estar*A_MLI*sigma*(T_4^4-T_MLI3^4)/&
(eps_MLI*A_MLI*sigma)

T_MLI4^4=Q_MLI4+estar*A_MLI*sigma*(T_4^4-T_MLI4^4)/&
(eps_MLI*A_MLI*sigma)

T_MLI5^4=Q_MLI5+estar*A_MLI*sigma*(T_4^4-T_MLI5^4)/&
(eps_MLI*A_MLI*sigma)

{Energy balance on SpaceCube}
g_dot+g_enc+eps_4b*F_rad_E*A_4b*EIR_Cold=estar*A_MLI*&
sigma*(T_4^4-T_MLI1^4)+estar*A_MLI*sigma*(T_4^4-&
T_MLI2^4)+estar*A_MLI*sigma*(T_4^4-T_MLI3^4)&
+estar*A_MLI*sigma*(T_4^4-T_MLI4^4)&
+estar*A_MLI*sigma*(T_4^4-T_MLI5^4)&
+eps_4b*A_4b*sigma*T_4^4

"Blackbody energy fluxes"
E_b1=sigma*(T_1^4)
E_b2=sigma*(T_2^4)
E_b3=sigma*(T_3^4)
E_b4=sigma*(T_4^4)
```

The temperature predictions for the structure and the E-Box are 251 K and 281.1 K respectively.

4.19 We can reference Section 4.7.3 for the basis of heat transfer with solar panels.

i) Based on the thermal-to-electric conversion efficiency, we know that

$$P_{out} = 0.35 \cdot \dot{Q}_{Solar}$$

where $\dot{Q}_{solar} = \alpha_S S A_{surf} \cos\beta$. Substituting in values we will have

$$P_{out} = 0.35 \cdot \alpha_S S A_{surf} \cos\beta$$

$$P_{out} = (0.35)(0.92)(1{,}361\ \text{W/m}^2)(4\ \text{W/m}^2)\cos(15°)$$

$$P_{out} = 1{,}693\ \text{W}$$

ii) Based on an energy balance applied to the solar panels, we know that

$$\dot{Q}_{solar} = P_{out} + \dot{Q}_{plate,rej} + \dot{Q}_{ref} + \dot{Q}_{waste}$$

we also know that

$$\dot{Q}_{total,rej} = \dot{Q}_{plate,rej} + \dot{Q}_{waste}$$

thus,

$$\dot{Q}_{solar} = P_{out} + \dot{Q}_{ref} + \dot{Q}_{total,rej}$$

Solving for the total heat rejection we would have

$$\dot{Q}_{tot,rej} = \dot{Q}_{solar} - P_{out} - \dot{Q}_{ref}$$

Substituting in values we would have

$$\dot{Q}_{tot,rej} = \dot{Q}_{solar} - 0.35 \cdot \dot{Q}_{solar} - (1-\varepsilon)\dot{Q}_{solar}$$

$$\dot{Q}_{tot,rej} = \dot{Q}_{solar}[1 - 0.35 - (1-\varepsilon)]$$

$$\dot{Q}_{tot,rej} = 0.5 \cdot \dot{Q}_{solar} = (0.5)(0.92)\left(1{,}361\ \text{W/m}^2\right)\left(4\ \text{W/m}^2\right)\cos(15°)$$

$$\dot{Q}_{tot,rej} = 2{,}419\ \text{W}$$

iii) From the energy balance relations we know that

$$\dot{Q}_{waste} = \dot{Q}_{solar} - P_{out} - \dot{Q}_{ref} - \dot{Q}_{plate,rej}$$

Expanding out the LHS we have

$$\varepsilon_{panel}\sigma A_{panel} T^4_{panel} = \dot{Q}_{solar} - 0.35 \cdot \dot{Q}_{solar} - (1-\varepsilon) \cdot \dot{Q}_{solar} - \varepsilon_{PV}\sigma A_{array} T^4_{array}$$

Substituting in values and solving for the panel temperature will result in a value of 124.1 K.

Chapter 5 Solutions

5.1 The R410a vapor compression cycle can be modeled in EES using the following

```
{Problem 5.1 R32/125 (50/50) Vapor compression cycle}
m_dot=0.045          [kg/s]
T_low=0              {Evaporator side temperature}
```

```
{Pressure ratio}
P_ratio=1.8
P_high=P_ratio*P_low

{State point 1}
x_1=1.0
P_low=pressure(R410A,T=T_low,x=x_1)
h_1=enthalpy(R410A,T=T_low,x=x_1)
s_1=entropy(R410A,T=T_low,x=x_1)

{State point 2}
s_2=s_1
h_2s=enthalpy(R410A,P=P_high,s=s_2)

{State point 3}
x_3=0.0
h_3=enthalpy(R410A,P=P_high,x=x_3)
T_high=temperature(R410A,P=P_high,x=x_3)

{State point 4}
h_4=h_3

{Compressor work input}
eta=0.9
W_comp_s=m_dot*(h_2s-h_1)
W_comp=W_comp_s/eta
h_2=W_comp/m_dot+h_1

{Evaporator cooling capacity}
Q_c=m_dot*(h_1-h_4)

{Condenser heat rejection}
Q_h=m_dot*(h_2-h_3)

{COPs}
T_low_K=T_low+273
T_high_K=T_high+273

COP_carnot=T_low_K/(T_high_K-T_low_K)

COP_actual=Q_c/W_comp
```

Upon execution of the code the answers to each part are as follows

i) 0.7869 kW

ii) 9.354 kW

iii) 10.89

5.2 The R32 vapor compression cycle can be modeled in EES using the following

```
{Problem 5.2 R32 vapor compression cycle}
m_dot=0.05          [kg/s]
T_low=0             {Evaporator-side temperature}

{Pressure ratio}
P_ratio=3
P_high=P_ratio*P_low

{State point 1}
x_1=1.0
P_low=pressure(R32,T=T_low,x=x_1)
h_1=enthalpy(R32,T=T_low,x=x_1)

{State point 2}
h_2=W_comp/m_dot+h_1

{State point 3}
x_3=0.0
h_3=enthalpy(R32,P=P_high,x=x_3)
T_high=temperature(R32,P=P_high,x=x_3)

{State point 4}
h_4=h_3

{Compressor work input}
W_comp=0.8

{Evaporator cooling capacity}
Q_c=m_dot*(h_1-h_4)

{Condenser heat rejection}
Q_h=m_dot*(h_2-h_3)

{COPs}
T_low_K=T_low+273
T_high_K=T_high+273

COP_carnot=T_low_K/(T_high_K-T_low_K)

COP_actual=Q_c/W_comp
```

Upon execution of the code the answer to part i) is

Condenser heat rejection = 12.85 kW.

The refrigeration coefficient of performance (actual) is 15.06.

For parts ii) and iii) the pressure ratio line should be commented out as shown below.

```
{P_ratio=3}
```

A table can then be established with a pressure ratio variation from 0 to 7. The following plots can then be attained

ii)

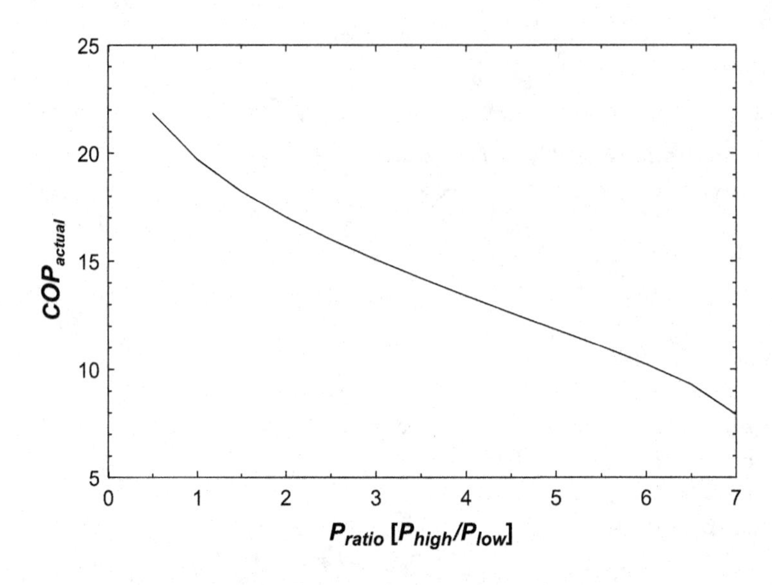

iii)

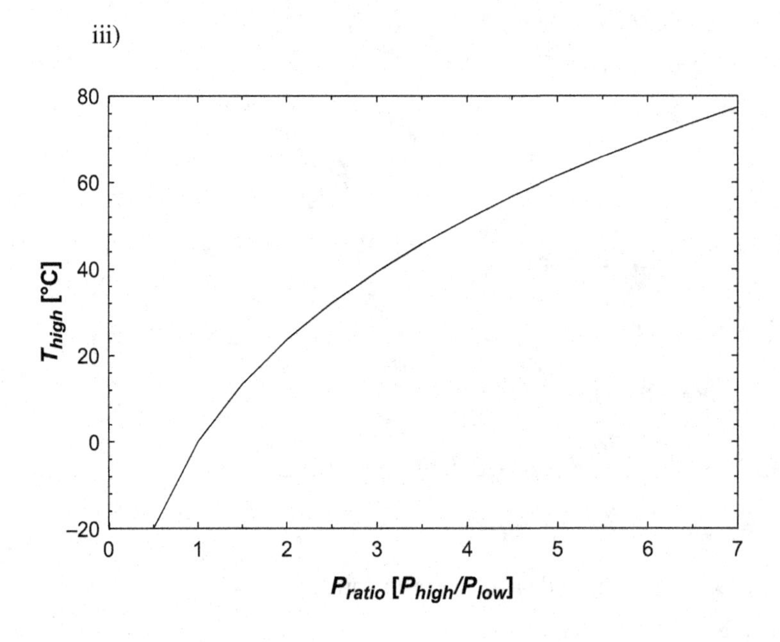

5.3 The R744 vapor compression cycle can be modeled in EES using the following

```
{ Problem 5.3 Carbon Dioxide vapor compression cycle with
condenser/radiator coupling}

sigma=5.67e-8    {Stefan-Boltzmann constant}
m_dot=0.002        [kg/s]

k_6061=167

T_low=-5             {Evaporator-side temperature}

{Radiator geometry and coating}
A_rad=1.0*1.0
th_rad=5/1000                    [m]

alpha_S=0.96; eps=0.88           {Carbon black NS-7 paint}

{Condenser geometry}
evap_base=0.05*0.10              [ m^2]
evap_height=0.05                 [m]

{Radiator and condenser temperatures}
R_evap_rad=(0.5*evap_height)/(k_6061*evap_base)+(0.5*th_rad)&
/(k_6061*A_rad)+1

T_rad_K=((Q_h*1000)/(eps*A_rad*sigma))^0.25

T_high=Q_h*R_evap_rad+T_rad    {Condenser-side temperature}

{State point 1}
x_1=1.0
P_low=pressure(CarbonDioxide,T=T_low,x=x_1)
h_1=enthalpy(CarbonDioxide,T=T_low,x=x_1)

{Pressure ratio}
P_ratio=2.25
P_high=P_ratio*P_low

{State point 2}
h_2=W_comp/m_dot+h_1

{State point 3}
x_3=0.0
h_3=enthalpy(CarbonDioxide,P=P_high,x=x_3)
T_3=temperature(CarbonDioxide,P=P_high,x=x_3)
```

```
PC=p_crit(CarbonDioxide)

{State point 4}
h_4=h_3

{Compressor work input}
W_comp=0.125

{Evaporator cooling capacity}
Q_c=m_dot*(h_1-h_4)

{Condenser heat rejection}
Q_h=m_dot*(h_2-h_3)

{COPs}
T_low_K=T_low+273
T_high_K=T_high+273
T_rad=T_rad_K-273

COP_carnot=T_low_K/(T_high_K-T_low_K)

COP_actual=Q_c/W_comp
```

Upon execution, the actual refrigeration COP is 2.323. The radiator temperature is 29.06°C.

5.4 The R744 (CO_2) vapor compression cycle can be modeled in EES using the following

```
{Problem 5.4 Carbon Dioxide vapor compression cycle with
condenser/radiator coupling}

sigma=5.67e-8        {Stefan–Boltzmann constant}
m_dot=0.012           [kg/s]

T_low=-10.0

{State point 1}
x_1=1.0
T_1=T_low
P_low=pressure(CarbonDioxide,T=T_low,x=x_1)
h_1=enthalpy(CarbonDioxide,T=T_low,x=x_1)
```

```
{Pressure ratio}
P_ratio=1.65
P_high=P_ratio*P_low

{State point 1p; LLSL heat exchanger relations}
P_1p=P_low
HX_Eff=0.98
HX_Eff=(T_1p-T_1)/(T_3-T_1)
h_1p=enthalpy(CarbonDioxide,P=P_1p,T=T_1p)

{State point 2}
h_2=W_comp/m_dot+h_1p

{State point 3}
x_3=0.0
h_3=enthalpy(CarbonDioxide,P=P_high,x=x_3)
T_3=temperature(CarbonDioxide,P=P_high,x=x_3)
T_high=T_3

{State point 4}
h_4=h_3p

{State point 3p}
h_3-h_3p=h_1p-h_1
T_3p=temperature(CarbonDioxide,P=P_high,h=h_3p)

{Compressor work input}
W_comp=0.190

{Evaporator cooling capacity}
Q_c=m_dot*(h_1-h_4)

{Condenser heat rejection}
Q_h=m_dot*(h_2-h_3)

{COPs}
T_low_K=T_low+273
T_high_K=T_high+273

COP_carnot=T_low_K/(T_high_K-T_low_K)

COP_actual=Q_c/W_comp
```

Upon execution, the heat rejection from the condenser is 3.036 kW and the actual refrigeration COP is 14.98.

5.5 The following EES code can be used to determine parts i) through iv)

```
{Problem 5.5 Thermal strap thermal coupling between an E-box
and a temperature sink}

{Sink temperature}
T_sink=-10                [C]        {Sink temperature}

{TS 1 dimensions}
delta_TS=5*0.0254/1000       [m]   {Foil thickness of 5 mil}
l_TS=7.0*2.54/100            [m]   {Length of TS1}
W_TS=2*2.54/100              [m]   {Width of TS foils, 2 inch}
N_TS=48

{TS efficiencies}
eta_S=0.99                          {Shape efficiency for TS 1}
eta_E=0.99                          {Endpoint efficiency for TS 1}

{Thermal conductivities}
k_OFHC=400                           [W/m-K]
k_Al_1100=220                        [W/m-K]

{TS conductance and resistance}
C_TS=(k_OFHC/l_TS)*(N_TS*delta_TS*W_TS)*eta_S*eta_E  {w/OFHC Foils}
R_TS=1/C_TS

{TS endpoint conductances}
A_end=(1.75*2.54/100)*(1.75*2.54/100) {Endpoint area for TS1}
C_interface=10000*A_end                 {Conductance for OFHC
interface at 5 MPa in vacuum}
R_interface=1/C_interface

{Energy dissipations}
g_dot=32                             {10 W produced at E-box 1}

{Energy balances}
g_dot=(T_1-T_sink)/(R_TS+2*R_interface)
```

i) Upon execution of the code, the TS's thermal conductance is 0.6828 W/K.
ii) Using OFHC foils, the E-box temperature is determined to be 40.1°C
iii) To determine the E-box temperature using Al-1100 foils as opposed to OFHC, the conductance equation in the "TS conduction and resistance" section of the code will be replaced with

```
C_TS=(k_Al_1100/l_TS)*(N_TS*delta_TS*W_TS)*eta_S*eta_E
{w/Al 1100 Foils}
```

Using 48 Al-1100 foils with the same geometry, the E-box temperature will be 78.5°C.

iv) An equation can be determined to calculate the number of foils required to match the performance of the OFHC foils by establishing an equality for the thermal conductances

$$C_{FCL,\text{Al-1100}} = C_{FCL,OFHC}$$

Expanding both sides of the equal sign out and eliminating common terms, the number of Al-1100 TS foils can be defined as

$$N_{\text{Al-1100}} = \frac{k_{OFHC}}{k_{\text{Al-1100}}} \cdot N_{TS}$$

The actual number can be calculated by adding it to the EES code using the following syntax

```
{Equivalent Al-1100 foils to match OFHC thermal conductance}
N_Al_1100=(k_OFHC/k_Al_1100)*N_TS            {# Al-1100 foils
equivalent to OFHC foil performance}
```

Upon execution, it is determined that a total of 88 Al-1100 foils will be required to match the conductance performance of the 48 OFHC foils.

5.6 i) $\eta_E = 0.99$; $\eta_S = 0.9$; $k|_{296\text{K}} = 398$ W/m · K

$$\dot{Q} = C_{TS} \cdot \Delta T$$

$$C_{TS} = \frac{\dot{Q}}{\Delta T} = \frac{10\text{ W}}{30\text{ K}} = 0.333\text{ W/K}$$

we know that $C_{TS} = \frac{1}{R_{TS}} = \frac{k}{l}(N_F \delta_F W_F)\eta_S \eta_E$
solving for N_F

$$N_F = \frac{l \cdot C_{TS}}{k \cdot \eta_S \cdot \eta_E \cdot \delta_F \cdot w_F}$$

$$N_F = \frac{(0.2\text{ m})(0.333\text{ W/K})}{(398\text{ W/m}\cdot\text{K})(0.9)(0.99)(7.62\times10^{-5}\text{ m})(0.0254\text{ m})}$$

$$N_F = 97.03\ \textit{layers}$$

ii)

$$k|_{20\text{K}} = 3{,}245\text{ W/m}\cdot\text{K}$$

$$C_{TS} = \left(\frac{3{,}245\text{ W/m}\cdot\text{K}}{0.2\text{ m}}\right)(97)(7.62\times10^{-5}\text{ m})(0.0254\text{ m})(0.9)(0.99)$$

$$C_{TS} = 2.714\text{ W/K}$$

Also, from $\dot{Q} = C_{TS} \cdot \Delta T$ we know that

$$\Delta T = \frac{\dot{Q}}{C_{TS}} = \frac{0.250\text{ W}}{2.714\text{ W/K}} = 0.092\text{ K}$$

5.7 The equation for thermal conductance of a braid must be used. This is

$$C_{Braid} = \frac{1}{R_{Braid}} = \frac{k}{l} \cdot N_{wire} \cdot \frac{\pi D^2}{4} \eta_S \eta_E$$

i) Using the geometry provided, the conductance can be determined. Given boundary temperatures, the heat transfer can be calculated using the following EES code.

```
{Problem 5.7 Braided copper wire with intermediate heater
for fine tuning control part i)}

l=12*2.54/100                    [m]
D_wire=2.0/1000                  [m]

k_OFHC=400                       [W/m-K]

eta_S=0.99                       {shape efficiency}
eta_E=0.99                       {endpoint efficiency}

N_wire=100                          {Number of wire in braid}

C_Braid=(k_OFHC/l)*N_wire*pi*((D_wire^2)/4)*eta_S*eta_E

T_1=30                              [C]

T_2=5                               [C]

Q_dot_1=C_Braid*(T_1-T_2)
```

The heat being transferred is 10.1 W.

ii) For part ii) an intermediate node at the collar must be added in to account for the temperature at that location. In addition, the braid conductance must be separated into two different sections. One section accounts for heat flow from node 1 to the collar. The other section accounts for heat flow from the collar to node 2. Balance equations accounting for the heat flow in each section must be specified. The EES code below can be used to determine the required heat input at the collar of 34.6 W.

```
{Problem 5.7 Braided copper wire with intermediate heater
for fine tuning control part ii)}

l=12*2.54/100                    [m]
D_wire=2.0/1000                  [m]
```

```
l_long=8.5                    {inches}
l_short=3.5                   {inches}

k_OFHC=400                    [W/m-K]

eta_S=0.99                    {shape efficiency}
eta_E=0.99                    {endpoint efficiency}

N_wire=100                    {Number of wires in braid}

C_Braid=(k_OFHC/l)*N_wire*pi*((D_wire^2)/4)*eta_S*eta_E

T_1=30

Q_dot_1=C_Braid*(12/l_long)*(T_1-T_collar)

T_collar=(Q_dot_1+Q_dot_collar)/(C_Braid*(12/l_short))+T_2

T_collar=30
T_2=5
```

5.8 The following EES code can be used to determine both parts i) and ii)

```
{Problem 5.8 Thermal conductance between an upper and lower
plate via multiple braids}

{Plate dimensions}
th_plate=1.9/100              [m]
side_plate=22.86/100        {length dimension of square plate}
A_plate=side_plate*side_plate  {area of upper and lower
                                plate mounting surfaces}

{Dimensions for a single wire}
l=18/100              [m]      {length of a single wire}
D_wire=2.0/1000       [m]      {dimensions of a single wire}

{Braid endpoint dimensions}
A_contact=(2.7/100)*(2.7/100)  {Area of endpoint mounting surface}

{Thermal conductivities}
k_OFHC=390        [W/m-K]          {OFHC copper conductivity}

k_6061_Al=155.3    [W/m-K]         {6061-aluminum}
```

```
{Braid efficiencies}
eta_S=0.99                              {Shape efficiency}
eta_E=0.95                              {Endpoint efficiency}

{Number of wires and braids}
N_wire=91
N_braids=6

{Conductance calculation for a single braid}
C_Braid=(k_OFHC/l)*N_wire*pi*((D_wire^2)/4)*eta_S*eta_E

{Braid contact conductance calculation}
C_interface=4900*A_contact
R_interface=1/C_interface

{Resistance for a single braid, multiple braids in parallel
and plate-to-plate}
R_Braid=1/C_Braid+2*R_interface {Thermal resistance of single
                                 braid}

R_parallel=R_Braid/N_braids    {Thermal resistance of braids
                               in parallel}

R_total=R_parallel+th_plate/(k_6061_Al*A_plate)

Area_wire=pi*0.25*D_wire^2

{Heat transfer across plates}
Q_dot=165

T_upper=308                                [K]

T_lower=T_upper-R_total*Q_dot
```

i) The lower plate's temperature is 245 K.
ii) In order to determine the number of braids required, the lower plate temperature can be set to 270 K while commenting out the N_{braids} variable. Executing the new program will result in 10 braids being required.

5.9 The following EES code can be used to determine the answers to part i) and part ii). Specific to part ii), the temperature for E-box 1 can be set to 40°C and the number of foils for TS 1 commented out of the code. Upon execution, N_{TS1} will be solved for.

```
{Problem 5.9 Two E-boxes coupled to a temperature sink via
TSs and a thin beam connection}

{Sink temperature}
T_sink=2                       [C]    {Sink temperature}
```

```
{TS1 dimensions}
delta_TS1=5*0.0254/1000    [m]    {Foil thickness of 5 mil}
l_TS1=12.75/100            [m]    {Length of TS1}
W_TS1=1*2.54/100           [m]    {Width of TS foils, 1 inch}
N_TS1=50

{ N_TS1=96}                        {Number of foils for TS1}

{TS 2 & 3 dimensions}
delta_TS2=5*0.0254/1000    [m]    {Foil thickness of 5 mils}
l_TS2=11/100               [m]    {Length of TS2}
W_TS2=1*2.54/100           [m]    {Width of TS2 foils, 1 inch}
N_TS2=50                          {Number of foils for TS2}

{Cross-beam dimensions}
th_beam=5/1000            [m]     {Thin beam thickness}
w_beam=20/100             [m]     {Beam width into page}
l_beam=40/100             [m]     {Beam length}

{ E-box dimensions}
side_1=20/100                  {length of one side of E-box 1}
side_2=20/100                  {length of one side of E-box 2}
th_Ebox=5/1000                 {Ebox thicknesses of 5 mm}

{Thermal conductivities}
k_OFHC=390                          [W/m-K]
k_6061_Al=155.3                     [W/m-K]

{ TS efficiencies}
eta_S_1=0.99                 {Shape efficiency for TS 1}
eta_E_1=0.99                 {Endpoint efficiency for TS 1}

eta_S_2=0.9                  {Shape efficiency for TS 2 & 3}
eta_E_2=0.99                 {Endpoint efficiency for TS 2 & 3}

{Shape factor determination for hollow E-boxes}
S_inf=3.391
ratio=side_1/(side_1-2*th_Ebox)
denom=(1+((pi^0.5)/6)*((ratio^3)-1))^(1/3)-1
S_star=(2*pi^0.5)/denom+S_inf

S=S_star*(side_1-2*th_Ebox)

{TS conductances and resistances}
C_TS1=(k_OFHC/l_TS1)*(N_TS1*delta_TS1*W_TS1)*eta_S_1*eta_E_1
R_TS1=1/C_TS1
```

```
C_TS2=(k_OFHC/l_TS2)*(N_TS2*delta_TS2*W_TS2)*eta_S_2*eta_E_2
R_TS2=1/C_TS2

{TS endpoint conductances}
A_end1=(3.3/100)*(3.3/100)     {Endpoint area for TS1}
C_interface1=12000*A_end1      {OFHC interface conductance at
                                5 MPa (vacuum)}
R_interface1=1/C_interface1

A_end2=(3.3/100)*(3.3/100)     {Endpoint area for TS2}
C_interface2=12000*A_end2      {OFHC interface conductance at
                                5 MPa (vacuum)}
R_interface2=1/C_interface2

{Beam thermal resistances}
R_beam=(0.5*l_beam)/(k_6061_Al*th_beam*w_beam)    {Between
                                the two beam segments}

R_beam_vert=(0.5*th_beam)/(k_6061_Al*l_beam*w_beam)
                             {From centerline to edge of beam}

R_ll_TS2s=0.5*(R_TS2+2*R_interface2)

{E-box resistance}
R_Ebox=1/(S*k_6061_Al)

{Energy dissipations}
g_dot_1=12                     {15 W produced at E-box 1}
g_dot_2=10                     {20 W produced at E-box 2}

{Energy balances}
T_Ebox2=T_Beam2+g_dot_2*(R_beam_vert+R_Ebox)

T_Beam2=T_Beam1+g_dot_2*R_beam

T_Beam1=T_sink+(g_dot_1+g_dot_2)*(R_beam_vert+R_ll_TS2s)

T_Ebox1=T_Beam1+g_dot_1*(R_beam_vert+R_TS1+2*R_interface1&
+R_Ebox)

{T_E-Box1=40}
```

i) The temperatures for E-box 1 and E-box 2 are 51.9°C and 38.2°C respectively.
ii) It will take at least 97 foils to have E-box 1 run at 40°C.

5.10 The following EES code can be used to determine the temperatures of the E-boxes

```
{Problem 5.10 Wishbone thermal strap/E-box configuration}

{Sink temperature}
T_sink=-11                    [C]    {Sink temperature}

{ TS 1 dimensions}
delta_TS1=5.5*0.0254/1000 [m]    {Foil thickness of 5 mil}
l_TS1=7.0*2.54/100            [m]    {Length of TS1}
W_TS1=2*2.54/100              [m]    {Width of TS foils, 2 inch}
N_TS1=50

{ TS 2 dimensions}
delta_TS2=5.5*0.0254/1000 [m]    {Foil thickness of 5 mils}
l_TS2=7.0*2.54/100            [m]    {Length of TS2}
W_TS2=2*2.54/100              [m]    {Width of TS foils, 2 inch}
N_TS2=60                              {Number of foils for TS2}

{ TS 3 wire braid dimensions}
l_wire=6.5*2.54/100          [m]    {Length of wire}
D_wire=0.5/1000             [m]    {Wire diameter}
N_wire=248                          {Number of wires in a braid}
N_Braids=8                          {Number of braids}

{ TS efficiencies}
eta_S_1=0.99                       {Shape efficiency for TS 1}
eta_E_1=0.99                       {Endpoint efficiency for TS 1}

eta_S_2=0.99                       {Shape efficiency for TS 2}
eta_E_2=0.99                       {Endpoint efficiency for TS 2}

eta_S_3=0.92                       {Shape efficiency for TS 3}
eta_E_3=0.95                       {Endpoint efficiency for TS 3}

{Thermal conductivities}
k_OFHC=390                          [W/m-K]
k_GRAPH=800                         [W/m-K]

{TS conductances and resistances}
C_TS1=(k_OFHC/l_TS1)*(N_TS1*delta_TS1*W_TS1)*eta_S_1*eta_E_1
R_TS1=1/C_TS1
```

```
C_TS2=(k_OFHC/l_TS2)*(N_TS2*delta_TS2*W_TS2)*eta_S_2*eta_E_2
R_TS2=1/C_TS2

C_Braid=(k_GRAPH/l_wire)*N_wire*pi*((D_wire^2)/4)*eta_S_3*eta_E_3
R_Braid=1/C_Braid

R_Braid_TS3=R_Braid/N_Braids

{TS endpoint conductances}
A_end1=(5.2/100)*(5.2/100)  {Endpoint area for TS1}
C_interface1=8000*A_end1    {Conductance for OFHC interface
                            at 5 MPa in vacuum}
R_interface1=1/C_interface1

A_end2=(5.2/100)*(5.2/100)  {Endpoint area for TS2}
C_interface2=8000*A_end2    {Conductance for OFHC interface
                            at 5 MPa in vacuum}
R_interface2=1/C_interface2

A_end3=(6.5/100)*(2/100)    {Endpoint area for TS3}
C_interface3=10000*A_end3   {Conductance for OFHC interface
                            at 5 MPa in vacuum}
R_interface3=1/C_interface3

{Energy dissipations}
g_dot_1=10                       {10 W produced at E-box 1}
g_dot_2=20                       {20 W produced at E-box 2}
g_dot_3=14                       {14 W produced at E-box 3}

{Energy balances}
g_dot_1=(T_1-T_2)/(R_TS1+2*R_interface1)

T_2=T_sink+(R_Braid_TS3+2*R_interface3)*(g_dot_1+g_dot_2&
+g_dot_3)

g_dot_3=(T_3-T_2)/(R_TS2+2*R_interface2)
```

5.11 We can perform a heat transfer comparison via the zero-g FOM

$$\frac{\sigma \rho h_{fg}}{\mu}$$

We will use liquid phase values at saturation conditions. After performing the calculations, the comparison plot will be

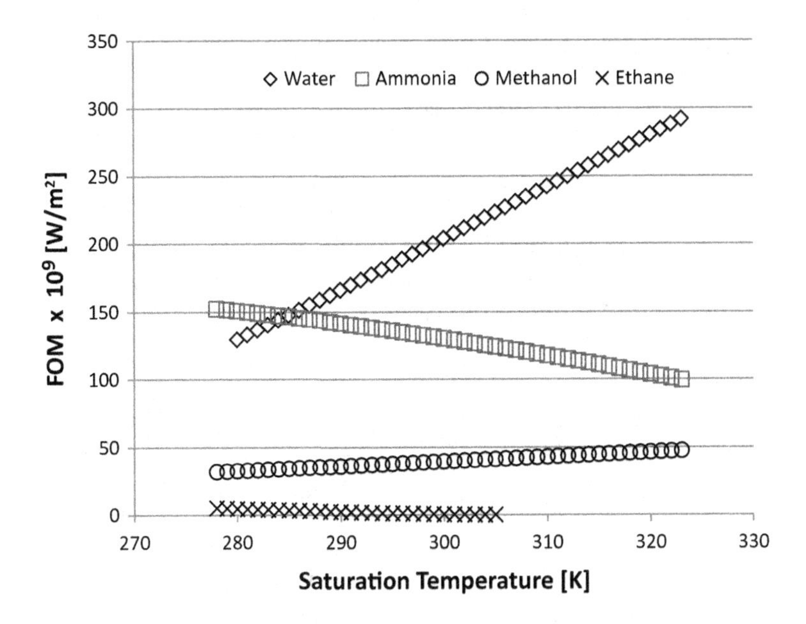

For the temperature range specified in the problem statement ammonia has the best heat transfer performance (in comparison to methanol and ethane).

5.12 EES can be used to determine the Bo numbers for each of these fluids under saturation conditions at 1 atm. The EES code is shown below.

```
{Problem 5.12: Bond number calculation for He, methane,
ammonia and water}

g=9.8   [m/s^2]
l_c=3.0/100 [m]
P_sat=101

{Saturation temperatures}
T_sat_He=t_sat(Helium,P=P_sat)
T_sat_CH4=t_sat(Methane,P=P_sat)
T_sat_NH3=t_sat(Ammonia,P=P_sat)
T_sat_H2O=t_sat(Water,P=P_sat)

{Liquid densities}
rho_l_He=density(Helium,T=T_sat_He,x=0.0)
rho_l_CH4=density(Methane,T=T_sat_CH4,x=0.0)
rho_l_NH3=density(Ammonia,T=T_sat_NH3,x=0.0)
rho_l_H2O=density(Water,T=T_sat_H2O,x=0.0)

{Vapor densities}
rho_v_He=density(Helium,T=T_sat_He,x=1.0)
```

```
rho_v_CH4=density(Methane,T=T_sat_CH4,x=1.0)
rho_v_NH3=density(Ammonia,T=T_sat_NH3,x=1.0)
rho_v_H2O=density(Water,T=T_sat_H2O,x=1.0)

{Surface tensions}
sigma_He=surfacetension(Helium,T=T_sat_He)
sigma_CH4=surfacetension(Methane,T=T_sat_CH4)
sigma_NH3=surfacetension(Ammonia,T=T_sat_NH3)
sigma_H2O=surfacetension(Water,T=T_sat_H2O)

{Bond numbers}
Bo_He=((rho_l_He-rho_v_He)*g*l_c^2)/sigma_He
Bo_CH4=((rho_l_CH4-rho_v_CH4)*g*l_c^2)/sigma_CH4
Bo_NH3=((rho_l_NH3-rho_v_NH3)*g*l_c^2)/sigma_NH3
Bo_H2O=((rho_l_H2O-rho_v_H2O)*g*l_c^2)/sigma_H2O
```

Upon execution of the code, it will result in the following values.

Bo He: 10,749; Bo CH_4: 277.1; Bo NH_3: 176; Bo H_2O: 143.4

As shown in the results, helium has the highest bond number in 1-g even through it has the smallest liquid-to-vapor density ratio of the liquids in question. Alternatively, water has the smallest bond number in 1-g while it has the largest liquid-to-vapor density ratio. Methane has the second highest Bo number in 1-g whereas ammonia is slight lower, yet greater than that for water.

5.13 EES can be used for the solution of this problem. The EES code is shown below.

```
{Problem 5.13, The heat pipe}

phi=0.33
L=0.33

Q_dot=120
T_sink=-20

k_Ni=90.7               [W/m-K]
k_6061=167              [W/m-K]
k_NH3=0.5               [W/m-K]

k_eff=(k_Ni*k_NH3)/((phi*k_Ni+k_NH3*(1-phi)))

r_o_wall=1.27/100
r_thick=0.12446/100
r_i_wall=r_o_wall-r_thick
```

```
r_o_wick=r_i_wall
r_wick_thick=2.5/1000
r_i_wick=r_o_wick-r_wick_thick

R_vapor=9.0e-4
R_interface=0.15

R_wick=(ln(r_o_wick/r_i_wick)/(2*pi*L*k_eff))
R_encasement=(ln(r_o_wall/r_i_wall)/(2*pi*L*k_6061))

R_Total=2*R_interface+2*R_encasement+2*R_wick+R_vapor

T_source=Q_dot*R_Total+T_sink
```

Upon execution of the code, the value for *T_source* is 35.2°C.

5.14 EES can be used to calculate the Entrainment, Boiling and Capillary Limit. The EES code is

```
{Problem 5.14. Heat pipe performance limits}
g=9.8 [ m/s^2]

{Saturation conditions}
T_sat=205 [K]
x_l=0.0
x_v=1.0

{Dimensions}
l=0.6; l_eff=0.5*l_evap+l_adia+0.5*l_cond
l_evap=0.22
l_cond=0.22
l_adia=l-l_evap-l_cond

d_o_wall=2.54/100
r_o_wall=0.5*d_o_wall
r_i_wall=r_o_wall-(1.5/1000)

r_o_wick=r_i_wall
r_i_wick=r_o_wick-(5.1/1000)

d_p=300*1e-6 [m]
r_eff=0.5*d_p

A_vapor=pi*r_i_wick^2

A_wick1=pi*(r_o_wick^2-r_i_wick^2)
```

```
{Thermophysical properties}
h_fg=enthalpy_vaporization(Ethane,T=T_sat)

rho_l=density(Ethane,T=T_sat,x=x_l)
rho_v=density(Ethane,T=T_sat,x=x_v)

mu_l=viscosity(Ethane,T=T_sat,x=x_l)

sigma_l=surfacetension(Ethane,T=T_sat)
k_l=conductivity(Ethane,T=T_sat,x=x_l)
k_wick=conductivity(Stainless_AISI304, T=T_sat)

{Entrainment limit}
phi=0.58
r_h_wick=(A_wick1)/(2*pi*(r_o_wick+r_i_wick)/2)
d_h_wick1=(4*A_wick1)/(2*pi*(r_o_wick+r_i_wick)/2)

Q_dot_ent1=A_vapor*h_fg*((sigma_l*rho_v/(d_p))^0.5)

{Boiling limit}
r_b=3.0e-7 [m]
k_eff=(1-phi)*k_wick+phi*k_l
Q_dot_boil=((4*pi*l_eff*k_eff*sigma_l*T_sat)/(h_fg*rho_v*&
(ln(r_i_wall/r_i_wick))))*(1/r_b-1/r_eff)

{Capillary limits}
angle=2
d_s=r_eff/0.41
d_h_check=2*(3*1e-4)*phi/(3*(1-phi))
K=((d_s^2)*(phi^3))/(150*(1-phi)^2)
Q_dot_cap_0=((rho_l*h_fg*K)/mu_l)*(((A_wick1)/l_eff))*&
((2*sigma_l)/r_eff)

Q_dot_cap_angle=((rho_l*h_fg*K)/mu_l)*(((A_wick1)/l_eff))&
*((2*sigma_l)/r_eff-rho_l*g*l*sin(angle))
```

i) 801.4 W
ii) 160 W
iii) 230.3 W
iv) 89.1 W

5.15 The schematic below is a plot of temperature vs. time for an LHP. There were a series of actions taken regarding the operation of the device. You have been called in to determine what took place with the LHP in the time frame specified on the plots.

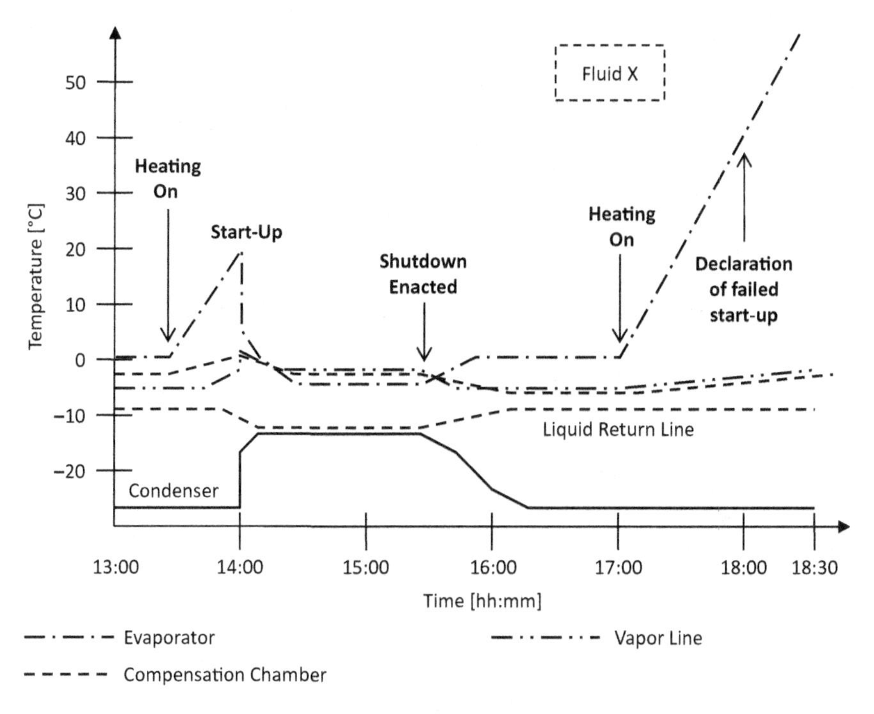

Figure Example LHP Temperature vs. Time Plot

i) Heating is initiated at approximately 13:25 and 17:00.
ii) Start-Up occurs at 14:00.
iii) The attempted start-up at 17:00 has failed. This is undeniably apparent at 18:00 where no signs of start-up have been observed.
iv) Shutdown occurs at approximately 15:25.
v) Remove heating to the evaporator and/or applying a moderate level of heating to the compensation chamber.

5.16 Answers to parts i) and ii) shown below.

i) Based on the LHP plot provided, the evaporator temperature is 4°C whereas the condenser temperature is –1°C. Thus, the effective ΔT across this LHP conductor is 5°C during steady-state operation. Given that the total heat dissipation is 125 W and 4% of the thermal energy is transferred through the E-box mounting surface (i.e., 5 W), the LHP is carrying 120 W towards to the radiator. The thermal resistance of the LHP then becomes

$$R_{LHP} = \frac{\Delta T}{\dot{Q}_{LHP}} = \frac{5\ \text{K}}{120\ \text{W}} = 0.042\ \text{K/W}$$

ii) Since we are neglecting thermal resistance through the LHP evaporator encasement structure, we simply take the thermal conductance from the center of the E-box to its interface with the evaporator

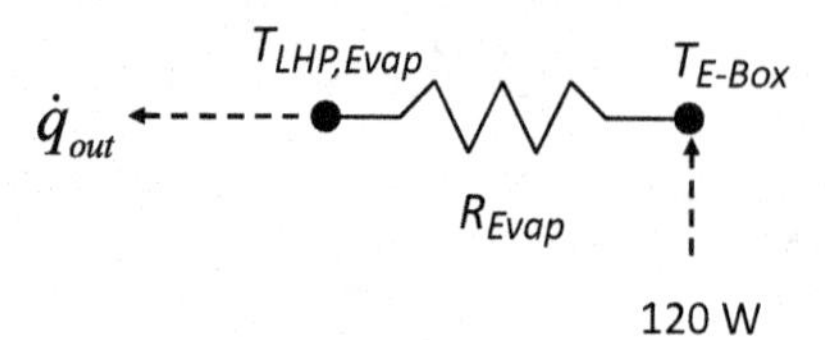

We know that the thermal resistance through the E-box is

$$R_{E\text{-}Box} = \frac{l}{k_{Al}A}$$

where
$l = 15$ cm $= 0.15$ m
$k = 167$ W/m · K
$A = 30$ cm · 30 cm
Thus,

$$R_{E\text{-}Box} = \frac{0.15\text{ m}}{167\frac{\text{W}}{\text{m}\cdot\text{K}}\cdot 0.09\text{ m}^2} = 0.01\text{ K/W}$$

We also know that

$$\dot{Q}_{out} = \frac{T_{E\text{-}Box} - T_{LHP,Evap}}{R_{E\text{-}Box}}$$

$$T_{E\text{-}Box} - T_{LHP,Evap} = \dot{Q}_{out}\cdot R_{E\text{-}Box}$$

$$T_{E\text{-}Box} = \dot{Q}_{out}\cdot R_{E\text{-}Box} + T_{LHP,Evap}$$

$$T_{E\text{-}Box} = 120\text{ W}\cdot 0.01\text{ K/W} + 277\text{ K}$$

$$T_{E\text{-}Box} = 1.2\text{ K} + 277\text{ K} = 278.2\text{ K}$$

5.17 Thermodynamic analysis of the LHP system can be performed using the following EES code.

```
{Problem 5.17 Propylene loop heat pipe model}

l_1=2.54*9.5/100             {primary wick length}
l_2=2.54*9.5/100             {secondary wick length}
l_eff=l_1/2                  {effective length of evaporator}
l_LRL=20                     {primary wick outer radius}
```

```
r_o_1=0.5*(2.54*2.0/100)     {primary wick outer radius}
r_i_1=r_o_1-(10/1000)        {primary wick inner radius}
A_wick_1=pi*(r_o_1^2-r_i_1^2)  {cross-sectional area of primary wick}

r_o_2=r_i_1                  {secondary wick outer radius}
r_i_2=r_o_2-(10/1000)        {secondary wick inner radius}
r_mid_2=(r_o_2+r_i_2)/2      {radial direction mid-point for
                              secondary wick}
A_wick_2=pi*(r_o_2^2-r_i_2^2) {cross-sectional area of secondary
                              wick}

d_p_1=50*1e-6 [m]            {primary wick pore diameter}
d_p_2=800*1e-6 [m]           {secondary wick pore diameter}

r_eff_1=0.5*d_p_1            {primary wick effective pore radius}
r_eff_2=0.5*d_p_2            {secondary wick effective pore radius}

d_s_p=r_eff_1/0.41           {diameter of primary wick sintered
                              spheres}
d_s_s=r_eff_2/0.41           {diameter of secondary wick sintered
                             spheres}

K_p=2.515e-12                {primary wick permeability}
K_s=1.128e-8                 {secondary wick permeability}

K_p=((d_s_p^2)*(phi_p^3))/(150*(1-phi_p)^2)   {permeability of
                                               primary wick}
K_s=((d_s_s^2)*(phi_s^3))/(150*(1-phi_s)^2)  {permeability
                              of secondary wick}

{Wick structure thermal conductivities}
k_s_p=conductivity(Nickel, T=T_3)      {Primary wick}
k_s_s=conductivity(Nickel, T=T_1)      {Secondary wick}

{Wick effective thermal conductivities}
k_eff_p=k_s_p*(((2+k_l_3/k_s_p-2*phi_p*(1-k_l_3/k_s_p)))/&
(2+k_l_3/k_s_p+phi_p*(1-k_l_3/k_s_p)))
k_eff_s=k_s_s*(((2+k_l_1/k_s_s-2*phi_s*(1-k_l_1/k_s_s)))/&
(2+k_l_1/k_s_s+phi_s*(1-k_l_1/k_s_s)))

{Wick resistances}
R_wall_p=(2*pi*l_eff*k_eff_p)/(ln(r_o_1/r_i_1))    {Radial
                               through primary wick}
R_1_2=(2*pi*l_eff*k_eff_s)/(ln(r_o_2/r_mid_2))   {Radial
                              through secondary wick}
R_long_2=l_eff/(k_eff_s*A_wick_2)
```

```
{Saturation conditions}
x_l=0.0
x_v=1.0

{State point 1: CC}
x_1=x_l
k_l_1=conductivity(Propylene,T=T_1,x=x_l)
c_p_1=cp(Propylene,T=T_1,x=x_l)
R_eff=R_wall_p+R_1_2+R_long_2
T_1=T_3-Q_dot_HL*R_eff

{State point 3: evaporator vapor groove}
x_3=x_v
T_3=295
h_3=enthalpy(Propylene,T=T_3,x=x_3)
k_l_3=conductivity(Propylene,T=T_3,x=x_3)
h_fg_3=enthalpy_vaporization(Propylene,T=T_3)

{State point 5: condenser entrance}
x_5=x_v
h_5=enthalpy(Propylene,T=T_5,x=x_5)
h_fg_5=enthalpy_vaporization(Propylene,T=T_5)

m_dot=Q_dot_VL/(h_3-h_5)

{State point 5p: end of condensation in condenser}
x_5p=x_l
T_5p=T_5
h_5p=enthalpy(Propylene,T=T_5p,x=x_5p)
c_p_5p=cp(Propylene,T=T_5,x=x_5p)

{State point 6: subcooler exit}
T_6=270
h_6=h_5p-Q_dot_out_SC/m_dot
c_p_6=cp('C3H6,propylene',T=T_6)

{ State point 7: entrance to CC on liquid return line}
h_7=Q_dot_LRL/m_dot+h_6
T_7=Q_dot_LRL/(m_dot*c_p_6)+T_6

{ Condenser/subcooler heat rejection}
Q_dot_out=203

{Energy balance for evaporator}
Q_dot_in_2ph=m_dot*h_fg_3
Q_dot_HL=Q_dot_in-Q_dot_in_2ph
```

```
{Energy balance for liquid return line}
Q_dot_flux=3.342e-2
A_LRL=2*pi*0.47625*l_LRL
Q_dot_LRL=Q_dot_flux*A_LRL

{Energy balance for CC}
Q_dot_CC=3.0
Q_dot_HL=m_dot*c_p_1*(T_1-T_7)-Q_dot_CC

{Energy balance for vapor line}
Q_dot_VL=0.75

{System-level energy balance}
Q_dot_in=Q_dot_out-Q_dot_LRL-Q_dot_CC+Q_dot_VL

{Energy balance for condenser}
Q_dot_out_2ph=m_dot*h_fg_5

{Energy balance for subcooler}
Q_dot_out_SC=Q_dot_out-Q_dot_out_2ph
```

5.18 The following EES program can be used to determine the cooling load and the COP.

```
{Problem 5.18  TEC cooling load determination}
i=6.8                          {Current in [A]}
N_couple=36                    {Number of couples in TEC module}
T_Cold=25            [C]
T_Hot=50             [C]

T_Cold_K=T_cold+273
T_Hot_K=T_Hot+273

{P-type & N-type element dimensions}
side=3.0                 {length of element side at base in [mm]}
Area=(side/1000)*(side/1000) {cross-sectional area of element in
                               [m^2]}
Height=4.2/1000                {height conversion from mm to m}

DELTAT=T_Hot-T_Cold

{Seebeck coefficients}
GAMMA_p_type=200e-6                              {V/K}
GAMMA_n_type=-215e-6                             {V/K}
GAMMA_pn=GAMMA_p_type-GAMMA_n_type               {V/K}
```

```
{Couple thermal resistance}
k_p_type=0.75                                         {W/m-K}
k_n_type=1.9                                          {W/m-K}

R_p_type=Height/(k_p_type*Area)                       {K/W}
R_n_type=Height/(k_n_type*Area)                       {K/W}

R_couple=1/((1/R_p_type)+(1/R_n_type))                {K/W}

{Couple electrical resistance}
sigma_p_type=(0.7e3)*100        {electrical conductivity in
                                (ohms-m)^-1}
sigma_n_type=(0.9e3)*100        {electrical conductivity in
                                (ohms-m)^-1}

rho_e_p_type=1/sigma_p_type     {electrical resistivity in
                                 ohms-m}
rho_e_n_type=1/sigma_n_type     {electrical resistivity in
                                 ohms-m}

R_e_p_type=rho_e_p_type*Height/Area                   {Ohms}
R_e_n_type=rho_e_n_type*Height/Area                   {Ohms}


R_e_couple=R_e_p_type+R_e_n_type

{Module-level resistances}
R_module=1/(N_couple/R_couple)
R_e_module=N_couple*R_e_couple

{Cooling load}
Q_dot_Cold=N_couple*GAMMA_pn*T_Cold_K*i-(DELTAT)/R_module&
-(i^2)*R_e_module/2

{Power input}
P_in=N_couple*GAMMA_pn*DELTAT*i+(i^2)*R_e_module

{Heat rejection load}
Q_dot_Hot=N_couple*GAMMA_pn*T_Hot_K*i-(DELTAT)/R_module&
+(i^2)*R_e_module/2

{Coefficient of performance}
COP=Q_dot_Cold/P_in
```

The cooling load is 15.3 W and the COP is 0.69.

5.19 The following EES program can be used to determine the cooling load and the COP.

i)

```
{Problem 5.19  TEC couplings, I_max and DeltaT_max determination}
i=7.578                           {Current in [ A]}
{ N_couple=56}                    {Number of couples in TEC module}
T_Cold=20                   [C]
T_Hot=62                    [C]

T_Cold_K=T_cold+273
T_Hot_K=T_Hot+273

{P-type & N-type element dimensions}
side=3.5                          {length of element side at base in
                                  [mm]}
Area=(side/1000)*(side/1000)      {cross-sectional area of element in
                                  [m^2]}
Height=5.0/1000

DELTAT=T_Hot-T_Cold

{Seebeck coefficients}
GAMMA_p_type=212e-6                                    {V/K}
GAMMA_n_type=-212e-6                                   {V/K}
GAMMA_pn=GAMMA_p_type-GAMMA_n_type                     {V/K}

{Couple thermal resistance}
k_p_type=1.0                                           {W/m-K}
k_n_type=1.0                                           {W/m-K}

R_p_type=Height/(k_p_type*Area)                        {K/W}
R_n_type=Height/(k_n_type*Area)                        {K/W}

R_couple=1/((1/R_p_type)+(1/R_n_type))                 {K/W}

{Couple electrical resistance}
sigma_p_type=(0.650e3)*100        {electrical conductivity in
                                  (ohms-m)^-1}
sigma_n_type=(0.650e3)*100        {electrical conductivity in
                                  (ohms-m)^-1}

rho_e_p_type=1/sigma_p_type       {electrical resistivity in
                                  ohms-m}
rho_e_n_type=1/sigma_n_type       {electrical resistivity in
                                  ohms-m}
```

```
R_e_p_type=rho_e_p_type*Height/Area          {Ohms}
R_e_n_type=rho_e_n_type*Height/Area          {Ohms}

R_e_couple=R_e_p_type+R_e_n_type

{Module-level resistances}
R_module=1/(N_couple/R_couple)
R_e_module=N_couple*R_e_couple

{Cooling load}
Q_dot_Cold=21
Q_dot_Cold=N_couple*GAMMA_pn*T_Cold_K*i-(DELTAT)/R_module&
-(i^2)*R_e_module/2

{Power input}
P_in=N_couple*GAMMA_pn*DELTAT*i+(i^2)*R_e_module

{Heat rejection load}
Q_dot_Hot=N_couple*GAMMA_pn*T_Hot_K*i-(DELTAT)/R_module&
+(i^2)*R_e_module/2

{Coefficient of performance}
COP=Q_dot_Cold/P_in

{Maximum}
i_max=N_couple*(GAMMA_pn*T_Cold_K)/(R_e_module)

DELTAT_max=0.5*((N_couple^2)*(GAMMA_pn^2)*(T_Cold_K^2*&
R_module))/R_e_module
```

The number of couples required is 56.

ii) First we must determine a relation for the maximum current. Referring to equation 5.60, the cooling load is a function of the current. As such, using Max/Min theory we can take the first derivative of the cooling load relation with respect to the current and set it to zero. This will be

$$\frac{d\dot{Q}_{C,\text{mod}}}{di} = N_{couple} \cdot \Gamma_{12} T_C - i_{max} R_{e,mod} = 0$$

Then solving for the current variable i_{max} we would have

$$i_{max} = \frac{N_{couple} \cdot \Gamma_{12} T_C}{R_{e,mod}}$$

Placing into the EES code and solving will result in a i_{max} value of 9.892 A.

iii) Using the hint provided in the problem statement, we shall return to equation 5.60, substitute in i_{max} for the current and set the total cooling load to zero

$$\dot{Q}_{C,\mathrm{mod}} = N_{couple} \cdot \Gamma_{12} T_C i_{max} - \frac{(T_H - T_C)_{max}}{R_{mod}} - \frac{1}{2} \cdot i^2_{max} R_{e,mod} = 0$$

$$N_{couple} \cdot \Gamma_{12} T_C \cdot \left[\frac{N_{couple} \cdot \Gamma_{12} T_C}{R_{e,mod}}\right] - \frac{\Delta T_{max}}{R_{mod}} - \frac{1}{2} \cdot \left[\frac{N_{couple} \cdot \Gamma_{12} T_C}{R_{e,mod}}\right]^2 \cdot R_{e,mod} = 0$$

$$\frac{N^2_{couple} \Gamma^2_{12} T^2_C}{R_{e,mod}} - \frac{\Delta T_{\max}}{R_{mod}} - \frac{1}{2} \cdot \frac{N^2_{couple} \Gamma^2_{12} T^2_C}{R_{e,mod}} = 0$$

Solving for ΔT_{max} would give

$$\Delta T_{max} = \frac{1}{2} \cdot \frac{N^2_{couple} \Gamma^2_{12} T^2_C R_{mod}}{R_{e,mod}}$$

Placing this into the EES code and solving will result in a ΔT_{max} value of 125.4 W.

5.20 A general program can be created in EES to produce the tables and associated data plots referenced in parts i), ii) and iii). In its general form the program will not execute using the solve command. There will be at least one variable remaining as undefined. This variable must be populated in the Tables data. Execution of the program can then be performed from the newly created table.

```
{Problem 5.20  TEC couplings COP performance and dependent
parameters}
N_couple=49

i=5.0                                      {Current in [A]}

T_Hot_K=T_Cold_K+DELTAT
DELTAT=30                         [K]

T_Cold_K=300                      [K]

T_avg=0.5*(T_Cold_K+T_Hot_K)

{P-type & N-type element dimensions}
ratio=(2.0)*0.001                 {A/l ratio in units of mm}

{Seebeck coefficients}
GAMMA_p_type=200*(1e-6)                              {V/K}
GAMMA_n_type=-200*(1e-6)                             {V/K}

GAMMA_pn=GAMMA_p_type-GAMMA_n_type                   {V/K}

{Couple thermal resistance}
k_p_type=(1.51e-2)*100                               {W/m-K}
k_n_type=(1.51e-2)*100                               {W/m-K}
```

```
R_p_type=1/(k_p_type*ratio)                              {K/W}
R_n_type=1/(k_n_type*ratio)                              {K/W}

R_couple=1/((1/R_p_type)+(1/R_n_type))                   {K/W}

{Couple electrical resistance}
sigma_p_type=1/rho_e_p_type     {electrical conductivity in
                                 (ohms-m)^-1}
sigma_n_type=1/rho_e_n_type     {electrical conductivity in
                                 (ohms-m)^-1}

rho_e_p_type=(1.01e-3)/100      {electrical resistivity in
                                 ohms-m}
rho_e_n_type=(1.01e-3)/100      {electrical resistivity in
                                 ohms-m}

R_e_p_type=rho_e_p_type/ratio  {Ohms}
R_e_n_type=rho_e_n_type/ratio  {Ohms}

R_e_couple=R_e_p_type+R_e_n_type

{Module-level resistances}
R_module=1/(N_couple/R_couple)
R_e_module=N_couple*R_e_couple

{Cooling load and power input}
Q_dot_Cold=N_couple*GAMMA_pn*T_Cold_K*i-(DELTAT)/R_module&
-(i^2)*R_e_module/2

Q_dot_Hot=N_couple*GAMMA_pn*T_Hot_K*i-(DELTAT)/R_module&
+(i^2)*R_e_module/2

P_in=N_couple*GAMMA_pn*DELTAT*i+(i^2)*R_e_module

{Coefficient of performance}
COP=Q_dot_Cold/P_in
```

i) Since the current is being varied, we will comment that line out as well as assign values to the temperature difference between the hot and cold side, as well as the temperature on the cold side. The following lines reflect these changes

```
{ i=5.0                                  {Current in [A]}}

DELTAT=30                                [K]

T_Cold_K=300                             [K]
```

The associated plot is shown below.

The current maximum location for each case is denoted by a peak in the curve, as well as a maximum value for the cooling load under the designated operating conditions. Based on inspection of the plot, the 0.5 mm case contains its maximum current at a value below 6 A. The 1.0 mm case has a current Max just under 6 A whereas the 1.5 mm and 2.0 mm curves are continuing to experience increases in cooling load at 6 A. The maximum cooling load will occur at a value greater than that shown on the plot for the 1.5 mm and 2.0 mm cases.

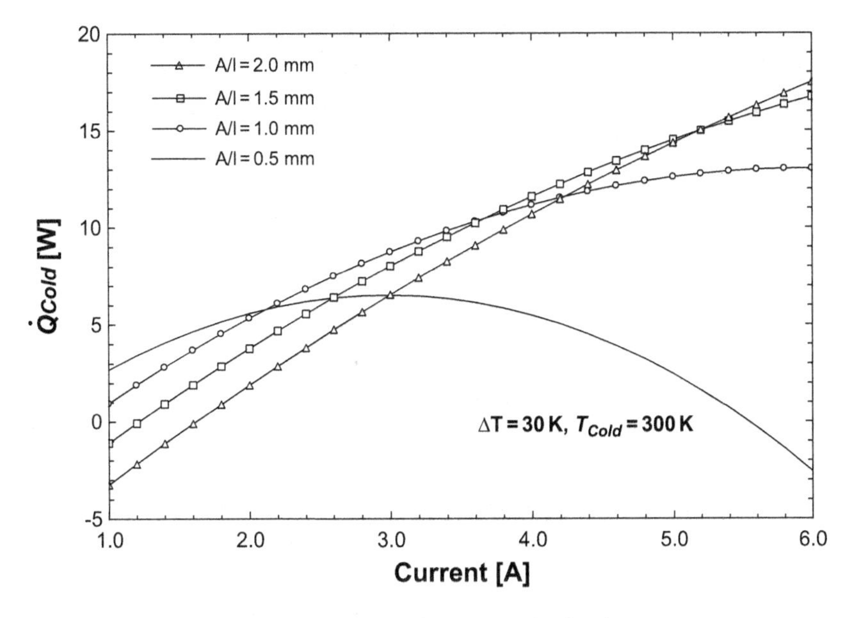

Figure S5.21a Cooling load as a function of current and *A/l* ratio

ii) Since the temperature difference is being varied, we will comment that line out and assign values to the current and the Cold side temperature. The following lines reflect these changes

```
i=5.0                                  {Current in [A]}
{DELTAT=30                             [K]}
T_Cold_K=300                           [K]
```

The plot for this case is shown below.

The 0.5 mm case has really low cooling load values because the current used for the calculation is noticeably greater than the maximum current for this case's geometry and conditions. Alternatively, the cooling load for the other cases (even though the 1.5 mm and 2.0 mm cases have not reached their maximum current values) is noticeably greater. In addition, the cooling load gradually decreases as the temperature difference increases.

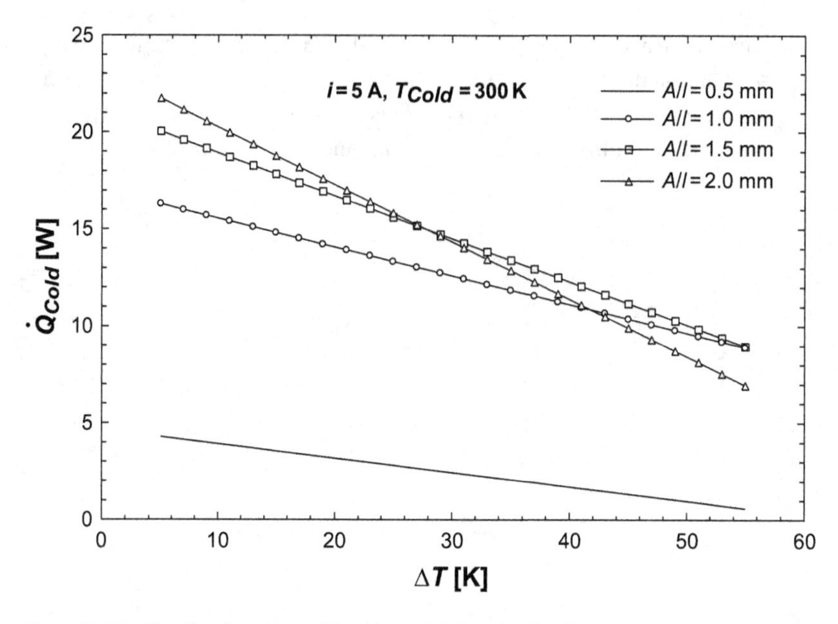

Figure S5.21b Cooling Load as a Function of ΔT and A/l ratio

iii) Since the cold-side temperature is being varied, we will comment that line out and assign values to the current and ΔT. The following lines reflect these changes

```
i=5.0                                   {Current in [A]}

DELTAT=30                               [K]

{T_Cold_K=300                           [K]}
```

The corresponding plot is shown below.

The COP tends to increase with A/l ratio and shows a small gradual increase with increasing T_cold.

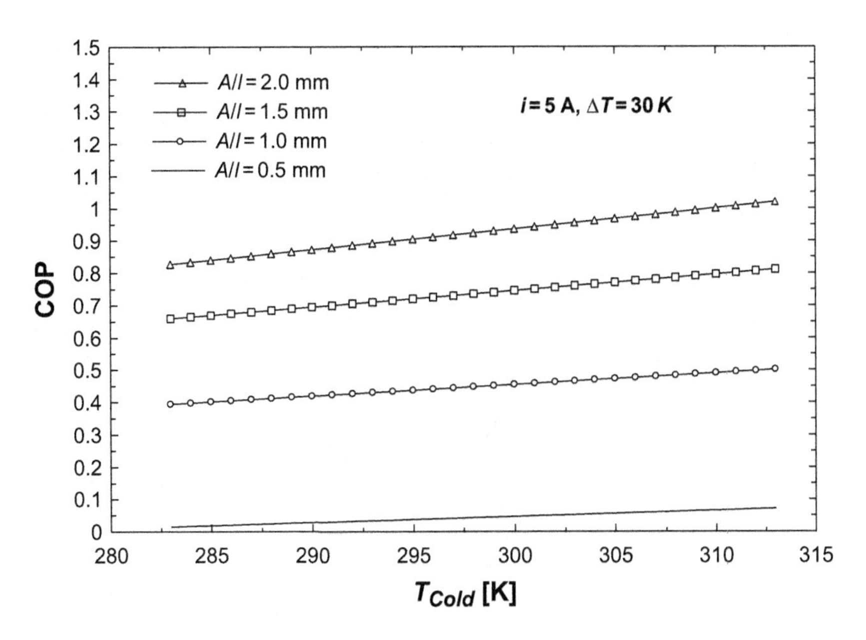

Figure S5.21c COP as a Function of T_{Cold} and A/l ratio

5.21 EES can be used to determine the current load required in both parts i) and ii)
i)

```
{Problem 5.21 TEC cooling load determination when coupled to
a remote temperature sink: part i)}
{FCL specifications}
N_FCL=47                                    {# of foil layers}
k_OFHC=398                        [W/m-K]
eta_S=0.99                                  {Shape efficiency}
eta_E=0.99                                  {Endpoint efficiency}
W_foil=1.8*0.0254               [m]         {Foil width}
delta_foil=8.5*0.001*0.0254                 {Foil thickness in meters}
l_foil=6.5*2.54/100             [m]         {Length of thermal strap}
C_foil=(k_OFHC/l_foil)*N_FCL*delta_foil*W_foil*eta_S*eta_E
R_foil=1/C_foil

{Interface resistances}
R_Hot_Strap=0.25                [K/W]
R_Strap_Sink=0.1                [K/W]
R_Total=R_Hot_Strap+R_foil+R_Strap_Sink

{TEC power and plate temperature specifications}
N_couple=49                       {Number of couples in TEC module}
T_Cold=22                     [C]
```

```
T_Cold_K=T_cold+273                      {TEC cold-side temperature}
T_Hot_K=Q_dot_Hot*R_Total+T_sink     {TEC hot-side temperature}
T_sink=3+273              [K]          { Temperature at the sink}

{P-type & N-type element dimensions}
side=3.5                              {length of element side at
                                       base in [ mm] }
Area=(side/1000)*(side/1000)         {cross-sectional area of element
                                      in [m^2] }
Height=4.6/1000

DELTAT=T_Hot_K-T_Cold_K

{Seebeck coefficients}
GAMMA_p_type=210e-6                                      {V/K}
GAMMA_n_type=-210e-6                                     {V/K}
GAMMA_pn=GAMMA_p_type-GAMMA_n_type                       {V/K}

{Couple Thermal Resistance}
k_p_type=0.8                                             {W/m-K}
k_n_type=0.8                                             {W/m-K}

R_p_type=Height/(k_p_type*Area)                          {K/W}
R_n_type=Height/(k_n_type*Area)                          {K/W}

R_couple=1/((1/R_p_type)+(1/R_n_type))                   {K/W}

{Couple electrical resistance}
sigma_p_type=(0.62e3)*100            {electrical conductivity
                                      in (ohms-m)^-1}
sigma_n_type=(0.62e3)*100            {electrical conductivity
                                      in (ohms-m)^-1}

rho_e_p_type=1/sigma_p_type         {electrical resistivity
                                      in ohms-m}
rho_e_n_type=1/sigma_n_type         {electrical resistivity
                                     in ohms-m}

R_e_p_type=rho_e_p_type*Height/Area                        {Ohms}
R_e_n_type=rho_e_n_type*Height/Area                        {Ohms}

R_e_couple=R_e_p_type+R_e_n_type

{Module-level resistances}
R_module=1/(N_couple/R_couple)
R_e_module=N_couple*R_e_couple
```

```
{Cooling load}
Q_dot_Cold=18
Q_dot_Cold=N_couple*GAMMA_pn*T_Cold_K*i-(DELTAT)/R_module&
-(i^2)*R_e_module/2

{Power input}
P_in=N_couple*GAMMA_pn*DELTAT*i+(i^2)*R_e_module

{Heat rejection load}
i=(Q_dot_Hot+(DELTAT)/R_module-(i^2)*R_e_module/2)/&
(N_couple*GAMMA_pn*T_Hot_K)
```

The required current is 5.523 A.

ii) Returning to the EES code, it can be modified as follows to determine the cooling load

```
{Problem 5-21  TEC cooling load determination when coupled
to a remote temperature sink: part ii)}
sigma=5.67e-8                        {Boltzmann's constant}

{Surface areas}
A_plate= (18*18)*(1.0e-4)           {TEC top plate surface area
                                     in m^2}
A_Encl=((19*19)+3*(3*19)+(2.6*19))*(1.0e-4)
                      {Dog-house interior surface area in m^2}

{Surface emissivities}
eps_TEC=0.92                        {ZOT w/Potassium Silicate}
eps_Encl=0.82                       {Black Anodize/Sample 1}

{View Factors}
F_TEC_Encl=1.0

{Radiation heat transfer to dog-house enclosure}
Term1=(1-eps_TEC)/(eps_TEC*A_plate)
Term2=1/(F_TEC_Encl*A_Encl)
Term3=(1-eps_Encl)/(eps_Encl*A_Encl)

T_Hot_K^4=(Q_dot_Hot*(Term1+Term2+Term3)/sigma)+T_sink^4

{TEC power and plate temperature specifications}
N_couple=49                    {Number of couples in TEC module}
T_Cold=22                           [C]
T_Hot=T_Hot_K-273
```

```
T_Cold_K=T_Cold+273                  {TEC cold-side temperature}
{T_Hot_K=356                         {TEC hot-side temperature}}

T_sink=3+273               [K]      {Temperature at the sink}

{P-type & N-type element dimensions}
side=3.5                      {length of element side at base
                              in [mm]}
Area=(side/1000)*(side/1000)  {cross-sectional area of element
                              in [m^2]}
Height=4.6/1000

DELTAT=T_Hot_K-T_Cold_K

{Seebeck Coefficients}
GAMMA_p_type=210e-6                     {V/K}
GAMMA_n_type=-210e-6                    {V/K}
GAMMA_pn=GAMMA_p_type-GAMMA_n_type      {V/K}

{Couple thermal resistance}
k_p_type=0.8                            {W/m-K}
k_n_type=0.8                            {W/m-K}

R_p_type=Height/(k_p_type*Area)         {K/W}
R_n_type=Height/(k_n_type*Area)         {K/W}

R_couple=1/((1/R_p_type)+(1/R_n_type)){K/W}

{Couple Electrical Resistance}
sigma_p_type=(0.62e3)*100       {electrical conductivity  in
                                (ohms-m)^-1}
sigma_n_type=(0.62e3)*100       {electrical conductivity in
                                (ohms-m)^-1}

rho_e_p_type=1/sigma_p_type      {electrical resistivity in
                                 ohms-m}
rho_e_n_type=1/sigma_n_type      {electrical resistivity in
                                 ohms-m}

R_e_p_type=rho_e_p_type*Height/Area                {Ohms}
R_e_n_type=rho_e_n_type*Height/Area                {Ohms}

R_e_couple=R_e_p_type+R_e_n_type
```

```
{Module-level resistances}
R_module=1/(N_couple/R_couple)
R_e_module=N_couple*R_e_couple

{Cooling load}
i=4.0
Q_dot_Cold=N_couple*GAMMA_pn*T_Cold_K*i-(DELTAT)/R_module&
-(i^2)*R_e_module/2

{Power input}
P_in=N_couple*GAMMA_pn*DELTAT*i+(i^2)*R_e_module

{Heat rejection Load}
Q_dot_Hot=N_couple*GAMMA_pn*T_Hot_K*i-(DELTAT)/R_module&
+(i^2)*R_e_module/2
```

The cooling load has now reduced to 8.045 W.

5.22 EES can be used to determine the power output and the resistance load using the following code.

```
{Problem 5.22  TEG panel arrangement}
{Surface area of plates}
A_surf=1.0                    [m^2]

{Hot plate temperature calculation}
T_Hot_K=420                                  [K]
T_Cold_K=263
DELTAT=T_Hot_K-T_Cold_K

{TEC Plate Temperature Specifications}
N_couple=36               {Number of couples in TEC module}
N_mod=16                  {Number of TEC modules}

{P-type & N-type element dimensions}
side=14.0                       {length of element side at
                                base in [mm]}
Area=(side/1000)*(side/1000)    {cross-sectional  area  of
                                element in [m^2]}
Height=8.6/1000

{Seebeck coefficients}
GAMMA_p_type=214e-6                          {V/K}
GAMMA_n_type=-214e-6                         {V/K}
GAMMA_pn=GAMMA_p_type-GAMMA_n_type           {V/K}
```

```
{Couple thermal resistance}
k_p_type=1.48                                         {W/m-K}
k_n_type=1.48                                         {W/m-K}

R_p_type=Height/(k_p_type*Area)                       {K/W}
R_n_type=Height/(k_n_type*Area)                       {K/W}

R_couple=1/((1/R_p_type)+(1/R_n_type))                {K/W}

{Couple electrical resistance}
sigma_p_type=1/rho_e_p_type            {electrical conductivity
                                       in (ohms-m)^-1}
sigma_n_type=1/rho_e_n_type            {electrical conductivity
                                       in (ohms-m)^-1}

rho_e_p_type=(4.5e-3)/100               {electrical resistivity
                                        in ohms-m}
rho_e_n_type=(4.5e-3)/100               {electrical resistivity
                                        in ohms-m}

R_e_p_type=rho_e_p_type*Height/Area                        {Ohms}
R_e_n_type=rho_e_n_type*Height/Area                        {Ohms}

R_e_couple=R_e_p_type+R_e_n_type                           {Ohms}

{Module-level resistances}
R_module=1/(N_couple/R_couple)
R_e_module=N_couple*R_e_couple

i=5.4

{Cooling load}
Q_dot_Cold_panel=N_mod*(N_couple*GAMMA_pn*T_Cold_K*i+&
(DELTAT)/R_module+(i^2)*R_e_module/2)

{Power Input}
P_out_panel=Q_dot_Hot_panel-Q_dot_Cold_panel
R_load=P_out_panel/i^2

{Heat rejection load}
Q_dot_Hot_panel=N_mod*(N_couple*GAMMA_pn*T_Hot_K*i+&
(DELTAT)/R_module-(i^2)*R_e_module/2)
```

The power output from the panel is 142.7 W with a resistive load on the circuit of 4.893 Ω.

5.23 EES can be used to determine the power output for part i).

```
{Problem 5-23  TEG panel arrangement in 1 AU orbit}
{Constants}
Solar=1361          [W/m^2]
sigma=5.67e-8       [W/m^2-K^4]

{Coatings}
alpha_S_1=0.9       {Maxorb UV absorptivity}
eps_1=0.1           {Maxorb IR emissivity}

eps_2=0.94          {Martin Black Velvet paint IR emissivity}

{Surface area of plates}
A_surf=1.0          [m^2]

{Environmental loading}
Q_dot_Solar=alpha_S_1*Solar*A_surf

{Hot plate temperature calculation}
DELTAT=T_Hot_K-T_Cold_K

{TEC plate temperature specifications}
N_couple=36         {Number of couples in TEC module}
N_mod=16            {Number of TEC modules}

{P-type & N-type element dimensions}
side=14.0                      {length of element side at base
                               in [ mm] }
Area=(side/1000)*(side/1000)   {cross-sectional area of element in
                               [m^2] }
Height=8.6/1000

{Seebeck coefficients}
GAMMA_p_type=214e-6                      {V/K}
GAMMA_n_type=-214e-6                     {V/K}
GAMMA_pn=GAMMA_p_type-GAMMA_n_type       {V/K}

{Couple Thermal Resistance}
k_p_type=1.48                            {W/m-K}
k_n_type=1.48                            {W/m-K}

R_p_type=Height/(k_p_type*Area)          {K/W}
R_n_type=Height/(k_n_type*Area)          {K/W}
```

```
R_couple=1/((1/R_p_type)+(1/R_n_type))              {K/W}

{Couple electrical resistance}
sigma_p_type=1/rho_e_p_type         {electrical conductivity
                                     in (ohms-m)^-1}
sigma_n_type=1/rho_e_n_type         {electrical conductivity
                                     in (ohms-m)^-1}

rho_e_p_type=(4.5e-3)/100            {electrical resistivity
                                      in ohms-m}
rho_e_n_type=(4.5e-3)/100            {electrical resistivity
                                      in ohms-m}

R_e_p_type=rho_e_p_type*Height/Area          {Ohms}
R_e_n_type=rho_e_n_type*Height/Area          {Ohms}

R_e_couple=R_e_p_type+R_e_n_type

{Module-level resistances}
R_module=1/(N_couple/R_couple)
R_e_module=N_couple*R_e_couple

i=1.2

{Cooling rejection load}

T_Cold_K=(eps_2*sigma*A_surf*T_Cold_K^4-N_mod*((DELTAT/&
R_module)+(i^2)*R_e_module/2))/&
(N_mod*N_couple*GAMMA_pn*i)

Q_dot_Cold_panel=N_mod*(N_couple*GAMMA_pn*T_Cold_K*i+&
(DELTAT)/R_module+(i^2)*R_e_module/2)

Q_dot_Cold_reject=eps_2*sigma*A_surf*T_Cold_K^4

{Power input}
P_out_panel=Q_dot_Hot_panel-Q_dot_Cold_panel

{Heat absorption load}
T_Hot_K=(Q_dot_Solar-eps_1*sigma*A_surf*T_Hot_K^4-N_mod*&
((DELTAT/R_module)-(i^2)*R_e_module/2))/&
(N_mod*N_couple*GAMMA_pn*i)

Q_dot_Hot_panel=N_mod*(N_couple*GAMMA_pn*T_Hot_K*i+&
(DELTAT)/R_module-(i^2)*R_e_module/2)

Q_dot_Hot_reject=eps_1*sigma*A_surf*T_Hot_K^4
```

The power output using environmental loading is 4.04 W.

ii) As shown in part i), the equilibrium temperatures on the plates resulting from environmental loads, heat rejection and thermoelectric effects fosters a small temperature difference. This results in a very low power output. Alternate thermo-electric materials could be opted for. However, improvements would not be significant given the small temperature difference occurring. In order to improve power output a larger temperature difference should be fostered across the plates. This can be performed through either a high temperature source or increasing the heat rejection surface area on plate 2.

5.24 i) Referencing Appendix B, one can determine the radiator surfaces' coating properties as being

$\alpha = 0.95$; $\varepsilon = 0.87$

Using the equation for radiator surface area (i.e., equation 5.85), the surface area would be

$$A_{surf} = \frac{\dot{g} \cdot \Delta t_{pulse}}{\sigma \varepsilon T_{melt}^4 \Delta t_{cycle}}$$

Substituting values in, the required surface area at the peak of pulsing can be calculated as

$$A_{surf} = \frac{300 \text{ W}}{(5.67e-8 \text{ W/m}^2 \cdot \text{K}^4)(0.87)\left((263 \text{ K})^4\right)} \cdot \left(\frac{45 \text{ min}}{90 \text{ min}}\right)$$

$$A_{surf} = 0.635 \text{ m}^2$$

ii) Referencing Table 5.4, n-Dodecane ($C_{12}H_{26}$) is the closest match for the temperature specified.

iii) Returning to equation 5.85, the net heat transfer on the radiator surface now becomes 350 W. Placing this thermal load in the equation would now give

$$A_{surf} = \frac{350 \text{ W}}{(5.67e-8 \text{ W/m}^2 \cdot \text{K}^4)(0.87)\left((263 \text{ K})^4\right)} \cdot \left(\frac{45 \text{ min}}{90 \text{ min}}\right)$$

$$A_{surf} = 0.742 \text{ m}^2$$

5.25 i) The solution to part i) can be determined using equation 5.88. The enthalpy of fusion for n-Eicosane can be referenced in Table 5.4.

$E_{max} = (1.5 \text{ kg})(246{,}000 \text{ J/kg})$

$E_{max} = 369{,}000$ Joules

ii) Since the radiator is nadir facing, we will have to take into account Earth-IR loading. This will be constant. However, because operation is occurring in eclipse, albedo can be neglected. The emissivity for Z-93 white paint can be referenced in Appendix B as $\varepsilon = 0.92$. The EES code for the solution is

```
{Example 5.25: PCM Melt/solidification cycle using n-eico-
sane}
Function heat(A,B,C,D,E)

   If (A<90) Then;  Y:=B+C*D*E; Endif
   If (A=90) or (A>90) Then;  Y:=C*D*E; Endif

   If (A>300) and (A<421) Then; Y:=B+C*D*E; Endif
   If (A=421) or (A>421) Then; Y:=C*D*E; Endif

   If (A>600) and (A<781) Then; Y:=B+C*D*E; Endif
   If (A=781) or (A>781) Then; Y:=C*D*E; Endif

   If (A>900) and (A<1051) Then; Y:=B+C*D*E; Endif
   If (A=1051) or (A>1051) Then; Y:=C*D*E; Endif

  heat:=Y
End

{Constants}
sigma=5.67e-8
h_sf_eicosane=246*1000                 {Enthalpy of fusion in
                                        J/kg, n-Tridecane}
T_melt=310                 [K]          {Melt temperature for
                                        n-tridecane}

{Pulse heat load}
Q_dot_pulse=500            [W]          {Pulse heat load}
Q_dot_Earth_IR=193                      {Earth-IR in W/m^2}

{Radiator features}
eps_rad=0.92                            {Z93 white paint}
A_rad=0.475                             {m^2}

{PCM features}
A_PCM=A_rad

rho_l=756                  [kg/m^3]     {density of liquid PCM}
rho_s=847                  [kg/m^3]     {density of solid PCM}

k_l=0.150                  [W/m-K]      {thermal conductivity of
                                        solid PCM}
k_s=0.150                  [W/m-K]      {thermal conductivity of
                                        solid PCM}
```

```
{Mass and maximum energy determination}
Q_dot_rej=eps_rad*A_rad*sigma*T_melt^4 {heat rejection from
                                        radiator at full
                                        melt}

mass=1.5                    {PCM mass in kg}

L=mass/(rho_l*A_PCM)        {Length required for PCM container}

E_max=mass*h_sf_eicosane

{Initial array settings}
g_dot_in[1]=Q_dot_pulse      {Thermal dissipation from E-box}

Time[1]=0.0                  {Initial time in seconds}

m_s[1]=L*A_PCM*rho_s         {Initial mass of solid PCM}

{m_l[1]=0.0                  {Initial mass of liquid PCM}}

E_sum[1]=DELTAt*Q_dot_out[1] {Energy accumulated in PCM}

z[1]=0.0                     {Initial location of melt line}

T_rad[1]=T_melt-((L-z[1])/(k_s*A_PCM))*eps_rad*sigma*&
A_rad*T_rad[1]^4              {Initial radiator temperature}

Q_dot_out[1]=eps_rad*A_rad*sigma*T_rad[1]^4  { Heat rejection from
                                              radiator}

{T_int[1]=T_melt             {Initial wall temperature}}

lapse[1]=0.0                 {Initial elapsed time in minutes}

x_MSR[1]=0.0             {Initial melt-to-solidification ratio}

{Al-6061 thermophysical properties}
c_p_plate=cp(Aluminum_6061, T=T_rad[1])     {Al-6061 specific
                                             heat}
rho_plate=density(Aluminum_6061, T=T_rad[1]) {Al-6061 density}
```

```
{Radiator plate mass determination}
th_plate=(5/1000)          { thickness of radiator plate in mm}
Vol_rad=th_plate*A_rad  { volume of radiator plate}
m_rad=rho_plate*Vol_rad { mass of radiator plate}

{Iterative loop}
DELTAt=2.0                          {Seconds}

ITER=4*10*30                        {Maximum iteration count}

Duplicate K=1,ITER

 T_rad[K+1]=T_rad[K]+(DELTAt/(m_rad*c_p_plate))*&
 ((k_s*A_PCM*(T_melt-T_rad[K+1]))/(L-z[K+1])-eps_rad*&
  sigma*A_rad*T_rad[K+1]^4)

 Q_dot_out[K+1]=eps_rad*A_rad*sigma*T_rad[K+1]^4

 g_dot_in[K+1]=heat(K,Q_dot_pulse,Q_dot_Earth_IR,eps_rad,&
 A_rad)        { E-Box dissipation}

 m_s[K+1]=m_s[K]-((g_dot_in[K+1]-Q_dot_out[K+1])*DELTAt)/&
 h_sf_eicosane

 Time[K+1]=K*DELTAt                 {Time in seconds}

 lapse[K+1]=K*DELTAt/60            {Total elapsed time in minutes}

 z[K+1]=L-m_s[K+1]/(A_PCM*rho_s)     {Update for melt-line}

 E_sum[K+1]=E_sum[K]+(g_dot_in[K+1]-Q_dot_out[K+1])*DELTAt

 x_MSR[K+1]=E_sum[K+1]/E_max    {melt-to-solidification ratio}

End
```

In the code, conditional statements to turn the E-box heat dissipation on and off are located in the function entitled "Heat." The plot for x_{MSR} and T_{rad} is shown below.

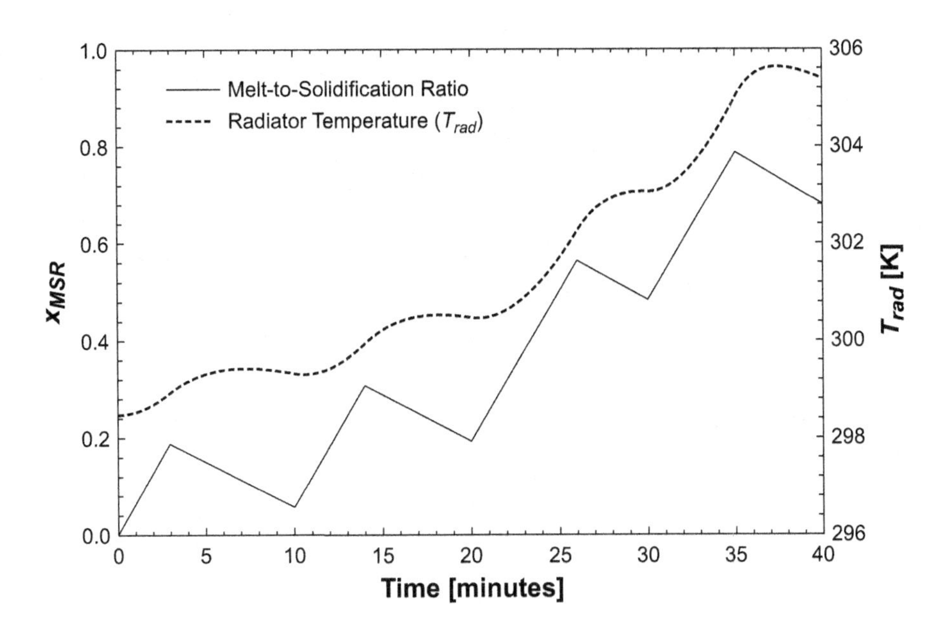

As shown in the plot, the maximum temperature observed on the radiator is just under 306 K. Thus, the melt temperature is not observed on the radiator. This is consistent with the melt-to-solidification ratio not reaching a value of 1.0 during the four cycles. In addition, the radiator temperature gradually increases from cycle to cycle due to the inability to completely reject the stored thermal energy acquired during each cycle. Two workarounds to achieve full reconditioning of the PCM would be to increase the radiator surface area or increase the cycle time and the time which the E-box is in the "off" state.

5.26 A computer code can be developed in EES to perform a time march starting at initial conditions. Bookkeeping of the melt-to-solidification ratio must be performed during program execution as well. The following EES code applies.

i)

```
{Example 5.26: PCM Melt/solidification cycle using n-Tetradecane}
Function heat(A,B)

    If (A<114) Then;  Y:=B; Endif
    If (A=114) or (A>114) Then;  Y:=0; Endif

    If (A>600) and (A<707)  Then;  Y:=B; Endif
    If (A=707) or (A>707) Then;  Y:=0; Endif

    heat:=Y
End

{Constants}
sigma=5.67e-8
```

```
h_sf_tetradecane=228*1000              {Enthalpy of fusion in
                                        J/kg, n-Tetradecane}
T_melt=279                [K]           {Melt temperature for
                                         n-Tetradecane}

{Pulse heat load}
Q_dot_pulse=250.2         [W]          {Pulse heat load}
Q_dot_Earth_IR=0.0                      {Earth-IR in W/m^2, not
                                        present in problem}

{Radiator features}
eps_rad=0.88                         {Chemglaze A276 White Paint}
A_rad=0.4*0.4                        {m^2}

{Area determination}
xi_source=Pulse/Cycle

{Time durations}
Pulse=696                            {seconds}
Pulse_min=Pulse/60

Cycle_min=60                         {seconds}
Cycle=60*Cycle_min

{PCM features}
A_PCM=A_rad

rho_l=771                [kg/m^3]      {density of liquid PCM}
rho_s=825                [kg/m^3]      {density of solid PCM}

k_l=0.150                [W/m-K]       {thermal conductivity of
                                        solid PCM}
k_s=0.150                [W/m-K]       {thermal conductivity of
                                        solid PCM}

{Mass and maximum energy determination}
Q_dot_rej=eps_rad*A_rad*sigma*T_melt^4
              {heat rejection from radiator at full melt}

mass=((Q_dot_pulse-Q_dot_rej)*Pulse)/h_sf_tetradecane
                                             {PCM mass in kg}

L=mass/(rho_l*A_PCM)
                      {Length required for PCM container}
```

```
E_max=mass*h_sf_tetradecane

{Initial array settings}
g_dot_in[1]=Q_dot_pulse  {Thermal dissipation from E-box}

m_s[1]=L*A_PCM*rho_s     {Initial mass of solid PCM}

E_sum[1]=DELTAt*Q_dot_out[1] {Energy accumulated in PCM}

z[1]=0.0                      {Initial location of melt line}

T_rad[1]=T_melt-((L-z[1])/(k_s*A_PCM))*eps_rad*sigma*&
A_rad*T_rad[1]^4
                              {Initial radiator temperature}

Q_dot_out[1]=eps_rad*A_rad*sigma*T_rad[1]^4
                              {Heat rejection from radiator}

T_int[1]=T_melt               {Initial wall temperature}

lapse[1]=0.0                 {Initial elapsed time in minutes}

x_MSR[1]=0.0            {Initial melt-to-solidification ratio}

{Al-6061 thermophysical properties}
c_p_plate=cp(Aluminum_6061, T=T_rad[1])
                                    {Al-6061 specific heat}
rho_plate=density(Aluminum_6061, T=T_rad[1])
                                   {Al-6061 density}

{Radiator plate mass determination}
th_plate=(5/1000)        { thickness of radiator plate in mm}
Vol_rad=th_plate*A_rad   { volume of radiator plate}
m_rad=rho_plate*Vol_rad  { mass of radiator plate}
m_int=m_rad              { mass of hot side interface plate}

{Iterative loop}
DELTAt=6.0               {Seconds}

ITER=1200                {Maximum iteration count}

Duplicate K=1,ITER

  T_rad[K+1]=T_rad[K]+(DELTAt/(m_rad*c_p_plate))*&
((k_s*A_PCM*(T_melt-T_rad[K+1]))/(L-z[K+1])-&
eps_rad*sigma*A_rad*T_rad[K+1]^4)
```

```
  T_int[K+1]=T_int[K]+(DELTAt/(m_int*c_p_plate))*&
((g_dot_in[K+1]-(k_l*A_PCM)*(T_int[K+1]-T_melt)/(z[K+1])))

 Q_dot_out[K+1]=eps_rad*A_rad*sigma*T_rad[K+1]^4

 g_dot_in[K+1]=heat(K,Q_dot_pulse)      {E-box dissipation}

 m_s[K+1]=m_s[K]-((g_dot_in[K+1]-Q_dot_out[K+1])*DELTAt)/&
 h_sf_tetradecane

 lapse[K+1]=K*DELTAt/60        {Total elapsed time in minutes}

 z[K+1]=L-m_s[K+1]/(A_PCM*rho_s) {Update for melt-line}

 E_sum[K+1]=E_sum[K]+(g_dot_in[K+1]-Q_dot_out[K+1])*DELTAt

 x_MSR[K+1]=E_sum[K+1]/E_max    {melt-to-solidification ratio}
End
```

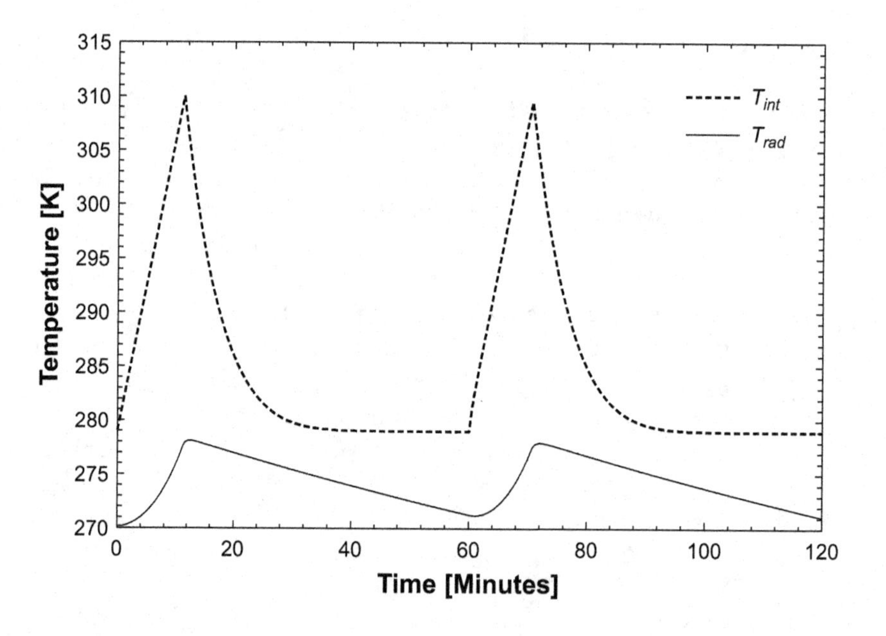

As shown in the plot, after initiation of heating the interface temperature gradually increases from 279 K to a peak of 310 K. Heat dissipation is reduced to zero immediately following the peak temperature. The interface temperature then gradually decreases back down to 279 K. The radiator increases from 270 K to a peak of 278 K. Following its peak there is an extrapolated cooling period over which the stored thermal energy is radiated away until. The second cycle begins at 60 minutes and is a repeat of the first cycle in terms of temperature profiles.

ii) Using an equivalent thermal mass of the E-box at the E-Box/PCM interface will require calculating the volume of the E-box and then multiplying by the material density in order to get the mass. The following EES code performs this calculation.

```
{E-box equivalent mass}
Vol_Outer=(40/100)^3          {E-box volume based on outer wall}
Vol_Inner=(39/100)^3          {E-box volume based on inner wall}
Vol_EBox=Vol_Outer-Vol_Inner           {Volume of E-box}
m_EBox=rho_plate*Vol_Ebox              {E-box mass}
```

In addition, the energy balance must be updated using the following syntax that includes the mass of the E-box

```
T_int[K+1]=T_int[K]+(DELTAt/(m_EBox*c_p_plate))*&
((g_dot_in[K+1]-(k_1*A_PCM)*(T_int[K+1]-T_melt)/(z[K+1])))
```

Upon execution, the temperature at the interface now peaks at 290 K when undergoing heating. When the pulse heat is turned off, the temperature gradually decreases back down to 279 K. However, the temperature decay is much more extrapolated across the "Off" portion of the cycle than before.

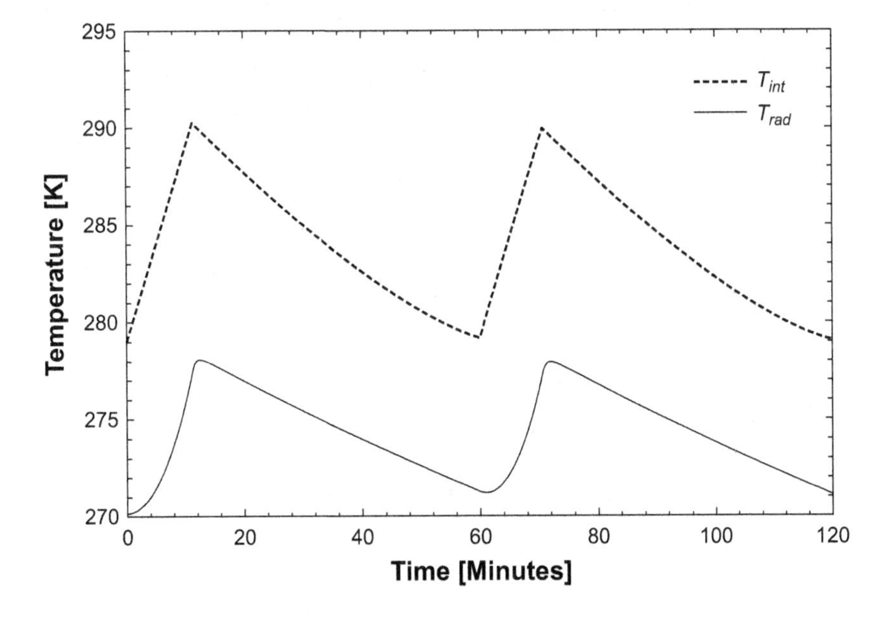

Chapter 6 Solutions

6.1 Cernox thermometers operate on resistance-based measurements and have a negligible sensitivity to magnetic fields. Their temperature-sensing range goes as low as 100 mK and their accuracy is ±5mK, which considerably exceeds the desired measurement accuracy. Use of Cernox thermometers would be the best option in this case.

6.2 Since the censor scheme is frozen for the telemetry sensors and changing it would have implications for the electrical wiring configuration and the test schedule, the one option to get measurements in the desired area would be to use TCs during testing to perform temperature measurements. Following the completion of testing (and prior to launch), the TCs (and/or wiring) would have to be removed or clipped and grounded to the structure.

6.3 The ideal locations for placement of the sensors are zones A, B, E and F. Since the E-boxes are operating at the same temperature as the planar zone they are attached to, placement of sensors in zones A and F will give insight into the planar zone temperature as well as the E-box temperature in each of these zones. The two planar locations receiving thermal energy (from the E-box dissipations) are in zones A and F. As such, zones C and D will have temperature values somewhere between the measured temperature of zones A and F. The coldest locations will be in zones B and E. The maximum planar temperature gradient can then be determined by taking the temperature difference between any of the four sensor-mounting locations selected.

6.4 Under ideal conditions five temperature sensors would be available for use. These would be placed on the CC ⑤, the evaporator ④, the vapor line ③, the condenser/subcooler ① and the liquid return line ②. However, since there are only three sensors available for placement the sensors should be selected in such a way as to insure that the LHP is either operating or not operating. The three sensor locations that can provide that certainty are the CC ⑤, the evaporator ④ and the condenser/subcooler ①.

6.5 Using the Steinhart–Hart equation shown in equation 6.1, and the temperature and resistance values from Table P6.5, three equations can be set up in EES, each having common constants of B_1, B_2 and B_3. Solving these relations simultaneously will result in values for each of the unknown constants. The final form of the Steinhart–Hart relation for the data shown in Table P6.5 then becomes.

$$\frac{1}{T} = -0.09642 + (0.02188) \cdot \ln(R_e) + \left(-1.526 \times 10^{-4}\right) \cdot \left[\ln(R_e)^3\right]$$

6.6 To solve this problem we shall perform a transient energy balance on the particle absorber in EES using milliseconds time increments. The following EES syntax may be used to predict the temperature of the absorber as a function of time.

```
{Problem 6.6  Photon absorber temperature rise calculation}

{Constants}
lambda=0.5*(0.01+10)*1.0e-9
```

```
N_photon_rate=100         {photons bombarding per millisecond}

T_abs[1] =2.0
T_sink=2.0 [ K]

{Elapsed time}
DELTAt=1/1000 [seconds]

{Time statistics}
t_max=25 [milliseconds]

{Thermal resistance and storage capacity}
mc_p=1.5e-11   [J/K]
R_link=(1/(7.5e-10))    {K/W}

Tau=mc_p/(1/R_link)

{Energy absorbed}
DELTAE_max=t_max*N_photon_rate*(1000*h#*C#/lambda){Joules}
DELTAE=DELTAE_max/t_max

{Temperature rise}
{DELT=(1/mc_p)*(DELTAE-DELTAt*((T_abs_final-T_sink)/&
R_link))}

Time[1] =0.0

Duplicate K=1,t_max

 T_abs[ K+1] =T_abs[ K] +(1/mc_p)*(DELTAE-DELTAt*((T_abs[ K] -&
 T_sink)/R_link))

 Time[K+1] =K* DELTAt          {Time in seconds}

End
```

In solving the EES program the total energy absorbed after 25 milliseconds is 3.969×10^{-15} J. After 25 milliseconds the absorber material has increased by 4 milli-Kelvin.

6.7 i) The duty cycle can be determined from the power dissipation of the heater, the length of time the heater is on and the time duration of the cycle using equation 6.4

$$Duty\ Cycle\,\% = \frac{12\ \text{min}}{12\ \text{min} + 18\ \text{min}} \cdot 100$$

$$Duty\ Cycle = 40\%$$

ii) The average power can be determined from the duty cycle and the ON power level using equation 6.5

$$P_{avg} = Duty\ Cycle \cdot P_{ON} = 0.4 \cdot 40\ \text{W}$$

$$P_{avg} = 16\ \text{W}$$

6.8 Based on the plot the cycle time is 2 hours, whereas the heater is on for a total of 30 minutes during the 2-hour span. Based on this observation the duty cycle can be determined using equation 6.4

$$Duty\ Cycle\,\% = \frac{30\ \text{min}}{30\ \text{min} + 90\ \text{min}} \cdot 100$$

$$Duty\ Cycle = 25\%$$

6.9 i) Using the relation for the average power while ON, the duty cycle can be solved for as

$$Duty\ Cycle = \frac{P_{avg}}{P_{ON}}$$

We also know that $P_{ON} = V^2/R_e$. Thus, the duty cycle relation can be recast as

$$Duty\ Cycle = \frac{P_{avg}}{V^2/R_e}$$

The voltage is given as 12 VDC and the electrical resistances are varied. Placing values in for each of the resistances gives duty cycles of 48.6%, 83.3% and 166% for resistances of 3.5 Ω, 6.0 Ω and 12 Ω respectively.

ii) The best heater option is the 3.5 Ω heater.

iii) The least practical heater option is the 12 Ω heater because it would result in a duty cycle of 166% in order to attain 20 W throughout the cycle.

6.10 i) There are 24 square inches of surface area available. All heaters selected must fit into this footprint. Heater HK6912 provides the largest dissipation (i.e., 53 W at 28 VDC) and is only 2.0 inches × 3.0 inches. At a dissipation of 7.8 W, HK6901 will require a total of 6 individual heaters to meet the 40 W requirement. This is somewhat cumbersome and impractical. Alternatively, HK6909 (dissipation of 33.3 W) would require two individual heaters. The two viable options are one HK6912 or two HK6909s.

ii) If using multiple heaters, ideally these should be in parallel. That way, if one fails then the entire circuit is not lost and you can still provide some heat. Also, as a precaution, an additional heater (providing heating margin) should be added as a back-up in case one does fail.

6.11 i) We know from equation 6.8 that the leak rate of a sealed volume can be determined from

$$X_{leak} = \frac{(P_{final} - P_{init})}{\Delta t} \cdot Vol$$

Using the information provided in the problem statement, we know the dimensions of the chamber, the initial and final pressure in the sealed chamber and the time between pressure measurements. Thus, we can calculate the leak rate. The code for determination of the leak in EES will be

```
{Problem 6.11 Part i: Leak rate on a sealed chamber}

d_chamber=0.75  [m]
l_chamber=1.0  [m]
r_chamber=d_chamber/2

Vol_chamber=pi*(r_chamber^2)*l_chamber

{Leak rate determination}

P_init=10e-3  [Pa]
P_final=0.1 [Pa]

DELTA_t=8*3600   {Factor of 3600 converts hours to seconds}

X_leak=((P_final-P_init)/DELTA_t)*Vol_chamber

X_leak_liters=X_leak*1000   {Factor of 1000 converts m^3 to liters}
```

Upon executing the code, the leak rate is determined to be 0.001381 $\text{Pa} \cdot \frac{\text{L}}{\text{s}}$

ii) Based on the problem statement, we can draw a relationship of

$$X_{leak} = \dot{Vol}_{sump} \cdot P_{equil}$$

As such, the volumetric flow rate can be determined by rearranging variables in the equation

$$\dot{Vol}_{sump} = \frac{X_{leak}}{P_{equil}}$$

We can then add the following lines of code to the previously created EES file

```
{Volumetric flow rate at chamber sump port}
P_equil=1.33e-4 [Pa]
Vol_dot_sump= X_leak/P_equil {volumetric flow rate in m^3/s}
```

Upon executing the code, the resultant volumetric flow rate will be 10.38 L/s

6.12 We know that the volumetric flow rate for the pump and the pump line, as well as the conductance on the vacuum line, are related through equation 6.9.

$$\frac{1}{\dot{Vol}_{sump}} = \frac{1}{\dot{Vol}_{pump}} + \frac{1}{C_{vac,path}}$$

The conductance along the vacuum path ($C_{vac,path}$) is determined via equation 6.13

$$C_{vac,path} = TP \cdot Vel_p \cdot \frac{A}{4}$$

Last but not least, the particle velocity in equation 6.13 can be determined using equation 6.12

$$Vel_{par} = \sqrt{\frac{8k_BT}{\pi m_{par}}}$$

Given these relations and the known variables given in the problem statement, an EES program can be written do determine the solution. The EES code is

```
{Problem 6.12 Determination of volumetric flow rate at
chamber sump line}

{Pump speed}
Vol_dot_pump_liters=300                    {liters/second}
Vol_dot_pump=Vol_dot_pump_liters*0.001     {cubic meters/second}

{Temperature}
T_C=22                 {Temperature in Celsius}
T=273+T_C              {Temperature in Kelvin}

{Particle average velocity}
R=1000*R#/molarmass(Air)
mass_air=molarmass(Air)/NA#

Vel_par=((8*1000*k#*T)/(pi*mass_air))^0.5

{Vacuum line dimensions}
D=0.5*2.54/100   [m]

A_aper=0.25*pi*(D^2)          {m^2}

{Vacuum line conductance}
TP=0.405                      {Transmission probability}

C_vac=TP*Vel_par*(A_aper/4)

Vol_dot_sump=1/(1/Vol_dot_pump+1/C_vac)

Vol_dot_sump_liters=Vol_dot_sump/0.001        {liters/second}
```

Upon executing the code the effective pump speed at the sump port is 5.84 liters/s.

6.13 i) The conductance at the aperture can be determined using equation 6.11

$$C_{aperture} = Vel_{par} \cdot \frac{A}{4}$$

However, we must first determine the particle velocity. This can be calculated using equation 6.12

$$Vel_{par} = \sqrt{\frac{8k_BT}{\pi m_{par}}}$$

We will use EES for this calculation. The corresponding EES code is

```
{Problem 6.13 Part i: Determination of aperture conductance
and pumpdown time}

{Pump Speed}
Vol_dot_pump_liters=45                          {liters/second}
Vol_dot_pump=Vol_dot_pump_liters*0.001          {cubic meters/second}

{Temperature}
T_C=21                    {Temperature in Celsius}
T=273+T_C                 {Temperature in Kelvin}

{Chamber dimensions}
d_chamber=1.5  [m]
l_chamber=2.5  [m]
r_chamber=d_chamber/2    [m]

Vol_chamber=pi*(r_chamber^2)*l_chamber

{Aperture area}
d_aper=1.5*2.54/100  [m]   {Conversion of aperture diameter
                           from inches to meters}
A=0.25*pi*d_aper^2 [ m^2]

{Particle average velocity}
mass_air=molarmass(Air)/NA#

Vel_par=((8*1000*k#*T)/(pi*mass_air))^0.5    [m/s]

{ Conductance calculation}
C_aper=Vel_par*A/4
```

```
{Pumpdown time between pressure levels}
P_init=0.1 [ Pa]
P_final=1e-4 [ Pa]

DELTAt_pumpdown=(Vol_chamber/Vol_dot_pump)*ln(P_init/P_final)

DELTAt_pumpdown_minutes=DELTAt_pumpdown/60
```

Execution of the code will result in a conductance value of 0.1321 m^3/s at the aperture and a pumpdown time of 11.3 minutes.

ii) The conductance at the aperture can be determined using equation 6.11, whereas the time for pumpdown can be determined using equation 6.7. In this case the speed will be taken as the speed at the sump port. The EES code for the calculation is

```
{Problem 6.13 Part ii: Determination of vacuum line con-
ductance and pumpdown time}

{Pump speed}
Vol_dot_pump_liters=45                          {liters/second}
Vol_dot_pump=Vol_dot_pump_liters*0.001          {cubic meters/second}

{Temperature}
T_C=21                  {Temperature in Celsius}
T=273+T_C               {Temperature in Kelvin}

{Chamber dimensions}
d_chamber=1.5  [m]
l_chamber=2.5  [m]
r_chamber=d_chamber/2    [m]

Vol_chamber=pi*(r_chamber^2)*l_chamber

{Aperture area}
d_aper=1.5*2.54/100   [m]   {Conversion of aperture diameter
                             from inches to meters}
A=0.25*pi*d_aper^2 [m^2]

{ Particle average velocity}
mass_air=molarmass(Air)/NA#

Vel_par=((8*1000*k#*T)/(pi*mass_air))^0.5    [ m/s]

{Conductance calculation for pump inlet location}
TP=0.05949            { Particle transmission probability}

C_aper=Vel_par*A/4
```

```
C_path=TP*C_aper

{Pumpdown time between pressure levels}
P_init=0.1 [Pa]
P_final=1e-4 [Pa]

{Volumetric flow rate determination at the pump inlet}
1/Vol_dot_sump=1/Vol_dot_pump+1/C_path
Vol_dot_sump_liters=Vol_dot_sump*1000

DELTAt_pumpdown=(Vol_chamber/Vol_dot_sump)*ln(P_init/P_final)

DELTAt_pumpdown_minutes=DELTAt_pumpdown/60
```

The solution for the code shows that the vacuum line conductance has a value of 0.00786 $\frac{m^3}{s}$ with a pumpdown time for the chamber of 76.01 minutes.

6.14 To calculate the pumpdown time in each case we will establish a program in EES. The syntax for the program is shown below.

```
{Problem 6.14 Variation in pumpdown time over molecular
dynamics regime}

{Temperature}
T=293 [K]

{Cylindrical chamber dimensions}
D=2.0               [m]              {Diameter}
L_cham=3.0          [m]              {Length}

Vol_cham=(0.25*pi*D^2)*L_cham     {Volume of T-Vac chamber}

Vol_cham_liters=Vol_cham/0.001

{Vacuum line dimensions}
D_line[1]=0.5*0.0254; D_line[2]=0.75*0.0254
D_line[3]=1.0*0.0254    {meters}

L_line=30   [m]

A_line[1]=0.25*pi*D_line[1]^2
A_line[2]=0.25*pi*D_line[2]^2
A_line[3]=0.25*pi*D_line[3]^2

{Pressure parameters}
P_init=0.1  [Pa]
P_final=1.0e-5 [Pa]
```

```
{Particle average velocity}
R=1000*R#/molarmass(Air)
mass_air=molarmass(Air)/NA#

Vel_par=((8*1000*k#*T)/(pi*mass_air))^0.5

{Vacuum line conductance}
Vol_dot_pump=100.0 [liters/s]

Duplicate K=1,3

a[K]=(4/3)*(D_line[K]/L_line)

C_vac_path[K]=((a[K]*Vel_par*A_line[K])/4)/0.001

Vol_dot_sump[K]=1/(1/Vol_dot_pump+1/(C_vac_path[K]/0.001))

{Pumpdown time}
DELTAt[K]=(Vol_cham_liters/Vol_dot_sump[K])*ln(P_init/P_final)/60

End
```

Using the program shown, a plot of the pumpdown times at different average speeds and vacuum line diameters can be produced. The resultant plot for the cases described in the problem is shown below.

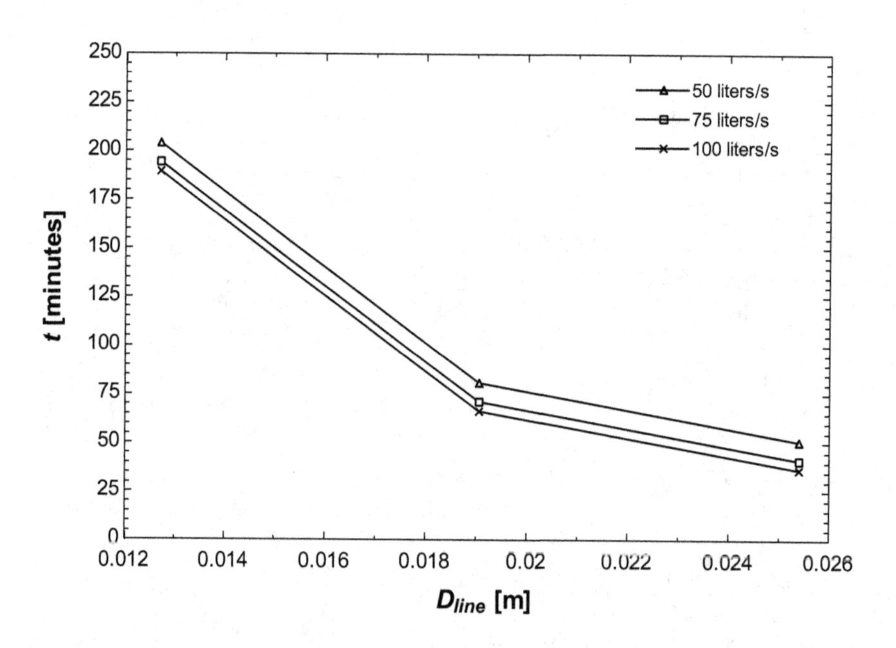

As shown in the plot, the pumpdown times are highly dependent upon the diameter of the vacuum line. While there is a reduction in the time with increasing pump speed, distinctions in pumpdown performance as a function of pump speed are minimal.

6.15 i) Since the distance between the measurement location and the signal-processing electronics is 1 m, two wires cut to this length will total 2 m. Using the resistance equation provided

$$R_{elec} = \rho_{elec}\frac{l}{A}$$

where

R_{elec} = Electrical resistance
ρ_{elec} = Electrical resistivity ($1.7 \times 10^{-8}\ \Omega \cdot$ m for copper)
l = Length of the conductor
A = Cross-sectional area of the conductor

we can calculate the electrical resistance of the 2-meter-long wire.
We know the area is

$$A = \pi(0.0004\ \text{m})^2 = 5.026 \times 10^{-7}\ \text{m}^2$$

Thus

$$R_{elec} = \frac{(1.7 \times 10^{-8}\ \Omega \cdot \text{m})(2.0\ \text{m})}{(5.026 \times 10^{-7}\ \text{m}^2)} = 0.0676\ \Omega$$

i) Using the RSS method, calculate the uncertainty in the resistance for the thermistor measurements.
To perform this calculation, the error for the resistance measurement of the leads must first be determined. Using the temperature correlation provided by the Omega website

$$\frac{1}{T} = a + b \cdot \ln(R) + c \cdot \ln^3(R)$$

and the provided constants for a, b and c, the resistance can be determined via a numerical solver for a temperature of 298.1 K and 297.9 K (i.e., ± 0.1 K relative to the nominal value of 298). Using this approach, it is determined that the corresponding error in the resistance of the leads is $\pm 41\ \Omega$. This value, along with the error for the wire (determined by the product of the 20% uncertainty specification and the nominal resistance calculated previously) can be used to calculate the total resistance error for the measurement that included the wire in series with the leads.

$$u_{R,Total} = \pm\sqrt{(u_{R,leads})^2 + (u_{R,wire})^2}$$

$$u_{R,Total} = \pm\sqrt{(41\ \Omega)^2 + (0.0676\ \Omega)^2} = \pm 41\ \Omega$$

ii) Determine the corresponding temperature uncertainty for the resistance uncertainty calculated.
The total uncertainty for the resistance is unchanged, therefore the corresponding uncertainty in the temperature measurement is unchanged as well (i.e., ± 0.1°C).

Chapter 7 Solutions

7.1 Based on the equation for linear expansion, we know that

$$\frac{\Delta L}{L_{init}} = \int_{T_1=ref}^{T_2} CTE\, dT$$

where $T_1 = ref$ is 293 K in EES. Since there are two boundary temperatures specified that are not the reference temperature, we can subtract integrals to determine a CTE value spanning 300 K to 55 K.

$$\int_{55\ \mathrm{K}}^{300\ \mathrm{K}} CTE\, dT = \int_{293\ \mathrm{K}}^{300\ \mathrm{K}} CTE\, dT - \int_{293\ \mathrm{K}}^{55\ \mathrm{K}} CTE\, dT$$

The change in length can then be calculated as

$$\Delta L = L_{init} \cdot \int_{55\ \mathrm{K}}^{300\ \mathrm{K}} CTE\, dT$$

The corresponding EES code is

```
{Problem 7.1  Thermal contraction In Ti 6Al-4V and stainless
steel 304 at 55K}

T_init=300  [K]
T_final=55  [K]

L_init=0.5

Delta_T=T_init-T_final

{Expansion for Ti-6Al-4V}
ER_Ti_1=totalthermalexp(Ti_6Al_4V, T=T_init)
ER_Ti_2=totalthermalexp(Ti_6Al_4V, T=T_final)

{Expansion for SS 304}
ER_SS_1=totalthermalexp(Stainless_AISI304, T=T_init)
ER_SS_2=totalthermalexp(Stainless_AISI304, T=T_final)

{Change in length}
DELTAL_Ti=L_init*(ER_Ti_2-ER_Ti_1)

DELTAL_SS=L_init*(ER_SS_2-ER_SS_1)
```

Execution of the code results in a change in length for SS304 of -1.54×10^{-3} and a change in length for Ti 6Al-4AV of -8.739×10^{-4} m.

7.2 i) Using equation 7.3, the STC value for each candidate material can be determined. The ratio for each material is as follows:

Material	STC ratio
Ti 6Al-4V	428.45
Stainless steel 316	57.12
T300	493.0
S-glass	1,480.3

ii) Based on the STC values for the candidate materials, S-glass has the largest value and is the best performer in this group.

7.3 For the grades of copper specified in the problem statement, we shall create a plot in EES. The EES code used to make the plot is

```
{Problem 7.3  Thermal conductivity performance of OFHC with variable RRR values}

k_RRR50=conductivity(Copper_RRR50, T=T)
k_RRR100=conductivity(Copper_RRR100, T=T)
k_RRR150=conductivity(Copper_RRR150, T=T)
k_RRR300=conductivity(Copper_RRR300, T=T)
k_RRR500=conductivity(Copper_RRR500, T=T)
```

where the temperature variable is varied in a table between 4 K and 300 K. The resultant plot is

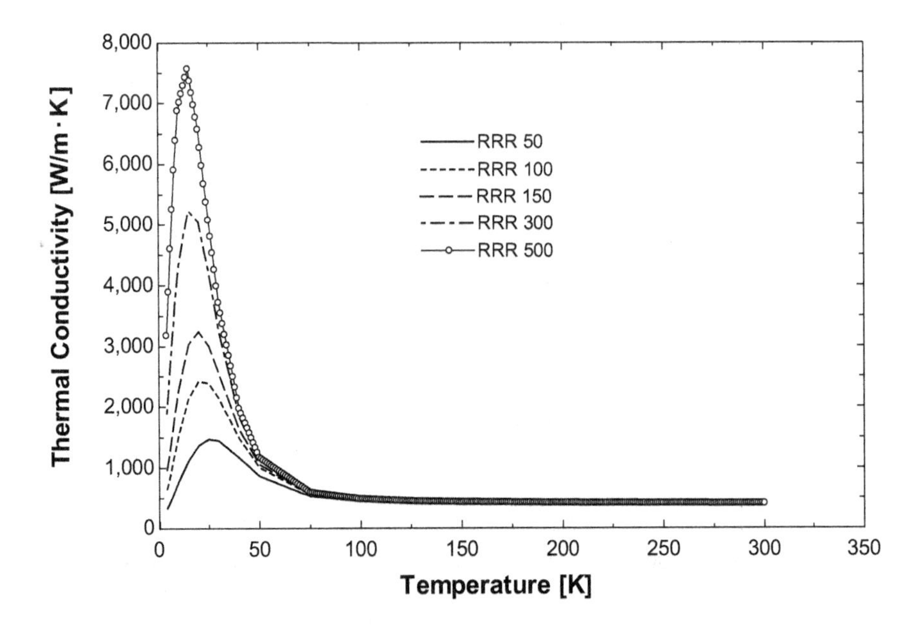

7.4 We know the wire has a diameter of 0.008 inches. The cross-sectional area of the wire can be calculated and the length is known. Using the conductivity integral function in EES, the heat transfer along the length of each wire can be determined using the following code

```
{Problem 7.4 Wire conductivity as a function of temperature}

{Endpoint temperatures}
T_1=250 [K]
T_2=30 [K]

{32 AWG wire dimensions}
D_in=0.008
D_meters=0.008*(2.54/100)

L=1

A_wire=pi*(D_meters^2)/4

{Copper RRR=50}
Cu_1=intk(Copper_RRR50, T=T_1)
Cu_2=intk(Copper_RRR50, T=T_2)

Q_dot_Cu=(A_wire/L)*(Cu_1-Cu_2)

{Nichrome}
Ni_1=intk(Nichrome, T=T_1)
Ni_2=intk(Nichrome, T=T_2)

Q_dot_Ni=(A_wire/L)*(Ni_1-Ni_2)

{ Platinum}
Plat_1=intk(Platinum, T=T_1)
Plat_2=intk(Platinum, T=T_2)

Q_dot_Plat=(A_wire/L)*(Plat_1-Plat_2)

{ Tungsten}
Tung_1=intk(Tungsten, T=T_1)
Tung_2=intk(Tungsten, T=T_2)

Q_dot_Tung=(A_wire/L)*(Tung_1-Tung_2)
```

Execution of the code shows that nichrome has the lowest conductive heat transfer with a value of 9.96×10^{-5} W.

7.5 An EES code will be developed to calculate the average resistivity across temperature boundaries and the joule heating in each segment of the wire. The EES code is

```
{Problem 7.5 Joule heating at variable temperatures}

{Temperature stages}
T0=300
T1=215
T2=120
T3=50

{32 AWG wire cross-sectional area}
D_wire=0.008*2.54/100 [m]
A_wire=0.25*pi*D_wire^2

{Wire segment lengths}
L_seg1=0.35 [m]
L_seg2=0.35 [m]
L_seg3=0.3 [m]

{Conductivites at temperature}
k0=conductivity(Copper_RRR100,  T=T0)
k1=conductivity(Copper_RRR100, T=T1)
k2=conductivity(Copper_RRR100, T=T2)
k3=conductivity(Copper_RRR100, T=T3)

{Resistivities}
Res_e0=electricalresistivity(Copper_RRR100, T=T0)
Res_e1=electricalresistivity(Copper_RRR100, T=T1)
Res_e2=electricalresistivity(Copper_RRR100, T=T2)
Res_e3=electricalresistivity(Copper_RRR100, T=T3)

{ Resistivity averages for each segment}
Res_avg_1=0.5* (Res_e0+Res_e1)
Res_avg_2=0.5* (Res_e1+Res_e2)
Res_avg_3=0.5* (Res_e2+Res_e3)

{ Resistances}
R_e_avg1=Res_avg_1* (L_seg1/A_wire)
R_e_avg2=Res_avg_2* (L_seg2/A_wire)
R_e_avg3=Res_avg_3* (L_seg3/A_wire)

{ Joule heating loads}
I=0.1                                   {Amps}
```

```
Q_dot_1=(I^2)*R_e_avg1
Q_dot_2=(I^2)*R_e_avg2
Q_dot_3=(I^2)*R_e_avg3

Q_Total=Q_dot_1+Q_dot_2+Q_dot_3
```

Upon execution, the dissipative heat in segments one, two and three are 1.57 mW, 0.909 mW and 0.27 mW respectively. The total joule heating load spanning the full length of the wire is 2.48 mW.

7.6 Since the polystyrene box has a cubic-shaped interior volume and a cubic-shaped exterior volume, we can use the hollow cubic shape factor relation from Chapter 2 (see p. 43). We will need to integrate the thermal conductivity of the polystyrene from the outer wall to the inner wall based on the temperatures in the problem statement. The thermal conductivity integral in EES can be used for the integration between two specific temperatures. As such, we will develop a code in EES to solve for the hold time. The EES code is

```
{Problem 7.6  Hold time for LAr in a polystyrene container}

{Temperatures and pressures}
T_o=293 [ K]

P_sat=101 [ kPa]
T_sat_Ar=t_sat(Argon,P=P_sat)

deltaT_sup=1 [ K]

T_i=T_sat_Ar+deltaT_sup

{Density and enthalpy of vaporization}
x_ar=0.0
rho_ar=density(Argon,T=T_sat_Ar,x=x_ar)

DELTAh_vap=enthalpy_vaporization(Argon,T=T_sat_Ar)*1000

{Argon volume}
Vol_ar=112.5*0.001  [m^3]

{Cubic shape factor determination}
Vol_in=0.125 [ m^3]

x_i=Vol_in^(1/3)            {Inner cubic side length.......m}

x_o=x_i+2*(3.81/100)  [m]  {Outer cubic side length......m}

Vol_out=x_o^3
```

```
A_i=6*x_i^2                    {Total surface area of inner cube}

s_inf=3.391

ratio=x_o/x_i

{Non-dimensional shape factor}
s_star=(2*(pi^0.5))/((1+((pi^0.5)/6)*(ratio^3-1))^(1/3)-1)+s_inf

S=s_star*(A_i^0.5)

{Conductivity integrals}
Poly_1=intk(Polystyrene, T=T_o)
Poly_2=intk(Polystyrene, T=T_i)

Cond_int=Poly_1-Poly_2

{Heat transfer through polystyrene}
Q_dot_leak=S*Cond_int

{Hold time}
Deltat=rho_ar*Vol_ar*DELTAh_vap/Q_dot_leak
Deltat_hours=Deltat/3600
```

Upon execution of the code, the hold time is approximately 25.4 hours.

7.7 To solve this problem we shall use EES. The code for the cycle is

```
{Problem 7.7  Joule-Thomson cryocooler cycle using Neon as
the refrigerant}
m_dot=0.00035                [kg/s]

T_c=t_crit(Neon)
T_high=150                        {warm-side temperature}

P_low=101*1.25             [kPa]
P_ratio=150
P_high=P_ratio*P_low

{State point 1}
x_1=1.0                       {Quality at exit of heat exchanger}
P_1=P_low
h_1=enthalpy(Neon, x=x_1, P=P_1)
T_1=temperature(Neon,x=x_1,P=P_1)
T_low=T_1
```

```
{State point 2}
P_2=P_low
T_2=T_high
h_2=enthalpy(Neon,T=T_2,P=P_2)
s_2=entropy(Neon,T=T_2,P=P_2)

{State point 3}
P_3=P_high
h_3=Q_dot_after/m_dot+h_3p
s_3=entropy(Neon, h=h_3,P=P_3)
T_3=temperature(Neon,h=h_3,P=P_3)

{Compressor}
T_avg=0.5*(T_2+T_3)
W_dot_comp=m_dot*T_avg*(s_2-s_3)-m_dot*(h_2-h_3)

{State point 3p; Recuperator high-pressure side inlet}
T_3p=154
P_3p=P_high
h_3p=enthalpy(Neon, T=T_3p, P=P_3p)

{State point 4}
P_4=P_high

{State point 5}
P_5=P_low
h_5=h_4
T_5=temperature(Neon,h=h_5,P=P_5)

{Recuperator}
h_2_max=enthalpy(Neon, T=T_3p, P=P_2)
Eff_recup=(h_2-h_1)/(h_2_max-h_1)

h_3p-h_4=h_2-h_1

{Aftercooler heat exchanger}
Q_dot_after=15

{Low-side heat exchanger cooling capacity}
Q_dot_in=m_dot*(h_1-h_5)

{Cycle performance metrics}
SP=W_dot_comp/Q_dot_in

COP_Thermo=1/SP
```

```
COP_Carnot=T_low/(T_high-T_low)

Percent=COP_Thermo/COP_Carnot
```

Upon executing the code, the following values are solved

$$Eff_{recup} = 0.969$$

$$\dot{W}_{comp} = 125.5 \text{ W}$$

$$S.P. = 202$$

$$\dot{Q}_{in} = 622 \text{ mW}$$

7.8 To solve this problem we shall use EES. The code for the cycle is

```
{Problem 7.8 Reverse-Brayton cryocooler cycle w/Helium as
the refrigerant}

m_dot=0.0013                [kg/s]

{Temperatures}
T_c=t_crit(Helium)
T_high=T_3                              {warm-side temperature}

T_sat=t_sat(Helium,P=P_low)

{Pressures}
P_low=101          [kPa]
P_ratio=1.6
P_high=P_ratio*P_low

{State point 2}
T_2=300        [K]
P_2=P_low
h_2=enthalpy(Helium, T=T_2, P=P_2)
s_2=entropy(Helium, T=T_2,P=P_2)

{Compressor}
eta=0.99
W_dot_comp_isen=m_dot*(h_3-h_2)
W_dot_comp=W_dot_comp_isen/eta

{State point 3}
P_3=P_high
s_3=s_2
h_3=enthalpy(Helium,s=s_3,P=P_3)
T_3=temperature(Helium,s=s_3,h=h_3)
```

```
{State point 3p}
P_3p=P_high
h_3p=W_dot_comp/m_dot+h_2
T_3p=temperature(Helium,h=h_3p,P=P_3p)

{State point 4}
P_4=P_high
T_4=302.8
h_4=enthalpy(Helium,T=T_4,P=P_4)

{Recuperator}
Eff_recup=0.99
h_2_max=enthalpy(Helium,T=T_4,P=P_2)
h_2-h_1=Eff_recup*(h_2_max-h_1)

h_5+h_2=h_1+h_4

{State point 5}
P_5=P_high
T_5=temperature(Helium,h=h_5,P=P_5)

{State point 6}
P_6=P_low
T_6=22.8     [K]
T_low=T_6
h_6=enthalpy(Helium,T=T_6,P=P_6)

{State point 1}
P_1=P_low
T_1=temperature(Helium,P=P_1,h=h_1)

{Cooling capacity}
Q_dot_in=m_dot*(h_1-h_6)

{Aftercooler}
Q_dot_rej=m_dot*(h_3p-h_4)

{Turbine}
W_dot_turb=m_dot*(h_5-h_6)

Q_dot_rej_check=Q_dot_in+W_dot_comp-W_dot_turb

{Overall work}
W_dot_net=W_dot_comp-W_dot_turb
```

```
{Performance metrics}
SP=W_dot_net/Q_dot_in

COP_Thermo=1/SP
COP_Carnot=T_low/(T_high-T_low)
Percent=COP_Thermo/COP_Carnot
```

Upon executing the code, the following values are solved

$$\dot{W}_{comp} = 423.4 \text{ W}$$

$$\dot{Q}_{in} = 1{,}007 \text{ mW}$$

$$S.P. = 400.6$$

% Carnot = 3.7%

7.9 To solve this problem we shall use EES. The code for the cycle is

```
{Problem 7.9 Reverse-Brayton cryocooler cycle w/heat-pipe
cooling at the compressor}

{Saturation conditions}
T_sat=319
x_l=0.0
x_v=1.0

{Dimensions}
l=0.75
l_evap=0.25
l_cond=0.25
l_adia=l-l_evap-l_cond

d_o_wall=2.54/100
r_o_wall=0.5*d_o_wall
r_i_wall=r_o_wall-(1.5/1000)

r_o_wick=r_i_wall
r_i_wick=r_o_wick-(5.1/1000)

d_p=250*1e-6 [m]
r_eff=0.5*d_p

A_vapor=pi*r_i_wick^2

A_wick1=pi*(r_o_wick^2-r_i_wick^2)
```

```
{ Thermophysical properties}
h_fg=enthalpy_vaporization(Ammonia,T=T_sat)

rho_l=density(Ammonia,T=T_sat,x=x_l)
rho_v=density(Ammonia,T=T_sat,x=x_v)

mu_l=viscosity(Ammonia,T=T_sat,x=x_l)

sigma_l=surfacetension(Ammonia,T=T_sat)
k_l=conductivity(Ammonia,T=T_sat,x=x_l)
k_wick=conductivity(Stainless_AISI304, T=T_sat)

{ Capillary limits}
phi=0.59
d_s=r_eff/0.41
d_h_check=2*(3*1e-4)*phi/(3*(1-phi))
K=((d_s^2)*(phi^3))/(150*(1-phi)^2)
Q_dot_cap=((rho_l*h_fg*K)/mu_l)*(((A_wick1)/l))*((2*sigma_l)/r_eff)

{T_avg=0.5*(T_3p+T_4)}

{Reverse-Brayton cryocooler cycle}
m_dot=0.0015            [kg/s]

{Temperatures}
T_high=T_3              {warm-side temperature}

{Pressures}
P_low=101       [kPa]
P_ratio=1.51
P_high=P_ratio*P_low

{State Point 2}
T_2=291.2       [K]
P_2=P_low
h_2=enthalpy(Helium, T=T_2, P=P_2)
s_2=entropy(Helium, T=T_2,P=P_2)

{Compressor}
eta=0.99
W_dot_comp_isen=m_dot*(h_3-h_2)
W_dot_comp=W_dot_comp_isen/eta

{State point 3}
P_3=P_high
s_3=s_2
```

```
h_3=enthalpy(Helium,s=s_3,P=P_3)
T_3=temperature(Helium,s=s_3,h=h_3)

{State point 3p}
P_3p=P_high
h_3p=W_dot_comp/m_dot+h_2
T_3p=temperature(Helium,h=h_3p,P=P_3p)

{State point 4}
P_4=P_high
T_4=temperature(Helium,h=h_4,P=P_4)

{Recuperator}
Eff_recup=0.99
h_2_max=enthalpy(Helium,T=T_4,P=P_2)
h_2-h_1=Eff_recup*(h_2_max-h_1)

h_5+h_2=h_1+h_4

{State point 5}
P_5=P_high
T_5=temperature(Helium,h=h_5,P=P_5)

{State point 6}
P_6=P_low
T_6=temperature(Helium,P=P_6,h=h_6)

{State point 1}
P_1=P_low
T_1=temperature(Helium,P=P_1,h=h_1)

{Cooling capacity}
Q_dot_in=0.3
h_6=h_1-Q_dot_in/m_dot

{Aftercooler}
eta_hp=0.97
h_4=h_3p-(eta_hp*Q_dot_cap/m_dot)

{Turbine}
W_dot_turb=m_dot*(h_5-h_6)

{Overall work}
W_dot_net=W_dot_comp-W_dot_turb
```

```
{Performance metrics}
SP=W_dot_net/Q_dot_in

COP_Thermo=1/SP

T_low=0.5*(T_1+T_6)

COP_Carnot=T_low/(T_high-T_low)
Percent=COP_Thermo/COP_Carnot
```

$$\dot{W}_{comp} = 410.8 \text{ W}$$

$$T_L = 4.915 \text{ K}$$

$$S.P. = 1{,}292$$

$$\% \text{ Carnot} = 5.32\%$$

7.10 i) We know that the specific power is defined as

$$S.P. = \frac{Total\ Input\ Power}{Cooling\ Power}$$

This relation can be rearranged to determine the cooling power

$$Cooling\ Power = \frac{Total\ Input\ Power}{S.P.}$$

Placing in values we would have

$$Cooling\ Power = \frac{85 \text{ W}}{300} = 0.283 \text{ W} = 283 \text{ mW}$$

ii) We know that the % of Carnot is defined as

$$\%COP_{Carnot} = \frac{COP_{Thermo}}{COP_{Carnot}}$$

Substituting in $1/S.P.$ for the thermodynamic COP, we can recast this relation as

$$\%COP_{Carnot} = \frac{1/S.P.}{COP_{Carnot}}$$

which is equal to

$$\%COP_{Carnot} = \frac{1/S.P.}{\left[\dfrac{T_L}{T_H - T_L}\right]}$$

Placing actual values in we have

$$\%COP_{Carnot} = \frac{1/300}{\left[\dfrac{23 \text{ K}}{300 \text{ K} - 23 \text{ K}}\right]} = \frac{0.0033}{0.083} = 0.04$$

Thus the cycle is operating at 4% of Carnot

iii) We will use EES to aid in determining the temperature at the dewar interface. The EES code is

```
{Problem 7.10, part iii, interface temperature determination}

{Temperatures}
T_cold=23  [K]
T_avg=0.5*(T_cold+T_int)       {Average temperature of thermal strap}

{Thermal strap specifications}
l=0.2  [m]                     {Length of foils}
W_F=2.54/100  [m]              {Width of foils}
delta_F=2.54e-5  [m]           {Thickness of foils}
eta_S=0.75                     {Shape efficiency}
eta_E=1.0                      {Endpoint efficiency}
N_F=20                         {Number of foils}

{Heat flow}
Q_dot=283/1000                 {Cooling capacity in mW}

{Thermal conductivity}
k_OFHC_RRR_150=conductivity(Aluminum_RRR150, T=T_avg)

{Thermal conductance calculation}
C_TS=(k_OFHC_RRR_150/l)*(N_F*delta_F*W_F)*eta_S*eta_E

R_TS=1/C_TS

{Interface temperature determination}
T_int=Q_dot*R_TS+T_cold
```

Upon execution of the code the interface temperature is 25.59 K.

iv) We will return to the EES code and add in a thermal coupling for the radiative heat transfer occurring between the aggregate perimeter surface of the thermal strap foils and the cylindrical surroundings defined in the problem statement. The EES code is

```
{Problem 7.10, part iv, interface temperature determination
with warm surroundings}

{constants}
sigma=5.67e-8                        {Stefan-Boltzmann constant}

{Temperatures}
T_cold=23  [K]
T_int=25.59
T_avg=0.5*(T_cold+T_int)       {Average temperature of thermal strap}
T_sur=200     [K]              {Temperature of surroundings}
```

```
{Surrounding enclosure specifications}
l_sur=20/100   [m]        {Length of cylindrical surrounding}
D_sur=5*2.54/100  [m] { Diameter of cylindrical surrounding}
A_sur=2*pi*(D_sur/2)*l_sur {Surface area of cylindrical surrounding}
eps_sur=0.03              {Emissivity of surrounding surface}

{Thermal strap specifications}
l=0.2  [m]                    {Length of foils}
W_F=2.54/100  [m]             {Width of foils}
delta_F=2.54e-5  [m]          {Thickness of foils}
eta_S=0.75                    {Shape efficiency}
eta_E=1.0                     {Endpoint efficiency}
N_F=20                        {Number of foils}
eps_TS=0.03                   {Emissivity of TS surfaces}

A_TS=2*W_F*l+2*N_F*delta_F    {Surface area of aggregate TS}

{Heat flow}
{Q_dot=283/1000               {Cooling capacity in mW}}

{Thermal conductivity}
k_OFHC_RRR_150=conductivity(Aluminum_RRR150, T=T_avg)

{Thermal conductance calculation}
C_TS=(k_OFHC_RRR_150/l)*(N_F*delta_F*W_F)*eta_S*eta_E

R_TS=1/C_TS

{Interface temperature determination}
F_TS_Sur=1.0                  {Thermal strap view to surrounding}

Rad_Res=(1-eps_TS)/(eps_TS*A_TS)+1/(A_TS*F_TS_Sur)+(1-&
eps_sur)/(eps_sur*A_sur)

Q_dot_rad=sigma*(T_sur^4-T_avg^4)/Rad_Res

Q_dot=(T_int-T_cold)/R_TS+Q_dot_rad
```

The newly determined cooling load requirement is 310 mW.

7.11 i) The thermal circuit diagram is

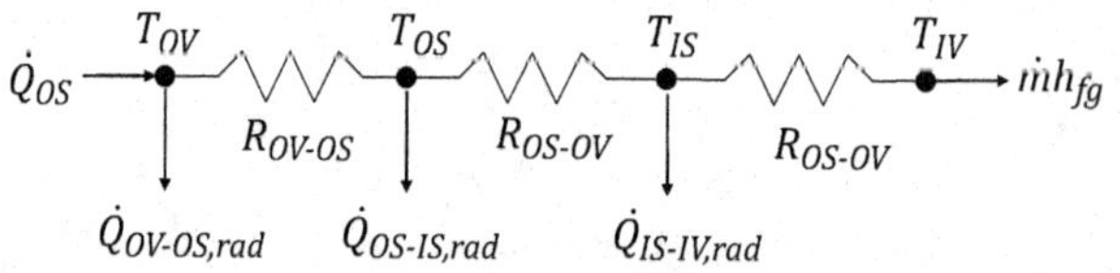

$\dot{Q}_{OS}$ → T_{OV} — $R_{OV\text{-}OS}$ — T_{OS} — $R_{OS\text{-}OV}$ — T_{IS} — $R_{OS\text{-}OV}$ — T_{IV} → $\dot{m}h_{fg}$

$\dot{Q}_{OV\text{-}OS,rad}$ $\dot{Q}_{OS\text{-}IS,rad}$ $\dot{Q}_{IS\text{-}IV,rad}$

ii) To solve this problem we shall use EES. The code for the dewar is

```
{Problem 7.11 part ii Shielded LH2 stored cryogen dewar system }

{constants}
sigma=5.67e-8          [W/m^2-K^4]

{Boundary temperatures}
T_OV=110   [K]

DELTAT_sup=4 [K]
T_IV=T_sat_LH2+DELTAT_sup

{Fluid properties}
P_LH2=101 [ kPa]
T_sat_LH2=t_sat(Hydrogen,P=P_LH2)
rho_LH2_liq=density(Hydrogen,T=T_sat_LH2,x=0.0)
h_fg=enthalpy_vaporization(Hydrogen,T=T_sat_LH2)

{Inner vessel}
side_IV=0.5 [m]
Vol_IV=0.5^3 [ m^3]

Area_IV=6*side_IV^2

{Inner shield}
d_IS=0.56 [m]
h_IS=0.6 [m]

Area_IS=2*pi*(d_IS/2)*h_IS+2*pi*(d_IS^2)/4  [m^2]

{Outer shield}
d_OS=0.62 [m]
h_OS=0.7 [m]

Area_OS=2*pi*(d_OS/2)*h_OS+2*pi*(d_OS^2)/4  [m^2]

{Outer vessel}
d_OV=0.68   [m]
h_OV=0.55   [m]
r_OV=d_OV/2 [m]
h_OV_cap=0.15 [m]

Area_OV=2*pi*(d_OV/2)*h_OV+2*pi*(r_OV^2+h_OV_cap^2)
```

```
{A strut: Ti 6Al-4V}
l_strut_A=3.5/100 [m]
OD_strut_A=1.9/100 [m]
ID_strut_A=OD_strut_A-2*(2.0/1000) [m]

Area_strut_A=0.25*pi*(OD_strut_A^2-ID_strut_A^2)

Area_contact=pi*0.5*OD_strut_A^2

{B strut: SS 304}
l_strut_B=3.5/100 [m]
OD_strut_B=1.9/100 [m]
ID_strut_B=OD_strut_B-2*(2.0/1000) [m]

Area_strut_B=0.25*pi*(OD_strut_B^2-ID_strut_B^2)

{Structure mounts}
S_tot=7.121e-3 [m]

{Conductivity Integrals}
{A struts}
k_A_1=intk(Ti_6Al_4V, T=T_IS)
k_A_2=intk(Ti_6Al_4V, T=T_IV)

k_A_avg=k_A_1-k_A_2

Q_A_Struts=(Area_strut_A/l_strut_A)*k_A_avg

{B struts}
k_B_1=intk(Ti_6Al_4V, T=T_OS)

k_B_2=intk(Ti_6Al_4V, T=T_IS)

k_B_avg=k_B_1-k_B_2

Q_B_Struts=(Area_strut_B/l_strut_B)*k_B_avg

{ Structure mounts}
k_SM_1=intk(Ti_6Al_4V, T=T_OV)

k_SM_2=intk(Ti_6Al_4V, T=T_OS)

k_SM_avg=k_SM_1-k_SM_2

Q_S_Mounts=S_tot*k_SM_avg
```

```
{Stored cryogen}
Vol_LH2=0.11875 [m^3]
Vol_LH2_liters=Vol_LH2/0.001

{Emissivities}
eps_OV=0.03
eps_OS=0.03
eps_IS=0.03
eps_IV=0.03

N=15

estar=1/((N+1)*((2-eps_IV)/eps_IV))

{FOVs}
F_OV_OS=1
F_OS_IS=1
F_IS_IV=1

{Energy balance relations}
{Outer shield}
T_OS^4=T_OV^4-((Q_dot_OS-6*Q_S_Mounts)*((1-eps_OS)/(eps_OS*&
Area_OS)+1/(F_OV_OS*Area_OV)+(1-eps_OV)/&
(eps_OV*Area_OV)))/sigma

{Inner shield}
T_IS^4=T_OS^4-((Q_dot_OS-6*Q_B_Struts)*((1-eps_OS)/(eps_OS*&
Area_OS)+1/(F_OS_IS*Area_OS)+(1-eps_IS)/&
(eps_IS*Area_IS)))/sigma

{Inner vessel}
Q_dot_OS=6*Q_A_Struts+(sigma*(T_IS^4-T_IV^4))/((1-eps_IS)/&
(eps_IS*Area_IS)+1/(F_IS_IV*Area_IS)+(1-estar)/&
(estar*Area_IV))

{Hold time calculations}
DELTA_t=rho_LH2_liq*Vol_LH2*(h_fg*1000)/(Q_dot_OS)

Time_days=DELTA_t/86400
Time_months=Time_days/30
```

The hold time is approximately 22 days.

iii) We can use the previously developed EES code to determine the hold time with a cryocooler heat intercept at the outer shield by reducing the cooling capacity by 1.26 W from the original heat load on that shield. The new hold time becomes 60.2 days.

Printed in the United States
by Baker & Taylor Publisher Services